Contents

Credits

CHAPTER 17 Quantum Mechanics

CHAPTER OUTLINE

We now begin the study of **quantum chemistry,** which applies quantum mechanics to chemistry. Chapter 17 deals with **quantum mechanics,** the laws governing the behavior of microscopic particles such as electrons and nuclei. Chapters 18 and 19 apply quantum mechanics to atoms and molecules. Chapter 20 applies quantum mechanics to spectroscopy, the study of the absorption and emission of electromagnetic radiation. Quantum mechanics is used in statistical mechanics (Chapter 21) and in theoretical chemical kinetics (Chapter 22).

Unlike thermodynamics, quantum mechanics deals with systems that are not part of everyday macroscopic experience, and the formulation of quantum mechanics is quite mathematical and abstract. This abstractness takes a while to get used to, and it is natural to feel somewhat uneasy when first reading Chapter 17.

In an undergraduate physical chemistry course, it is not possible to give a full presentation of quantum mechanics. Derivations of results that are given without proof may be found in quantum chemistry texts listed in the Bibliography.

Sections 17.1 to 17.4 give the historical background of quantum mechanics. Section 17.5 discusses the uncertainty principle, a key concept that underlies the differences between quantum mechanics and classical (Newtonian) mechanics. Quantum mechanics describes the state of a system using a state function (or wave function) Ψ. Sections 17.6 and 17.7 describe the meaning of Ψ and the time-dependent and time-independent Schrödinger equations used to find Ψ. Sections 17.8, 17.9, 17.10, 17.12, 17.13, and 17.14 consider the Schrödinger equation, the wave functions, and the allowed quantum-mechanical energy levels for several systems. Sections 17.11 and 17.16 discuss operators, which are used extensively in quantum mechanics. Section 17.15 introduces some of the approximation methods used to apply quantum mechanics to chemistry.

Essentially all of chemistry is a consequence of the laws of quantum mechanics. If we want to understand chemistry at the fundamental level of electrons, atoms, and molecules, we must understand quantum mechanics. Quantities such as the heat of combustion of octane, the 25°C entropy of liquid water, the reaction rate of N_2 and H_2 gases at specified conditions, the equilibrium constants of chemical reactions, the absorption spectra of coordination compounds, the NMR spectra of organic compounds, the nature of the products formed when organic compounds react, the shape a protein molecule folds into when it is formed in a cell, the structure and function of DNA are all a consequence of quantum mechanics.

In 1929, Dirac, one of the founders of quantum mechanics, wrote that "The general theory of quantum mechanics is now almost complete . . . . The underlying physical laws necessary for the mathematical theory of . . . the whole of chemistry are thus completely known, and the difficulty is only that the exact application of these laws leads to equations much too complicated to be soluble." After its discovery, quantum mechanics was used to develop many concepts that helped explain chemical properties. However, because of the very difficult calculations needed to apply quantum mechanics to chemical systems, quantum mechanics was of little practical value in accurately calculating the properties of chemical systems for many years after its discovery. Nowadays, however, the extraordinary computational power of modern computers allows quantum-mechanical calculations to give accurate chemical predictions in many systems of real chemical interest. As computers become even more powerful and applications of quantum mechanics in chemistry increase, the need for all chemists to be familiar with quantum mechanics will increase.

17.1 BLACKBODY RADIATION AND ENERGY QUANTIZATION

Classical physics is the physics developed before 1900. It consists of classical mechanics (Sec. 2.1), Maxwell's theory of electricity, magnetism, and electromagnetic radiation (Sec. 20.1), thermodynamics, and the kinetic theory of gases (Chapters 14 and 15). In the late nineteenth century, some physicists believed that the theoretical structure of physics was complete, but in the last quarter of the nineteenth century, various experimental results were obtained that could not be explained by classical physics. These results led to the development of quantum theory and the theory of relativity. An understanding of atomic structure, chemical bonding, and molecular spectroscopy must be based on quantum theory, which is the subject of this chapter.

One failure of classical physics was the incorrect $C_{V,m}$ values of polyatomic molecules predicted by the kinetic theory of gases (Sec. 14.10). A second failure was the inability of classical physics to explain the observed frequency distribution of radiant energy emitted by a hot solid.

When a solid is heated, it emits light. Classical physics pictures light as a wave consisting of oscillating electric and magnetic fields, an **electromagnetic wave.** (See Sec. 20.1 for a fuller discussion.) The frequency ν (nu) and wavelength λ (lambda) of an electromagnetic wave traveling through vacuum are related by

$$\lambda\nu = c \qquad \textbf{(17.1)*}$$

where $c = 3.0 \times 10^8$ m/s is the speed of light in vacuum. The human eye is sensitive to electromagnetic waves whose frequencies lie in the range 4×10^{14} to 7×10^{14} cycles/s. However, electromagnetic radiation can have any frequency (see Fig. 20.2). We shall use the term "light" as synonymous with electromagnetic radiation, not restricting it to visible light.

Different solids emit radiation at different rates at the same temperature. To simplify things, one deals with the radiation emitted by a blackbody. A **blackbody** is a body that absorbs all the electromagnetic radiation that falls on it. A good approximation to a blackbody is a cavity with a tiny hole. Radiation that enters the hole is repeatedly reflected within the cavity (Fig. 17.1a). At each reflection, a certain fraction

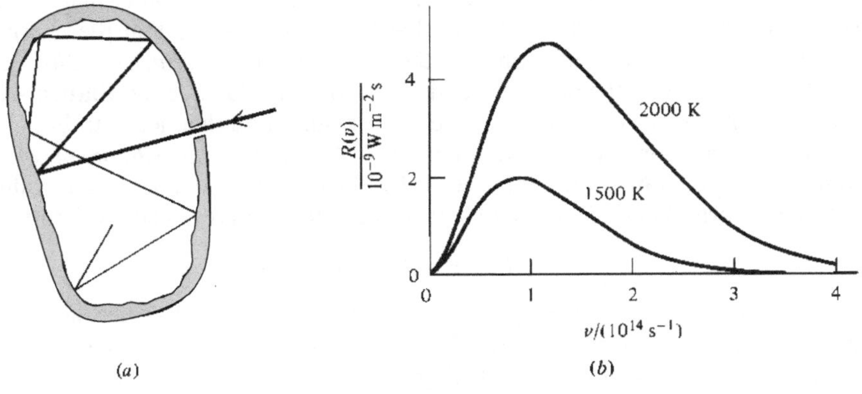

Figure 17.1

(a) A cavity acting as a blackbody. (b) Frequency distribution of blackbody radiation at two temperatures. (The visible region is from 4×10^{14} to 7×10^{14} s^{-1}.)

of the radiation is absorbed by the cavity walls, and the large number of reflections causes virtually all the incident radiation to be absorbed. When the cavity is heated, its walls emit light, a tiny portion of which escapes through the hole. It can be shown that the rate of radiation emitted per unit surface area of a blackbody is a function of only its temperature and is independent of the material of which the blackbody is made. (See *Zemansky and Dittman,* sec. 4-14, for a proof.)

By using a prism to separate the various frequencies emitted by the cavity, one can measure the amount of blackbody radiant energy emitted in a given narrow frequency range. Let the *frequency distribution* of the emitted blackbody radiation be described by the function $R(\nu)$, where $R(\nu)\,d\nu$ is the energy with frequency in the range ν to $\nu + d\nu$ that is radiated per unit time and per unit surface area. (Recall the discussion of distribution functions in Sec. 14.4.) Figure 17.1*b* shows some experimentally observed $R(\nu)$ curves. As T increases, the maximum in $R(\nu)$ shifts to higher frequencies. When a metal rod is heated, it first glows red, then orange-yellow, then white, then blue-white. (White light is a mixture of all colors.) Our bodies are not hot enough to emit visible light, but we do emit infrared radiation.

In June 1900, Lord Rayleigh attempted to derive the theoretical expression for the function $R(\nu)$. Using the equipartition-of-energy theorem (Sec. 14.10), he found that classical physics predicted $R(\nu) = (2\pi kT/c^2)\nu^2$, where k and c are Boltzmann's constant and the speed of light. But this result is absurd, since it predicts that the amount of energy radiated would increase without limit as ν increases. In actuality, $R(\nu)$ reaches a maximum and then falls off to zero as ν increases (Fig. 17.1*b*). Thus, classical physics fails to predict the spectrum of blackbody radiation.

On October 19, 1900, the physicist Max Planck announced to the German Physical Society his discovery of a formula that gave a highly accurate fit to the observed curves of blackbody radiation. Planck's formula was $R(\nu) = a\nu^3/(e^{b\nu/T} - 1)$, where a and b are constants with certain numerical values. Planck had obtained this formula by trial and error and at that time had no theory to explain it. On December 14, 1900, Planck presented to the German Physical Society a theory that yielded the blackbody-radiation formula he had found empirically a few weeks earlier. Planck's theory gave the constants a and b as $a = 2\pi h/c^2$ and $b = h/k$, where h was a new constant in physics and k is Boltzmann's constant [Eq. (3.57)]. Planck's theoretical expression for the frequency distribution of blackbody radiation is then

$$R(\nu) = \frac{2\pi h}{c^2}\frac{\nu^3}{e^{h\nu/kT} - 1} \qquad (17.2)$$

Planck considered the walls of the blackbody to contain electric charges that oscillated (vibrated) at various frequencies [Maxwell's electromagnetic theory of light (Sec. 20.1) shows that electromagnetic waves are produced by accelerated electric charges. A charge oscillating at frequency ν will emit radiation at that frequency.] In order to derive (17.2), Planck found that he had to assume that the energy of each oscillating charge could take on only the possible values 0, $h\nu$, $2h\nu$, $3h\nu$, . . . , where ν is the frequency of the oscillator and h is a constant (later called **Planck's constant**) with the dimensions of energy × time. This assumption then leads to Eq. (17.2). (For Planck's derivation, see M. Jammer, *The Conceptual Development of Quantum Mechanics,* McGraw-Hill, 1966, sec. 1.2.) Planck obtained a numerical value of h by fitting the formula (17.2) to the observed blackbody curves. The modern value is

$$h = 6.626 \times 10^{-34}\ \text{J}\cdot\text{s} \qquad \textbf{(17.3)*}$$

In classical physics, energy takes on a continuous range of values, and a system can lose or gain any amount of energy. In direct contradiction to classical physics, Planck restricted the energy of each oscillating charge to a whole-number multiple of $h\nu$ and hence restricted the amount of energy each oscillator could gain or lose to an integral multiple of $h\nu$. Planck called the quantity $h\nu$ a **quantum** of energy (the Latin word *quantum* means "how much"). *In classical physics, energy is a continuous variable. In quantum physics, the energy of a system is* **quantized,** meaning that the energy can take on only certain values. Planck introduced the idea of energy quantization in one case, the emission of blackbody radiation. In the years 1900–1926, the concept of energy quantization was gradually extended to all microscopic systems.

Planck was not very explicit in his derivation and several historians of science have argued that Planck's apparent introduction of energy quantization was solely for mathematical convenience to allow him to evaluate a certain quantity needed in the derivation, and Planck was not actually proposing energy quantization as a physical reality. See S. G. Brush, *Am. J. Phys.,* **70,** 119 (2002); O. Darrigol, *Centaurus,* **43,** 219 (2001)—available at www.mpiwg-berlin.mpg.de/Preprints/P150.PDF.

17.2 THE PHOTOELECTRIC EFFECT AND PHOTONS

The person who recognized the value of Planck's idea was Einstein, who applied the concept of energy quantization to electromagnetic radiation and showed that this explained the experimental observations in the photoelectric effect.

In the **photoelectric effect,** a beam of electromagnetic radiation (light) shining on a metal surface causes the metal to emit electrons; electrons absorb energy from the light beam, thereby acquiring enough energy to escape from the metal. A practical application is the photoelectric cell, used to measure light intensities, to prevent elevator doors from crushing people, and in smoke detectors (light scattered by smoke particles causes electron emission, which sets off an alarm).

Experimental work around 1900 had shown that (*a*) Electrons are emitted only when the frequency of the light exceeds a certain minimum frequency ν_0 (the *threshold frequency*). The value of ν_0 differs for different metals and lies in the ultraviolet for most metals. (*b*) Increasing the intensity of the light increases the number of electrons emitted but does not affect the kinetic energy of the emitted electrons. (*c*) Increasing the frequency of the radiation increases the kinetic energy of the emitted electrons.

These observations on the photoelectric effect cannot be understood using the classical picture of light as a wave. The energy in a wave is proportional to its intensity but is independent of its frequency, so one would expect the kinetic energy of the emitted electrons to increase with an increase in light intensity and to be independent of the light's frequency. Moreover, the wave picture of light would predict the photoelectric effect to occur at any frequency, provided the light is sufficiently intense.

In 1905 Einstein explained the photoelectric effect by extending Planck's concept of energy quantization to electromagnetic radiation. (Planck had applied energy quantization to the oscillators in the blackbody but had considered the electromagnetic radiation to be a wave.) Einstein proposed that in addition to having wavelike properties, light could also be considered to consist of particlelike entities (quanta), each quantum of light having an energy $h\nu$, where h is Planck's constant and ν is the frequency of the light. These entities were later named **photons,** and the energy of a photon is

$$E_{\text{photon}} = h\nu \qquad \textbf{(17.4)*}$$

The energy in a light beam is the sum of the energies of the individual photons and is therefore quantized.

Let electromagnetic radiation of frequency ν fall on a metal. The photoelectric effect occurs when an electron in the metal is hit by a photon. The photon disappears, and its energy $h\nu$ is transferred to the electron. Part of the energy absorbed by the electron is used to overcome the forces holding the electron in the metal, and the remainder appears as kinetic energy of the emitted electron. Conservation of energy therefore gives

$$h\nu = \Phi + \tfrac{1}{2}mv^2 \qquad (17.5)$$

where the *work function* Φ is the minimum energy needed by an electron to escape the metal and $\frac{1}{2}mv^2$ is the kinetic energy of the free electron. The valence electrons in metals have a distribution of energies (Sec. 23.11), so some electrons need more energy than others to leave the metal. The emitted electrons therefore show a distribution of kinetic energies, and $\frac{1}{2}mv^2$ in (17.5) is the maximum kinetic energy of emitted electrons.

Einstein's equation (17.5) explains all the observations in the photoelectric effect. If the light frequency is such that $h\nu < \Phi$, a photon does not have enough energy to allow an electron to escape the metal and no photoelectric effect occurs. The minimum frequency ν_0 at which the effect occurs is given by $h\nu_0 = \Phi$. (The work function Φ differs for different metals, being lowest for the alkali metals.) Equation (17.5) shows the kinetic energy of the emitted electrons to increase with ν and to be independent of the light intensity. An increase in intensity with no change in frequency increases the energy of the light beam and hence increases the number of photons per unit volume in the light beam, thereby increasing the rate of emission of electrons.

Einstein's theory of the photoelectric effect agreed with the qualitative observations, but it wasn't until 1916 that R. A. Millikan made an accurate quantitative test of Eq. (17.5). The difficulty in testing (17.5) is the need to maintain a very clean surface of the metal. Millikan found accurate agreement between (17.5) and experiment.

At first, physicists were very reluctant to accept Einstein's hypothesis of photons. Light shows the phenomena of diffraction and interference (Sec. 17.5), and these effects are shown only by waves, not by particles. Eventually, physicists became convinced that the photoelectric effect could be understood only by viewing light as being composed of photons. However, diffraction and interference can be understood only by viewing light as a wave and not as a collection of particles.

Thus, light seems to exhibit a dual nature, behaving like waves in some situations and like particles in other situations. This apparent duality is logically contradictory, since the wave and particle models are mutually exclusive. Particles are localized in space, but waves are not. The photon picture gives a quantization of the light energy, but the wave picture does not. In Einstein's equation $E_{\text{photon}} = h\nu$, the quantity E_{photon} is a particle concept, but the frequency ν is a wave concept, so this equation is, in a sense, self-contradictory. An explanation of these apparent contradictions is given in Sec. 17.4.

In 1907 Einstein applied the concept of energy quantization to the vibrations of the atoms in a solid, thereby showing that the heat capacity of a solid goes to zero as T goes to zero, a result in agreement with experiment but in disagreement with the classical equipartition theorem. See Sec. 23.12 for details.

17.3 THE BOHR THEORY OF THE HYDROGEN ATOM

The next major application of energy quantization was the Danish physicist Niels Bohr's 1913 theory of the hydrogen atom. A heated gas of hydrogen atoms emits electromagnetic radiation containing only certain distinct frequencies (Fig. 20.36). During 1885 to 1910, Balmer, Rydberg, and others found that the following empirical formula correctly reproduces the observed H-atom spectral frequencies:

$$\frac{\nu}{c} = \frac{1}{\lambda} = R\left(\frac{1}{n_b^2} - \frac{1}{n_a^2}\right) \qquad n_b = 1, 2, 3, \ldots; \quad n_a = 2, 3, \ldots; \quad n_a > n_b \quad (17.6)$$

where the *Rydberg constant* R equals 1.096776×10^5 cm^{-1}. There was no explanation for this formula until Bohr's work.

If one accepts Einstein's equation $E_{\text{photon}} = h\nu$, the fact that only certain frequencies of light are emitted by H atoms indicates that contrary to classical ideas, a hydrogen atom can exist only in certain energy states. Bohr therefore postulated that the energy of a hydrogen atom is quantized: (1) An atom can take on only certain distinct energies $E_1, E_2, E_3, \ldots$. Bohr called these allowed states of constant energy the *stationary states* of the atom. This term is not meant to imply that the electron is at rest in a stationary state. Bohr further assumed that (2) An atom in a stationary state does not emit electromagnetic radiation. To explain the line spectrum of hydrogen, Bohr assumed that (3) When an atom makes a transition from a stationary state with energy E_{upper} to a lower-energy stationary state with energy E_{lower}, it emits a photon of light. Since $E_{\text{photon}} = h\nu$, conservation of energy gives

$$E_{\text{upper}} - E_{\text{lower}} = h\nu \qquad \textbf{(17.7)*}$$

where $E_{\text{upper}} - E_{\text{lower}}$ is the energy difference between the atomic states involved in the transition and ν is the frequency of the light emitted. Similarly, an atom can make a transition from a lower-energy to a higher-energy state by absorbing a photon of frequency given by (17.7). The Bohr theory provided no description of the transition process between two stationary states. Of course, transitions between stationary states can occur by means other than absorption or emission of electromagnetic radiation. For example, an atom can gain or lose electronic energy in a collision with another atom.

Equations (17.6) and (17.7) with upper and lower replaced by a and b give $E_a - E_b = Rhc(1/n_b^2 - 1/n_a^2)$, which strongly indicates that the energies of the H-atom stationary states are given by $E = -Rhc/n^2$, with $n = 1, 2, 3, \ldots$. Bohr then introduced further postulates to derive a theoretical expression for the Rydberg constant. He assumed that (4) The electron in an H-atom stationary state moves in a circle around the nucleus and obeys the laws of classical mechanics. The energy of the electron is the sum of its kinetic energy and the potential energy of the electron–nucleus electrostatic attraction. Classical mechanics shows that the energy depends on the radius of the orbit. Since the energy is quantized, only certain orbits are allowed. Bohr used one final postulate to select the allowed orbits. Most books give this postulate as (5) The allowed orbits are those for which the electron's angular momentum $m_e vr$ equals $nh/2\pi$, where m_e and v are the electron's mass and speed, r is the radius of the orbit, and $n = 1, 2, 3, \ldots$. Actually, Bohr used a different postulate which is less arbitrary than 5 but less simple to state. The postulate Bohr used is equivalent to 5 and is omitted here. (If you're curious, see *Karplus and Porter,* sec. 1.4.)

With his postulates, Bohr derived the following expression for the H-atom energy levels: $E = -m_e e^4/8\varepsilon_0^2 h^2 n^2$, where e is the proton charge and the electric constant ε_0 occurs in Coulomb's law (13.1). Therefore, Bohr predicted that $Rhc = m_e e^4/8\varepsilon_0^2 h^2$ and $R = m_e e^4/8\varepsilon_0^2 h^3 c$. Substitution of the values of m_e, e, h, ε_0, and c gave a result in good agreement with the experimental value of the Rydberg constant, indicating that the Bohr model gave the correct energy levels of H.

Although the Bohr theory is historically important for the development of quantum theory, postulates 4 and 5 are in fact *false,* and the Bohr theory was superseded in 1926 by the Schrödinger equation, which provides a correct picture of electronic behavior in atoms and molecules. Although postulates 4 and 5 are false, postulates 1, 2, and 3 are consistent with quantum mechanics.

17.4 THE DE BROGLIE HYPOTHESIS

In the years 1913 to 1925, attempts were made to apply the Bohr theory to atoms with more than one electron and to molecules. However, all attempts to derive the spectra of such systems using extensions of the Bohr theory failed. It gradually became clear

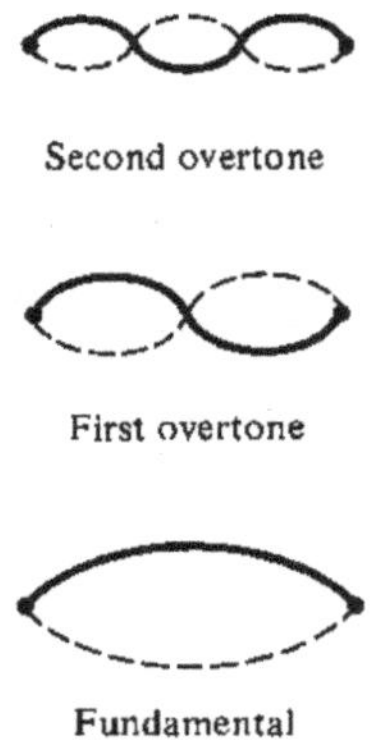

Figure 17.2

Fundamental and overtone vibrations of a string.

that there was a fundamental error in the Bohr theory. The fact that the Bohr theory works for H is something of an accident.

A key idea toward resolving these difficulties was advanced by the French physicist Louis de Broglie (1892–1987) in 1923. The fact that a heated gas of atoms or molecules emits radiation of only certain frequencies shows that the energies of atoms and molecules are quantized, only certain energy values being allowed. Quantization of energy does not occur in classical mechanics; a particle can have any energy in classical mechanics. Quantization does occur in wave motion. For example, a string held fixed at each end has quantized modes of vibration (Fig. 17.2). The string can vibrate at its fundamental frequency ν, at its first overtone frequency 2ν, at its second overtone frequency 3ν, etc. Frequencies lying between these integral multiples of ν are not allowed.

De Broglie therefore proposed that just as light shows both wave and particle aspects, matter also has a "dual" nature. As well as showing particlelike behavior, an electron could also show wavelike behavior, the wavelike behavior manifesting itself in the quantized energy levels of electrons in atoms and molecules. Holding the ends of a string fixed quantizes its vibrational frequencies. Similarly, confining an electron in an atom quantizes its energies.

De Broglie obtained an equation for the wavelength λ to be associated with a material particle by reasoning in analogy with photons. We have $E_{\text{photon}} = h\nu$. Einstein's special theory of relativity gives the energy of a photon as $E_{\text{photon}} = pc$, where p is the momentum of the photon and c is the speed of light. Equating these two expressions for E_{photon}, we get $h\nu = pc$. But $\nu = c/\lambda$, so $hc/\lambda = pc$ and $\lambda = h/p$ for a photon. By analogy, de Broglie proposed that a material particle with momentum p would have a wavelength λ given by

$$\lambda = h/p \tag{17.8}$$

The momentum of a particle with a speed v much less than the speed of light is $p = mv$, where m is the particle's rest mass.

The de Broglie wavelength of an electron moving at 1.0×10^6 m/s is

$$\lambda = \frac{6.6 \times 10^{-34}\text{ J s}}{(9.1 \times 10^{-31}\text{ kg})(1.0 \times 10^6\text{ m/s})} = 7 \times 10^{-10}\text{ m} = 7\text{ Å}$$

This wavelength is on the order of magnitude of molecular dimensions and indicates that wave effects are important in electronic motions in atoms and molecules. For a macroscopic particle of mass 1.0 g moving at 1.0 cm/s, a similar calculation gives $\lambda = 7 \times 10^{-27}$ cm. The extremely small size of λ (which results from the smallness of Planck's constant h in comparison with mv) indicates that quantum effects are unobservable for the motion of macroscopic objects.

De Broglie's bold hypothesis was experimentally confirmed in 1927 by Davisson and Germer, who observed diffraction effects when an electron beam was reflected from a crystal of Ni; G. P. Thomson observed diffraction effects when electrons were passed through a thin sheet of metal. See Fig. 17.3. Similar diffraction effects have been observed with neutrons, protons, helium atoms, and hydrogen molecules, indicating that the de Broglie hypothesis applies to all material particles, not just electrons. An application of the wavelike behavior of microscopic particles is the use of electron diffraction and neutron diffraction to obtain molecular structures (Secs. 23.9 and 23.10).

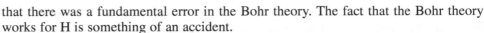

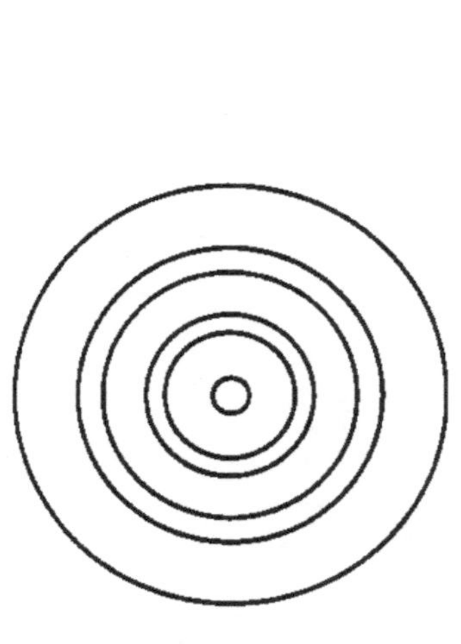

Figure 17.3

Diffraction rings observed when electrons are passed through a thin polycrystalline metal sheet.

Electrons show particlelike behavior in some experiments (for example, the cathode-ray experiments of J. J. Thomson, Sec. 18.2) and wavelike behavior in other experiments. As noted in Sec. 17.2, the wave and particle models are incompatible with each other. An entity cannot be both a wave and a particle. How can we explain the apparently contradictory behavior of electrons? The source of the difficulty is the attempt to describe microscopic entities like electrons by using concepts developed from our experience in the macroscopic world. The particle and wave concepts were developed from observations on large-scale objects, and there is no guarantee that they will be fully applicable on the microscopic scale. Under certain experimental conditions, an electron behaves like a particle. Under other conditions, it behaves like a wave. However, an electron is neither a particle nor a wave. It is something that cannot be adequately described in terms of a model we can visualize.

A similar situation holds for light, which shows wave properties in some situations and particle properties in others. Light originates in the microscopic world of atoms and molecules and cannot be fully understood in terms of models visualizable by the human mind.

Although both electrons and light exhibit an apparent "wave–particle duality," there are significant differences between these entities. Light travels at speed c in vacuum, and photons have zero rest mass. Electrons always travel at speeds less than c and have a nonzero rest mass.

17.5 THE UNCERTAINTY PRINCIPLE

The apparent wave–particle duality of matter and of radiation imposes certain limitations on the information we can obtain about a microscopic system. Consider a microscopic particle traveling in the y direction. Suppose we measure the x coordinate of the particle by having it pass through a narrow slit of width w and fall on a fluorescent screen (Fig. 17.4). If we see a spot on the screen, we can be sure the particle passed through the slit. Therefore, we have measured the x coordinate at the time of passing the slit to an accuracy w. Before the measurement, the particle had zero velocity v_x and zero momentum $p_x = mv_x$ in the x direction. Because the microscopic particle has wavelike properties, it will be diffracted at the slit. Photographs of electron-diffraction patterns at a single slit and at multiple slits are given in C. Jönsson, *Am. J. Phys.*, **42,** 4 (1974).

Diffraction is the bending of a wave around an obstacle. A classical particle would go straight through the slit, and a beam of such particles would show a spread of length w in where they hit the screen. A wave passing through the slit will spread out to give a diffraction pattern. The curve in Fig. 17.4 shows the intensity of the wave at various points on the screen. The maxima and minima result from constructive and destructive interference between waves originating from various parts of the slit. **Interference** results from the superposition of two waves traveling through the same region of space. When the waves are in phase (crests occurring together), constructive interference occurs, with the amplitudes adding to give a stronger wave. When the waves are out of phase (crests of one wave coinciding with troughs of the second wave), destructive interference occurs and the intensity is diminished.

The first minima (points P and Q) in the single-slit diffraction pattern occur at places on the screen where waves originating from the top of the slit travel one-half wavelength less or more than waves originating from the middle of the slit. These waves are then exactly out of phase and cancel each other. Similarly, waves originating from a distance d below the top of the slit cancel waves originating a distance d below the center of the slit. The condition for the first diffraction minimum is then $\overline{DP} - \overline{AP} = \frac{1}{2}\lambda = \overline{CD}$ in Fig. 17.4, where C is located so that $\overline{CP} = \overline{AP}$. Because the distance from the slit to the screen is much greater than the slit width, angle APC is nearly zero and angles PAC and ACP are each nearly 90°. Hence, angle ACD is essentially 90°. Angles PDE and DAC each equal 90° minus angle ADC. These

Figure 17.4

Diffraction at a slit.

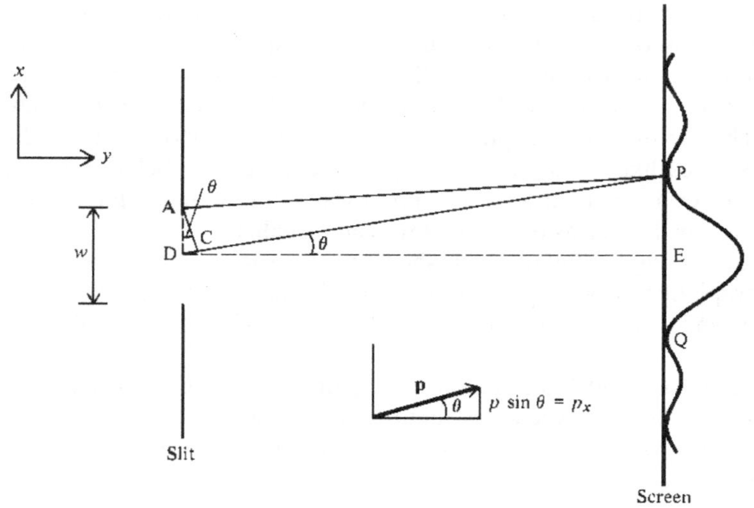

two angles are therefore equal and have been marked θ. We have $\sin\theta = \overline{\text{DC}}/\overline{\text{AD}} = \frac{1}{2}\lambda/\frac{1}{2}w = \lambda/w$. The angle θ at which the first diffraction minimum occurs is given by $\sin\theta = \lambda/w$.

For a microscopic particle passing through the slit, diffraction at the slit will change the particle's direction of motion. A particle diffracted by angle θ and hitting the screen at P or Q will have an x component of momentum $p_x = p \sin\theta$ at the slit (Fig. 17.4), where p is the particle's momentum. The intensity curve in Fig. 17.4 shows that the particle is most likely to be diffracted by an angle lying in the range $-\theta$ to $+\theta$, where θ is the angle to the first diffraction minimum. Hence, the position measurement produces an uncertainty in the p_x value given by $p\sin\theta - (-p\sin\theta) = 2p\sin\theta$. We write $\Delta p_x = 2p\sin\theta$, where Δp_x gives the uncertainty in our knowledge of p_x at the slit. We saw in the previous paragraph that $\sin\theta = \lambda/w$, so $\Delta p_x = 2p\lambda/w$. The de Broglie relation (17.8) gives $\lambda = h/p$, so $\Delta p_x = 2h/w$. The uncertainty in our knowledge of the x coordinate is given by the slit width, so $\Delta x = w$. Therefore, $\Delta x\, \Delta p_x = 2h$.

Before the measurement, we had no knowledge of the particle's x coordinate, but we knew that it was traveling in the y direction and so had $p_x = 0$. Thus, before the measurement, $\Delta x = \infty$ and $\Delta p_x = 0$. The slit of width w gave the x coordinate to an uncertainty w ($\Delta x = w$) but introduced an uncertainty $\Delta p_x = 2h/w$ in p_x. By reducing the slit width w, we can measure the x coordinate as accurately as we please, but as $\Delta x = w$ becomes smaller, $\Delta p_x = 2h/w$ becomes larger. The more we know about x, the less we know about p_x. The measurement introduces an uncontrollable and unpredictable disturbance in the system, changing p_x by an unknown amount.

Although we have analyzed only one experiment, analysis of many other experiments leads to the same conclusion: the product of the uncertainties in x and p_x of a particle is on the order of magnitude of Planck's constant or greater:

$$\Delta x\, \Delta p_x \gtrsim h \qquad \textbf{(17.9)*}$$

This is the **uncertainty principle,** discovered by Heisenberg in 1927. A general quantum-mechanical proof of (17.9) was given by Robertson in 1929. Similarly we have $\Delta y\, \Delta p_y \gtrsim h$ and $\Delta z\, \Delta p_z \gtrsim h$.

The small size of h makes the uncertainty principle of no consequence for macroscopic particles.

17.6 QUANTUM MECHANICS

The fact that electrons and other microscopic "particles" show wavelike as well as particlelike behavior indicates that electrons do not obey classical mechanics. Classical mechanics was formulated from the observed behavior of macroscopic objects and does not apply to microscopic particles. The form of mechanics obeyed by microscopic systems is called **quantum mechanics,** since a key feature of this mechanics is the quantization of energy. The laws of quantum mechanics were discovered by Heisenberg, Born, and Jordan in 1925 and by Schrödinger in 1926. Before discussing these laws, we consider some aspects of *classical* mechanics.

Classical Mechanics

The motion of a one-particle, one-dimensional classical-mechanical system is governed by Newton's second law $F = ma = m\, d^2x/dt^2$. To obtain the particle's position x as a function of time, this differential equation must be integrated twice with respect to time. The first integration gives dx/dt, and the second integration gives x. Each integration introduces an arbitrary integration constant. Therefore, integration of $F = ma$ gives an equation for x that contains two unknown constants c_1 and c_2; we have $x = f(t, c_1, c_2)$, where f is some function. To evaluate c_1 and c_2, we need two pieces of information about the system. If we know that at a certain time t_0, the particle was at the position x_0 and had speed v_0, then c_1 and c_2 can be evaluated from the equations $x_0 = f(t_0, c_1, c_2)$ and $v_0 = f'(t_0, c_1, c_2)$, where f' is the derivative of f with respect to t. Thus, provided we know the force F and the particle's initial position and velocity (or momentum), we can use Newton's second law to predict the position of the particle at any future time. A similar conclusion holds for a three-dimensional many-particle classical system.

The **state** of a system in classical mechanics is defined by specifying all the forces acting and all the positions and velocities (or momenta) of the particles. We saw in the preceding paragraph that knowledge of the present state of a classical-mechanical system enables its future state to be predicted with certainty.

The Heisenberg uncertainty principle, Eq. (17.9), shows that simultaneous specification of position and momentum is impossible for a microscopic particle. Hence, the very knowledge needed to specify the classical-mechanical state of a system is unobtainable in quantum theory. The state of a quantum-mechanical system must therefore involve less knowledge about the system than in classical mechanics.

Quantum Mechanics

In quantum mechanics, the **state** of a system is defined by a mathematical function Ψ (capital psi) called the **state function** or the **time-dependent wave function.** (As part of the definition of the state, the potential-energy function V must also be specified.) Ψ is a function of the coordinates of the particles of the system and (since the state may change with time) is also a function of time. For example, for a two-particle system, $\Psi = \Psi(x_1, y_1, z_1, x_2, y_2, z_2, t)$, where x_1, y_1, z_1 and x_2, y_2, z_2 are the coordinates of particles 1 and 2, respectively. The state function is in general a complex quantity; that is, $\Psi = f + ig$, where f and g are real functions of the coordinates and time and $i \equiv \sqrt{-1}$. The state function is an abstract entity, but we shall later see how Ψ is related to physically measurable quantities.

The state function changes with time. For an n-particle system, quantum mechanics postulates that the equation governing how Ψ changes with t is

$$-\frac{\hbar}{i}\frac{\partial \Psi}{\partial t} = -\frac{\hbar^2}{2m_1}\left(\frac{\partial^2\Psi}{\partial x_1^2} + \frac{\partial^2\Psi}{\partial y_1^2} + \frac{\partial^2\Psi}{\partial z_1^2}\right) - \cdots - \frac{\hbar^2}{2m_n}\left(\frac{\partial^2\Psi}{\partial x_n^2} + \frac{\partial^2\Psi}{\partial y_n^2} + \frac{\partial^2\Psi}{\partial z_n^2}\right) + V\Psi \qquad (17.10)$$

In this equation, $\hbar$ (**h-bar**) is Planck's constant divided by 2π,

$$\hbar \equiv h/2\pi \tag{17.11*}$$

i is $\sqrt{-1}$; $m_1, \ldots, m_n$ are the masses of particles $1, \ldots, n$; x_1, y_1, z_1 are the spatial coordinates of particle 1; and V is the potential energy of the system. Since the potential energy is energy due to the particles' positions, V is a function of the particles' coordinates. Also, V can vary with time if an externally applied field varies with time. Hence, V is in general a function of the particles' coordinates and the time. V is derived from the forces acting in the system; see Eq. (2.17). The dots in Eq. (17.10) stand for terms involving the spatial derivatives of particles $2, 3, \ldots, n - 1$.

Equation (17.10) is a complicated partial differential equation. For most of the problems dealt with in this book, it will not be necessary to use (17.10), so don't panic.

The concept of the state function Ψ and Eq. (17.10) were introduced by the Austrian physicist Erwin Schrödinger (1887–1961) in 1926. Equation (17.10) is the **time-dependent Schrödinger equation.** Schrödinger was inspired by the de Broglie hypothesis to search for a mathematical equation that would resemble the differential equations that govern wave motion and that would have solutions giving the allowed energy levels of a quantum system. Using the de Broglie relation $\lambda = h/p$ and certain plausibility arguments, Schrödinger proposed Eq. (17.10) and the related time-independent equation (17.24) below. These plausibility arguments have been omitted in this book. It should be emphasized that these arguments can at best make the Schrödinger equation seem plausible. They can in no sense be used to derive or prove the Schrödinger equation. The Schrödinger equation is a fundamental postulate of quantum mechanics and cannot be derived. The reason we believe it to be true is that its predictions give excellent agreement with experimental results. "One could argue that the Schrödinger equation has had more to do with the evolution of twentieth-century science and technology than any other discovery in physics." (Jeremy Bernstein, *Cranks, Quarks, and the Cosmos,* Basic Books, 1993, p. 54.)

In 1925, several months before Schrödinger's work, Werner Heisenberg (1901–1976), Max Born (1882–1970), and Pascual Jordan (1902–1980) developed a form of quantum mechanics based on mathematical entities called matrices. A matrix is a rectangular array of numbers; matrices are added and multiplied according to certain rules. The *matrix mechanics* of these workers turns out to be fully equivalent to the Schrödinger form of quantum mechanics (which is often called *wave mechanics*). We shall not discuss matrix mechanics.

Schrödinger also contributed to statistical mechanics, relativity, and the theory of color vision and was deeply interested in philosophy. In an epilogue to his 1944 book, *What Is Life?*, Schrödinger wrote: "So let us see whether we cannot draw the correct, non-contradictory conclusion from the following two premises: (i) My body functions as a pure mechanism according to the Laws of Nature. (ii) Yet I know, by incontrovertible direct experience, that I am directing its motions . . . . The only possible inference from these two facts is, I think, that I—I in the widest meaning of the word, that is to say, every conscious mind that has ever said or felt 'I'—am the person, if any, who controls the 'motion of the atoms' according to the Laws of Nature." Schrödinger's life and loves are chronicled in W. Moore, *Schrödinger, Life and Thought,* Cambridge University Press, 1989.

The time-dependent Schrödinger equation (17.10) contains the first derivative of Ψ with respect to t, and a single integration with respect to time gives us Ψ. Integration of (17.10) therefore introduces only one integration constant, which can be evaluated if Ψ is known at some initial time t_0. Therefore, knowing the initial quantum-mechanical state $\Psi(x_1, \ldots, z_n, t_0)$ and the potential energy V, we can use (17.10) to predict the future quantum-mechanical state. The time-dependent Schrödinger equation is the quantum-mechanical analog of Newton's second law, which allows the future state of a classical-mechanical system to be predicted from its present state. We shall soon see, however, that knowledge of the state in quantum mechanics usually involves a knowledge of only probabilities, rather than certainties, as in classical mechanics.

What is the relation between quantum mechanics and classical mechanics? Experiment shows that macroscopic bodies obey classical mechanics (provided their speed is much less than the speed of light). We therefore expect that in the classical-mechanical limit of taking $h \to 0$, the time-dependent Schrödinger equation ought to reduce to Newton's second law. This was shown by Ehrenfest in 1927; for Ehrenfest's proof, see *Park,* sec. 3.3.

Physical Meaning of the State Function Ψ

Schrödinger originally conceived of Ψ as the amplitude of some sort of wave that was associated with the system. It soon became clear that this interpretation was wrong. For example, for a two-particle system, Ψ is a function of the six spatial coordinates x_1, y_1, z_1, x_2, y_2, and z_2, whereas a wave moving through space is a function of only three spatial coordinates. The correct physical interpretation of Ψ was given by Max Born in 1926. Born postulated that $|\Psi|^2$ gives the **probability density** for finding the particles at given locations in space. (Probability densities for molecular speeds were discussed in Sec. 14.4.) To be more precise, suppose a one-particle system has the state function $\Psi(x, y, z, t')$ at time t'. Consider the probability that a measurement of the particle's position at time t' will find the particle with its x, y, and z coordinates in the infinitesimal ranges x_a to $x_a + dx$, y_a to $y_a + dy$, and z_a to $z_a + dz$, respectively. This is the probability of finding the particle in a tiny rectangular-box-shaped region located at point (x_a, y_a, z_a) in space and having edges dx, dy, and dz (Fig. 17.5). Born's postulate is that the probability is given by

$$\Pr(x_a \le x \le x_a + dx, y_a \le y \le y_a + dy, z_a \le z \le z_a + dz)$$
$$= |\Psi(x_a, y_a, z_a, t')|^2 \, dx \, dy \, dz \qquad \textbf{(17.12)*}$$

where the left side of (17.12) denotes the probability the particle is found in the box of Fig. 17.5.

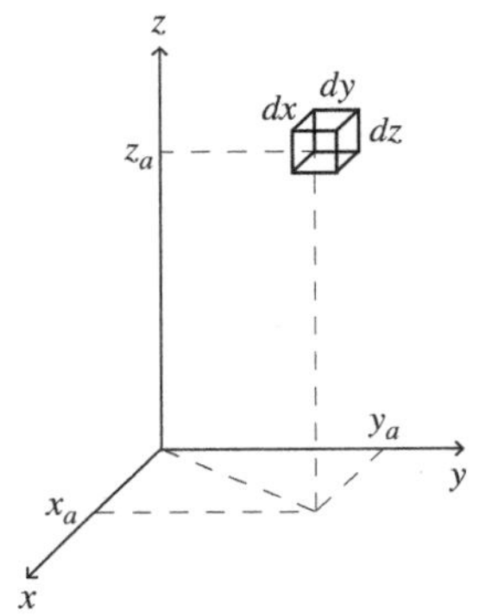

Figure 17.5

An infinitesimal box in space.

EXAMPLE 17.1 Probability for finding a particle

Suppose that at time t' the state function of a one-particle system is

$$\Psi = (2/\pi c^2)^{3/4} e^{-(x^2+y^2+z^2)/c^2} \qquad \text{where } c = 2 \text{ nm}$$

[One nanometer (nm) $\equiv 10^{-9}$ m.] Find the probability that a measurement of the particle's position at time t' will find the particle in the tiny cubic region with its center at $x = 1.2$ nm, $y = -1.0$ nm, and $z = 0$ and with edges each of length 0.004 nm.

The distance 0.004 nm is much less than the value of c and a change of 0.004 nm in one or more of the coordinates will not change the probability density $|\Psi|^2$ significantly. It is therefore a good approximation to consider the interval 0.004 nm as infinitesimal and to use (17.12) to write the desired probability as

$$|\Psi|^2 \, dx \, dy \, dz = (2/\pi c^2)^{3/2} e^{-2(x^2+y^2+z^2)/c^2} \, dx \, dy \, dz$$
$$= [2/(4\pi \text{ nm}^2)]^{3/2} e^{-2[(1.2)^2+(-1)^2+0^2]/4} (0.004 \text{ nm})^3$$
$$= 1.200 \times 10^{-9}$$

Exercise

(*a*) At what point is the probability density a maximum for the Ψ of this example? Answer by simply looking at $|\Psi|^2$. (*b*) Redo the calculation with x changed to its minimum value in the tiny cubic region and then with x changed to its maximum value in the region. Compare the results with that found when the central value of x is used. [*Answers:* (*a*) At the origin. (*b*) 1.203×10^{-9}, 1.197×10^{-9}.]

The state function Ψ is a complex quantity, and $|\Psi|$ is the absolute value of Ψ. Let $\Psi = f + ig$, where f and g are real functions and $i \equiv \sqrt{-1}$. The **absolute value** of Ψ is defined by $|\Psi| \equiv (f^2 + g^2)^{1/2}$. For a real quantity, g is zero, and the absolute value becomes $(f^2)^{1/2}$, which is the usual meaning of absolute value for a real quantity. The **complex conjugate** Ψ^* of Ψ is defined by

$$\Psi^* \equiv f - ig, \qquad \text{where } \Psi = f + ig \qquad \textbf{(17.13)*}$$

To get Ψ^*, we replace i by $-i$ wherever it occurs in Ψ. Note that

$$\Psi^*\Psi = (f - ig)(f + ig) = f^2 - i^2g^2 = f^2 + g^2 = |\Psi|^2 \qquad (17.14)$$

since $i^2 = -1$. Therefore, instead of $|\Psi|^2$, we can write $\Psi^*\Psi$. The quantity $|\Psi|^2 = \Psi^*\Psi = f^2 + g^2$ is real and nonnegative, as a probability density must be.

In a two-particle system, $|\Psi(x_1, y_1, z_1, x_2, y_2, z_2, t')|^2\, dx_1\, dy_1\, dz_1\, dx_2\, dy_2\, dz_2$ is the probability that, at time t', particle 1 is in a tiny rectangular-box-shaped region located at point (x_1, y_1, z_1) and having dimensions dx_1, dy_1, and dz_1, and particle 2 is simultaneously in a box-shaped region at (x_2, y_2, z_2) with dimensions dx_2, dy_2, and dz_2. Born's interpretation of Ψ gives results fully consistent with experiment.

For a one-particle, one-dimensional system, $|\Psi(x, t)|^2\, dx$ is the probability that the particle is between x and $x + dx$ at time t. The probability that it is in the region between a and b is found by summing the infinitesimal probabilities over the interval from a to b to give the definite integral $\int_a^b |\Psi|^2\, dx$. Thus

$$\Pr(a \le x \le b) = \int_a^b |\Psi|^2\, dx \qquad \text{one-particle, one-dim. syst.} \qquad \textbf{(17.15)*}$$

The probability of finding the particle somewhere on the x axis must be 1. Hence, $\int_{-\infty}^{\infty} |\Psi|^2\, dx = 1$. When Ψ satisfies this equation, it is said to be **normalized.** The normalization condition for a one-particle, three-dimensional system is

$$\int_{-\infty}^{\infty}\int_{-\infty}^{\infty}\int_{-\infty}^{\infty} |\Psi(x, y, z, t)|^2\, dx\, dy\, dz = 1 \qquad (17.16)$$

For an n-particle, three-dimensional system, the integral of $|\Psi|^2$ over all $3n$ coordinates $x_1, \ldots, z_n$, each integrated from $-\infty$ to ∞, equals 1.

The integral in (17.16) is a multiple integral. In a double integral like $\int_a^b \int_c^d f(x, y)\, dx\, dy$, one first integrates $f(x, y)$ with respect to x (while treating y as a constant) between the limits c and d, and then integrates the result with respect to y. For example, $\int_0^1 \int_0^4 (2xy + y^2)\, dx\, dy = \int_0^1 (x^2y + xy^2)\big|_0^4\, dy = \int_0^1 (16y + 4y^2)\, dy = 28/3$. To evaluate a triple integral like (17.16), we first integrate with respect to x while treating y and z as constants, then integrate with respect to y while treating z as constant, and finally integrate with respect to z.

The normalization requirement is often written

$$\int |\Psi|^2\, d\tau = 1 \qquad \textbf{(17.17)*}$$

where $\int d\tau$ is a shorthand notation that stands for the *definite* integral over the full ranges of all the spatial coordinates of the system. For a one-particle, three-dimensional system, $\int d\tau$ implies a triple integral over x, y, and z from $-\infty$ to ∞ for each coordinate [Eq. (17.16)].

By substitution, it is easy to see that, if Ψ is a solution of (17.10), then so is $c\Psi$, where c is an arbitrary constant. Thus, there is always an arbitrary multiplicative constant in each solution to (17.10). The value of this constant is chosen so as to satisfy the normalization requirement (17.17).

From the state function Ψ, we can calculate the probabilities of the various possible outcomes when a measurement of position is made on the system. In fact, Born's work is more general than this. It turns out that Ψ gives information on the outcome of a measurement of *any* property of the system, not just position. For example, if Ψ is known, we can calculate the probability of each possible outcome when a measurement of p_x, the x component of momentum, is made. The same is true for a measurement of energy, or angular momentum, etc. (The procedure for calculating these probabilities from Ψ is discussed in *Levine,* sec. 7.6.)

The state function Ψ is not to be thought of as a physical wave. Instead Ψ is an abstract mathematical entity that gives information about the state of the system. Everything that can be known about the system in a given state is contained in the state function Ψ. Instead of saying "the state described by the function Ψ," we can just as well say "the state Ψ." The information given by Ψ is the probabilities for the possible outcomes of measurements of the system's physical properties.

The state function Ψ describes a physical system. In Chapters 17 to 20, the system will usually be a particle, atom, or molecule. One can also consider the state function of a system that contains a large number of molecules, for example, a mole of some compound; this will be done in Chapter 21 on statistical mechanics.

Classical mechanics is a deterministic theory in that it allows us to predict the exact paths taken by the particles of the system and tells us where they will be at any future time. In contrast, quantum mechanics gives only the probabilities of finding the particles at various locations in space. The concept of a path for a particle becomes rather fuzzy in a time-dependent quantum-mechanical system and disappears in a time-independent quantum-mechanical system.

Some philosophers have used the Heisenberg uncertainty principle and the nondeterministic nature of quantum mechanics as arguments in favor of human free will.

The probabilistic nature of quantum mechanics disturbed many physicists, including Einstein, Schrödinger, and de Broglie. (Einstein wrote in 1926: "Quantum mechanics . . . says a lot, but does not really bring us any closer to the secret of the Old One. I, at any rate, am convinced that He does not throw dice." When someone pointed out to Einstein that Einstein himself had introduced probability into quantum theory when he interpreted a light wave's intensity in each small region of space as being proportional to the probability of finding a photon in that region, Einstein replied, "A good joke should not be repeated too often.") These scientists believed that quantum mechanics does not furnish a complete description of physical reality. However, attempts to replace quantum mechanics by an underlying deterministic theory have failed. There appears to be a fundamental randomness in nature at the microscopic level.

Summary

The state of a quantum-mechanical system is described by its state function Ψ, which is a function of time and the spatial coordinates of the particles of the system. The state function provides information on the probabilities of the outcomes of measurements on the system. For example, when a position measurement is made on a one-particle system at time t', the probability that the particle's coordinates are found to be in the ranges x to $x + dx$, y to $y + dy$, z to $z + dz$ is given by $|\Psi(x, y, z, t')|^2\, dx\, dy\, dz$. The function $|\Psi|^2$ is the probability density for position. Because the total probability of finding the particles somewhere is 1, the state function is normalized, meaning that the definite integral of $|\Psi|^2$ over the full range of all the spatial coordinates is equal to 1. The state function Ψ changes with time according to the time-dependent Schrödinger equation (17.10), which allows the future state (function) to be calculated from the present state (function).

17.7 THE TIME-INDEPENDENT SCHRÖDINGER EQUATION

For an isolated atom or molecule, the forces acting depend only on the coordinates of the charged particles of the system and are independent of time. Therefore, the potential energy V is independent of t for an isolated system. For systems where V is independent of time, the time-dependent Schrödinger equation (17.10) has solutions of the form $\Psi(x_1, \ldots, z_n, t) = f(t)\psi(x_1, \ldots, z_n)$, where ψ (lowercase psi) is a function of the $3n$ coordinates of the n particles and f is a certain function of time. We shall demonstrate this for a one-particle, one-dimensional system.

For a one-particle, one-dimensional system with V independent of t, Eq. (17.10) becomes

$$-\frac{\hbar^2}{2m}\frac{\partial^2\Psi}{\partial x^2} + V(x)\Psi = -\frac{\hbar}{i}\frac{\partial\Psi}{\partial t} \tag{17.18}$$

Let us look for those solutions of (17.18) that have the form

$$\Psi(x, t) = f(t)\psi(x) \tag{17.19}$$

We have $\partial^2\Psi/\partial x^2 = f(t)\, d^2\psi/dx^2$ and $\partial\Psi/\partial t = \psi(x)\, df/dt$. Substitution into (17.18) followed by division by $f\psi = \Psi$ gives

$$-\frac{\hbar^2}{2m}\frac{1}{\psi(x)}\frac{d^2\psi}{dx^2} + V(x) = -\frac{\hbar}{i}\frac{1}{f(t)}\frac{df(t)}{dt} \equiv E \tag{17.20}$$

where the parameter E was defined as $E \equiv -(\hbar/i)f'(t)/f(t)$.

From the definition of E, it is equal to a function of t only and hence is independent of x. However, (17.20) shows that $E = -(\hbar^2/2m)\psi''(x)/\psi(x) + V(x)$, which is a function of x only and is independent of t. Hence, E is independent of t as well as independent of x and must therefore be a constant. Since the constant E has the same dimensions as V, it has the dimensions of energy. Quantum mechanics postulates that E is in fact the energy of the system.

Equation (17.20) gives $df/f = -(iE/\hbar)\, dt$, which integrates to $\ln f = -iEt/\hbar + C$. Therefore $f = e^C e^{-iEt/\hbar} = Ae^{-iEt/\hbar}$, where $A \equiv e^C$ is an arbitrary constant. The constant A can be included as part of the $\psi(x)$ factor in (17.19), so we omit it from f. Thus

$$f(t) = e^{-iEt/\hbar} \tag{17.21}$$

Equation (17.20) also gives

$$-\frac{\hbar^2}{2m}\frac{d^2\psi(x)}{dx^2} + V(x)\psi(x) = E\psi(x) \tag{17.22}$$

which is the **(time-independent) Schrödinger equation** for a one-particle, one-dimensional system. Equation (17.22) can be solved for ψ when the potential-energy function $V(x)$ has been specified.

For an n-particle, three-dimensional system, the same procedure that led to Eqs. (17.19), (17.21), and (17.22) gives

$$\Psi = e^{-iEt/\hbar}\psi(x_1, y_1, z_1, \ldots, x_n, y_n, z_n) \tag{17.23}$$

where the function ψ is found by solving

$$-\frac{\hbar^2}{2m_1}\left(\frac{\partial^2\psi}{\partial x_1^2}+\frac{\partial^2\psi}{\partial y_1^2}+\frac{\partial^2\psi}{\partial z_1^2}\right)-\cdots-\frac{\hbar^2}{2m_n}\left(\frac{\partial^2\psi}{\partial x_n^2}+\frac{\partial^2\psi}{\partial y_n^2}+\frac{\partial^2\psi}{\partial z_n^2}\right)+V\psi=E\psi$$

(17.24)*

The solutions ψ to the time-independent Schrödinger equation (17.24) are the **(time-independent) wave functions.** States for which Ψ is given by (17.23) are called **stationary states.** We shall see that for a given system there are many different solutions to (17.24), different solutions corresponding to different values of the energy E. In general, quantum mechanics gives only probabilities and not certainties for the outcome of a measurement. However, when a system is in a stationary state, a measurement of its energy is certain to give the particular energy value that corresponds to the wave function ψ of the system. Different systems have different forms for the potential-energy function $V(x_1, \ldots, z_n)$, and this leads to different sets of allowed wave functions and energies when (17.24) is solved for different systems. All this will be made clearer by the examples in the next few sections.

For a stationary state, the probability density $|\Psi|^2$ becomes

$$|\Psi|^2 = |f\psi|^2 = (f\psi)^*f\psi = f^*\psi^*f\psi = e^{iEt/\hbar}\psi^*e^{-iEt/\hbar}\psi = e^0\psi^*\psi = |\psi|^2 \quad (17.25)$$

where we used (17.19), (17.21) and the identity

$$(f\psi)^* = f^*\psi^*$$

(Prob. 17.19). Hence, for a stationary state, $|\Psi|^2 = |\psi|^2$, which is independent of time. For a stationary state, the probability density and the energy are constant with time. There is no implication, however, that the particles of the system are at rest in a stationary state.

It turns out that the probabilities for the outcomes of measurements of any physical property involve $|\Psi|$, and since $|\Psi| = |\psi|$, these probabilities are independent of time for a stationary state. Thus, the $e^{-iEt/\hbar}$ factor in (17.23) is of little consequence, and *the essential part of the state function for a stationary state is the time-independent wave function* $\psi(x_1, \ldots, z_n)$. For a stationary state, the normalization condition (17.17) becomes $\int |\psi|^2\, d\tau = 1$, where $\int d\tau$ denotes the definite integral over all space.

The wave function ψ of a stationary state of energy E must satisfy the time-independent Schrödinger equation (17.24). However, quantum mechanics postulates that not all functions that satisfy (17.24) are allowed as wave functions for the system. In addition to being a solution of (17.24), a wave function must meet the following three conditions: (*a*) The wave function must be **single-valued.** (*b*) The wave function must be **continuous.** (*c*) The wave function must be **quadratically integrable.** Condition (*a*) means that ψ has one and only one value at each point in space. The function of Fig. 17.6*a*, which is multiple-valued at some points, is not a possible wave function for a one-particle, one-dimensional system. Condition (*b*) means that ψ makes no sudden jumps in value. A function like that in Fig. 17.6*b* is ruled out. Condition (*c*) means that the integral over all space $\int |\psi|^2\, d\tau$ is a finite number. The function x^2 (Fig. 17.6*c*) is not quadratically integrable, since $\int_{-\infty}^{\infty} x^4\, dx = (x^5/5)|_{-\infty}^{\infty} = \infty - (-\infty) = \infty$. Condition (*c*) allows the wave function to be multiplied by a constant that normalizes it, that is, that makes $\int |\psi|^2\, d\tau = 1$. [If ψ is a solution of the Schrödinger equation (17.24), then so is $k\psi$, where k is any constant; see Prob. 17.20.] A function obeying conditions (*a*), (*b*), and (*c*) is said to be **well-behaved.**

Since E occurs as an undetermined parameter in the Schrödinger equation (17.24), the solutions ψ that are found by solving (17.24) will depend on E as a parameter: $\psi = \psi(x_1, \ldots, z_n; E)$. It turns out that ψ *is well-behaved only for certain particular values of E, and it is these values that are the allowed energy levels.* An example is given in the next section.

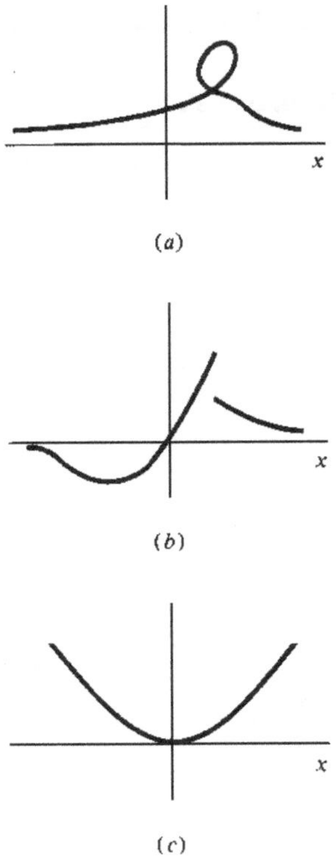

Figure 17.6

(*a*) A multivalued function. (*b*) A discontinuous function. (*c*) A function that is not quadratically integrable.

We shall mainly be interested in the stationary states of atoms and molecules, since these give the allowed energy levels. For a collision between two molecules or for a molecule exposed to the time-varying electric and magnetic fields of electromagnetic radiation, the potential energy V depends on time, and one must deal with the time-dependent Schrödinger equation and with nonstationary states.

Summary
In an isolated atom or molecule, the potential energy V is independent of time and the system can exist in a stationary state, which is a state of constant energy and time-independent probability density. For a stationary state, the probability density for the particles' locations is given by $|\psi|^2$, where the time-independent wave function ψ is a function of the coordinates of the particles of the system. The possible stationary-state wave functions and energies for a system are found by solving the time-independent Schrödinger equation (17.24) and picking out only those solutions that are single-valued, continuous, and quadratically integrable.

17.8 THE PARTICLE IN A ONE-DIMENSIONAL BOX

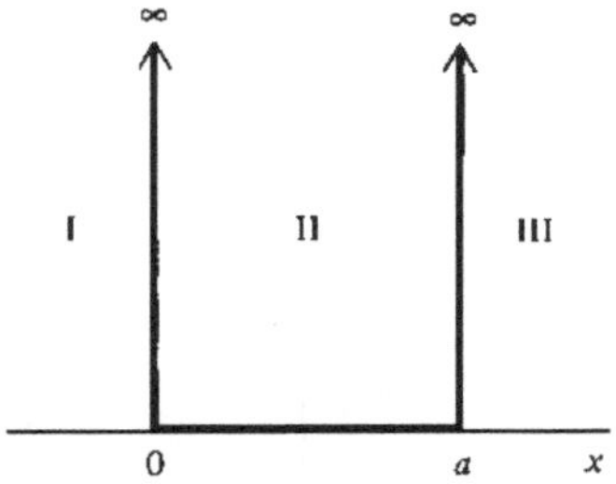

Figure 17.7

Potential-energy function for a particle in a one-dimensional box.

The introduction to quantum mechanics in the last two sections is quite abstract. To help make the ideas of quantum mechanics more understandable, this section examines the stationary states of a simple system, **a particle in a one-dimensional box.** By this is meant a single microscopic particle of mass m moving in one dimension x and subject to the potential-energy function of Fig. 17.7. The potential energy is zero for x between 0 and a (region II) and is infinite elsewhere (regions I and III):

$$V = \begin{cases} 0 & \text{for } 0 \leq x \leq a \\ \infty & \text{for } x < 0 \text{ and for } x > a \end{cases}$$

This potential energy confines the particle to move in the region between 0 and a on the x axis. No real system has a V as simple as Fig. 17.7, but the particle in a box can be used as a crude model for dealing with pi electrons in conjugated molecules (Sec. 19.11).

We restrict ourselves to considering the states of constant energy, the stationary states. For these states, the (time-independent) wave functions ψ are found by solving the Schrödinger equation (17.24), which for a one-particle, one-dimensional system is

$$-\frac{\hbar^2}{2m}\frac{d^2\psi}{dx^2} + V\psi = E\psi \tag{17.26}$$

Since a particle cannot have infinite energy, there must be zero probability of finding the particle in regions I and III, where V is infinite. Therefore, the probability density $|\psi|^2$ and hence ψ must be zero in these regions: $\psi_{\text{I}} = 0$ and $\psi_{\text{III}} = 0$, or

$$\psi = 0 \quad \text{for} \quad x < 0 \text{ and for } x > a \tag{17.27}$$

Inside the box (region II), V is zero and (17.26) becomes

$$\frac{d^2\psi}{dx^2} = -\frac{2mE}{\hbar^2}\psi \quad \text{for} \quad 0 \leq x \leq a \tag{17.28}$$

To solve this equation, we need a function whose second derivative gives us the same function back again, but multiplied by a constant. Two functions that behave this way are the sine function and the cosine function, so let us try as a solution

$$\psi = A \sin rx + B \cos sx$$

where A, B, r, and s are constants. Differentiation of ψ gives $d^2\psi/dx^2 = -Ar^2 \sin rx - Bs^2 \cos sx$. Substitution of the trial solution in (17.28) gives

$$-Ar^2 \sin rx - Bs^2 \cos sx = -2mE\hbar^{-2}A \sin rx - 2mE\hbar^{-2}B \cos sx \qquad (17.29)$$

If we take $r = s = (2mE)^{1/2}\hbar^{-1}$, Eq. (17.29) is satisfied. The solution of (17.28) is therefore

$$\psi = A \sin[(2mE)^{1/2}\hbar^{-1}x] + B \cos[(2mE)^{1/2}\hbar^{-1}x] \qquad \text{for } 0 \le x \le a \qquad (17.30)$$

A more formal derivation than we have given shows that (17.30) is indeed the general solution of the differential equation (17.28).

As noted in Sec. 17.7, not all solutions of the Schrödinger equation are acceptable wave functions. Only well-behaved functions are allowed. The solution of the particle-in-a-box Schrödinger equation is the function defined by (17.27) and (17.30), where A and B are arbitrary constants of integration. For this function to be continuous, the wave function inside the box must go to zero at the two ends of the box, since ψ equals zero outside the box. We must require that ψ in (17.30) go to zero as $x \to 0$ and as $x \to a$. Setting $x = 0$ and $\psi = 0$ in (17.30), we get $0 = A \sin 0 + B \cos 0 = A \cdot 0 + B \cdot 1$, so $B = 0$. Therefore

$$\psi = A \sin[(2mE)^{1/2}\hbar^{-1}x] \qquad \text{for } 0 \le x \le a \qquad (17.31)$$

Setting $x = a$ and $\psi = 0$ in (17.31), we get $0 = \sin[(2mE)^{1/2}\hbar^{-1}a]$. The function $\sin w$ equals zero when w is $0, \pm\pi, \pm 2\pi, \ldots, \pm n\pi$, so we must have

$$(2mE)^{1/2}\hbar^{-1}a = \pm n\pi \qquad (17.32)$$

Substitution of (17.32) in (17.31) gives $\psi = A \sin(\pm n\pi x/a) = \pm A \sin(n\pi x/a)$, since $\sin(-z) = -\sin z$. The use of $-n$ instead of n multiplies ψ by -1. Since A is arbitrary, this doesn't give a solution different from the $+n$ solution, so there is no need to consider the $-n$ values. Also, the value $n = 0$ must be ruled out, since it would make $\psi = 0$ everywhere (Prob. 17.26), meaning there is no probability of finding the particle in the box. The allowed wave functions are therefore

$$\psi = A \sin(n\pi x/a) \qquad \text{for } 0 \le x \le a, \qquad \text{where } n = 1, 2, 3, \ldots \qquad (17.33)$$

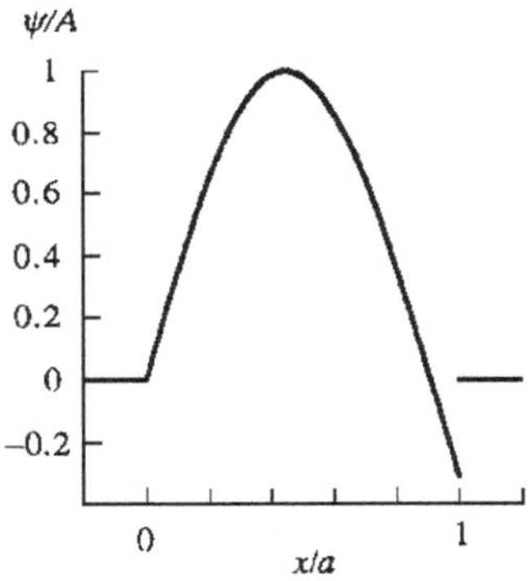

Figure 17.8

Plot of the solution to the particle-in-a-box Schrödinger equation for $E = (1.1)^2h^2/8ma^2$. This solution is discontinuous at $x = a$.

The allowed energies are found by solving (17.32) for E to get

$$E = \frac{n^2h^2}{8ma^2}, \qquad n = 1, 2, 3, \ldots \qquad \mathbf{(17.34)*}$$

where $\hbar \equiv h/2\pi$ was used. Only these values of E make ψ a well-behaved (continuous) function. For example, Fig. 17.8 plots ψ of (17.27) and (17.31) for $E = (1.1)^2h^2/8ma^2$. Because of the discontinuity at $x = a$, this is not an acceptable wave function.

Confining the particle to be between 0 and a requires that ψ be zero at $x = 0$ and $x = a$, and this quantizes the energy. An analogy is the quantization of the vibrational modes of a string that occurs when the string is held fixed at both ends. The energy levels (17.34) are proportional to n^2, and the separation between adjacent levels increases as n increases (Fig. 17.9).

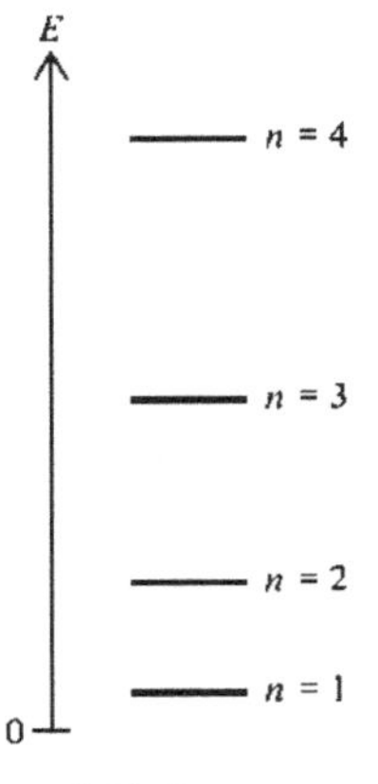

Figure 17.9

Lowest four energy levels of a particle in a one-dimensional box.

The magnitude of the constant A in ψ in (17.33) is found from the normalization condition (17.17) and (17.25): $\int |\psi|^2\, d\tau = 1$. Since $\psi = 0$ outside the box, we need only integrate from 0 to a, and

$$1 = \int_{-\infty}^{\infty} |\psi|^2\, dx = \int_0^a |\psi|^2\, dx = |A|^2 \int_0^a \sin^2\left(\frac{n\pi x}{a}\right) dx$$

608

Chapter 17
Quantum Mechanics

A table of integrals gives $\int \sin^2 cx\, dx = x/2 - (1/4c) \sin 2cx$, and we find $|A| = (2/a)^{1/2}$. The **normalization constant** A can be taken as any number having absolute value $(2/a)^{1/2}$. We could take $A = (2/a)^{1/2}$, or $A = -(2/a)^{1/2}$, or $A = i(2/a)^{1/2}$ (where $i = \sqrt{-1}$), etc. Choosing $A = (2/a)^{1/2}$, we get

$$\psi = \left(\frac{2}{a}\right)^{1/2} \sin \frac{n\pi x}{a} \quad \text{for } 0 \le x \le a, \quad \text{where } n = 1, 2, 3, \ldots \tag{17.35}$$

For a one-particle, one-dimensional system, $|\psi(x)|^2\, dx$ is a probability. Since probabilities have no units, $\psi(x)$ must have dimensions of length$^{-1/2}$, as is true for ψ in (17.35).

The state functions for the stationary states of the particle in a box are given by (17.19), (17.21), and (17.35) as $\Psi = e^{-iEt/\hbar}(2/a)^{1/2} \sin(n\pi x/a)$, for $0 \le x \le a$, where $E = n^2h^2/8ma^2$ and $n = 1, 2, 3, \ldots$.

EXAMPLE 17.2 Calculation of a transition wavelength

Find the wavelength of the light emitted when a 1×10^{-27} g particle in a 3-Å one-dimensional box goes from the $n = 2$ to the $n = 1$ level.

The wavelength λ can be found from the frequency ν. The quantity $h\nu$ is the energy of the emitted photon and equals the energy *difference* between the two levels involved in the transition [Eq. (17.7)]:

$$h\nu = E_{\text{upper}} - E_{\text{lower}} = 2^2h^2/8ma^2 - 1^2h^2/8ma^2 \quad \text{and} \quad \nu = 3h/8ma^2$$

where (17.34) was used. Use of $\lambda = c/\nu$ and 1 Å ≡ 10^{-10} m [Eq. (2.87)] gives

$$\lambda = \frac{8ma^2c}{3h} = \frac{8(1 \times 10^{-30}\ \text{kg})(3 \times 10^{-10}\ \text{m})^2(3 \times 10^8\ \text{m/s})}{3(6.6 \times 10^{-34}\ \text{J s})} = 1 \times 10^{-7}\ \text{m}$$

(The mass m is that of an electron, and the wavelength lies in the ultraviolet.)

Exercise

(*a*) For a particle of mass 9.1×10^{-31} kg in a certain one-dimensional box, the $n = 3$ to $n = 2$ transition occurs at $\nu = 4.0 \times 10^{14}\ \text{s}^{-1}$. Find the length of the box. (*Answer:* 1.07 nm.) (*b*) Show that the frequency of the $n = 3$ to 2 particle-in-a-one-dimensional-box transition is 5/3 times the frequency of the 2 to 1 transition.

Let us contrast the quantum-mechanical and classical pictures. Classically, the particle can rattle around in the box with any nonnegative energy; $E_{\text{classical}}$ can be any number from zero on up. (The potential energy is zero in the box, so the particle's energy is entirely kinetic. Its speed v can have any nonnegative value, so $\frac{1}{2}mv^2$ can have any nonnegative value.) Quantum-mechanically, the energy can take on only the values (17.34). The energy is quantized in quantum mechanics, whereas it is continuous in classical mechanics.

Classically, the minimum energy is zero. Quantum-mechanically, the particle in a box has a minimum energy that is greater than zero. This energy, $h^2/8ma^2$, is the **zero-point energy.** Its existence is a consequence of the uncertainty principle. Suppose the particle could have zero energy. Since its energy is entirely kinetic, its speed v_x and momentum $mv_x = p_x$ would then be zero. With p_x known to be zero, the uncertainty Δp_x is zero, and the uncertainty principle $\Delta x\, \Delta p_x \gtrsim h$ gives $\Delta x = \infty$. However, we know the particle to be somewhere between $x = 0$ and $x = a$, so Δx cannot exceed a. Hence, a zero energy is impossible for a particle in a box.

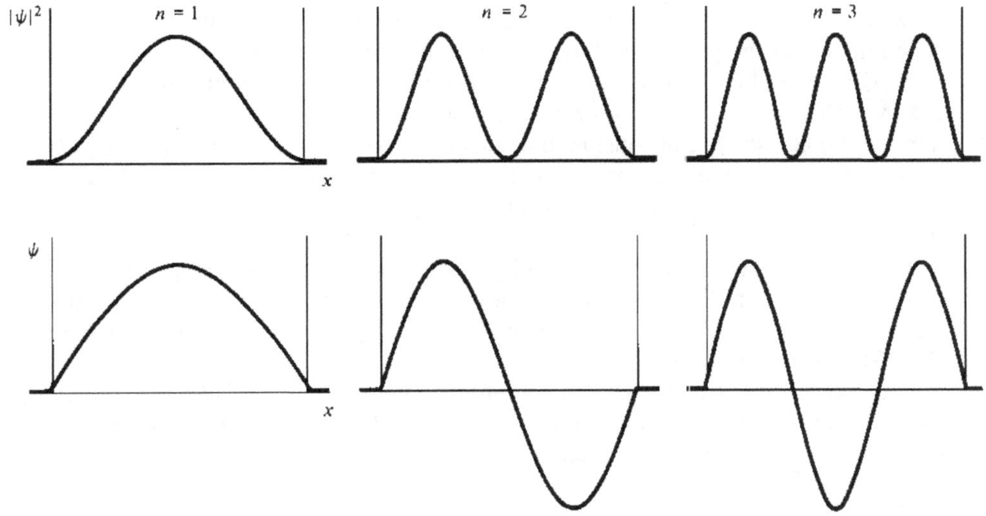

Figure 17.10

Wave functions and probability densities for the lowest three particle-in-a-box stationary states.

The stationary states of a particle in a box are specified by giving the value of the integer n in (17.35). n is called a **quantum number.** The lowest-energy state ($n = 1$) is the **ground state.** States higher in energy than the ground state are **excited states.**

Figure 17.10 plots the wave functions ψ and the probability densities $|\psi|^2$ for the first three particle-in-a-box stationary states. For $n = 1$, $n\pi x/a$ in the wave function (17.35) goes from 0 to π as x goes from 0 to a, so ψ is half of one cycle of a sine function.

Classically, all locations for the particle in the box are equally likely. Quantum-mechanically, the probability density is not uniform along the length of the box, but shows oscillations. In the limit of a very high quantum number n, the oscillations in $|\psi|^2$ come closer and closer together and ultimately become undetectable; this corresponds to the classical result of uniform probability density. The relation $8ma^2E/h^2 = n^2$ shows that for a macroscopic system (E, m, and a having macroscopic magnitudes), n is very large, so the limit of large n is the classical limit.

A point at which $\psi = 0$ is called a **node.** The number of nodes increases by 1 for each increase in n. The existence of nodes is surprising from a classical viewpoint. For example, for the $n = 2$ state, it is hard to understand how the particle can be found in the left half of the box or in the right half but never at the center. The behavior of microscopic particles (which have a wave aspect) cannot be rationalized in terms of a visualizable model.

The wave functions ψ and probability densities $|\psi|^2$ are spread out over the length of the box, much like a wave (compare Figs. 17.10 and 17.2). However, quantum mechanics does not assert that the particle itself is spread out like a wave; a measurement of position will give a definite location for the particle. It is the wave function ψ (which gives the probability density $|\psi|^2$) that is spread out in space and obeys a wave equation.

EXAMPLE 17.3 Probability calculations

(*a*) For the ground state of a particle in a one-dimensional box of length a, find the probability that the particle is within $\pm 0.001a$ of the point $x = a/2$. (*b*) For the particle-in-a-box stationary state with quantum number n, write down (but do not evaluate) an expression for the probability that the particle will be found between $a/4$ and $a/2$. (*c*) For a particle-in-a-box stationary state, what is the probability that the particle will be found in the left half of the box?

(*a*) The probability density (the probability per unit length) equals $|\psi|^2$. Figure 17.10 shows that $|\psi|^2$ for $n = 1$ is essentially constant over the very small interval $0.002a$, so we can consider this interval to be infinitesimal and take $|\psi|^2\,dx$ as the desired probability. For $n = 1$, Eq. (17.35) gives $|\psi|^2 = (2/a)\sin^2(\pi x/a)$. With $x = a/2$ and $dx = 0.002a$, the probability is $|\psi|^2\,dx = (2/a)\sin^2(\pi/2) \times 0.002a = 0.004$.

(*b*) From Eq. (17.15), the probability that the particle is between points c and d is $\int_c^d |\Psi|^2\,dx$. But $|\Psi|^2 = |\psi|^2$ for a stationary state [Eq. (17.25)], so the probability is $\int_c^d |\psi|^2\,dx$. The desired probability is $\int_{a/4}^{a/2} (2/a)\sin^2(n\pi x/a)\,dx$, where (17.35) was used for ψ.

(*c*) For each particle-in-a-box stationary state, the graph of $|\psi|^2$ is symmetric about the midpoint of the box, so the probabilities of being in the left and right halves are equal and are each equal to 0.5.

Exercise

For the $n = 2$ state of a particle in a box of length a, (*a*) find the probability the particle is within $\pm 0.0015a$ of $x = a/8$; (*b*) find the probability the particle is between $x = 0$ and $x = a/8$. (*Answers:* (*a*) 0.0030; (*b*) $1/8 - 1/4\pi = 0.0454$.)

If ψ_i and ψ_j are particle-in-a-box wave functions with quantum numbers n_i and n_j, one finds (Prob. 17.29) that

$$\int_0^a \psi_i^* \psi_j\,d\tau = 0 \qquad \text{for } n_i \neq n_j \tag{17.36}$$

where $\psi_i = (2/a)^{1/2}\sin(n_i\pi x/a)$ and $\psi_j = (2/a)^{1/2}\sin(n_j\pi x/a)$. The functions f and g are said to be **orthogonal** when $\int f^*g\,d\tau = 0$, where the integral is a definite integral over the full range of the spatial coordinates. One can show that two wave functions that correspond to different energy levels of a quantum-mechanical system are orthogonal (Sec. 17.16).

17.9 THE PARTICLE IN A THREE-DIMENSIONAL BOX

The particle in a three-dimensional box is a single particle of mass m confined to remain within the volume of a box by an infinite potential energy outside the box. The simplest box shape to deal with is a rectangular parallelepiped. The potential energy is therefore $V = 0$ for points such that $0 \le x \le a$, $0 \le y \le b$, and $0 \le z \le c$ and $V = \infty$ elsewhere. The dimensions of the box are a, b, and c. In Secs. 20.3 and 21.6, this system will be used to give the energy levels for translational motion of ideal-gas molecules in a container.

Let us solve the time-independent Schrödinger equation for the stationary-state wave functions and energies. Since $V = \infty$ outside the box, ψ is zero outside the box, just as for the corresponding one-dimensional problem. Inside the box, $V = 0$, and the Schrödinger equation (17.24) becomes

$$-\frac{\hbar^2}{2m}\left(\frac{\partial^2\psi}{\partial x^2} + \frac{\partial^2\psi}{\partial y^2} + \frac{\partial^2\psi}{\partial z^2}\right) = E\psi \tag{17.37}$$

Let us assume that solutions of (17.37) exist that have the form $X(x)Y(y)Z(z)$, where $X(x)$ is a function of x only and Y and Z are functions of y and z. For an arbitrary partial differential equation, it is not in general possible to find solutions in which the variables are present in separate factors. However, it can be proved mathematically that, if we succeed in finding well-behaved solutions to (17.37) that have the form $X(x)Y(y)Z(z)$, then there are no other well-behaved solutions, so we shall have found the general solution of (17.37). Our assumption is then

$$\psi = X(x)Y(y)Z(z) \tag{17.38}$$

Partial differentiation of (17.38) gives

$$\partial^2\psi/\partial x^2 = X''(x)Y(y)Z(z), \quad \partial^2\psi/\partial y^2 = X(x)Y''(y)Z(z), \quad \partial^2\psi/\partial z^2 = X(x)Y(y)Z''(z)$$

Substitution in (17.37) followed by division by $X(x)Y(y)Z(z) = \psi$ gives

$$-\frac{\hbar^2}{2m}\frac{X''(x)}{X(x)} - \frac{\hbar^2}{2m}\frac{Y''(y)}{Y(y)} - \frac{\hbar^2}{2m}\frac{Z''(z)}{Z(z)} = E \tag{17.39}$$

Let $E_x \equiv -(\hbar^2/2m)X''(x)/X(x)$. Then (17.39) gives

$$E_x \equiv -\frac{\hbar^2}{2m}\frac{X''(x)}{X(x)} = E + \frac{\hbar^2}{2m}\frac{Y''(y)}{Y(y)} + \frac{\hbar^2}{2m}\frac{Z''(z)}{Z(z)} \tag{17.40}$$

From its definition, E_x is a function of x only. However, the relation $E_x = E + \hbar^2Y''/2mY + \hbar^2Z''/2mZ$ in (17.40) shows E_x to be independent of x. Therefore E_x is a constant, and we have from (17.40)

$$-(\hbar^2/2m)X''(x) = E_xX(x) \qquad \text{for } 0 \le x \le a \tag{17.41}$$

Equation (17.41) is the same as the Schrödinger equation (17.28) for a particle in a one-dimensional box if X and E_x in (17.41) are identified with ψ and E, respectively, in (17.28). Moreover, the condition that $X(x)$ be continuous requires that $X(x) = 0$ at $x = 0$ and at $x = a$, since the three-dimensional wave function is zero outside the box. These are the same requirements that ψ in (17.28) must satisfy. Therefore, the well-behaved solutions of (17.41) and (17.28) are the same. Replacing ψ and E in (17.34) and (17.35) by X and E_x, we get

$$X(x) = \left(\frac{2}{a}\right)^{1/2}\sin\frac{n_x\pi x}{a}, \qquad E_x = \frac{n_x^2h^2}{8ma^2}, \qquad n_x = 1, 2, 3, \ldots \tag{17.42}$$

where the quantum number is called n_x.

Equation (17.39) is symmetric with respect to x, y, and z, so the same reasoning that gave (17.42) gives

$$Y(y) = \left(\frac{2}{b}\right)^{1/2}\sin\frac{n_y\pi y}{b}, \qquad E_y = \frac{n_y^2h^2}{8mb^2}, \qquad n_y = 1, 2, 3, \ldots \tag{17.43}$$

$$Z(z) = \left(\frac{2}{c}\right)^{1/2}\sin\frac{n_z\pi z}{c}, \qquad E_z = \frac{n_z^2h^2}{8mc^2}, \qquad n_z = 1, 2, 3, \ldots \tag{17.44}$$

where, by analogy to (17.40),

$$E_y \equiv -\frac{\hbar^2}{2m}\frac{Y''(y)}{Y(y)}, \qquad E_z \equiv -\frac{\hbar^2}{2m}\frac{Z''(z)}{Z(z)} \tag{17.45}$$

We assumed in Eq. (17.38) that the wave function ψ is the product of separate factors $X(x)$, $Y(y)$, and $Z(z)$ for each coordinate. Having found X, Y, and Z [Eqs. (17.42), (17.43), and (17.44)], we have as the stationary-state wave functions for a particle in a three-dimensional rectangular box

$$\psi = \left(\frac{8}{abc}\right)^{1/2}\sin\frac{n_x\pi x}{a}\sin\frac{n_y\pi y}{b}\sin\frac{n_z\pi z}{c} \qquad \text{inside the box} \tag{17.46}$$

Outside the box, $\psi = 0$.

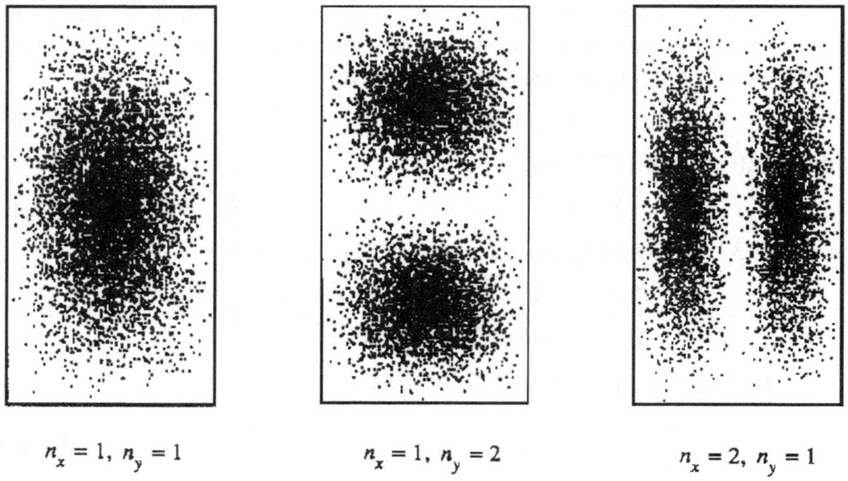

Figure 17.11

Probability densities for three states of a particle in a two-dimensional box whose dimensions have a 2:1 ratio. The states are the ψ_{11}, ψ_{12}, and ψ_{21} states, where the subscripts give the n_x and n_y values.

Equations (17.39), (17.40), and (17.45) give $E = E_x + E_y + E_z$, and use of (17.42) to (17.44) for E_x, E_y, and E_z gives the allowed energy levels as

$$E = \frac{h^2}{8m}\left(\frac{n_x^2}{a^2} + \frac{n_y^2}{b^2} + \frac{n_z^2}{c^2}\right) \qquad (17.47)$$

The quantities E_x, E_y, and E_z are the kinetic energies associated with motion in the x, y, and z directions.

The procedure used to solve (17.37) is called **separation of variables.** The conditions under which it works are discussed in Sec. 17.11.

The wave function has three quantum numbers because this is a three-dimensional problem. The quantum numbers n_x, n_y, and n_z vary independently of one another. The state of the particle in the box is specified by giving the values of n_x, n_y, and n_z. The ground state is $n_x = 1$, $n_y = 1$, and $n_z = 1$.

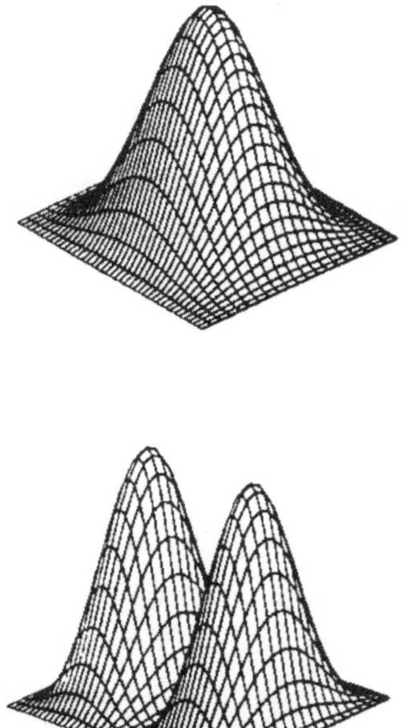

Figure 17.12

Three-dimensional plot of $|\psi|^2$ for the ψ_{11} and ψ_{12} states of a two-dimensional box with $b = 2a$.

The Particle in a Two-Dimensional Box

For a particle in a two-dimensional rectangular box with sides a and b, the same procedure that gave (17.46) and (17.47) gives

$$\psi = (4/ab)^{1/2} \sin\,(n_x\pi x/a) \sin\,(n_y\pi y/b) \qquad \text{for } 0 \le x \le a,\ 0 \le y \le b \qquad (17.48)$$

and $E = (h^2/8m)\,(n_x^2/a^2 + n_y^2/b^2)$. For a two-dimensional box with $b = 2a$, Fig. 17.11 shows the variation of the probability density $|\psi|^2$ in the box for three states. The greater the density of dots in a region, the greater the value of $|\psi|^2$. Figure 17.12 shows three-dimensional graphs of $|\psi|^2$ for the lowest two states. The height of the surface above the xy plane gives the value of $|\psi|^2$ at point (x, y). Figure 17.13 is a three-dimensional graph of ψ for the $n_x = 1$, $n_y = 2$ state; ψ is positive in half the box, negative in the other half, and zero on the line that separates these two halves. Figure 17.14 shows contour plots of constant $|\psi|$ for the $n_x = 1$, $n_y = 2$ state; the contours shown are those for which $|\psi|/|\psi|_{max} = 0.9$ (the two innermost loops), 0.7, 0.5, 0.3, and 0.1, where $|\psi|_{max}$ is the maximum value of $|\psi|$. These contours correspond to $|\psi|^2/|\psi^2|_{max} = 0.81$, 0.49, 0.25, 0.09, and 0.01.

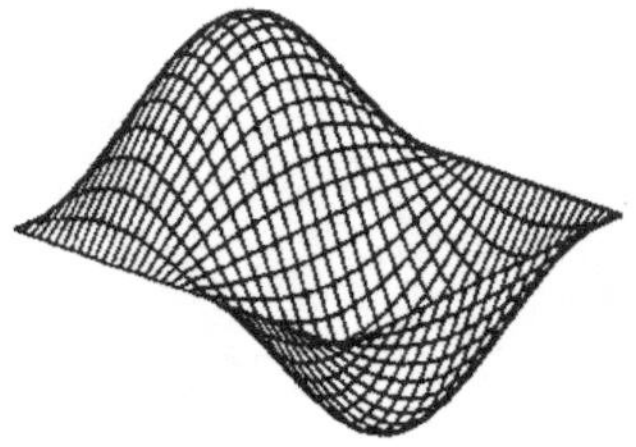

Figure 17.13

Three-dimensional plot of ψ_{12} for a particle in a two-dimensional box with $b = 2a$.

17.10 DEGENERACY

Suppose the sides of the three-dimensional box of the last section have equal lengths: $a = b = c$. Then (17.46) and (17.47) become

$$\psi = (2/a)^{3/2} \sin\,(n_x\pi x/a) \sin\,(n_y\pi y/a) \sin\,(n_z\pi z/a) \qquad (17.49)$$

$$E = (n_x^2 + n_y^2 + n_z^2)h^2/8ma^2 \quad (17.50)$$

Let us use numerical subscripts on ψ to specify the n_x, n_y, and n_z values. The lowest-energy state is ψ_{111} with $E = 3h^2/8ma^2$. The states ψ_{211}, ψ_{121}, and ψ_{112} each have energy $6h^2/8ma^2$. Even though they have the same energy, these are different states. With $n_x = 2, n_y = 1$, and $n_z = 1$ in (17.49), we get a different wave function than with $n_x = 1, n_y = 2$, and $n_z = 1$. The ψ_{211} state has zero probability density of finding the particle at $x = a/2$ (see Fig. 17.10), but the ψ_{121} state has a maximum probability density at $x = a/2$.

The terms "state" and "energy level" have different meanings in quantum mechanics. A **stationary state** is specified by giving the wave function ψ. Each different ψ is a different state. An **energy level** is specified by giving the value of the energy. Each different value of E is a different energy level. The three different particle-in-a-box states ψ_{211}, ψ_{121}, and ψ_{112} belong to the same energy level, $6h^2/8ma^2$. Figure 17.15 shows the lowest few stationary states and energy levels of a particle in a cubic box.

An energy level that corresponds to more than one state is said to be **degenerate.** The number of different states belonging to the level is the **degree of degeneracy** of the level. The particle-in-a-cubic-box level $6h^2/8ma^2$ is threefold degenerate. The particle-in-a-box degeneracy arises when the dimensions of the box are made equal. Degeneracy usually arises from the symmetry of the system.

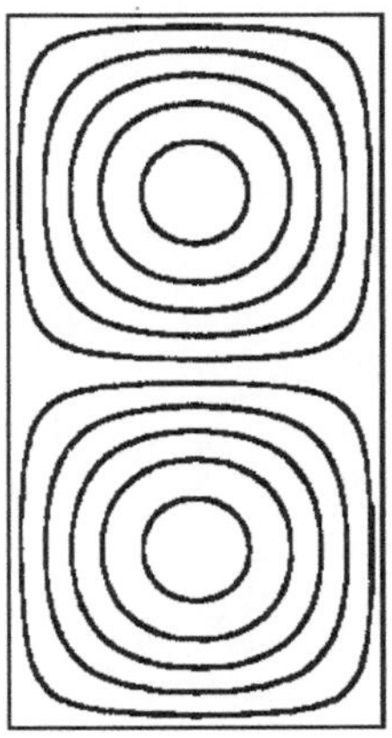

Figure 17.14

Contour plot of constant $|\psi|$ for the state of Fig. 17.13.

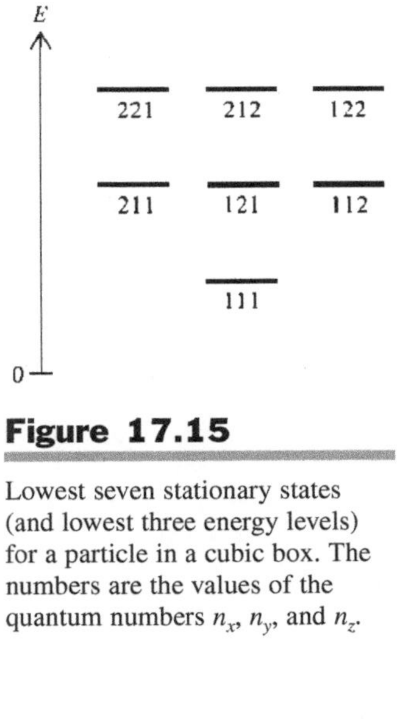

Figure 17.15

Lowest seven stationary states (and lowest three energy levels) for a particle in a cubic box. The numbers are the values of the quantum numbers n_x, n_y, and n_z.

17.11 OPERATORS

Operators

Quantum mechanics is most conveniently formulated in terms of operators. An **operator** is a rule for transforming a given function into another function. For example, the operator d/dx transforms a function into its first derivative: $(d/dx)f(x) = f'(x)$. Let $\hat{A}$ symbolize an arbitrary operator. (We shall use a circumflex to denote an operator.) If $\hat{A}$ transforms the function $f(x)$ into the function $g(x)$, we write $\hat{A}f(x) = g(x)$. If $\hat{A}$ is the operator d/dx, then $g(x) = f'(x)$. If $\hat{A}$ is the operator "multiplication by $3x^2$," then $g(x) = 3x^2 f(x)$. If $\hat{A} = \log$, then $g(x) = \log f(x)$.

The **sum** of two operators $\hat{A}$ and $\hat{B}$ is defined by

$$(\hat{A} + \hat{B})f(x) \equiv \hat{A}f(x) + \hat{B}f(x) \quad \mathbf{(17.51)^*}$$

For example, $(\ln + d/dx)f(x) = \ln f(x) + (d/dx)f(x) = \ln f(x) + f'(x)$. Similarly, $(\hat{A} - \hat{B})f(x) \equiv \hat{A}f(x) - \hat{B}f(x)$.

The **square** of an operator is defined by $\hat{A}^2 f(x) \equiv \hat{A}[\hat{A}f(x)]$. For example,

$$(d/dx)^2 f(x) = (d/dx)[(d/dx)f(x)] = (d/dx)[f'(x)] = f''(x) = (d^2/dx^2)f(x)$$

Therefore, $(d/dx)^2 = d^2/dx^2$.

The **product** of two operators is defined by

$$(\hat{A}\hat{B})f(x) \equiv \hat{A}[\hat{B}f(x)] \quad \mathbf{(17.52)^*}$$

The notation $\hat{A}[\hat{B}f(x)]$ means that we first apply the operator $\hat{B}$ to the function $f(x)$ to get a new function, and then we apply the operator $\hat{A}$ to this new function.

Two operators are **equal** if they produce the same result when operating on an arbitrary function: $\hat{B} = \hat{C}$ if and only if $\hat{B}f = \hat{C}f$ for every function f.

EXAMPLE 17.4 Operator algebra

Let the operators $\hat{A}$ and $\hat{B}$ be defined as $\hat{A} \equiv x\cdot$ and $\hat{B} \equiv d/dx$. (*a*) Find $(\hat{A} + \hat{B})(x^3 + \cos x)$. (*b*) Find $\hat{A}\hat{B}f(x)$ and $\hat{B}\hat{A}f(x)$. Are the operators $\hat{A}\hat{B}$ and $\hat{B}\hat{A}$ equal? (*c*) Find $\hat{A}\hat{B} - \hat{B}\hat{A}$.

(*a*) Using the definition (17.51) of the sum of operators, we have

$$(\hat{A} + \hat{B})(x^3 + \cos x) = (x + d/dx)(x^3 + \cos x)$$
$$= x(x^3 + \cos x) + (d/dx)(x^3 + \cos x)$$
$$= x^4 + x \cos x + 3x^2 - \sin x$$

(*b*) The definition (17.52) of the operator product gives

$$\hat{A}\hat{B}f(x) \equiv \hat{A}[\hat{B}f(x)] = x[(d/dx)f(x)] = x[f'(x)] = xf'(x)$$
$$\hat{B}\hat{A}f(x) \equiv \hat{B}[\hat{A}f(x)] = (d/dx)[xf(x)] = xf'(x) + f(x)$$

In this example, $\hat{A}\hat{B}$ and $\hat{B}\hat{A}$ produce different results when they operate on $f(x)$, so $\hat{A}\hat{B}$ and $\hat{B}\hat{A}$ are not equal in this case. In multiplication of numbers, the order doesn't matter. In multiplication of operators, the order may matter.

(*c*) To find the operator $\hat{A}\hat{B} - \hat{B}\hat{A}$, we examine the result of applying it to an arbitrary function $f(x)$. We have $(\hat{A}\hat{B} - \hat{B}\hat{A})f(x) = \hat{A}\hat{B}f - \hat{B}\hat{A}f = xf' - (xf' + f) = -f$, where the definition of the difference of operators and the results of (*b*) were used. Since $(\hat{A}\hat{B} - \hat{B}\hat{A})f(x) = -1 \cdot f(x)$ for all functions $f(x)$, the definition of equality of operators gives

$$\hat{A}\hat{B} - \hat{B}\hat{A} = -1$$

where the multiplication sign after the -1 is omitted, as is customary.

The operator $\hat{A}\hat{B} - \hat{B}\hat{A}$ is called the **commutator** of $\hat{A}$ and $\hat{B}$ and is symbolized by $[\hat{A}, \hat{B}]$;

$$[\hat{A}, \hat{B}] \equiv \hat{A}\hat{B} - \hat{B}\hat{A}$$

Exercise

Let $\hat{R} \equiv x^2$ and $\hat{S} \equiv d^2/dx^2$. (*a*) Find $(\hat{R} + \hat{S})(x^4 + 1/x)$. (*b*) Find $\hat{R}\hat{S}f(x)$ and $\hat{S}\hat{R}f(x)$. (*c*) Find $[\hat{R}, \hat{S}]$. [*Answers:* (*a*) $x^6 + 12x^2 + x + 2/x^3$; (*b*) $x^2f''(x)$, $2f(x) + 4xf'(x) + x^2f''(x)$; (*c*) $-2 - 4x\,(d/dx)$.]

Operators in Quantum Mechanics

In quantum mechanics, each physical property of a system has a corresponding operator. The operator that corresponds to p_x, the x component of momentum of a particle, is postulated to be $(\hbar/i)(\partial/\partial x)$, with similar operators for p_y and p_z:

$$\hat{p}_x = \frac{\hbar}{i}\frac{\partial}{\partial x}, \qquad \hat{p}_y = \frac{\hbar}{i}\frac{\partial}{\partial y}, \qquad \hat{p}_z = \frac{\hbar}{i}\frac{\partial}{\partial z} \qquad \textbf{(17.53)*}$$

where $\hat{p}_x$ is the quantum-mechanical operator for the property p_x and $i \equiv \sqrt{-1}$. The operator that corresponds to the x coordinate of a particle is multiplication by x, and the operator that corresponds to $f(x, y, z)$, where f is any function, is multiplication by that function. Thus,

$$\hat{x} = x\times, \qquad \hat{y} = y\times, \qquad \hat{z} = z\times, \qquad \hat{f}(x, y, z) = f(x, y, z)\times \qquad \textbf{(17.54)*}$$

To find the operator that corresponds to any other physical property, we write down the classical-mechanical expression for that property as a function of cartesian coordinates and corresponding momenta and then replace the coordinates and momenta by their corresponding operators (17.53) and (17.54). For example, the energy of a one-particle system is the sum of its kinetic and potential energies:

$$E = K + V = \tfrac{1}{2}m(v_x^2 + v_y^2 + v_z^2) + V(x, y, z, t)$$

To express E as a function of the momenta and coordinates, we note that $p_x = mv_x$, $p_y = mv_y$, $p_z = mv_z$. Therefore,

$$E = \frac{1}{2m}(p_x^2 + p_y^2 + p_z^2) + V(x, y, z, t) \equiv H \qquad \textbf{(17.55)*}$$

The expression for the energy as a function of coordinates and momenta is called the system's **Hamiltonian** H [after W. R. Hamilton (1805–1865), who reformulated Newton's second law in terms of H]. The use of Eq. (17.53) and $i^2 = -1$ gives

$$\hat{p}_x^2 f(x, y, z) = (\hbar/i)(\partial/\partial x)[(\hbar/i)(\partial/\partial x)f] = (\hbar^2/i^2)\,\partial^2 f/\partial x^2 = -\hbar^2\,\partial^2 f/\partial x^2$$

so $\hat{p}_x^2 = -\hbar^2\,\partial^2/\partial x^2$ and $\hat{p}_x^2/2m = -(\hbar^2/2m)\,\partial^2/\partial x^2$. From (17.54), the potential-energy operator is simply multiplication by $V(x, y, z, t)$. (Time is a parameter in quantum mechanics, and there is no time operator.) Replacing p_x^2, p_y^2, p_z^2, and V in (17.55) by their operators, we get as the energy operator, or **Hamiltonian operator,** for a one-particle system

$$\hat{E} = \hat{H} = -\frac{\hbar^2}{2m}\left(\frac{\partial^2}{\partial x^2} + \frac{\partial^2}{\partial y^2} + \frac{\partial^2}{\partial z^2}\right) + V(x, y, z, t)\,\times \qquad (17.56)$$

To save time in writing, we define the **Laplacian operator** ∇^2 (read as "del squared") by $\nabla^2 \equiv \partial^2/\partial x^2 + \partial^2/\partial y^2 + \partial^2/\partial z^2$ and write the one-particle Hamiltonian operator as

$$\hat{H} = -(\hbar^2/2m)\nabla^2 + V \qquad (17.57)$$

where the multiplication sign after V is understood.

For a many-particle system, we have $\hat{p}_{x,1} = (\hbar/i)\,\partial/\partial x_1$ for particle 1, and the Hamiltonian operator is readily found to be

$$\hat{H} = -\frac{\hbar^2}{2m_1}\nabla_1^2 - \frac{\hbar^2}{2m_2}\nabla_2^2 - \cdots - \frac{\hbar^2}{2m_n}\nabla_n^2 + V(x_1, \ldots, z_n, t) \qquad \textbf{(17.58)*}$$

$$\nabla_1^2 \equiv \frac{\partial^2}{\partial x_1^2} + \frac{\partial^2}{\partial y_1^2} + \frac{\partial^2}{\partial z_1^2} \qquad \textbf{(17.59)*}$$

with similar definitions for $\nabla_2^2, \ldots, \nabla_n^2$. The terms in (17.58) are the operators for the kinetic energies of particles 1, 2, . . . , n and the potential energy of the system.

From (17.58), we see that the time-dependent Schrödinger equation (17.10) can be written as

$$-\frac{\hbar}{i}\frac{\partial \Psi}{\partial t} = \hat{H}\Psi \qquad (17.60)$$

and the time-independent Schrödinger equation (17.24) can be written as

$$\hat{H}\psi = E\psi \qquad \textbf{(17.61)*}$$

where V in (17.61) is independent of time. Since there is a whole set of allowed stationary-state wave functions and energies, (17.61) is often written as $\hat{H}\psi_j = E_j\psi_j$, where the subscript j labels the various wave functions (states) and their energies.

When an operator $\hat{B}$ applied to the function f gives the function back again but multiplied by the constant c, that is, when

$$\hat{B}f = cf$$

one says that f is an **eigenfunction** of $\hat{B}$ with **eigenvalue** c. (However, the function $f = 0$ everywhere is not allowed as an eigenfunction.) The wave functions ψ in (17.61) are eigenfunctions of the Hamiltonian operator $\hat{H}$, the eigenvalues being the allowed energies E.

Operator algebra differs from ordinary algebra. From $\hat{H}\psi = E\psi$ [Eq. (17.61)], one cannot conclude that $\hat{H} = E$. $\hat{H}$ is an operator and E is a number, and the two are not equal. Note, for example, that $(d/dx)e^{2x} = 2e^{2x}$, but $d/dx \neq 2$. In Example 17.4, we found that $(\hat{A}\hat{B} - \hat{B}\hat{A})f(x) = -1 \cdot f(x)$ (for $\hat{A} = x\cdot$ and $\hat{B} = d/dx$) and concluded that $\hat{A}\hat{B} - \hat{B}\hat{A} = -1\cdot$. Because this equation applies to all functions $f(x)$, it is valid to delete the $f(x)$ here. However, the relation $(d/dx)e^{2x} = 2e^{2x}$ applies only to the function e^{2x}, and this function cannot be deleted.

EXAMPLE 17.5 Eigenfunctions

Verify directly that $\hat{H}\psi = E\psi$ for the particle in a one-dimensional box.

Inside the box (Fig. 17.7), $V = 0$ and Eq. (17.56) gives $\hat{H} = -(\hbar^2/2m)\, d^2/dx^2$ for this one-dimensional problem. The wave functions are given by (17.35) as $\psi = (2/a)^{1/2} \sin(n\pi x/a)$. We have, using (1.27) and (17.11):

$$\hat{H}\psi = -\frac{\hbar^2}{2m}\frac{d^2}{dx^2}\left[\left(\frac{2}{a}\right)^{1/2}\sin\frac{n\pi x}{a}\right] = -\frac{h^2}{4\pi^2(2m)}\left(\frac{2}{a}\right)^{1/2}\left(-\frac{n^2\pi^2}{a^2}\right)\sin\frac{n\pi x}{a}$$

$$= \frac{n^2h^2}{8ma^2}\left(\frac{2}{a}\right)^{1/2}\sin\frac{n\pi x}{a} = E\psi$$

since $E = n^2h^2/8ma^2$ [Eq. (17.34)].

Exercise

Verify that the function Ae^{ikx}, where A and k are constants, is an eigenfunction of the operator $\hat{p}_x$. What is the eigenvalue? (*Answer:* $k\hbar$.)

The operators that correspond to physical quantities in quantum mechanics are linear. A **linear operator** $\hat{L}$ is one that satisfies the following two equations for all functions f and g and all constants c:

$$\hat{L}(f + g) = \hat{L}f + \hat{L}g \quad \text{and} \quad \hat{L}(cf) = c\hat{L}f$$

The operator $\partial/\partial x$ is linear, since $(\partial/\partial x)(f + g) = \partial f/\partial x + \partial g/\partial x$ and $(\partial/\partial x)(cf) = c\,\partial f/\partial x$. The operator $\sqrt{\ }$ is nonlinear, since $\sqrt{f + g} \neq \sqrt{f} + \sqrt{g}$.

If the function ψ satisfies the time-independent Schrödinger equation $\hat{H}\psi = E\psi$, then so does the function $c\psi$, where c is any constant. Proof of this follows from the fact that the Hamiltonian operator $\hat{H}$ is a linear operator. We have $\hat{H}(c\psi) = c\hat{H}\psi = cE\psi = E(c\psi)$. The freedom to multiply ψ by a constant enables us to normalize ψ.

Measurement

Multiplication of $\hat{H}\psi = E\psi$ [Eq. (17.61)] by $e^{-iEt/\hbar}$ gives $e^{-iEt/\hbar}\hat{H}\psi = Ee^{-iEt/\hbar}\psi$. For a stationary state, $\hat{H}$ does not involve time and $e^{-iEt/\hbar}\hat{H}\psi = \hat{H}(e^{-iEt/\hbar}\psi)$. Using $\Psi = e^{-iEt/\hbar}\psi$ [Eq. (17.23)], we have

$$\hat{H}\Psi = E\Psi$$

so Ψ is an eigenfunction of $\hat{H}$ with eigenvalue E for a stationary state. A stationary state has a definite energy, and measurement of the system's energy will always give a single predictable value when the system is in a stationary state. For example, for the $n = 2$ particle-in-a-box stationary state, measurement of the energy will always give the result $2^2h^2/8ma^2$ [Eq. (17.34)].

What about properties other than the energy? Let the operator $\hat{M}$ correspond to the property M. Quantum mechanics postulates that *if the system's state function Ψ happens to be an eigenfunction of $\hat{M}$ with eigenvalue c (that is, if $\hat{M}\Psi = c\Psi$), then a* *measurement of M is certain to give the value c as the result.* (Examples will be given when we consider angular momentum in Sec. 18.4). If Ψ is not an eigenfunction of $\hat{M}$, then the result of measuring M cannot be predicted. (However, the probabilities of the various possible outcomes of a measurement of M can be calculated from Ψ, but discussion of how this is done is omitted.) For stationary states, the essential part of Ψ is the time-independent wave function ψ, and ψ replaces Ψ in the italicized statement in this paragraph.

Average Values

From (14.38), the average value of x for a one-particle, one-dimensional quantum-mechanical system equals $\int_{-\infty}^{\infty} xg(x)\,dx$, where $g(x)$ is the probability density for finding the particle between x and $x + dx$. But the Born postulate (Sec. 17.6) gives $g(x) = |\Psi(x)|^2$. Hence, $\langle x\rangle = \int_{-\infty}^{\infty} x|\Psi(x)|^2\,dx$. Since $|\Psi|^2 = \Psi^*\Psi$, we have $\langle x\rangle = \int_{-\infty}^{\infty} \Psi^* x\Psi\,dx = \int_{-\infty}^{\infty} \Psi^*\hat{x}\Psi\,dx$, where (17.54) was used.

What about the average value of an arbitrary physical property M for a general quantum-mechanical system? Quantum mechanics *postulates* that the average value of any physical property M in a system whose state function is Ψ is given by

$$\langle M\rangle = \int \Psi^*\hat{M}\Psi\,d\tau \tag{17.62}$$

where $\hat{M}$ is the operator for the property M and the integral is a definite integral over all space. In (17.62), $\hat{M}$ operates on Ψ to produce the result $\hat{M}\Psi$, which is a function. The function $\hat{M}\Psi$ is then multiplied by Ψ^*, and the resulting function $\Psi^*\hat{M}\Psi$ is integrated over the full range of the spatial coordinates of the system. For example, Eq. (17.53) gives the p_x operator as $\hat{p}_x = (\hbar/i)\,\partial/\partial x$, and the average value of p_x for a one-particle, three-dimensional system whose state function is Ψ is $\langle p_x\rangle = (\hbar/i)\int_{-\infty}^{\infty}\int_{-\infty}^{\infty}\int_{-\infty}^{\infty} \Psi^*(\partial\Psi/\partial x)\,dx\,dy\,dz$.

The average value of M is the average of the results of a very large number of measurements of M made on identical systems, each of which is in the same state Ψ just before the measurement.

If Ψ happens to be an eigenfunction of $\hat{M}$ with eigenvalue c, then $\hat{M}\Psi = c\Psi$ and (17.62) becomes $\langle M\rangle = \int \Psi^*\hat{M}\Psi\,d\tau = \int \Psi^* c\Psi\,d\tau = c\int \Psi^*\Psi\,d\tau = c$, since Ψ is normalized. This result makes sense since, as noted in the last subsection, c is the only possible result of a measurement of M if $\hat{M}\Psi = c\Psi$.

For a stationary state, Ψ equals $e^{-iEt/\hbar}\psi$ [Eq. (17.23)]. Since $\hat{M}$ doesn't affect the $e^{-iEt/\hbar}$ factor, we have

$$\Psi^*\hat{M}\Psi = e^{iEt/\hbar}\psi^*\hat{M}e^{-iEt/\hbar}\psi = e^{iEt/\hbar}e^{-iEt/\hbar}\psi^*\hat{M}\psi = \psi^*\hat{M}\psi$$

Therefore, for a stationary state,

$$\langle M\rangle = \int \psi^*\hat{M}\psi\,d\tau \tag{17.63*}$$

EXAMPLE 17.6 Average value

For a particle in a one-dimensional-box stationary state, give the expression for $\langle x^2\rangle$.

For a one-particle, one-dimensional problem, $d\tau = dx$. Since $\hat{x}^2 = x^2\,\cdot$, we have

$$\langle x^2\rangle = \int_{-\infty}^{\infty} \psi^* x^2\psi\,dx = \int_{-\infty}^{0} x^2|\psi|^2\,dx + \int_{0}^{a} x^2|\psi|^2\,dx + \int_{a}^{\infty} x^2|\psi|^2\,dx$$

since $\psi^*\psi = |\psi|^2$ [Eq. (17.14)]. For $x < 0$ and $x > a$, we have $\psi = 0$ [Eq. (17.27)] and inside the box $\psi = (2/a)^{1/2} \sin (n\pi x/a)$ [Eq. (17.35)]. Therefore

$$\langle x^2 \rangle = \frac{2}{a} \int_0^a x^2 \sin^2 \frac{n\pi x}{a}\, dx$$

Evaluation of the integral is left as a homework problem (Prob. 17.42).

Exercise

Evaluate $\langle p_x \rangle$ for a particle in a one-dimensional-box stationary state. [*Answer:* $(2n\pi\hbar/ia^2) \int_0^a \sin (n\pi x/a) \cos (n\pi x/a)\, dx = 0$.]

Separation of Variables

Let $q_1, q_2, \ldots, q_r$ be the coordinates of a system. For example, for a two-particle system, $q_1 = x_1, q_2 = y_1, \ldots, q_6 = z_2$. Suppose the Hamiltonian operator has the form

$$\hat{H} = \hat{H}_1 + \hat{H}_2 + \cdots + \hat{H}_r \tag{17.64}$$

where the operator $\hat{H}_1$ involves only q_1, the operator $\hat{H}_2$ involves only q_2, etc. An example is the particle in a three-dimensional box, where one has $\hat{H} = \hat{H}_x + \hat{H}_y + \hat{H}_z$, with $\hat{H}_x \equiv -(\hbar^2/2m)\, \partial^2/\partial x^2$, etc. We saw in Sec. 17.9 that, for this case, $\psi = X(x)Y(y)Z(z)$ and $E = E_x + E_y + E_z$, where $\hat{H}_x X(x) = E_x X(x)$, $\hat{H}_y Y(y) = E_y Y(y)$, $\hat{H}_z Z(z) = E_z Z(z)$ [Eqs. (17.41) and (17.45)].

The same type of argument used in Sec. 17.9 shows (Prob. 17.43) that when $\hat{H}$ is the sum of separate terms for each coordinate, as in (17.64), then each stationary-state wave function is the product of separate factors for each coordinate and each stationary-state energy is the sum of energies for each coordinate:

$$\psi = f_1(q_1) f_2(q_2) \cdots f_r(q_r) \qquad \textbf{(17.65)*}$$

$$E = E_1 + E_2 + \cdots + E_r \qquad \textbf{(17.66)*}$$

where $E_1, E_2, \ldots$ and the functions $f_1, f_2, \ldots$ are found by solving

$$\hat{H}_1 f_1 = E_1 f_1, \quad \hat{H}_2 f_2 = E_2 f_2, \quad \ldots, \quad \hat{H}_r f_r = E_r f_r \tag{17.67}$$

The equations in (17.67) are, in effect, separate Schrödinger equations, one for each coordinate.

Noninteracting Particles

An important case where separation of variables applies is a system of n noninteracting particles, meaning that the particles exert no forces on one another. For such a system, the classical-mechanical energy is the sum of the energies of the individual particles, so the classical Hamiltonian H and the quantum-mechanical Hamiltonian operator $\hat{H}$ have the forms $H = H_1 + H_2 + \cdots + H_n$ and $\hat{H} = \hat{H}_1 + \hat{H}_2 + \cdots + \hat{H}_n$, where $\hat{H}_1$ involves only the coordinates of particle 1, $\hat{H}_2$ involves only particle 2, etc. Here, by analogy to (17.65) to (17.67), we have

$$\psi = f_1(x_1, y_1, z_1) f_2(x_2, y_2, z_2) \cdots f_n(x_n, y_n, z_n) \qquad \textbf{(17.68)*}$$

$$E = E_1 + E_2 + \cdots + E_n \qquad \textbf{(17.69)*}$$

$$\hat{H}_1 f_1 = E_1 f_1, \quad \hat{H}_2 f_2 = E_2 f_2, \quad \ldots, \quad \hat{H}_n f_n = E_n f_n \qquad \textbf{(17.70)*}$$

For a system of noninteracting particles, there is a separate Schrödinger equation for each particle, the wave function is the product of wave functions of the individual particles, and the energy is the sum of the energies of the individual particles. (For noninteracting particles, the probability density $|\psi|^2$ is the product of probability densities for each particle: $|\psi|^2 = |f_1|^2|f_2|^2 \cdots |f_n|^2$. This is in accord with the theorem that the probability that several independent events will all occur is the product of the probabilities of the separate events.)

17.12 THE ONE-DIMENSIONAL HARMONIC OSCILLATOR

The one-dimensional **harmonic oscillator** is a useful model for treating the vibration of a diatomic molecule (Sec. 20.3) and is also relevant to vibrations of polyatomic molecules (Sec. 20.8) and crystals (Sec. 23.12).

Classical Treatment

Before examining the quantum mechanics of a harmonic oscillator, we review the classical treatment. Consider a particle of mass m that moves in one dimension and is attracted to the coordinate origin by a force proportional to its displacement from the origin: $F = -kx$, where k is called the **force constant.** When x is positive, the force is in the $-x$ direction, and when x is negative, F is in the $+x$ direction. A physical example is a mass attached to a frictionless spring, x being the displacement from the equilibrium position. From (2.17), $F = -dV/dx$, where V is the potential energy. Hence $-dV/dx = -kx$, and $V = \frac{1}{2}kx^2 + c$. The choice of zero of potential energy is arbitrary. Choosing the integration constant c as zero, we have (Fig. 17.16)

$$V = \tfrac{1}{2}kx^2 \tag{17.71}$$

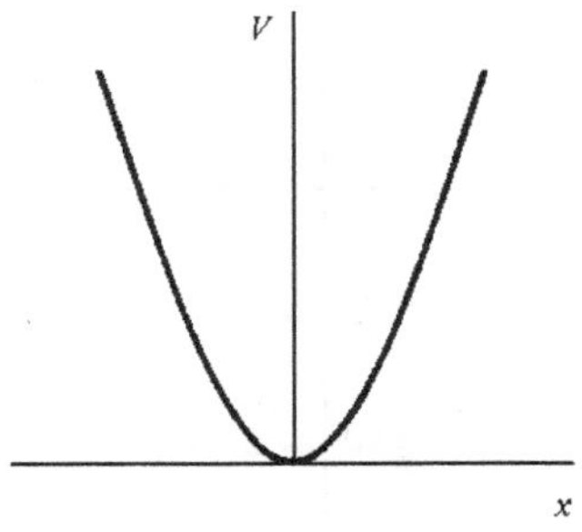

Figure 17.16

The potential-energy function for a one-dimensional harmonic oscillator.

Newton's second law $F = ma$ gives $m\, d^2x/dt^2 = -kx$. The solution to this differential equation is

$$x = A \sin\left[(k/m)^{1/2}t + b\right] \tag{17.72}$$

as can be verified by substitution in the differential equation (Prob. 17.52). In (17.72), A and b are integration constants. The maximum and minimum values of the sine function are $+1$ and -1, so the particle's x coordinate oscillates back and forth between $+A$ and $-A$. A is the *amplitude* of the motion.

The *period* τ (tau) of the oscillator is the time required for one complete cycle of oscillation. For one cycle of oscillation, the argument of the sine function in (17.72) must increase by 2π, since 2π is the period of a sine function. Hence the period satisfies $(k/m)^{1/2}\tau = 2\pi$, and $\tau = 2\pi(m/k)^{1/2}$. The **frequency** ν is the reciprocal of the period and equals the number of vibrations per second ($\nu = 1/\tau$); thus

$$\nu = \frac{1}{2\pi}\left(\frac{k}{m}\right)^{1/2} \tag{17.73*}$$

The energy of the harmonic oscillator is $E = K + V = \frac{1}{2}mv_x^2 + \frac{1}{2}kx^2$. The use of (17.72) for x and of $v_x = dx/dt = (k/m)^{1/2}A \cos\left[(k/m)^{1/2}t + b\right]$ leads to (Prob. 17.52)

$$E = \tfrac{1}{2}kA^2 \tag{17.74}$$

Equation (17.74) shows that the classical energy can have any nonnegative value. As the particle oscillates, its kinetic energy and potential energy continually change, but the total energy remains constant at $\frac{1}{2}kA^2$.

Classically, the particle is limited to the region $-A \le x \le A$. When the particle reaches $x = A$ or $x = -A$, its speed is zero (since it reverses its direction of motion at $+A$ and $-A$) and its potential energy is a maximum, being equal to $\frac{1}{2}kA^2$. If the particle were to move beyond $x = \pm A$, its potential energy would increase above $\frac{1}{2}kA^2$. This is impossible for a classical particle. The total energy is $\frac{1}{2}kA^2$ and the kinetic energy is nonnegative, so the potential energy ($V = E - K$) cannot exceed the total energy.

Quantum-Mechanical Treatment

Now for the quantum-mechanical treatment. Substitution of $V = \frac{1}{2}kx^2$ in (17.26) gives the time-independent Schrödinger equation as

$$-\frac{\hbar^2}{2m}\frac{d^2\psi}{dx^2} + \tfrac{1}{2}kx^2\psi = E\psi \tag{17.75}$$

Solution of the harmonic-oscillator Schrödinger equation (17.75) is complicated and is omitted in this book (see any quantum chemistry text). Here, we examine the results. One finds that quadratically integrable (Sec. 17.7) solutions to (17.75) exist only for the following values of E:

$$E = (v + \tfrac{1}{2})h\nu \qquad \text{where } v = 0, 1, 2, \ldots \qquad \mathbf{(17.76)^*}$$

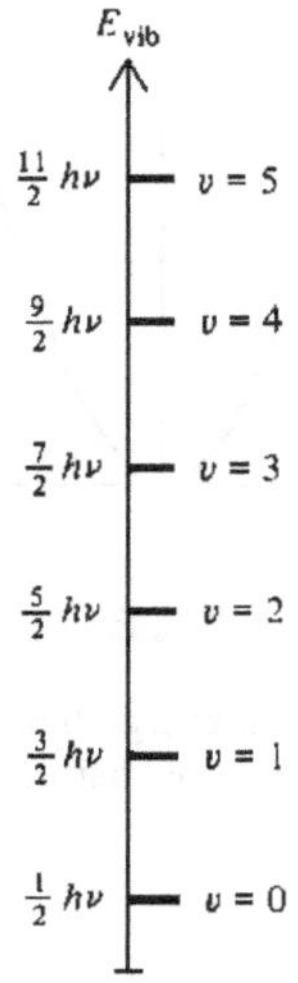

Figure 17.17

Energy levels of a one-dimensional harmonic oscillator.

where the vibrational frequency ν is given by (17.73) and the quantum number v takes on nonnegative integral values. [Don't confuse the typographically similar symbols ν (nu) and v (vee).] The energy is quantized. The allowed energy levels (Fig. 17.17) are equally spaced (unlike the particle in a box). The zero-point energy is $\frac{1}{2}h\nu$. (For a collection of harmonic oscillators in thermal equilibrium, all the oscillators will fall to the ground state as the temperature goes to absolute zero; hence the name *zero-point energy*.) For all values of E other than (17.76), one finds that the solutions to (17.75) go to infinity as x goes to $\pm\infty$, so these solutions are not quadratically integrable and are not allowed as wave functions.

The well-behaved solutions to (17.75) turn out to have the form

$$\psi_v = \begin{cases} e^{-\alpha x^2/2}(c_0 + c_2x^2 + \cdots + c_vx^v) & \text{for } v \text{ even} \\ e^{-\alpha x^2/2}(c_1x + c_3x^3 + \cdots + c_vx^v) & \text{for } v \text{ odd} \end{cases}$$

where $\alpha \equiv 2\pi\nu m/\hbar$. The polynomial that multiplies $e^{-\alpha x^2/2}$ contains only even powers of x or only odd powers, depending on whether the quantum number v is even or odd. The explicit forms of the lowest few wave functions ψ_0, ψ_1, ψ_2, and ψ_3 (where the subscript on ψ gives the value of the quantum number v) are given in Fig. 17.18, which plots these ψ's. As with the particle in a one-dimensional box, the number of nodes increases by one for each increase in the quantum number. Note the qualitative resemblance of the wave functions in Figs. 17.18 and 17.10.

The harmonic-oscillator wave functions fall off exponentially to zero as $x \to \pm\infty$. Note, however, that even for very large values of x, the wave function ψ and the probability density $|\psi|^2$ are not zero. There is some probability of finding the particle at an indefinitely large value of x. For a classical-mechanical harmonic oscillator with energy $(v + \frac{1}{2})h\nu$, Eq. (17.74) gives $(v + \frac{1}{2})h\nu = \frac{1}{2}kA^2$, and $A = [(2v + 1)h\nu/k]^{1/2}$. A classical oscillator is confined to the region $-A \le x \le A$. However, a quantum-mechanical oscillator has some probability of being found in the *classically forbidden* regions $x > A$ and $x < -A$, where the potential energy is greater than the particle's total energy. This penetration into classically forbidden regions is called **tunneling.** Tunneling occurs more readily the smaller the particle's mass and is most important in chemistry for electrons, protons, and H atoms. Tunneling influences the rates of reactions involving these species (see Secs. 22.3 and 22.4). Electron tunneling is the basis for the scanning tunneling microscope, a remarkable device that gives pictures of the atoms on the surface of a solid (Sec. 23.10). Tunneling makes possible the fusion of hydrogen nuclei to helium nuclei in the sun, despite the electrical repulsion between two hydrogen nuclei.

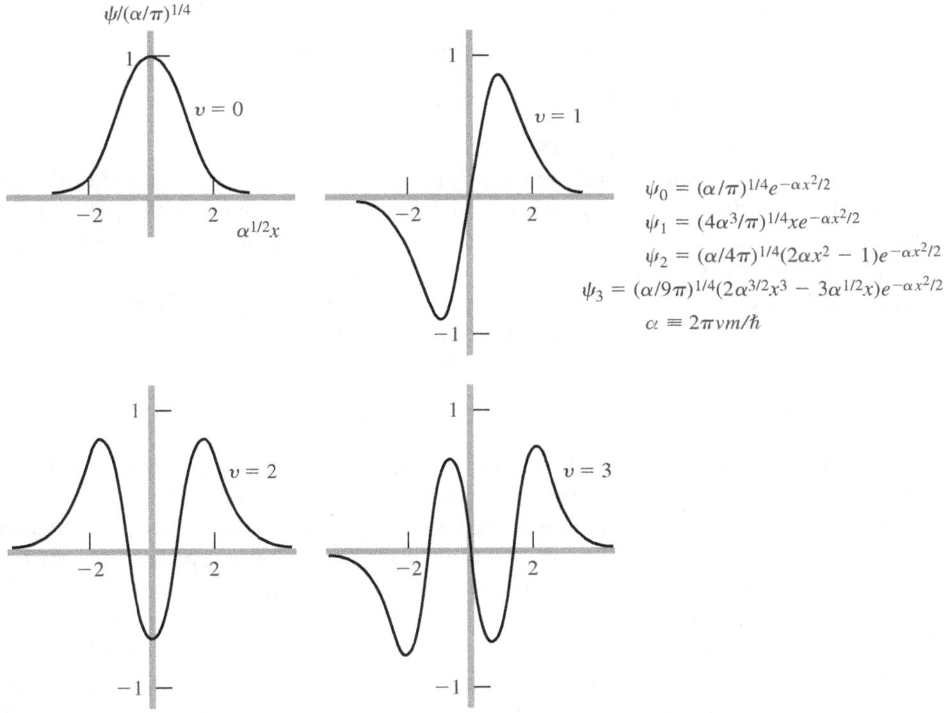

Figure 17.18

Wave functions for the lowest four harmonic-oscillator stationary states.

$$\psi_0 = (\alpha/\pi)^{1/4}e^{-\alpha x^2/2}$$
$$\psi_1 = (4\alpha^3/\pi)^{1/4}xe^{-\alpha x^2/2}$$
$$\psi_2 = (\alpha/4\pi)^{1/4}(2\alpha x^2 - 1)e^{-\alpha x^2/2}$$
$$\psi_3 = (\alpha/9\pi)^{1/4}(2\alpha^{3/2}x^3 - 3\alpha^{1/2}x)e^{-\alpha x^2/2}$$

$$\alpha \equiv 2\pi\nu m/\hbar$$

17.13 TWO-PARTICLE PROBLEMS

Consider a two-particle system where the coordinates of the particles are x_1, y_1, z_1 and x_2, y_2, z_2. The **relative** (or **internal**) **coordinates** x, y, z are defined by

$$x \equiv x_2 - x_1, \qquad y \equiv y_2 - y_1, \qquad z \equiv z_2 - z_1 \qquad (17.77)$$

These are the coordinates of particle 2 in a coordinate system whose origin is attached to particle 1 and moves with it.

In most cases, the potential energy V of the two-particle system depends only on the relative coordinates x, y, and z. For example, if the particles are electrically charged, the Coulomb's law potential energy of interaction between the particles depends only on the distance r between them, and $r = (x^2 + y^2 + z^2)^{1/2}$. Let us assume that $V = V(x, y, z)$. Let X, Y, and Z be the coordinates of the center of mass of the system; X is given by $(m_1x_1 + m_2x_2)/(m_1 + m_2)$, where m_1 and m_2 are the masses of the particles (*Halliday and Resnick*, sec. 9-1). If one expresses the classical energy (that is, the classical Hamiltonian) of the system in terms of the internal coordinates x, y, and z and the center-of-mass coordinates X, Y, and Z, instead of x_1, y_1, z_1, x_2, y_2, and z_2, it turns out (see Prob. 17.55) that

$$H = \left[\frac{1}{2\mu}(p_x^2 + p_y^2 + p_z^2) + V(x, y, z)\right] + \left[\frac{1}{2M}(p_X^2 + p_Y^2 + p_Z^2)\right] \qquad (17.78)$$

where M is the system's total mass ($M = m_1 + m_2$), the **reduced mass** μ is defined by

$$\mu \equiv \frac{m_1m_2}{m_1 + m_2} \qquad (17.79)*$$

and the momenta in (17.78) are defined by

$$p_x \equiv \mu v_x, \qquad p_y \equiv \mu v_y, \qquad p_z \equiv \mu v_z$$
$$p_X \equiv M v_X, \qquad p_Y \equiv M v_Y, \qquad p_Z \equiv M v_Z \qquad (17.80)$$

where $v_x = dx/dt$, etc., and $v_X = dX/dt$, etc.

Equation (17.55) shows that the Hamiltonian (17.78) is the sum of a Hamiltonian for a fictitious particle of mass μ and coordinates x, y, and z that has the potential energy $V(x, y, z)$ and a Hamiltonian for a second fictitious particle of mass $M = m_1 + m_2$ and coordinates X, Y, and Z that has $V = 0$. Moreover, there is no term for any interaction between these two fictitious particles. Hence, Eqs. (17.69) and (17.70) show that the quantum-mechanical energy E of the two-particle system is given by $E = E_\mu + E_M$, where E_μ and E_M are found by solving

$$\hat{H}_\mu \psi_\mu(x, y, z) = E_\mu \psi_\mu(x, y, z) \quad \text{and} \quad \hat{H}_M \psi_M(X, Y, Z) = E_M \psi_M(X, Y, Z)$$

The Hamiltonian operator $\hat{H}_\mu$ is formed from the terms in the first pair of brackets in (17.78), and $\hat{H}_M$ is formed from the terms in the second pair of brackets.

Introduction of the relative coordinates x, y, and z and the center-of-mass coordinates X, Y, and Z reduces the two-particle problem to two separate one-particle problems. We solve a Schrödinger equation for a fictitious particle of mass μ moving subject to the potential energy $V(x, y, z)$, and we solve a separate Schrödinger equation for a fictitious particle whose mass is M $(= m_1 + m_2)$ and whose coordinates are the system's center-of-mass coordinates X, Y, and Z. The Hamiltonian $\hat{H}_M$ involves only kinetic energy. If the two particles are confined to a box, we can use the particle-in-a-box energies (17.47) for E_M. The energy E_M is translational energy of the two-particle system as a whole. The Hamiltonian $\hat{H}_\mu$ involves the kinetic energy and potential energy of motion of the particles relative to each other, so E_μ is the energy associated with this relative or "internal" motion.

The system's total energy E is the sum of its translational energy E_M and its internal energy E_μ. For example, the energy of a hydrogen atom in a box is the sum of the atom's translational energy through space and the atom's internal energy, which is composed of potential energy of interaction between the electron and the proton and kinetic energy of motion of the electron relative to the proton.

17.14 THE TWO-PARTICLE RIGID ROTOR

The **two-particle rigid rotor** consists of particles of masses m_1 and m_2 constrained to remain a fixed distance d from each other. This is a useful model for treating the rotation of a diatomic molecule; see Sec. 20.3. The system's energy is wholly kinetic, and $V = 0$. Since $V = 0$ is a special case of V being a function of only the relative coordinates of the particles, the results of the last section apply. The quantum-mechanical energy is the sum of the translational energy of the system as a whole and the energy of internal motion of one particle relative to the other. The interparticle distance is constant, so the internal motion consists entirely of changes in the spatial orientation of the interparticle axis. The internal motion is a rotation of the two-particle system.

Solution of the Schrödinger equation for internal motion is complicated, so we shall just quote the results without proof. (For a derivation, see, for example, *Levine,* sec. 6.4.) The allowed rotational energies turn out to be

$$E_{\text{rot}} = J(J+1)\frac{\hbar^2}{2I} \quad \text{where } J = 0, 1, 2, \ldots \qquad \textbf{(17.81)*}$$

where the rotor's **moment of inertia** I is

$$I = \mu d^2 \qquad \textbf{(17.82)*}$$

with $\mu = m_1m_2/(m_1 + m_2)$. The spacing between adjacent rotational energy levels increases with increasing quantum number J (Fig. 17.19). There is no zero-point rotational energy.

The rotational wave functions are most conveniently expressed in terms of the angles θ and ϕ that give the spatial orientation of the rotor (Fig. 17.20). One finds $\psi_{\text{rot}} = \Theta_{JM_J}(\theta)\Phi_{M_J}(\phi)$, where Θ_{JM_J} is a function of θ whose form depends on the two quantum numbers J and M_J, and Φ_{M_J} is a function of ϕ whose form depends on M_J. These functions won't be given here but will be discussed in Sec. 18.3.

Ordinarily, the wave function for internal motion of a two-particle system is a function of three coordinates. However, since the interparticle distance is held fixed in this problem, ψ_{rot} is a function of only two coordinates, θ and ϕ. Since there are two coordinates, there are two quantum numbers, J and M_J. The possible values of M_J turn out to range from $-J$ to J in steps of 1:

$$M_J = -J, -J + 1, \ldots, J - 1, J \qquad \textbf{(17.83)*}$$

For example, if $J = 2$, then $M_J = -2, -1, 0, 1, 2$. For a given J, there are $2J + 1$ values of M_J. The quantum numbers J and M_J determine the rotational wave function, but E_{rot} depends only on J. Hence, each rotational level is $(2J + 1)$-fold degenerate. For example, the value $J = 1$ corresponds to one energy level ($E_{\text{rot}} = \hbar^2/I$) and corresponds to the three M_J values $-1, 0, 1$. Therefore for $J = 1$ there are three different ψ_{rot} functions, that is, three different rotational states.

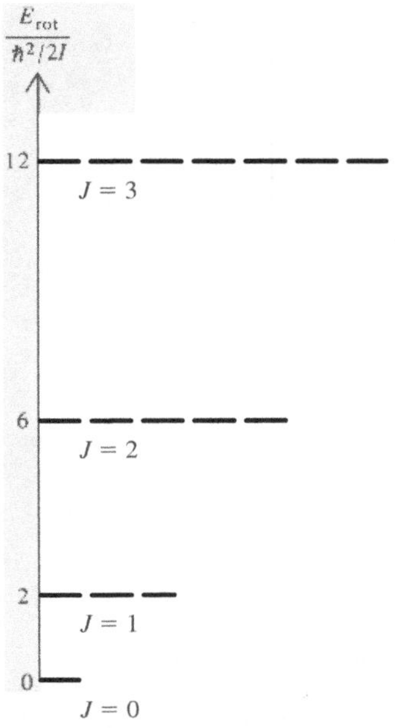

Figure 17.19

Lowest four energy levels of a two-particle rigid rotor. Each energy level consists of $2J + 1$ states.

EXAMPLE 17.7 Rotational energy levels

Find the two lowest rotational energy levels of the $^1H^{35}Cl$ molecule, treating it as a rigid rotor. The bond distance is 1.28 Å in HCl. Atomic masses are listed in a table inside the back cover.

The rotational energy [Eqs. (17.81) and (17.82)] depends on the reduced mass μ of Eq. (17.79). The atomic mass m_1 in μ equals the molar mass M_1 divided by the Avogadro constant N_A. Using the table of atomic masses, we have

$$\mu = \frac{m_1m_2}{m_1 + m_2} = \frac{[(1.01\text{ g/mol})/N_A][(35.0\text{ g/mol})/N_A]}{[(1.01\text{ g/mol}) + (35.0\text{ g/mol})]/N_A} = \frac{0.982\text{ g/mol}}{6.02 \times 10^{23}/\text{mol}}$$

$$= 1.63 \times 10^{-24}\text{ g}$$

$$I = \mu d^2 = (1.63 \times 10^{-27}\text{ kg})(1.28 \times 10^{-10}\text{ m})^2 = 2.67 \times 10^{-47}\text{ kg m}^2$$

The two lowest rotational levels have $J = 0$ and $J = 1$, and (17.81) gives $E_{J=0} = 0$ and

$$E_{J=1} = \frac{J(J + 1)\hbar^2}{2I} = \frac{1(2)(6.63 \times 10^{-34}\text{ J s})^2}{2(2\pi)^2(2.67 \times 10^{-47}\text{ kg m}^2)} = 4.17 \times 10^{-22}\text{ J}$$

Exercise

The separation between the two lowest rotational levels of $^{12}C^{32}S$ is 3.246×10^{-23} J. Calculate the bond distance in $^{12}C^{32}S$. (*Answer:* 1.538 Å.)

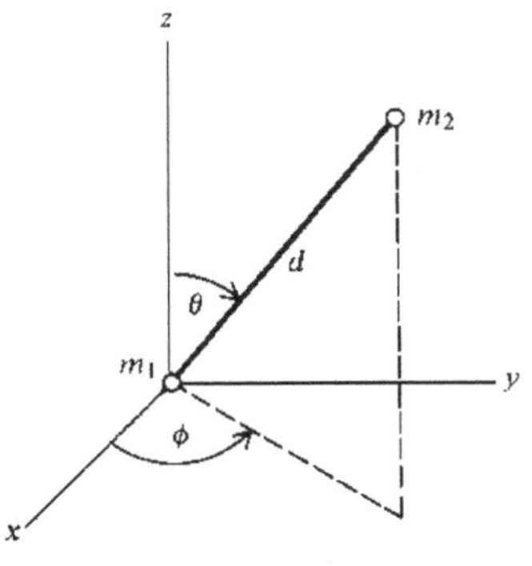

Figure 17.20

A two-particle rigid rotor.

17.15 APPROXIMATION METHODS

For a many-electron atom or molecule, the interelectronic repulsion terms in the potential energy V make it impossible to solve the Schrödinger equation (17.24) exactly. One must resort to approximation methods.

The Variation Method

The most widely used approximation method is the **variation method.** From the postulates of quantum mechanics, one can deduce the following theorem (for the proof, see Prob. 17.68). Let $\hat{H}$ be the time-independent Hamiltonian operator of a quantum-mechanical system. If ϕ is any *normalized, well-behaved* function of the coordinates of the particles of the system, then

$$\int \phi^* \hat{H} \phi \, d\tau \geq E_{gs} \qquad \text{for } \phi \text{ normalized} \tag{17.84}$$

where E_{gs} is the system's true ground-state energy and the definite integral goes over all space. (Do not confuse the variation *function* ϕ with the *angle* ϕ in Fig. 17.20.)

To apply the variation method, one takes many different normalized, well-behaved functions $\phi_1, \phi_2, \ldots$, and for each of them one computes the **variational integral** $\int \phi^* \hat{H} \phi \, d\tau$. The variation theorem (17.84) shows that the function giving the lowest value of $\int \phi^* \hat{H} \phi \, d\tau$ provides the closest approximation to the ground-state energy (Fig. 17.21). This function can serve as an approximation to the true ground-state wave function and can be used to compute approximations to ground-state molecular properties in addition to the energy (for example, the dipole moment).

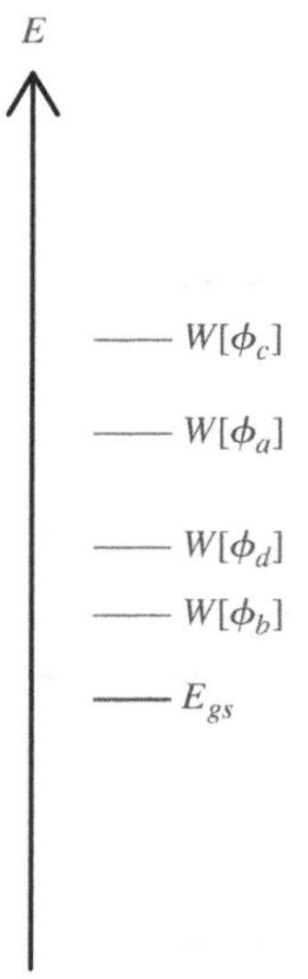

Figure 17.21

The variational integral cannot be less than the true ground-state energy E_{gs}. The quantities $W[\phi_a]$, $W[\phi_b]$, $W[\phi_c]$, and $W[\phi_d]$ are the values of the variational integral in (17.84) for the normalized functions ϕ_a, ϕ_b, ϕ_c, and ϕ_d. Of these functions, ϕ_b gives the lowest W and so its W is closest to E_{gs}.

Suppose we were lucky enough to guess the true ground-state wave function ψ_{gs}. Substitution of $\phi = \psi_{gs}$ in (17.84) and the use of (17.61) and (17.17) give the variational integral as $\int \psi_{gs}^* \hat{H} \psi_{gs} \, d\tau = \int \psi_{gs}^* E_{gs} \psi_{gs} \, d\tau = E_{gs} \int \psi_{gs}^* \psi_{gs} \, d\tau = E_{gs}$. We would then get the true ground-state energy.

If the variation function ϕ is not normalized, it must be multiplied by a normalization constant N before being used in (17.84). The normalization condition is $1 = \int |N\phi|^2 \, d\tau = |N|^2 \int |\phi|^2 \, d\tau$. Hence,

$$|N|^2 = \frac{1}{\int |\phi|^2 \, d\tau} \tag{17.85}$$

Use of the normalized function $N\phi$ in place of ϕ in (17.84) gives $\int N^* \phi^* \hat{H}(N\phi) \, d\tau = |N|^2 \int \phi^* \hat{H} \phi \, d\tau \geq E_{gs}$, where we used the linearity of $\hat{H}$ (Sec. 17.11) to write $\hat{H}(N\phi) = N\hat{H}\phi$. Substitution of (17.85) into the last inequality gives

$$\frac{\int \phi^* \hat{H} \phi \, d\tau}{\int \phi^* \phi \, d\tau} \geq E_{gs} \qquad \textbf{(17.86)*}$$

where ϕ need not be normalized but must be well behaved.

EXAMPLE 17.8 Trial variation function

Devise a trial variation function for the particle in a one-dimensional box and use it to estimate E_{gs}.

The particle in a box is exactly solvable, and there is no need to resort to an approximate method. For instructional purposes, let's pretend we don't know how to solve the particle-in-a-box Schrödinger equation. We know that the true ground-state wave function is zero outside the box, so we take the variation function ϕ to be zero outside the box. Equations (17.84) and (17.86) are valid only if ϕ is a well-behaved function, and this requires that ϕ be continuous. For ϕ to be continuous at the ends of the box, it must be zero at $x = 0$ and at $x = a$, where a is the box length. Perhaps the simplest way to get a function that vanishes at 0 and a is to take $\phi = x(a - x)$ for the region inside the box. As noted above, $\phi = 0$

outside the box. Since we did not normalize ϕ, Eq. (17.86) must be used. For the particle in a box, $V = 0$ and $\hat{H} = -(\hbar^2/2m)\, d^2/dx^2$ inside the box. We have

$$\int \phi^* \hat{H} \phi \, d\tau = \int_0^a x(a-x)\left(\frac{-\hbar^2}{2m}\right)\frac{d^2}{dx^2}[x(a-x)]\, dx$$

$$= \frac{-\hbar^2}{2m}\int_0^a x(a-x)(-2)\, dx = \frac{\hbar^2 a^3}{6m}$$

Also, $\int \phi^*\phi \, d\tau = \int_0^a x^2(a-x)^2 \, dx = a^5/30$. The variation theorem (17.86) becomes $(\hbar^2 a^3/6m) \div (a^5/30) \geq E_{gs}$, or

$$E_{gs} \leq 5h^2/4\pi^2 ma^2 = 0.12665h^2/ma^2$$

From (17.34), the true ground-state energy is $E_{gs} = h^2/8ma^2 = 0.125h^2/ma^2$. The variation function $x(a-x)$ gives a 1.3% error in E_{gs}.

Figure 17.22 plots the normalized variation function $(30/a^5)^{1/2}x(a-x)$ and the true ground-state wave function $(2/a)^{1/2}\sin(\pi x/a)$. Figure 17.22 also plots the percent deviation of the variation function from the true wave function versus x.

Exercise

Which of the following functions could be used as trial variation functions for the particle in a box? All functions are zero outside the box and the expression given applies only inside the box. *(a)* $-x^2(a-x)^2$; *(b)* x^2; *(c)* x^3; *(d)* $\sin(\pi x/a)$; *(e)* $\cos(\pi x/a)$; *(f)* $x(a-x)\sin(\pi x/a)$. [*Answer*: (a), (d), (f).]

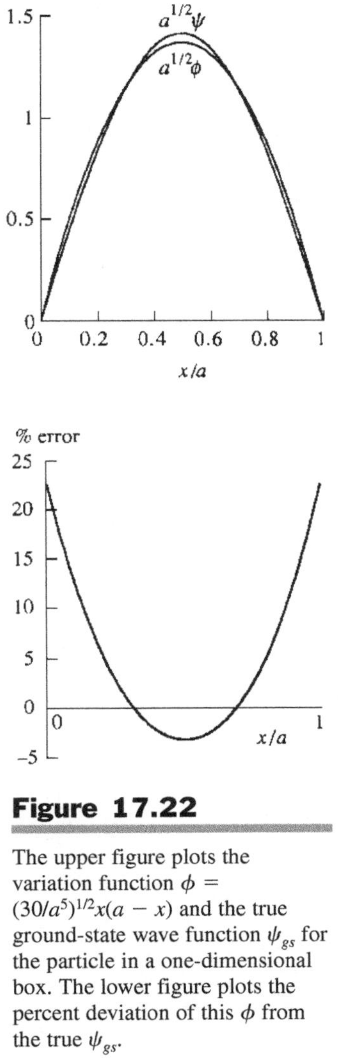

Figure 17.22

The upper figure plots the variation function $\phi = (30/a^5)^{1/2}x(a-x)$ and the true ground-state wave function ψ_{gs} for the particle in a one-dimensional box. The lower figure plots the percent deviation of this ϕ from the true ψ_{gs}.

If the normalized variation function ϕ contains the parameter c, then the variational integral $W \equiv \int \phi^* \hat{H}\phi\, d\tau$ will be a function of c, and one minimizes W by setting $\partial W/\partial c = 0$.

EXAMPLE 17.9 Variation function with a parameter

Apply the variation function e^{-cx^2} to the harmonic oscillator, where c is a parameter whose value is chosen to minimize the variational integral.

The harmonic-oscillator potential energy (17.71) is $\frac{1}{2}kx^2$ and the Hamiltonian operator (17.56) is $\hat{H} = -(\hbar^2/2m)(d^2/dx^2) + \frac{1}{2}kx^2$. We have

$$\hat{H}\phi = -\frac{\hbar^2}{2m}\frac{d^2(e^{-cx^2})}{dx^2} + \frac{1}{2}kx^2e^{-cx^2} = -\frac{\hbar^2}{2m}(4c^2x^2 - 2c)e^{-cx^2} + \frac{1}{2}kx^2e^{-cx^2}$$

$$\int \phi^*\hat{H}\phi\, d\tau = \int_{-\infty}^{\infty}\left[-\frac{\hbar^2}{2m}(4c^2x^2 - 2c)e^{-2cx^2} + \frac{1}{2}kx^2e^{-2cx^2}\right]dx$$

$$= \frac{\hbar^2}{m}\left(\frac{\pi c}{8}\right)^{1/2} + \frac{k}{4}\left(\frac{\pi}{8c^3}\right)^{1/2}$$

where Table 14.1 of Sec. 14.4 was used to evaluate the integrals. Also,

$$\int \phi^*\phi\, d\tau = \int_{-\infty}^{\infty} e^{-2cx^2}\, dx = \left(\frac{\pi}{2c}\right)^{1/2}$$

$$W \equiv \frac{\int \phi^*\hat{H}\phi\, d\tau}{\int \phi^*\phi\, d\tau} = \frac{\hbar^2 c}{2m} + \frac{k}{8c}$$

We now find the value of c that minimizes W:

$$0 = \frac{\partial W}{\partial c} = \frac{\hbar^2}{2m} - \frac{k}{8c^2}$$

We have $c^2 = mk/4\hbar^2$ and $c = \pm(mk)^{1/2}/2\hbar$. The negative value for c would give a positive exponent in the variation function $\phi = e^{-cx^2}$; ϕ would go to infinity as x goes to $\pm\infty$ and ϕ would not be quadratically integrable. We therefore reject the negative value of c. With $c = (mk)^{1/2}/2\hbar$, the variational integral W becomes

$$W = \frac{\hbar^2 c}{2m} + \frac{k}{8c} = \frac{\hbar k^{1/2}}{4m^{1/2}} + \frac{\hbar k^{1/2}}{4m^{1/2}} = \frac{hk^{1/2}}{4\pi m^{1/2}} = \frac{h\nu}{2}$$

where $\nu = (1/2\pi)(k/m)^{1/2}$ [Eq. (17.73)] was used. The value $h\nu/2$ is the true ground-state energy of the harmonic oscillator [Eq. (17.76)], and with $c = (mk)^{1/2}/2\hbar = \pi\nu m/\hbar$, the trial function e^{-cx^2} is the same as the unnormalized ground-state wave function of the harmonic oscillator (Fig. 17.18).

Exercise

Verify the integration results in this example.

A common form for variational functions in quantum mechanics is the **linear variation function**

$$\phi = c_1 f_1 + c_2 f_2 + \cdots + c_n f_n$$

where $f_1, \ldots, f_n$ are functions and $c_1, \ldots, c_n$ are variational parameters whose values are determined by minimizing the variational integral. Let W be the left side of (17.86). Then the conditions for a minimum in W are $\partial W/\partial c_1 = 0$, $\partial W/\partial c_2 = 0, \ldots,$ $\partial W/\partial c_n = 0$. These conditions lead to a set of equations that allows the c's to be found. It turns out that there are n different sets of coefficients $c_1, \ldots, c_n$ that satisfy $\partial W/\partial c_1 = \cdots = \partial W/\partial c_n = 0$, so we end up with n different variational functions $\phi_1, \ldots, \phi_n$ and n different values for the variational integral $W_1, \ldots, W_n$, where $W_1 = \int \phi_1^* \hat{H}\phi_1\, d\tau / \int \phi_1^*\phi_1\, d\tau$, etc. If these W's are numbered in order of increasing energy, it can be shown that $W_1 \geq E_{gs}$, $W_2 \geq E_{gs+1}$, etc., where $E_{gs}, E_{gs+1}, \ldots$ are the true energies of the ground state, the next-lowest state, etc. Thus, use of the linear variation function $c_1 f_1 + \cdots + c_n f_n$ gives us approximations to the energies and wave functions of the lowest n states in the system. (In using this method, one deals separately with wave functions of different symmetry.)

Perturbation Theory

In recent years, the perturbation-theory approximation method has become important in molecular electronic structure calculations. Let $\hat{H}$ be the time-independent Hamiltonian operator of a system whose Schrödinger equation $\hat{H}\psi_n = E_n\psi_n$ we seek to solve. In the perturbation-theory approximation, one divides $\hat{H}$ into two parts:

$$\hat{H} = \hat{H}^0 + \hat{H}' \tag{17.87}$$

where $\hat{H}^0$ is the Hamiltonian operator of a system whose Schrödinger equation can be solved exactly and $\hat{H}'$ is a term whose effects one hopes are small. The system with Hamiltonian $\hat{H}^0$ is called the *unperturbed system,* $\hat{H}'$ is called the *perturbation,* and the system with Hamiltonian $\hat{H} = \hat{H}^0 + \hat{H}'$ is called the *perturbed system.* One finds that the energy E_n of state n of the perturbed system can be written as

$$E_n = E_n^{(0)} + E_n^{(1)} + E_n^{(2)} + \cdots \tag{17.88}$$

where $E_n^{(0)}$ is the energy of state n of the unperturbed system, and $E_n^{(1)}, E_n^{(2)}, \ldots$ are called the first-order, second-order, . . . corrections to the energy. (For derivations of this and other perturbation-theory equations, see a quantum chemistry text.) If the problem is suitable for perturbation theory, the quantities $E_n^{(1)}, E_n^{(2)}, E_n^{(3)}, \ldots$ decrease as the order of the perturbation correction increases.

To find $E_n^{(0)}$we solve the Schrödinger equation $\hat{H}^0\psi_n^{(0)} = E_n^{(0)}\psi_n^{(0)}$ of the unperturbed system.

Perturbation theory shows that the first-order energy correction $E_n^{(1)}$ is given by

$$E_n^{(1)} = \int \psi_n^{(0)*}\hat{H}'\psi_n^{(0)}\,d\tau \qquad (17.89)$$

Since $\psi_n^{(0)}$ is known, $E_n^{(1)}$is easily calculated. The formulas for $E_n^{(2)}, E_n^{(3)}, \ldots$ are complicated and are omitted.

EXAMPLE 17.10 Perturbation theory

Suppose a one-particle, one-dimensional system has

$$\hat{H} = -(\hbar^2/2m)\, d^2/dx^2 + \tfrac{1}{2}kx^2 + bx^4$$

where b is small. Apply perturbation theory to obtain an approximation to the stationary-state energies of this system.

If we take $\hat{H}^0 = -(\hbar^2/2m)\, d^2/dx^2 + \frac{1}{2}kx^2$ and $\hat{H}' = bx^4$, then the unperturbed system is a harmonic oscillator, whose energies and wave functions are known (Sec. 17.12). From (17.76), we have $E_n^{(0)} = (n + \frac{1}{2})h\nu$, with $n = 0, 1, 2, \ldots$ and $\nu = (1/2\pi)(k/m)^{1/2}$, where the quantum-number symbol was changed from v to n to conform with the notation of this section. Including only the first-order correction to E_n, we have from (17.76) and (17.89)

$$E_n \approx E_n^{(0)} + E_n^{(1)} = (n + \tfrac{1}{2})h\nu + b\int_{-\infty}^{\infty} \psi_{n,\text{ho}}^{*}x^4\psi_{n,\text{ho}}\,dx \qquad (17.90)$$

where $\psi_{n,\text{ho}}$ is the harmonic-oscillator wave function with quantum number n. Substitution of the known $\psi_{n,\text{ho}}$ functions (Fig. 17.18) enables $E_n^{(1)}$ to be found.

Exercise

Evaluate $E_n^{(1)}$ in (17.90) for the ground state. Use Table 14.1. (*Answer:* $3bh^2/64\pi^4\nu^2m^2$.)

17.16 HERMITIAN OPERATORS

Section 17.11 noted that operators in quantum mechanics are linear. Quantum-mechanical operators that correspond to a physical property must have another property besides linearity, namely, they must be Hermitian. This section discusses Hermitian operators and their properties. The material of this section is important for a thorough understanding of quantum mechanics, but is not essential to understanding the material in the remaining chapters of this book and so may be omitted if time does not allow its inclusion. The abstract material of this section can induce dizziness in susceptible individuals and is best studied in small doses.

Hermitian Operators

The quantum-mechanical average value $\langle M\rangle$ of the physical quantity M must be a real number. To take the complex conjugate of a number, we replace i by $-i$ wherever it occurs. A real number does not contain i, so a real number equals its complex conjugate: $z = z^*$ if z is real. Hence $\langle M\rangle = \langle M\rangle^*$. We have $\langle M\rangle = \int \Psi^*\hat{M}\Psi\,d\tau$ [Eq. (17.62)] and

$\langle M\rangle^* = \int (\Psi^*\hat{M}\Psi)^*\, d\tau = \int (\Psi^*)^*(\hat{M}\Psi)^*\, d\tau = \int \Psi(\hat{M}\Psi)^*\, d\tau$, where the result of Prob. 17.19 was used. Therefore

$$\int \Psi^*\hat{M}\Psi\, d\tau = \int \Psi(\hat{M}\Psi)^*\, d\tau \tag{17.91}$$

Equation (17.91) must hold for all possible state functions Ψ, that is, for all functions that are continuous, single-valued, and quadratically integrable. A linear operator that obeys (17.91) for all well-behaved functions is called a **Hermitian operator.** If $\hat{M}$ is a Hermitian operator, it follows from (17.91) that (Prob. 17.63)

$$\int f^*\hat{M}g\, d\tau = \int g(\hat{M}f)^*\, d\tau \qquad \textbf{(17.92)*}$$

where f and g are arbitrary well-behaved functions (not necessarily eigenfunctions of any operator) and the integrals are definite integrals over all space. Although (17.92) looks like a more stringent requirement than (17.91), it is actually a consequence of (17.91). Thus a Hermitian operator obeys (17.92). The Hermitian property (17.92) is readily verified for the quantum-mechanical operators $x\cdot$ and $(\hbar/i)(\partial/\partial x)$ (Prob. 17.64).

Eigenvalues of Hermitian Operators

Section 17.11 noted that when Ψ is an eigenfunction of $\hat{M}$ with eigenvalue c, a measurement of M will give the value c. Since measured values are real, we expect c to be a real number. We now prove that *the eigenvalues of a Hermitian operator are real numbers.* To prove the theorem, we take the special case of (17.92) where f and g are the same function and this function is an eigenfunction of $\hat{M}$ with eigenvalue b. With $f = g$ and $\hat{M}f = bf$, (17.92) becomes

$$\int f^*bf\, d\tau = \int f(bf)^*\, d\tau$$

Using $(bf)^* = b^*f^*$ and taking the constants outside the integrals, we get $b \int f^*f\, d\tau = b^* \int ff^*\, d\tau$ or

$$(b - b^*)\int |f|^2\, d\tau = 0 \tag{17.93}$$

The quantity $|f|^2$ is never negative. The only way the definite integral $\int |f|^2\, d\tau$ (which is the infinite sum of the nonnegative infinitesimal quantities $|f|^2\, d\tau$) could be zero would be if the function f were zero everywhere. However, the function $f = 0$ is not allowed as an eigenfunction (Sec. 17.11). Therefore (17.93) requires that $b - b^* = 0$ and $b = b^*$. Only a real number is equal to its complex conjugate, so the eigenvalue b must be real.

Orthogonality of Eigenfunctions

We noted in Eq. (17.36) that the particle-in-a-one-dimensional-box stationary-state wave functions, which are eigenfunctions of $\hat{H}$, are orthogonal, meaning that $\int \psi_i^*\psi_j\, d\tau = 0$ when $i \neq j$. This is an example of
the theorem that *two eigenfunctions of a Hermitian operator that correspond to different eigenvalues are orthogonal.* The proof is as follows.

The Hermitian property (17.92) holds for all well-behaved functions. In particular, it holds if we take f and g as two of the eigenfunctions of the Hermitian operator $\hat{M}$. With $\hat{M}f = bf$ and $\hat{M}g = cg$, the Hermitian property $\int f^*\hat{M}g\, d\tau = \int g(\hat{M}f)^*\, d\tau$ becomes

$$c\int f^*g\, d\tau = \int g(bf)^*\, d\tau = \int gb^*f^*\, d\tau = b\int gf^*\, d\tau$$

since a Hermitian operator has real eigenvalues. We have

$$(c - b)\int f^*g\,d\tau = 0$$

If the eigenvalues c and b are different ($c \neq b$), then $\int f^*g\,d\tau = 0$, and the theorem is proved.

If the eigenvalues b and c happen to be equal, then orthogonality need not necessarily hold. Recall that we saw examples of different eigenfunctions of $\hat{H}$ having the same eigenvalue when we discussed the degenerate energy levels of the particle in a three-dimensional cubic box and the rigid two-particle rotor (Secs. 17.10 and 17.14). Because the quantum-mechanical operator $\hat{M}$ is linear, one can show (Prob. 17.65) that if the functions f_1 and f_2 are eigenfunctions of $\hat{M}$ with the same eigenvalue, that is, if $\hat{M}f_1 = bf_1$ and $\hat{M}f_2 = bf_2$, then any linear combination $c_1f_1 + c_2f_2$ (where c_1 and c_2 are constants) is an eigenfunction of $\hat{M}$ with eigenvalue b. This freedom to take linear combinations of eigenfunctions with the same eigenvalue enables us to choose the constants c_1 and c_2 so as to give orthogonal eigenfunctions (Prob. 17.66). From here on, we shall assume that this has been done, so that all eigenfunctions of a Hermitian operator that we deal with will be orthogonal.

Let the set of functions $g_1, g_2, g_3, \ldots$ be the eigenfunctions of a Hermitian operator. Since these functions are (or can be chosen to be) orthogonal, we have $\int g_j^*g_k\,d\tau = 0$ when $j \neq k$ (that is, when g_j and g_k are different eigenfunctions). We shall always normalize eigenfunctions of operators, so $\int g_j^*g_j\,d\tau = 1$. These two equations expressing orthogonality and normalization can be written as the single equation

$$\int g_j^*g_k\,d\tau = \delta_{jk} \qquad (17.94)$$

where the **Kronecker delta** δ_{jk} is a special symbol defined to equal 1 when $j = k$ and to equal 0 when j and k differ:

$$\delta_{jk} \equiv 1 \quad \text{when } j = k, \qquad \delta_{jk} \equiv 0 \quad \text{when } j \neq k \qquad (17.95)$$

A set of functions that are orthogonal and normalized is an **orthonormal set.**

Complete Sets of Eigenfunctions

A set of functions $g_1, g_2, g_3, \ldots$ is said to be a **complete set** if every well-behaved function that depends on the same variables as the g's and obeys the same boundary conditions as the g's can be expressed as the sum $\sum_i c_ig_i$, where the c's are constants whose values depend on the function being expressed. The sets of eigenfunctions of many of the Hermitian operators that occur in quantum mechanics have been proved to be complete, and quantum mechanics assumes that *the set of eigenfunctions of a Hermitian operator that represents a physical quantity is a complete set.* If F is a well-behaved function and the set $g_1, g_2, g_3, \ldots$ is the set of eigenfunctions of the Hermitian operator $\hat{R}$ that corresponds to the physical property R, then

$$F = \sum_k c_kg_k \qquad (17.96)$$

and one says that F has been *expanded* in terms of the set of g's.

How do we find the coefficients c_k in the expansion (17.96)? Multiplication of (17.96) by g_j^* gives $g_j^*F = \sum_k c_kg_j^*g_k$. Integration of this equation over the full range of all the coordinates gives

$$\int g_j^*\,F\,d\tau = \int \sum_k c_kg_j^*g_k\,d\tau = \sum_k \int c_kg_j^*g_k\,d\tau = \sum_k c_k \int g_j^*g_k\,d\tau = \sum_k c_k\delta_{jk}$$

where the orthonormality of the eigenfunctions of a Hermitian operator [Eq. (17.94)] and the fact that the integral of a sum equals the sum of the integrals were used. The

Kronecker delta δ_{jk} is always zero except when k equals j. Therefore every term in the sum $\Sigma_k c_k \delta_{jk}$ is zero except for the one term where k becomes equal to j: thus, $\Sigma_k c_k \delta_{jk} = c_j \delta_{jj} = c_j$ [Eq. (17.95)]. Therefore

$$c_j = \int g_j^* F \, d\tau$$

Changing j to k in this equation and substituting in the expansion (17.96), we have

$$F = \sum_k \left(\int g_k^* F \, d\tau \right) g_k \tag{17.97}$$

where the definite integrals $\int g_j^* F \, d\tau$ are constants. Equation (17.97) shows how to expand any function F in terms of a known complete set of functions $g_1, g_2, g_3, \ldots$.

Suppose we are unable to solve the Schrödinger equation for a system we are interested in. We can express the unknown ground-state wave function as $\psi_{gs} = \Sigma_k c_k g_k$, where the g's are a known complete set of functions. We then use the linear variation method (Sec. 17.15) to solve for the coefficients c_k, thereby obtaining ψ_{gs}. The difficulty with this approach is that a complete set of functions usually contains an infinite number of functions. We are therefore forced to limit ourselves to a finite number of functions in the expansion sum, thereby introducing error into our determination of ψ_{gs}. Most methods of calculating wave functions for molecules use expansions, as we shall see in Chapter 19.

Consider an example. Let the function F be defined as $F = x^2(a - x)$ for x between 0 and a and $F = 0$ elsewhere. Could we use the particle-in-a-box stationary-state wave functions $\psi_n = (2/a)^{1/2} \sin (n\pi x/a)$ [Eq. (17.35)] to expand F? The function F is well-behaved and satisfies the same boundary conditions as ψ_n, namely, F is zero at the ends of the box. The functions ψ_n are the eigenfunctions of a Hermitian operator (the particle-in-a-box Hamiltonian $\hat{H}$) and so are a complete set. Therefore we can express F as $F = \Sigma_{n=1}^{\infty} c_n \psi_n$, where the coefficients c_n are given in (17.97) as

$$c_n = \int \psi_n^* F \, d\tau = \int_0^a \left(\frac{2}{a}\right)^{1/2} \sin \frac{n\pi x}{a} x^2(a - x) \, dx \tag{17.98}$$

Problem 17.67 evaluates c_n and shows how the sum $\Sigma_n c_n \psi_n$ becomes a more and more accurate representation of F as more terms are included in the sum.

The proof of the variation theorem (17.84) uses an expansion of the variation function ϕ, as in Eq. (17.96). The proof is outlined in Prob. 17.68.

Summary

Quantum-mechanical operators that correspond to physical properties are Hermitian, meaning that they satisfy (17.92) for all well-behaved functions f and g. The eigenvalues of a Hermitian operator are real. The eigenfunctions of a Hermitian operator are (or can be chosen to be) orthogonal. The eigenfunctions of a Hermitian operator form a complete set, meaning that any well-behaved function can be expanded in terms of them.

17.17 SUMMARY

Electromagnetic waves of frequency ν and wavelength λ travel at speed $c = \lambda\nu$ in vacuum. Processes involving absorption or emission of electromagnetic radiation (for example, blackbody radiation, the photoelectric effect, spectra of atoms and molecules) can be understood by viewing the electromagnetic radiation to be composed of photons, each photon having an energy $h\nu$, where h is Planck's constant. When an atom or molecule absorbs or emits a photon, it makes a transition between two energy levels E_a and E_b whose energy difference is $h\nu$; $E_a - E_b = h\nu$.

De Broglie proposed that microscopic particles such as electrons have wavelike properties, and this was confirmed by observation of electron diffraction. Because of this wave–particle duality, simultaneous measurement of the precise position and momentum of a microscopic particle is impossible (the Heisenberg uncertainty principle).

The state of a quantum-mechanical system is described by the state function Ψ, which is a function of the particles' coordinates and the time. The change in Ψ with time is governed by the time-dependent Schrödinger equation (17.10) [or (17.60)], which is the quantum-mechanical analog of Newton's second law in classical mechanics. The probability density for finding the system's particles is $|\Psi|^2$. For example, for a two-particle, one-dimensional system, $|\Psi(x_1, x_2, t)|^2\, dx_1\, dx_2$ is the probability of simultaneously finding particle 1 between x_1 and $x_1 + dx_1$ and particle 2 between x_2 and $x_2 + dx_2$ at time t.

When the system's potential energy V is independent of time, the system can exist in one of many possible stationary states. For a stationary state, the state function is $\Psi = e^{-iEt/\hbar}\psi$. The (time-independent) wave function ψ is a function of the particles' coordinates and is one of the well-behaved solutions of the (time-independent) Schrödinger equation $\hat{H}\psi = E\psi$, where E is the energy and the Hamiltonian operator $\hat{H}$ is the quantum-mechanical operator that corresponds to the classical quantity E. To find the operator corresponding to a classical quantity M, one writes down the classical-mechanical expression for M in terms of cartesian coordinates and momenta and then replaces the coordinates and momenta by their corresponding quantum-mechanical operators: $\hat{x}_1 = x_1 \times$, $\hat{p}_{x,1} = (\hbar/i)\, \partial/\partial x_1$, etc. For a stationary state, $|\Psi|^2 = |\psi|^2$ and the probability density and energy are independent of time.

In accord with the probability interpretation, the state function is normalized to satisfy $\int |\Psi|^2\, d\tau = 1$, where $\int d\tau$ denotes the definite integral over the full range of the particles' coordinates. For a stationary state, the normalization condition becomes $\int |\psi|^2\, d\tau = 1$.

The average value of property M for a system in stationary state ψ is $\langle M \rangle = \int \psi^* \hat{M} \psi\, d\tau$, where $\hat{M}$ is the quantum-mechanical operator for property M.

The stationary-state wave functions and energies were found for the following systems.

(*a*) Particle in a one-dimensional box ($V = 0$ for x between 0 and a; $V = \infty$ elsewhere): $E = n^2h^2/8ma^2$, $\psi = (2/a)^{1/2} \sin (n\pi x/a)$, $n = 1, 2, 3, \ldots$.

(*b*) Particle in a three-dimensional rectangular box with dimensions a, b, c: $E = (h^2/8m) \cdot (n_x^2/a^2 + n_y^2/b^2 + n_z^2/c^2)$.

(*c*) One-dimensional harmonic oscillator ($V = \frac{1}{2}kx^2$): $E = (v + \frac{1}{2})h\nu$, $\nu = (1/2\pi)(k/m)^{1/2}$, $v = 0, 1, 2, \ldots$.

(*d*) Two-particle rigid rotor (particles at fixed distance d and energy entirely kinetic): $E = J(J + 1)\hbar^2/2I$, $I = \mu d^2$, $J = 0, 1, 2, \ldots$; $\mu \equiv m_1m_2/(m_1 + m_2)$ is the reduced mass.

When more than one state function corresponds to the same energy level, that energy level is said to be degenerate. There is degeneracy for the particle in a cubic box and for the two-particle rigid rotor.

For a system of noninteracting particles, the stationary-state wave functions are products of wave functions for each particle and the energy is the sum of the energies of the individual particles.

The variation theorem states that for any well-behaved trial variation function ϕ, one has $\int \phi^* \hat{H} \phi\, d\tau / \int \phi^* \phi\, d\tau \geq E_{gs}$, where $\hat{H}$ is the system's Hamiltonian operator and E_{gs} is its true ground-state energy.

Important kinds of calculations discussed in this chapter include:

- Use of $\lambda\nu = c$ to calculate the wavelength of light from the frequency, and vice versa.
- Use of $E_{upper} - E_{lower} = h\nu$ to calculate the frequency of the photon emitted or absorbed when a quantum-mechanical system makes a transition between two states.

- Use of energy-level formulas such as $E = n^2h^2/8ma^2$ for the particle in a box or $E = (v + \frac{1}{2})h\nu$ for the harmonic oscillator to calculate energy levels of quantum-mechanical systems.
- For a one-particle, one-dimensional, stationary-state system, use of $|\psi|^2\,dx$ to calculate the probability of finding the particle between x and $x + dx$ and of $\int_a^b |\psi|^2\,dx$ to calculate the probability of finding the particle between a and b.
- Use of $\langle M\rangle = \int \psi^*\hat{M}\psi\,d\tau$ to calculate average values.
- Use of the variation theorem to estimate the ground-state energy of a quantum-mechanical system.

FURTHER READING

Hanna, chap. 3; *Karplus and Porter,* chap. 2; *Levine,* chaps. 1–4, 8, 9; *Lowe and Peterson,* chaps. 1–3, 7; *McQuarrie* (1983), chaps. 1–5; *Atkins and Friedman,* chaps. 1–3, 6.

PROBLEMS

Section 17.1

17.1 (*a*) Let ν_{max} be the frequency at which the blackbody-radiation function (17.2) is a maximum. Show that $\nu_{max} = kTx/h$, where x is the nonzero solution of $x + 3e^{-x} = 3$. Since x is a constant, ν_{max} increases linearly with T. (*b*) Use a calculator with an e^x key to solve the equation in (*a*) by trial and error. To save time, use interpolation after you have found the successive integers that x lies between. Alternatively, use the Solver in Excel. (*c*) Calculate ν_{max} for a blackbody at 300 K and at 3000 K. Refer to Fig. 20.2 to state in which portions of the electromagnetic spectrum these frequencies lie. (*d*) The light emitted by the sun conforms closely to the blackbody radiation law and has $\nu_{max} = 3.5 \times 10^{14}\ s^{-1}$. Estimate the sun's surface temperature. (*e*) The skin temperature of humans is 33°C, and the emission spectrum of human skin at this temperature conforms closely to blackbody radiation. Find ν_{max} for human skin at 33°C. What region of the electromagnetic spectrum is this in?

17.2 (*a*) Use the fact that $\int_0^\infty [z^3/(e^z - 1)]\,dz = \pi^4/15$ to show that the total radiant energy emitted per second by unit area of a blackbody is $2\pi^5k^4T^4/15c^2h^3$. Note that this quantity is proportional to T^4 (*Stefan's law*). (*b*) The sun's diameter is 1.4×10^9 m and its effective surface temperature is 5800 K. Assume the sun is a blackbody and estimate the rate of energy loss by radiation from the sun. (*c*) Use $E = mc^2$ to calculate the relativistic mass of the photons lost by radiation from the sun in 1 year.

Section 17.2

17.3 The work function of K is 2.2 eV and that of Ni is 5.0 eV, where 1 eV = 1.60×10^{-19} J. (*a*) Calculate the threshold frequencies and wavelengths for these two metals. (*b*) Will violet light of wavelength 4000 Å cause the photoelectric effect in K? In Ni? (*c*) Calculate the maximum kinetic energy of the electrons emitted in (*b*).

17.4 Calculate the energy of a photon of red light of wavelength 700 nm. (1 nm = 10^{-9} m.)

17.5 A 100-W sodium-vapor lamp emits yellow light of wavelength 590 nm. Calculate the number of photons emitted per second.

17.6 Millikan found the following data for the photoelectric effect in Na:

$10^{12}K_{max}$/ergs	3.41	2.56	1.95	0.75
λ/Å	3125	3650	4047	5461

where K_{max} is the maximum kinetic energy of emitted electrons and λ is the wavelength of the incident radiation. Plot K_{max} versus ν. From the slope and intercept, calculate h and the work function for Na.

Section 17.4

17.7 Calculate the de Broglie wavelength of (*a*) a neutron moving at 6.0×10^6 cm/s; (*b*) a 50-g particle moving at 120 cm/s.

Section 17.5

17.8 A beam of electrons traveling at 6.0×10^8 cm/s falls on a slit of width 2400 Å. The diffraction pattern is observed on a screen 40 cm from the slit. The x and y axes are defined as in Fig. 17.4. Find (*a*) the angle θ to the first diffraction minimum; (*b*) the width of the central maximum of the diffraction pattern on the screen; (*c*) the uncertainty Δp_x at the slit.

17.9 Estimate the minimum uncertainty in the x component of velocity of an electron whose position is measured to an uncertainty of 1×10^{-10} m.

Section 17.6

17.10 For a system containing three particles, what are the variables on which the state function Ψ depends?

17.11 True or false? (*a*) In the equation $\int |\Psi|^2\,d\tau = 1$, the integral is an indefinite integral. (*b*) The state function Ψ takes on only real values. (*c*) If z is a complex number, then $zz^* = |z|^2$. (*d*) If z is a complex number, then $z + z^*$ is always a real number. (*e*) If $z = a + bi$, where a and b are real numbers, and we plot a on the x axis and b on the y axis, the distance of the point (a, b) from the origin in the xy plane is equal to $|z|$. (*f*) The absolute value $|z|$ of a complex number must be a real nonnegative number.

17.12 Find the complex conjugate and the absolute value of (*a*) -2; (*b*) $3 - 2i$; (*c*) $\cos\theta + i\sin\theta$; (*d*) $-3e^{-i\pi/5}$.

17.13 Verify that, if Ψ is a solution of the time-dependent Schrödinger equation (17.10), then $c\Psi$ is also a solution, where c is any constant.

17.14 Show that

$$\int_a^b\int_c^d\int_s^t f(r)g(\theta)h(\phi)\,dr\,d\theta\,d\phi = \int_s^t f(r)\,dr\int_c^d g(\theta)\,d\theta\int_a^b h(\phi)\,d\phi$$

where the limits are constants.

17.15 Verify that Ψ in Example 17.1 is normalized.

Section 17.7

17.16 For a system consisting of three particles, on what variables does the time-independent wave function ψ depend?

17.17 True or false? For a stationary state, (*a*) $|\psi| = |\Psi|$; (*b*) $\psi = \Psi$; (*c*) the probability density is independent of time; (*d*) the energy is a constant.

17.18 Which is more general, the time-dependent Schrödinger equation or the time-independent Schrödinger equation?

17.19 Prove that $(fg)^* = f^*g^*$, where f and g are complex quantities.

17.20 Verify that if ψ is a solution of the time-independent Schrödinger equation (17.24), then $k\psi$ is also a solution, where k is any constant.

17.21 State whether each of the functions (*a*) to (*d*) is quadratically integrable. (a and b are positive constants.) (*a*) e^{-ax^2} (*Hint:* See Table 14.1.) (*b*) e^{-bx}. (*c*) $1/x$. (*Hint:* In (*c*) and (*d*), write the integral as the sum of two integrals.) (*d*) $1/|x|^{1/4}$. (*e*) True or false? A function that becomes infinite at a point must not be quadratically integrable.

Section 17.8

17.22 Calculate the wavelength of the photon emitted when a 1.0×10^{-27} g particle in a box of length 6.0 Å goes from the $n = 5$ to the $n = 4$ level.

17.23 (*a*) For a particle in the stationary state n of a one-dimensional box of length a, find the probability that the particle is in the region $0 \le x \le a/4$. (*b*) Calculate this probability for $n = 1$, 2, and 3.

17.24 For a 1.0×10^{-26} g particle in a box whose ends are at $x = 0$ and $x = 2.000$ Å, calculate the probability that the particle's x coordinate is between 1.6000 and 1.6001 Å if (*a*) $n = 1$; (*b*) $n = 2$.

17.25 Sketch ψ and $|\psi|^2$ for the $n = 4$ and $n = 5$ states of a particle in a one-dimensional box.

17.26 Solve Eq. (17.28) for the special case $E = 0$. Then apply the continuity requirement at each end of the box to evaluate the two integration constants and thus show that $\psi = 0$ for $E = 0$. From (17.32), $E = 0$ corresponds to $n = 0$, so $n = 0$ is not allowed.

17.27 For an electron in a certain one-dimensional box, the lowest observed transition frequency is $2.0 \times 10^{14}\ s^{-1}$. Find the length of the box.

17.28 If the $n = 3$ to 4 transition for a certain particle-in-a-box system occurs at $4.00 \times 10^{13}\ s^{-1}$, find the frequency of the $n = 6$ to 9 transition in this system.

17.29 Verify the orthogonality equation (17.36) for particle-in-a-box wave functions.

17.30 For the particle in a box, check that the wave functions (17.35) satisfy the Schrödinger equation (17.28) by substituting (17.35) into (17.28).

Section 17.9

17.31 For a particle in a two-dimensional box with sides of equal lengths, draw rough sketches of contours of constant $|\psi|$ for the states (*a*) $n_x = 2$, $n_y = 1$; (*b*) $n_x = 2$, $n_y = 2$. At what points in the box is $|\psi|$ a maximum for each state? (*Hint:* The maximum value of $|\sin\theta|$ is 1.)

Section 17.10

17.32 For a particle in a cubic box of length a, give the degree of degeneracy of the energy level with energy (*a*) $21h^2/8ma^2$; (*b*) $24h^2/8ma^2$.

17.33 For a particle in a cubic box of edge a: (*a*) How many states have energies in the range 0 to $16h^2/8ma^2$? (*b*) How many energy levels lie in this range?

Section 17.11

17.34 True or false? (*a*) $(\hat{A} + \hat{B})f(x)$ is always equal to $\hat{A}f(x) + \hat{B}f(x)$. (*b*) $\hat{A}[f(x) + g(x)]$ is always equal to $\hat{A}f(x) + \hat{A}g(x)$. (*c*) $\hat{B}\hat{C}f(x)$ is always equal to $\hat{C}\hat{B}f(x)$. (*d*) $[\hat{A}f(x)]/f(x)$ is always equal to $\hat{A}$, provided $f(x) \ne 0$. (*e*) $3x$ is an eigenvalue of $\hat{x}$. (*f*) $3x$ is an eigenfunction of $\hat{x}$. (*g*) $e^{ikx/\hbar}$, where k is a constant, is an eigenfunction of $\hat{p}_x$ with eigenvalue k.

17.35 If f is a function, state whether or not each of the following expressions is equal to $f^*\hat{B}f$. (*a*) $f^*(\hat{B}f)$. (*b*) $\hat{B}(f^*f)$. (*c*) $(\hat{B}f)f^*$. (*d*) $f^*f\hat{B}$.

17.36 If the energy of a particle in the $n = 5$ stationary state of a particle in a one-dimensional box of length a is measured, state the possible result(s).

17.37 Let $\hat{A} = d^2/dx^2$ and $\hat{B} = x\times$. (*a*) Find $\hat{A}\hat{B}f(x) - \hat{B}\hat{A}f(x)$. (*b*) Find $(\hat{A} + \hat{B})(e^{x^2} + \cos 2x)$.

17.38 (*a*) Classify each of these operators as linear or nonlinear: $\partial^2/\partial x^2$, $2\,\partial/\partial z$, $3z^2\times$, $(\ )^2$, $(\ )^*$. (*b*) Verify that $\hat{H}$ in (17.58) is linear.

17.39 State whether each of the following entities is an operator or a function: (*a*) $\hat{A}\hat{B}g(x)$; (*b*) $\hat{A}\hat{B} + \hat{B}\hat{A}$; (*c*) $\hat{B}^2f(x)$; (*d*) $g(x)\hat{A}$, (*e*) $g(x)\hat{A}f(x)$.

17.40 Find the quantum-mechanical operator for (*a*) p_x^3; (*b*) p_z^4.

17.41 (*a*) Which of the functions sin $3x$, 6 cos $4x$, $5x^3$, $1/x$, $3e^{-5x}$, ln $2x$ are eigenfunctions of d^2/dx^2? (*b*) For each eigenfunction, state the eigenvalue.

17.42 For a particle in a one-dimensional-box stationary state, show that (*a*) $\langle p_x \rangle = 0$; (*b*) $\langle x \rangle = a/2$; (*c*) $\langle x^2 \rangle = a^2(1/3 - 1/2n^2\pi^2)$.

17.43 Verify the separation-of-variables equations of Sec. 17.11 as follows. For the Hamiltonian (17.64), write the time-independent Schrödinger equation. Assume solutions of the form (17.65) and substitute into the Schrödinger equation to derive (17.66) and (17.67).

17.44 For a system of two noninteracting particles of masses m_1 and m_2 in a one-dimensional box of length a, give the formulas for the stationary-state wave functions and energies.

Section 17.12

17.45 For each of the following, state whether it is a nu or a vee and whether it is a quantum number or a frequency. (*a*) ν; (*b*) v.

17.46 Calculate the frequency of radiation emitted when a harmonic oscillator of frequency 6.0×10^{13} s^{-1} goes from the $v = 8$ to the $v = 7$ level.

17.47 Draw rough sketches of ψ^2 for the $v = 0, 1, 2,$ and 3 harmonic-oscillator states.

17.48 Find the most probable value(s) of x for a harmonic oscillator in the state (*a*) $v = 0$; (*b*) $v = 1$.

17.49 Verify that ψ_0 in Fig. 17.18 is a solution of the Schrödinger equation (17.75).

17.50 Verify that the harmonic-oscillator ψ_1 in Fig. 17.18 is normalized. (See Table 14.1.)

17.51 For the ground state of a harmonic oscillator, calculate (*a*) $\langle x \rangle$; (*b*) $\langle x^2 \rangle$; (*c*) $\langle p_x \rangle$. See Table 14.1.

17.52 (*a*) Verify Eq. (17.74). (*b*) Verify by substitution that (17.72) satisfies the differential equation $m\, d^2x/dt^2 = -kx$.

17.53 A mass of 45 g on a spring oscillates at the frequency of 2.4 vibrations per second with an amplitude 4.0 cm. (*a*) Calculate the force constant of the spring. (*b*) What would be the quantum number v if the system were treated quantum-mechanically?

17.54 For a three-dimensional harmonic oscillator, $V = \frac{1}{2}k_x x^2 + \frac{1}{2}k_y y^2 + \frac{1}{2}k_z z^2$, where the three force constants k_x, k_y, k_z are not necessarily equal. (*a*) Write down the expression for the energy levels of this system. Define all symbols. (*b*) What is the zero-point energy?

Section 17.13

17.55 Substitute (17.79), (17.80), and $M = m_1 + m_2$ into (17.78) and verify that H reduces to $p_1^2/2m_1 + p_2^2/2m_2 + V$, where p_1 is the momentum of particle 1.

Section 17.14

17.56 Consider the $^{12}C^{16}O$ molecule to be a two-particle rigid rotor with m_1 and m_2 equal to the atomic masses and the interparticle distance fixed at the CO bond length 1.13 Å. (*a*) Find the reduced mass. (*b*) Find the moment of inertia. (*c*) Find the energies of the four lowest rotational levels and give the degeneracy of each of these levels. (*d*) Calculate the frequency of the radiation absorbed when a $^{12}C^{16}O$ molecule goes from the $J = 0$ level to the $J = 1$ level. Repeat for $J = 1$ to $J = 2$.

Section 17.15

17.57 The unnormalized particle-in-a-box variation function $x(a - x)$ for x between 0 and a was used in Example 17.8. (*a*) Use work in that example to show that $(30/a^5)^{1/2}x(a - x)$ is the normalized form of this function. (*b*) Calculate $\langle x^2 \rangle$ using the normalized function in (*a*) and compare the result with the exact ground-state $\langle x^2 \rangle$ (Prob. 17.42).

17.58 For the particle in a one-dimensional box, one finds that use of the normalized variation function $\phi = Nx^k(a - x)^k$, where k is a parameter, gives $\int \phi^* \hat{H} \phi \, d\tau = (\hbar^2/ma^2) \cdot (4k^2 + k)/(2k - 1)$. Find the value of k that minimizes the variational integral W and find the value of W for this k value. Compare the percent error in the ground-state energy with that for the function $Nx(a - x)$ used in Example 17.8.

17.59 (*a*) Apply the variation function $x^2(a - x)^2$ for x between 0 and a to the particle in a box and estimate the ground-state energy. Calculate the percent error in E_{gs}. (*b*) Explain why the function x^2 (for x between 0 and a) cannot be used as a variation function for the particle in a box.

17.60 Consider a one-particle, one-dimensional system with $V = \infty$ for $x < 0$, $V = \infty$ for $x > a$, and $V = kx$ for $0 \le x \le a$, where k is small. Treat the system as a perturbed particle in a box and find $E_n^{(0)} + E_n^{(1)}$ for the state with quantum number n. Use a table of integrals.

Section 17.16

17.61 True or false? (*a*) All the eigenvalues of a Hermitian operator are real numbers. (*b*) Two eigenfunctions of the same Hermitian operator are always orthogonal. (*c*) $\delta_{jk} = \delta_{kj}$. (*d*) A Hermitian operator cannot contain the imaginary number i. (*e*) $\sum_n b_m c_m \delta_{mn} = b_n c_n$.

17.62 Prove that the sum of two Hermitian operators is a Hermitian operator.

17.63 Use the following procedure to show that for the Hermitian operator $\hat{M}$, Eq. (17.92) is a consequence of (17.91). (*a*) Set $\Psi = f + cg$ in (17.91), where c is an arbitrary constant. Use (17.91) to cancel some terms in the resulting equation, thereby getting

$$c^* \int g^* \hat{M} f \, d\tau + c \int f^* \hat{M} g \, d\tau$$

$$= c \int g (\hat{M} f)^* \, d\tau + c^* \int f (\hat{M} g)^* \, d\tau \qquad (17.99)$$

(*b*) First set $c = 1$ in (17.99). Then set $c = i$ in (17.99) and divide the resulting equation by i. Add these two equations, thereby proving (17.92).

17.64 Verify that if f and g are functions of x and $d\tau = dx$, then (*a*) the Hermitian property (17.92) holds for $\hat{x}$; (*b*) (17.92) holds for $\hat{p}_x$. [*Hint:* For part (*b*), use integration by parts and the fact that a quadratically integrable function must go to zero as x goes to $\pm\infty$.]

17.65 Given that $\hat{M}$ is a linear operator and that $\hat{M}f_1 = bf_1$ and $\hat{M}f_2 = bf_2$, prove that $c_1f_1 + c_2f_2$, where c_1 and c_2 are constants, is an eigenfunction of $\hat{M}$ with eigenvalue b.

17.66 If $\hat{M}$ is a linear operator with $\hat{M}f_1 = bf_1$ and $\hat{M}f_2 = bf_2$, and we define g_1 and g_2 as $g_1 \equiv f_1$ and $g_2 \equiv f_2 + kf_1$, where $k \equiv -\int f_1^* f_2\, d\tau / \int f_1^* f_1\, d\tau$, verify that g_1 and g_2 are orthogonal.

17.67 Let $F \equiv x^2(a - x)$ for x between 0 and a. Let $G \equiv \sum_{n=1}^{m} c_n\psi_n$, where ψ_n is a particle-in-a-one-dimensional box wave function with quantum number n and c_n is given by (17.98). To make things a bit simpler, take the length a as equal to 1. (*a*) Use a table of integrals to find c_n. (*b*) Use a spreadsheet to calculate F and G for $m = 3$ and plot them on the same graph. (*c*) Repeat (*b*) for $m = 5$ and comment on the results.

17.68 For a system with Hamiltonian operator $\hat{H}$ and stationary-state wave functions and energies ψ_n and E_n, we have $\hat{H}\psi_n = E_n\psi_n$, where n labels the various states. Let ϕ be a normalized, well-behaved variation function that depends on the coordinates of the system's particles and let $W \equiv \int \phi^*\hat{H}\phi\, d\tau$ be the variational integral. Since the eigenfunctions ψ_n form a complete set, we can use (17.96) to expand ϕ as $\phi = \sum_k c_k\psi_k$. (*a*) Explain why $\phi^* = \sum_j c_j^*\psi_j^*$. (*b*) Substitute these expansions for ϕ^* and ϕ into the normalization condition $\int \phi^*\phi\, d\tau = 1$ and use the orthonormality of the eigenfunctions ψ_n to show that $1 = \sum_j \sum_k c_j^* c_k \delta_{jk}$. Explain why this last equation becomes $\sum_k |c_k|^2 = 1$. (*c*) Substitute the expansions for ϕ^* and ϕ into W and show that $W = \sum_k |c_k|^2 E_k$. (*d*) We have $E_k \geq E_{gs}$, where E_{gs} is the ground-state energy. Since $|c_k|^2$ is never negative, we can multiply the inequality by $|c_k|^2$ to get $|c_k|^2 E_k \geq |c_k|^2 E_{gs}$. Explain why $W \geq \sum_k |c_k|^2 E_{gs}$. Then use the result of part (*b*) to conclude that $W \geq E_{gs}$. (*e*) In the expansions of ϕ^* and ϕ, we used different letters j and k for the summation indices. Explain why this was done. *Hint:* Express the product $(a_1 + a_2)(b_1 + b_2)$ using summation notation and move both sum signs to the left of everything.

General

17.69 What are the SI units of a stationary-state wave function ψ for (*a*) a one-particle, one-dimensional system; (*b*) a one-particle, three-dimensional system; (*c*) a two-particle, three-dimensional system?

17.70 By fitting experimental blackbody-radiation curves using Eq. (17.2), Planck not only obtained a value for h, but also obtained the first reasonably accurate values of k, N_A, and the proton charge e. Explain how Planck obtained values for these constants.

17.71 State quantitatively the effect on the system's energy levels of each of the following: (*a*) doubling the box length for a particle in a one-dimensional box; (*b*) doubling the interparticle distance of a two-particle rigid rotor; (*c*) doubling the mass of a harmonic oscillator.

17.72 True or false? (*a*) In classical mechanics, knowledge of the present state of an isolated system allows the future state to be predicted with certainty. (*b*) In quantum mechanics, knowledge of the present state of an isolated system allows the future state to be predicted with certainty. (*c*) For a stationary state, Ψ is the product of a function of time and a function of the coordinates. (*d*) An increase in the particle mass would decrease the ground-state energy of both the particle in a box and the harmonic oscillator. (*e*) For a system of noninteracting particles, each stationary-state wave function is equal to the sum of wave functions for each particle. (*f*) The one-dimensional harmonic-oscillator energy levels are nondegenerate. (*g*) Ψ must be real. (*h*) The energies of any two photons must be equal. (*i*) In the variation method, the variational function ϕ must be an eigenfunction of $\hat{H}$.

17.73 *Scientist trivia question.* Name the scientist referred to in each of the following descriptions. The same name can be used more than once and all names except (*f*) appear in Chapter 17. (*a*) On Christmas vacation in Arosa, Switzerland, this 38-year-old professor of theoretical physics at the University of Zurich began work on a series of papers titled "Quantisierung als Eigenwertproblem" ("Quantization as an Eigenvalue Problem"), papers described by Born as "of a grandeur unsurpassed in theoretical physics." (*b*) On the tiny North Sea island of Helgoland, where he had gone to recover from a severe attack of hay fever, this 23-year-old lecturer at the University of Göttingen conceived the ideas embodied in his paper "Über Quantentheoretische Umdeutung Kinematischer und Mechanischer Beziehungen" ("On a Quantum-Theoretical Interpretation of Kinematical and Mechanical Relations") that marks the birth of modern quantum mechanics. He later wrote: "it was almost three o'clock in the morning before the final result of my computations lay before me . . . I could no longer doubt the mathematical consistency and coherence of the kind of quantum mechanics to which my calculations pointed. At first, I was deeply alarmed. I had the feeling that, through the surface of atomic phenomena, I was looking at a strangely beautiful interior . . . I was far too excited to sleep, and so, as a new day dawned, I made for the southern tip of the island, where I had been longing to climb a rock jutting out into the sea. I now did so without too much trouble, and waited for the sun to rise." (*c*) He headed Germany's atom-bomb project during World War II. (*d*) In September 1943, under cover of darkness, he crossed the Öresund strait by boat from German-occupied Denmark to Sweden. A few days later, a British bomber flew him to Scotland. The plane had no passenger seat and he flew in the bomb bay. He failed to use the oxygen mask supplied him and lost consciousness during the flight, but recovered when the plane landed. (*e*) In a discussion with Bohr about quantum mechanics, he said "If we are still going to have to put up with these damn quantum jumps, I am sorry I ever had anything to do with quantum theory." (*f*) Besides founding a journal of physical chemistry, he also founded a journal of philosophy. When the philosopher Ludwig Wittgenstein had trouble finding a publisher for his first book, this physical chemist published the book in 1921 in his philosophy journal. (Wittgenstein was chosen by *Time* magazine as one of the 100 most influential people of the 20th century.) (*g*) He spent 1940–1956 working mainly on a unified field theory at the Dublin Institute for Advanced Studies; during the early years of this period he lived in a two-story house with his wife Anny, his mistress Hilde March (wife of the physicist Arthur March), and his daughter Ruth (who did not find out she was his daughter until she was 17). (*h*) This nineteenth-century mathematician and physicist reformulated classical mechanics in a form especially suitable for formulating quantum mechanics. He was appointed Professor of Astronomy at Trinity College while still an undergraduate student. (*i*) He used the pseudonym Nicholas Baker when he worked on the atomic bomb in Los Alamos, New Mexico. (*j*) In a 1999 poll of physicists he was chosen as the greatest physicist of all time and was chosen by *Time* magazine as "Person of the Twentieth Century." (*k*) His son Erwin was found guilty of complicity in the 1944 plot to kill Hitler and was executed in February 1945. Shortly thereafter, when the town he was living in became a battlefield, this 87-year-old physicist had to hide in the woods and sleep in haystacks. (*l*) He believed in the Hindu philosophy of Vedanta, which he summarized as "we living beings all belong to one another . . . we are all actually members or aspects of a single being, which we may in western terminology call God, while in the Upanishads it is called Brahman." (*m*) His 1919 divorce settlement with his first wife provided that in addition to providing child support, if he were to win a Nobel Prize, the prize money would be given to his ex-wife. She received the prize money in 1923. In a letter he praised a friend's ability to have a happy, long-lasting marriage, "an undertaking in which I failed twice rather disgracefully." (*n*) He helped develop quantum mechanics and from 1957 to 1961 was a member of West Germany's parliament. (*o*) In a letter to a father whose child had died, he wrote: "A human being is a part of the whole, called by us "Universe," a part limited in time and space. He experiences himself, his thoughts and feelings as something separated from the rest—a kind of optical delusion of his consciousness. This delusion is a prison for us, restricting us to our personal desires and to affection for a few persons nearest to us. Our task must be to free ourselves from this prison by widening our circle of compassion to embrace all living creatures and the whole of nature in its beauty."

Chapter 17

17.1 **(a)** $dR/d\nu = 0 = \frac{2\pi h}{c^2}\left[\frac{3\nu^2}{e^{h\nu/kT}-1} - \frac{\nu^3(h/kT)e^{h\nu/kT}}{(e^{h\nu/kT}-1)^2}\right]$

So $3 = (h\nu_{max}/kT)e^{h\nu_{max}/kT}/(e^{h\nu_{max}/kT} - 1) = xe^x/(e^x - 1)$, where $x = h\nu_{max}/kT$. Then $3e^x - 3 = xe^x$ and multiplication by e^{-x} gives $x + 3e^{-x} = 3$.

(b) For $x = 0, 1, 2, 3$, the function $x + 3e^{-x}$ equals 3, 2.104, 2.406, and 3.149. So the nonzero root lies between 2 and 3. We have $(3 - 2.406)/(3.149 - 2.406) = 0.80$, so interpolation gives $x \approx 2.80$. For $x = 2.80, 2.81, 2.82, 2.83$, we find $x + 3e^{-x} = 2.98243, 2.99061, 2.99882, 3.00704$. The root lies between 2.82 and 2.83, and interpolation gives $x = 2.821_4$. The Excel Solver gives 2.82143937… . (Use Options to change the Precision to 10^{-15}.)

(c) At 300 K, $\nu_{max} = kTx/h = (1.3807 \times 10^{-23}\text{ J/K})(300\text{ K})(2.821)/(6.626 \times 10^{-34}\text{ J s}) = 1.76 \times 10^{13}\text{ s}^{-1}$. At 3000 K, ν_{max} is 10 times as large, namely, $1.76 \times 10^{14}\text{ s}^{-1}$. From Fig. 20.2, these frequencies lie in the infrared.

(d) $T = h\nu_{max}/xk = (6.626 \times 10^{-34}\text{ J s})(3.5 \times 10^{14}\text{ s}^{-1})/2.821(1.38 \times 10^{-23}\text{ J/K}) = 6000\text{ K}$.

(e) $\nu_{max} = kTx/h = (1.38 \times 10^{-23}\text{ J/K})(306\text{ K})2.821/(6.626 \times 10^{-34}\text{ J s}) = 1.80 \times 10^{13}\text{ s}^{-1}$. Infrared.

17.2 **(a)** The total emission per unit time and per unit area is $\int_0^\infty R(\nu)\,d\nu = (2\pi h/c^2)\int_0^\infty [\nu^3/(e^{h\nu/kT} - 1)]\,d\nu$. Let $z = h\nu/kT$; then $dz = (h/kT)\,d\nu$. We have $\int_0^\infty R(\nu)\,d\nu = (2\pi h/c^2)(kT/h)^4 \int_0^\infty [z^3/(e^z - 1)]\,dz = (2\pi h/c^2)(kT/h)^4\pi^4/15 = 2\pi^5k^4T^4/15c^2h^3$.

(b) The emission rate is $(2\pi^5k^4T^4/15c^2h^3)(4\pi r^2) =$

$$\frac{8\pi^6(1.381\times10^{-23}\text{ J/K})^4(5800\text{ K})^4(0.7\times10^9\text{ m})^2}{15(2.998\times10^8\text{ m/s})^2(6.626\times10^{-34}\text{ J s})^3} = 3.9_6 \times 10^{26}\text{ J/s}$$

(similar to the value given in Prob. 16.113).

(c) In 1 year, $\Delta E = (3.9_6 \times 10^{26}\text{ J/s})(365.25 \times 24 \times 60 \times 60\text{ s}) = 1.2_5 \times 10^{34}\text{ J}$. So $\Delta m = \Delta E/c^2 = (1.2_5 \times 10^{34}\text{ J})/(2.998 \times 10^8\text{ m/s})^2 = 1.4 \times 10^{17}\text{ kg}$.

17.3 **(a)** $h\nu = \Phi + \frac{1}{2}m\upsilon^2$ and $h\nu_{thr} = \Phi$. For K we have $\nu_{thr} = \Phi/h =$ $(2.2 \text{ eV})(1.60 \times 10^{-19} \text{ J/eV})/(6.626 \times 10^{-34} \text{ J s}) = 5.3 \times 10^{14} \text{ s}^{-1}$ and $\lambda_{thr} = c/\nu_{thr} = (3.0 \times 10^{10} \text{ cm/s})/(5.3 \times 10^{14} \text{ s}^{-1}) = 5.7 \times 10^{-5}$ cm. For Ni we find $\nu_{thr} = 1.2 \times 10^{15} \text{ s}^{-1}$ and $\lambda_{thr} = 2.5 \times 10^{-5}$ cm.

(b) K-yes; Ni-no.

(c) $\frac{1}{2}m\upsilon^2 = h\nu - \Phi = (6.63 \times 10^{-34} \text{ J s})(3.00 \times 10^8 \text{ m/s})/(4.00 \times 10^{-7} \text{ m}) -$ $(2.2 \text{ eV})(1.60 \times 10^{-19} \text{ J/eV}) = 1.4 \times 10^{-19} \text{ J} = 0.9$ eV.

17.4 $E = h\nu = hc/\lambda = (6.626 \times 10^{-34} \text{ J s})(2.998 \times 10^8 \text{ m/s})/(700 \times 10^{-9} \text{ m}) =$ 2.8×10^{-19} J.

17.5 $E_{photon} = h\nu = hc/\lambda = (6.626 \times 10^{-34} \text{ J s})(2.998 \times 10^8 \text{ m/s})/(590 \times 10^{-9} \text{ m}) =$ 3.37×10^{-19} J. Then 100 J/s $= N(3.37 \times 10^{-19} \text{ J})$ and $N = 2.97 \times 10^{20}$ photons/s.

17.6 $h\nu = \Phi + \frac{1}{2}m\upsilon^2 = \Phi + K_{max}$ and $K_{max} = h\nu - \Phi$.

$10^{12}K_{max}$/ergs	3.41	2.56	1.95	0.75
$10^{-14}\nu/\text{s}^{-1}$	9.593	8.213	7.408	5.490

where we used $\nu = c/\lambda$. The slope is $6.5_3 \times 10^{-27}$ erg s $= 6.5_3 \times 10^{-34}$ J s $= h$. Φ can be found from the graph as the value of $h\nu$ at $K_{max} = 0$ or as the negative of the intercept at $\nu = 0$. We find $\Phi = 2.8_5 \times 10^{-12}$ erg $= 1.8$ eV.

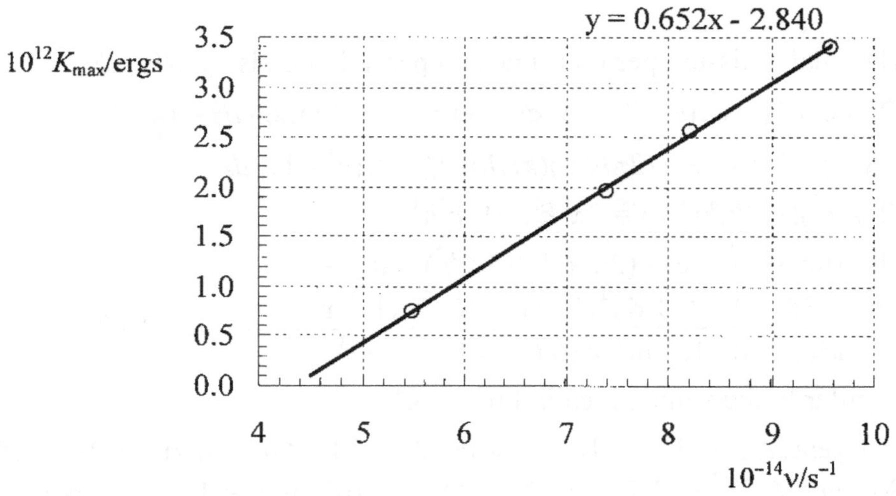

17.7 **(a)** $\lambda = h/m\upsilon = (6.626 \times 10^{-34}\text{ J s})/(1.67 \times 10^{-27}\text{ kg})(6.0 \times 10^{4}\text{ m/s}) = 6.6 \times 10^{-12}\text{ m}$.

(b) $\lambda = (6.626 \times 10^{-34}\text{ J s})/(0.050\text{ kg})(1.20\text{ m/s}) = 1.1 \times 10^{-32}\text{ m}$.

17.8 **(a)** $\sin\theta = \lambda/w$ and $\lambda = h/m\upsilon$, so $\sin\theta = h/m\upsilon w =$

$$\frac{6.26\times10^{-34}\text{ J}\cdot\text{s}}{(9.11\times10^{-31}\text{ kg})(6.0\times10^{6}\text{ m/s})(2400\times10^{-10}\text{ m})} =$$

5.05×10^{-4} and $\theta = 0.0289° = 5.05 \times 10^{-4}$ rad.

(b) Let z be the width of the central maximum. Figure 17.4 gives $\tan\theta = \overline{PE}/\overline{DE} = \frac{1}{2}z/(40\text{ cm})$ and $z = 2(40\text{ cm})\tan 0.0289° = 0.040$ cm.

(c) $\Delta p_x = 2h/w = 2(6.63 \times 10^{-34}\text{ J s})/(2400 \times 10^{-10}\text{ m}) = 5.5 \times 10^{-27}\text{ kg m/s}$.

17.9 $\Delta x\,\Delta p_x \gtrsim h$ and $\Delta p_x \gtrsim h/\Delta x = (6.6 \times 10^{-34}\text{ J s})/(1 \times 10^{-10}\text{ m}) = 6._6 \times 10^{-24}\text{ kg m s}^{-1}$. We have $\Delta p_x = \Delta(m\upsilon_x) = m\,\Delta\upsilon_x$ and $\Delta\upsilon_x = \Delta p_x/m \gtrsim (6._6 \times 10^{-24}\text{ kg m s}^{-1})/(9.1 \times 10^{-31}\text{ kg}) = 7 \times 10^{6}$ m/s, which is very large.

17.10 The time t and the 9 spatial coordinates $x_1, y_1, z_1, x_2, y_2, z_2, x_3, y_3, z_3$ of the three particles.

17.11 **(a)** F; **(b)** F; **(c)** T; **(d)** T; **(e)** T; **(f)** T.

17.12 $|z|^2 = zz^*$ and $|z| = (zz^*)^{1/2}$.

(a) $|-2| = 2$;

(b) $|3 - 2i| = [(3 - 2i)(3 + 2i)]^{1/2} = (9 + 4)^{1/2} = 13^{1/2}$;

(c) $|\cos\theta + i\sin\theta| = [(\cos\theta + i\sin\theta)(\cos\theta - i\sin\theta)]^{1/2} = (\cos^2\theta + \sin^2\theta)^{1/2} = 1^{1/2} = 1$;

(d) $|-3e^{-i\pi/5}| = [(-3e^{-i\pi/5})(-3e^{i\pi/5})]^{1/2} = (9e^{0})^{1/2} = 3$.

17.13 Let $\text{ls}_{17.10}$ and $\text{rs}_{17.10}$ denote the left side and right side of (17.10), respectively. We have $\text{ls}_{17.10} = \text{rs}_{17.10}$. To see if $c\Psi$ is a solution of (17.10), we replace Ψ in (17.10) by $c\Psi$ and see if (17.10) is satisfied. With $c\Psi$ as the proposed solution, the left side of (17.10) becomes $(-\hbar/i)(\partial/\partial t)(c\Psi) = c(-\hbar/i)(\partial\Psi/\partial t) = c\,\text{ls}_{17.10}$.

The right side becomes
$-(\hbar^2/2m_1)[\partial^2(c\Psi)/\partial x_1^2 + \cdots] - \cdots -(\hbar^2/2m_n)[\partial^2(c\Psi)/\partial x_n^2 + \cdots] + Vc\Psi =$
$c\ \mathrm{rs}_{17.10}$. Since Ψ is a solution of (17.10), we have $\mathrm{ls}_{17.10} = \mathrm{rs}_{17.10}$. Then
$c\ \mathrm{ls}_{17.10} = c\ \mathrm{rs}_{17.10}$ and so $c\Psi$ satisfies (17.10).

17.14 $\int_a^b \int_c^d [\int_s^t f(r)g(\theta)h(\phi)\,dr]\,d\theta\,d\phi = \int_a^b \int_c^d g(\theta)h(\phi)[\int_s^t f(r)\,dr]\,d\theta\,d\phi =$
$\int_s^t f(r)\,dr \int_a^b [\int_c^d g(\theta)h(\phi)\,d\theta]\,d\phi = \int_s^t f(r)\,dr \int_c^d g(\theta)\,d\theta \int_a^b h(\phi)\,d\phi$

17.15 The result of Prob. 17.14 with r, θ, ϕ replaced by x, y, z, gives $\int |\Psi|^2\,d\tau =$
$\int_{-\infty}^{\infty} (2/\pi c^2)^{1/2} e^{-2x^2/c^2} (2/\pi c^2)^{1/2} e^{-2y^2/c^2} (2/\pi c^2)^{1/2} e^{-2z^2/c^2}\,dx\,dy\,dz =$
$(2/\pi c^2)^{1/2} \int_{-\infty}^{\infty} e^{-2x^2/c^2}\,dx \cdot (2/\pi c^2)^{1/2} \int_{-\infty}^{\infty} e^{-2y^2/c^2}\,dy \cdot (2/\pi c^2)^{1/2} \int_{-\infty}^{\infty} e^{-2z^2/c^2}\,dz$
Use of integrals 1 (with $n = 0$) and 2 in Table 14.1 gives
$(2/\pi c^2)^{1/2} \int_{-\infty}^{\infty} e^{-2x^2/c^2}\,dx = 2(2/\pi c^2)^{1/2} \int_0^{\infty} e^{-2x^2/c^2}\,dx =$
$2(2/\pi c^2)^{1/2}\pi^{1/2}(1/2)(c^2/2)^{1/2} = 1$. By symmetry the y and z integrals also equal 1 and Ψ is normalized.

17.16 The 9 spatial coordinates $x_1, y_1, z_1, x_2, y_2, z_2, x_3, y_3, z_3$ of the three particles.

17.17 **(a)** T; **(b)** F; **(c)** T; **(d)** T.

17.18 The time-dependent Schrödinger equation is more general, since the time-independent equation applies only to stationary states.

17.19 Let $f = f_1 + if_2$ and $g = g_1 + ig_2$, where f_1 is the real part of f, and f_2 is the coefficient of the imaginary part of f. Then $(fg)^* = [(f_1 + if_2)(g_1 + ig_2)]^* =$
$[f_1g_1 - f_2g_2 + i(f_2g_1 + f_1g_2)]^* = f_1g_1 - f_2g_2 - i(f_2g_1 + f_1g_2)$. Also, $f^*g^* =$
$(f_1 + if_2)^*(g_1 + ig_2)^* = (f_1 - if_2)(g_1 - ig_2) = f_1g_1 - f_2g_2 - i(f_2g_1 + f_1g_2) = (fg)^*$.

17.20 Let $\mathrm{ls}_{17.24}$ and $\mathrm{rs}_{17.24}$ denote the left and right side of (17.24). To see if $k\psi$ is a solution of (17.24) we replace ψ in (17.24) by $k\psi$ and see if (17.24) is satisfied. With $k\psi$ as the proposed solution, the left side of (17.24) becomes
$(-\hbar^2/2m_1)[\partial^2(k\psi)/\partial x_1^2 + \cdots] - \cdots + V(k\psi) =$

$k[(-\hbar^2/2m_1)(\partial^2\psi/\partial x_1^2 + \cdots) - \cdots + V\psi] = k\,\text{ls}_{17.24}$. The right side becomes $Ek\psi = k(E\psi) = k\,\text{rs}_{17.24}$. Equation (17.24) is $\text{ls}_{17.24} = \text{rs}_{17.24}$ and multiplication by k gives $k\,\text{ls}_{17.24} = k\,\text{rs}_{17.24}$, so $k\psi$ is a solution of (17.24).

17.21 **(a)** Yes. The integral $\int_{-\infty}^{\infty} e^{-2ax^2}\,dx$ with $a > 0$ is finite, as shown by integral 1 with $n = 0$ and integral 2 in Table 14.1.

(b) No. The integral of the square of this function is infinite.

(c) No. $\int_{-\infty}^{\infty} x^{-2}dx = \int_{-\infty}^{0} x^{-2}dx + \int_{0}^{\infty} x^{-2}dx = -x^{-1}\,|_{-\infty}^{0} - x^{-1}\,|_{0}^{\infty} = \infty - 0 - (0 - \infty) = \infty$.

(d) No. For $x < 0$, $|x| = -x$, and the integral from $-\infty$ to ∞ of the square of this function is $\int_{-\infty}^{0} [1/(-x)^{1/2}]\,dx + \int_{0}^{\infty} (1/x^{1/2})\,dx = -2(-x)^{1/2}\Big|_{-\infty}^{0} + 2x^{1/2}\Big|_{0}^{\infty} = 0 - (-2)(\infty) + 2(\infty) - 0 = \infty$.

(e) False. Consider the function $f \equiv e^{-ax^2}/|x|^{1/4}$, which is the product of the functions in (a) and (d). The integrand is infinite at $x = 0$. Despite this, the integral $\int_{-\infty}^{\infty} |f|^2\,dx$ is not infinite. For x between -1 and 1, the integrand is less than $|x|^{-1/2}$ (except at $x = 0$), so $\int_{-1}^{1} e^{-2ax^2}|x|^{-1/2}\,dx < \int_{-1}^{1} |x|^{-1/2}\,dx = \int_{-1}^{0} |-x|^{-1/2}\,dx + \int_{0}^{1} |x|^{-1/2}\,dx = -2(-x)^{1/2}\Big|_{-1}^{0} + 2x^{1/2}\Big|_{0}^{1} = 2 + 2 = 4$. So $0 < \int_{-1}^{1} |f|^2\,dx < 4$. Over the rest of the integration range we deal with $\int_{-\infty}^{-1} e^{-2ax^2}|x|^{-1/2}\,dx$ and $\int_{1}^{\infty} e^{-2ax^2}|x|^{-1/2}\,dx$. Here, the $1/|x|^{1/2}$ factor makes the integrand less than e^{-2ax^2} and, since integration of e^{-2ax^2} gives a finite area under the curve (see part a), the integrals $\int_{-\infty}^{-1} e^{-2ax^2}|x|^{-1/2}\,dx$ and $\int_{1}^{\infty} e^{-2ax^2}|x|^{-1/2}\,dx$ are finite. Hence, $\int_{-\infty}^{\infty} |f|^2\,dx$ is finite and f is quadratically integrable even though it is infinite at the origin.

17.22 $E = n^2h^2/8ma^2$; $|\Delta E| = h\nu = hc/\lambda$; $\lambda = hc/|\Delta E| = hc[8ma^2/h^2|n_2^2 - n_1^2|] = 8ma^2c/h|n_2^2 - n_1^2| = \dfrac{8(1.0\times10^{-30}\text{ kg})(6.0\times10^{-10}\text{ m})^2(3.0\times10^{8}\text{ m/s})}{(6.63\times10^{-34}\text{ J}\cdot\text{s})(25-16)} = 1.4_5\times10^{-7}$ m.

17.23 **(a)** $\int_0^{a/4} |\psi|^2\, dx = (2/a) \int_0^{a/4} \sin^2 (n\pi x/a)\, dx =$
$(2/a)[\frac{1}{2}x - \frac{1}{4}(a/n\pi) \sin (2n\pi x/a)]\Big|_0^{a/4} = \frac{1}{4} - (1/2n\pi) \sin \frac{1}{2}n\pi$ where we used $\int \sin^2 cx\, dx = \frac{1}{2}x - \frac{1}{4}c^{-1} \sin 2cx$.

(b) For $n = 1, 2, 3$, we get 0.091, 0.250, and 0.303, respectively.

17.24 The interval 0.0001 Å is much, much smaller than the box length, so we can consider this to be an "infinitesimal" interval. The probability is then $|\psi|^2\, dx = (2/a) \sin^2 (n\pi x/a)\, dx$.

(a) For $n = 1$, we get $[2/(2\text{ Å})] \sin^2 [\pi(1.6\text{ Å})/(2.0\text{ Å})](0.0001\text{ Å}) = 3.45 \times 10^{-5}$.

(b) For $n = 2$, we get 9.05×10^{-5}.

17.25

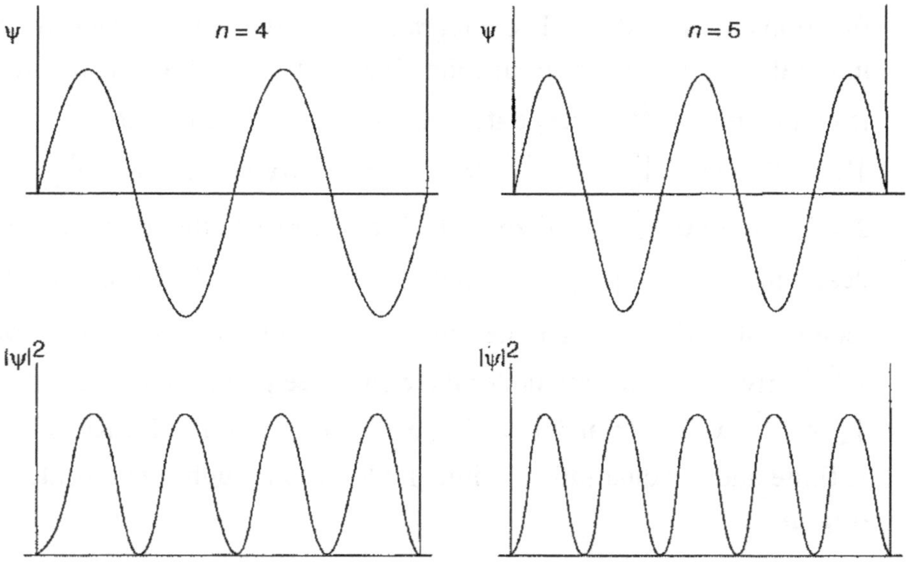

Note that ψ^2 has zero slope at the nodes.

17.26 $d^2\psi/dx^2 = 0$. Integration with respect to x gives $d\psi/dx = c$ and a second integration gives $\psi = cx + d$, where c and d are integration constants. The continuity condition at $x = 0$ requires that $\psi = 0$ at $x = 0$, so $0 = c(0) + d$ and

$d = 0$. Then $\psi = cx$. The continuity condition at $x = a$ requires that $\psi = 0$ at $x = a$, and $0 = ca$, so $c = 0$. Then $\psi = 0$.

17.27 The lowest frequency transition corresponds to $n = 1 \rightarrow 2$. Use of (17.7) gives $2^2h^2/8ma^2 - 1^2h^2/8ma^2 = h\nu$ and $a = (3h/8m\nu)^{1/2} =$ $[3(6.63 \times 10^{-34}\text{ J s})]^{1/2}/[8(9.11 \times 10^{-31}\text{ kg})(2.0 \times 10^{14}\text{ s}^{-1})]^{1/2} = 1.2 \times 10^{-9}$ m.

17.28 $h\nu = E_{\text{upper}} - E_{\text{lower}} = (n_u^2 - n_\ell^2)h^2/8ma^2$, where u and ℓ stand for upper and lower. $\nu = (n_u^2 - n_\ell^2)h/8ma^2$. $\nu_{3\rightarrow4} = (4^2 - 3^2)h/8ma^2 = 7h/8ma^2$. $\nu_{6\rightarrow9} = (9^2 - 6^2)h/8ma^2 = 45h/8ma^2$. So $\nu_{6\rightarrow9}/\nu_{3\rightarrow4} = 45/7$ and $\nu_{6\rightarrow9} = (45/7)(4.00 \times 10^{13}\text{ s}^{-1}) = 2.57 \times 10^{14}\text{ s}^{-1}$.

17.29 $\int_0^a \psi_i^* \psi_j\, dx = 2a^{-1} \int_0^a \sin(n_i\pi x/a) \sin(n_j\pi x/a)\, dx$, $n_i \neq n_j$. A table of integrals gives $\int \sin cx \sin bx\, dx = [1/2(c - b)] \sin[(c - b)x] - [1/2(c + b)] \sin[(c + b)x]$, provided $c^2 \neq b^2$. So $\int_0^a \psi_i^* \psi_j\, dx =$

$$\frac{2}{a}\left[\frac{\sin[(n_i - n_j)\pi x/a]}{2(n_i - n_j)\pi/a} - \frac{\sin[(n_i + n_j)\pi x/a]}{2(n_i + n_j)\pi/a}\right]\Bigg|_0^a = 0$$

since $\sin[(n_i - n_j)\pi] = 0$, $\sin[(n_i + n_j)\pi] = 0$, and $\sin 0 = 0$.

17.30 The left side of (17.28) becomes $d^2\psi/dx^2 = (2/a)^{1/2}(-1)(n^2\pi^2/a^2)\sin(n\pi x/a)$. With use of (17.34), the right side of (17.28) becomes $-(2m/\hbar^2)(n^2h^2/8ma^2) \times (2/a)^{1/2}\sin(n\pi x/a) = -(\pi^2n^2/a^2)(2/a)^{1/2}\sin(n\pi x/a)$, which equals the left side.

17.31 For (a), $|\psi|$ is a maximum at $(\frac{1}{4}a, \frac{1}{2}a)$ and at $(\frac{3}{4}a, \frac{1}{2}a)$, where a is the box length. For (b), $|\psi|$ is a maximum at $(\frac{1}{4}a, \frac{1}{4}a)$, $(\frac{1}{4}a, \frac{3}{4}a)$, $(\frac{3}{4}a, \frac{1}{4}a)$, $(\frac{3}{4}a, \frac{3}{4}a)$.

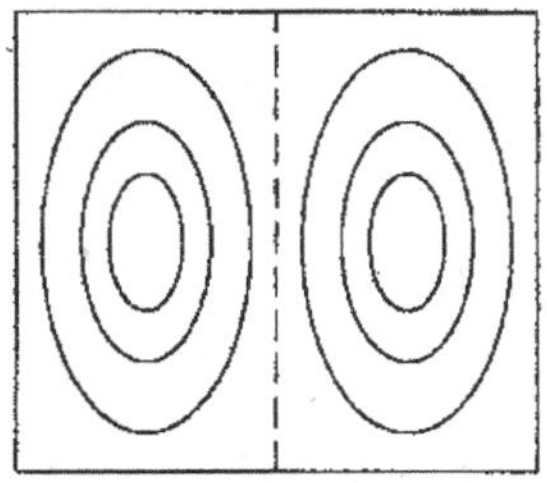
(a)

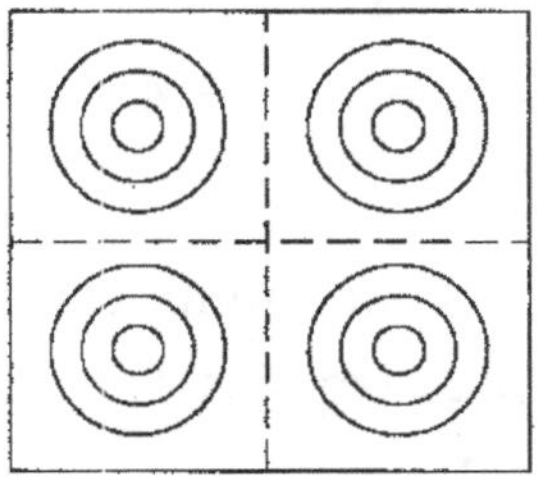
(b)

where the dashed lines are nodal lines and the x axis is horizontal.

17.32 **(a)** $n_x^2 + n_y^2 + n_z^2 = 21$ is satisfied by $n_x n_y n_z$ = 124, 142, 214, 241, 412, 421 and no others, so the degeneracy is 6.

(b) $n_x^2 + n_y^2 + n_z^2 = 24$ is satisfied only by $n_x n_y n_z$ = 422, 242, 224, and the degeneracy is 3.

17.33 **(a)** $E = (h^2/8ma^2)(n_x^2 + n_y^2 + n_z^2)$. There are 17 states with $n_x^2 + n_y^2 + n_z^2 \le$ 16, namely: $n_x n_y n_z$ = 111, 211, 121, 112, 122, 212, 221, 311, 131, 113, 222, 123, 132, 213, 231, 312, 321.

(b) These states give a total of 6 different values for $n_x^2 + n_y^2 + n_z^2$, namely, $n_x^2 + n_y^2 + n_z^2 = 3, 6, 9, 11, 12, 14$, so there are 6 energy levels in the given range.

17.34 **(a)** T; **(b)** F; **(c)** F; **(d)** F; **(e)** F; **(f)** F; **(g)** T.

17.35 **(a)** Yes; **(b)** no; **(c)** yes; **(d)** no.

17.36 Since the wave function is an eigenfunction of the Hamiltonian (energy) operator, we must get the eigenvalue $25h^2/8ma^2$.

17.37 **(a)** $\hat{A}\hat{B}f(x) - \hat{B}\hat{A}f(x) = (d^2/dx^2)[xf(x)] - x[(d^2/dx^2)f(x)] =$ $(d/dx)[xf'(x) + f(x)] - xf''(x) = xf''(x) + f'(x) + f'(x) - xf''(x) = 2f'(x)$.

(b) $(\hat{A} + \hat{B})(e^{x^2} + \cos 2x) = (d^2/dx^2 + x)[e^{x^2} + \cos 2x] =$ $(d^2/dx^2)(e^{x^2} + \cos 2x) + x(e^{x^2} + \cos 2x) = d^2(e^{x^2})/dx^2 + d^2(\cos 2x)/dx^2 +$ $xe^{x^2} + x\cos 2x = 2e^{x^2} + 4x^2e^{x^2} - 4\cos 2x + xe^{x^2} + x\cos 2x$.

17.38 **(a)** $(\ \ )^2$ and $(\ \ )^*$ are nonlinear; the others are linear.

(b) $\hat{H}(f+g) = [(-\hbar^2/2m_1)\nabla_1^2 - \cdots + V](f+g) =$ $[(-\hbar^2/2m_1)\nabla_1^2 - \cdots](f+g) + V(f+g) = (-\hbar^2/2m_1)\nabla_1^2 f - (\hbar^2/2m_1)\nabla_1^2 g$ $- \cdots + Vf + Vg = [-(\hbar^2/2m_1)\nabla_1^2 - \cdots + V]f + [-(\hbar^2/2m_1)\nabla_1^2 - \cdots + V]g$

$= \hat{H}f + \hat{H}g$, where the definition of the sum of operators and the fact that $(\partial^2/\partial x_1^2)(f+g) = \partial^2 f/\partial x_1^2 + \partial^2 g/\partial x_1^2$ were used. Similarly, one finds $\hat{H}(cf) = c\hat{H}f$. So $\hat{H}$ is linear.

17.39 **(a)** When $\hat{B}$ operates on $g(x)$, it turns g into another function, which we shall call $f(x)$. When $\hat{A}$ operates on $f(x)$, we get another function, so $\hat{A}\hat{B}g(x)$ is a function. **(b)** Operator. **(c)** Function. **(d)** Operator. **(e)** Function.

17.40 **(a)** $\hat{p}_x^3 = [(\hbar/i)(\partial/\partial x)]^3 = -(\hbar^3/i)(\partial^3/\partial x^3)$.

(b) $\hat{p}_z^4 = [(\hbar/i)(\partial/\partial z)]^4 = \hbar^4(\partial^4/\partial z^4)$.

17.41 $(d^2/dx^2)(\sin 3x) = (d/dx)(3\cos 3x) = -9(\sin 3x)$, so $\sin 3x$ is an eigenfunction of d^2/dx^2 with eigenvalue -9. $(d^2/dx^2)(6\cos 4x) = -96\cos 4x = -16(6\cos 4x)$, so $6\cos 4x$ is an eigenfunction of d^2/dx^2 with eigenvalue -16. $(d^2/dx^2)(5x^3) = 30x$, which does not equal a constant times $5x^3$; so $5x^3$ is not an eigenfunction of d^2/dx^2. $(d^2/dx^2)x^{-1} = 2x^{-3} \neq (\text{const.})x^{-1}$. $(d^2/dx^2)(3e^{-5x}) = 75e^{-5x} = 25(3e^{-5x})$, and the eigenvalue is 25. $(d^2/dx^2)\ln 2x = -1/x^2 \neq (\text{const.})\ln 2x$.

17.42 **(a)**

$\langle p_x\rangle = \int_{-\infty}^{\infty} \psi^* \hat{p}_x\psi\, dx = \int_{-\infty}^{\infty} \psi^*(\hbar/i)(\partial/\partial x)\psi\, dx =$
$(\hbar/i)\int_{-\infty}^{0} \psi^*(\partial\psi/\partial x)\, dx + (\hbar/i)\int_0^a \psi^*(\partial\psi/\partial x)\, dx +$
$(\hbar/i)\int_a^{\infty} \psi^*(\partial\psi/\partial x)\, dx =$
$(\hbar/i)\int_0^a \psi^*(\partial\psi/\partial x)\, dx = (\hbar/i)(2/a)(n\pi/a)\int_0^a \sin(n\pi x/a)\cos(n\pi x/a)\, dx =$
$(2n\pi\hbar/ia^2)(a/n\pi)\tfrac{1}{2}\sin^2(n\pi x/a)\big|_0^a = 0$, since $\sin n\pi = 0$ for $n =$ 1, 2, 3, ... (We used the fact that $\psi^* = 0$ outside the box.)

(b) $\langle x\rangle = \int_0^a \psi^* x\psi\, dx = \int_0^a (2/a)^{1/2}\sin(n\pi x/a)\, x\,(2/a)^{1/2}\sin(n\pi x/a)\, dx =$
$(2/a)\int_0^a x\sin^2(n\pi x/a)\, dx$. A table of integrals gives
$\int x\sin^2 cx\, dx = \tfrac{1}{4}x^2 - (x/4c)\sin 2cx - (1/8c^2)\cos 2cx$, so $\langle x\rangle =$
$(2/a)[\tfrac{1}{4}x^2 - (ax/4n\pi)\sin(2n\pi x/a) - (a^2/8n^2\pi^2)\cos(2n\pi x/a)]\big|_0^a =$
$(2/a)(\tfrac{1}{4}a^2 - a^2/8n^2\pi^2 + a^2/8n^2\pi^2) = a/2$, since $\sin 2n\pi = 0$ and $\cos 2n\pi = 1$ for $n = 1, 2, \ldots$.

(c) $\langle x^2 \rangle = \int_0^a \psi^* x^2 \psi \, dx = (2/a) \int_0^a x^2 \sin^2 (n\pi x/a) \, dx$. A table of integrals gives $\int x^2 \sin^2 cx \, dx = x^3/6 - (x^2/4c - 1/8c^3) \sin 2cx - (x/4c^2) \cos 2cx$, so $\langle x^2 \rangle = (2/a)[x^3/6 - (ax^2/4n\pi - a^3/8n^3\pi^3) \sin (2n\pi x/a) - (a^2x/4n^2\pi^2) \cos (2n\pi x/a)] \big|_0^a = (2/a)(a^3/6 - a^3/4n^2\pi^2) = a^2/3 - a^2/2n^2\pi^2$.

17.43 The time-independent Schrödinger equation $\hat{H}\psi = E\psi$ for (17.64) is $(\hat{H}_1 + \hat{H}_2 + \cdots + \hat{H}_r)\psi = E\psi$ and $(\hat{H}_1\psi + \hat{H}_2\psi + \cdots + \hat{H}_r\psi) = E\psi$ (1). Taking $\psi = f_1(q_1)f_2(q_2) \cdots f_r(q_r)$, we have $\hat{H}_1\psi = \hat{H}_1[f_1(q_1)f_2(q_2) \cdots f_r(q_r)] = (f_2 \cdots f_r)\hat{H}_1 f_1$, since $\hat{H}_1$ involves only q_1. Equation (1) becomes $(f_2 \cdots f_r)\hat{H}_1 f_1 + (f_1 f_3 \cdots f_r)\hat{H}_2 f_2 + \cdots + (f_1 \cdots f_{r-1})\hat{H}_r f_r = E f_1 f_2 \cdots f_r$. Division by $f_1 f_2 \cdots f_r$ gives $(1/f_1)\hat{H}_1 f_1 + (1/f_2)\hat{H}_2 f_2 + \cdots + (1/f_r)\hat{H}_r f_r = E$ (2). By the same kind of argument used after Eq. (17.39), each term on the left side of equation (2) must be a constant. Calling these constants $E_1, E_2, \ldots, E_r$, we have $(1/f_1)\hat{H}_1 f_1 = E_1$ or $\hat{H}_1 f_1 = E_1 f_1$, etc., and equation (2) gives $E_1 + E_2 + \cdots + E_r = E$.

17.44 $\psi = (2/a) \sin (n_1\pi x_1/a) \sin (n_2\pi x_2/a)$. $E = n_1^2 h^2/8m_1a^2 + n_2^2 h^2/8m_2a^2$.

17.45 **(a)** nu, frequency; **(b)** vee, quantum number.

17.46 $\nu_{\text{light}} = (E_{\text{upper}} - E_{\text{lower}})/h = [(\upsilon_{\text{upper}} + \frac{1}{2})h\nu_{\text{osc}} - (\upsilon_{\text{lower}} + \frac{1}{2})h\nu_{\text{osc}}]/h = (\upsilon_{\text{upper}} - \upsilon_{\text{lower}})\nu_{\text{osc}} = (8 - 7)\nu_{\text{osc}} = \nu_{\text{osc}} = 6.0 \times 10^{13}\ \text{s}^{-1}$.

17.47 Squaring the curves in Fig. 17.18, we get the following curves (note the unequal peak heights):

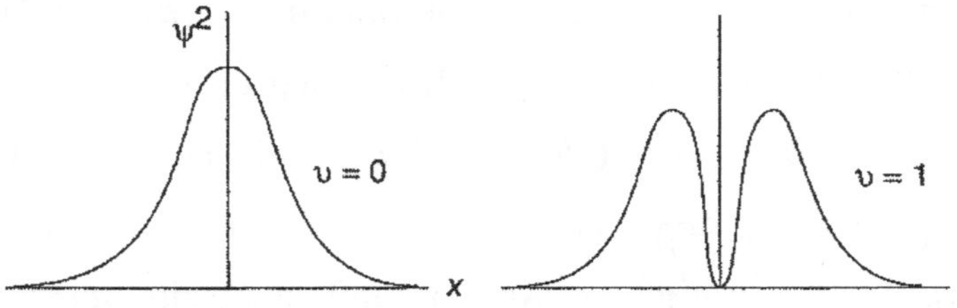

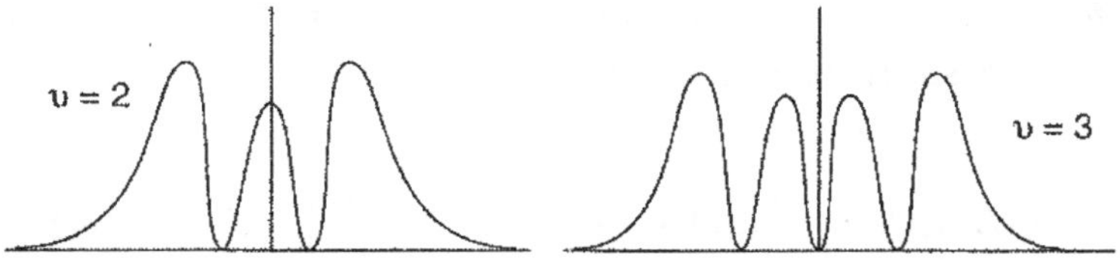

17.48 **(a)** From Fig. 17.18, ψ_0 is a maximum at $x = 0$; likewise, ψ_0^2 is a maximum at $x = 0$ and this is the most probable value of x.

(b) $d\psi_1^2/dx = 0 = (4\alpha^3/\pi)^{1/2}(2xe^{-\alpha x^2} - 2\alpha x^3 e^{-\alpha x^2})$, so $x = \pm 1/\alpha^{1/2} = \pm(\hbar/2\pi\nu m)^{1/2}$. ($x = 0$ is a minimum.)

17.49 $d^2\psi_0/dx^2 = (\alpha/\pi)^{1/4}(d^2/dx^2)e^{-\alpha x^2/2} = (\alpha/\pi)^{1/4}(d/dx)(-\alpha x e^{-\alpha x^2/2}) = (\alpha/\pi)^{1/4}(-\alpha e^{-\alpha x^2/2} + \alpha^2 x^2 e^{-\alpha x^2/2}) = (\alpha^2 x^2 - \alpha)\psi_0 = (16\pi^4\nu^2 m^2 x^2/h^2 - 4\pi^2\nu m/h)\psi_0$. Equation (17.73) gives $k = 4\pi^2\nu^2 m$, so $\frac{1}{2}kx^2\psi_0 = 2\pi^2\nu^2 m x^2\psi_0$. Then $-(\hbar^2/2m)(d^2\psi_0/dx^2) + \frac{1}{2}kx^2\psi_0 = (-2\pi^2\nu^2 m x^2 + \frac{1}{2}h\nu)\psi_0 + 2\pi^2\nu^2 m x^2\psi_0 = \frac{1}{2}h\nu\psi_0 = E_0\psi_0$.

17.50 $\int_{-\infty}^{\infty}\psi_1^*\psi_1\, dx = (4\alpha^3/\pi)^{1/2}\int_{-\infty}^{\infty}x^2 e^{-\alpha x^2}\, dx = 2(4\alpha^3/\pi)^{1/2}\int_0^{\infty}x^2 e^{-\alpha x^2}\, dx = 2(4\alpha^3/\pi)^{1/2}(2\pi^{1/2}/2^3\alpha^{3/2}) = 1$, See integrals 1 and 3 (with $n = 1$) in Table 14.1.

17.51 **(a)** $\langle x\rangle = \int_{-\infty}^{\infty}\psi^* x\psi\, dx = (\alpha/\pi)^{1/2}\int_{-\infty}^{\infty}xe^{-\alpha x^2}\, dx = 0$, where integral 4 (with $n = 0$) in Table 14.1 was used. This result is obvious from Fig. 17.18.

(b) $\langle x^2\rangle = \int_{-\infty}^{\infty}\psi^* x^2\psi\, dx = (\alpha/\pi)^{1/2}\int_{-\infty}^{\infty}x^2 e^{-\alpha x^2}\, dx = 2(\alpha/\pi)^{1/2}\int_0^{\infty}x^2 e^{-\alpha x^2}\, dx = 2(\alpha/\pi)^{1/2}(2\pi^{1/2}/2^3\alpha^{3/2}) = 1/2\alpha = h/8\pi^2\nu m$, where we used integrals 1 and 3 in Table 14.1.

(c) $\langle p_x\rangle = \int_{-\infty}^{\infty}\psi^*\hat{p}_x\psi\, dx = (\alpha/\pi)^{1/2}\int_{-\infty}^{\infty}e^{-\alpha x^2/2}(\hbar/i)(\partial/\partial x)e^{-\alpha x^2/2}\, dx = (\alpha/\pi)^{1/2}(\hbar/i)(-\alpha)\int_{-\infty}^{\infty}xe^{-\alpha x^2}\, dx = 0$ (from integral 4).

17.52 **(a)** $E = K + V = \frac{1}{2}m(dx/dt)^2 + \frac{1}{2}kx^2 = \frac{1}{2}m\{(k/m)^{1/2}A\cos[(k/m)^{1/2}t + b]\}^2 + \frac{1}{2}kA^2\sin^2[(k/m)^{1/2}t + b] = \frac{1}{2}kA^2\{\cos^2[(k/m)^{1/2}t + b] + \sin^2[(k/m)^{1/2}t + b]\} = \frac{1}{2}kA^2$.

(b) $m\, d^2x/dt^2 = m(d^2/dt^2)\{A \sin [(k/m)^{1/2}t + b]\} = -mA(k/m) \sin [(k/m)^{1/2}t + b]$
$= -k\{A \sin [(k/m)^{1/2}t + b]\} = -kx$

17.53 (a) $\nu = (1/2\pi)(k/m)^{1/2}$ and $k = 4\pi^2\nu^2 m = 4\pi^2(2.4\ \text{s}^{-1})^2(0.045\ \text{kg}) = 10.2$ N/m.

(b) $E = \frac{1}{2}kA^2 = 0.5(10.2\ \text{N/m})(0.04\ \text{m})^2 = 0.0082\ \text{J} = (\upsilon + \frac{1}{2})h\nu$, so $\upsilon + \frac{1}{2} =$ $(0.0082\ \text{J})/(6.626 \times 10^{-34}\ \text{J}\cdot\text{s})(2.4\ \text{s}^{-1}) = 5.2 \times 10^{30} = \upsilon$.

17.54 (a) The Hamiltonian is the sum of three one-dimensional harmonic-oscillator Hamiltonians, one for each coordinate; the separation-of-variables theorem [Eqs. (17.65) and (17.66)] gives the energy as the sum of three one-dimensional-harmonic-oscillator energies:
$E = E_x + E_y + E_z = (\upsilon_x + ½)h\nu_x + (\upsilon_y + ½)h\nu_y + (\upsilon_z + ½)h\nu_z$, where
$\upsilon_x = 0, 1, 2, \ldots, \upsilon_y = 0, 1, 2, \ldots, \upsilon_z = 0, 1, 2, \ldots$, and
$\nu_x = (1/2\pi)(k_x/m)^{1/2}$, $\nu_y = (1/2\pi)(k_y/m)^{1/2}$, $\nu_z = (1/2\pi)(k_z/m)^{1/2}$, where m is the particle's mass.

(b) The lowest level has $\upsilon_x = \upsilon_y = \upsilon_z = 0$ and $E = \frac{1}{2}h(\nu_x + \nu_y + \nu_z)$.

17.55 $H = (1/2\mu)(\mu^2\upsilon_x^2 + \mu^2\upsilon_y^2 + \mu^2\upsilon_z^2) + V + (1/2M)(M^2\upsilon_X^2 + M^2\upsilon_Y^2 + M^2\upsilon_Z^2) =$
$V + \frac{1}{2}\mu(\upsilon_x^2 + \upsilon_y^2 + \upsilon_z^2) + \frac{1}{2}M(\upsilon_X^2 + \upsilon_Y^2 + \upsilon_Z^2)$. Using Eq. (17.77), we have
$\upsilon_x = dx/dt = dx_2/dt - dx_1/dt = \upsilon_{x,2} - \upsilon_{x,1}$. Similarly, $\upsilon_y = \upsilon_{y,2} - \upsilon_{y,1}$ and
$\upsilon_z = \upsilon_{z,2} - \upsilon_{z,1}$. Since $X = (m_1x_1 + m_2x_2)/M$, we have $\upsilon_X = dX/dt =$
$[m_1(dx_1/dt) + m_2(dx_2/dt)]/M = (m_1\upsilon_{x,1} + m_2\upsilon_{x,2})/(m_1 + m_2)$; similar equations hold for υ_Y and υ_Z. So $H = V + \frac{1}{2}[m_1m_2/(m_1 + m_2)] \times$
$(\upsilon_{x,1}^2 - 2\upsilon_{x,1}\upsilon_{x,2} + \upsilon_{x,2}^2 + \cdots) + \frac{1}{2}(m_1 + m_2)(m_1 + m_2)^{-2} \times$
$(m_1^2\upsilon_{x,1}^2 + 2m_1m_2\upsilon_{x,1}\upsilon_{x,2} + m_2^2\upsilon_{x,2}^2 + \cdots) =$
$V + \frac{1}{2}(m_1 + m_2)^{-1}[(m_1 + m_2)m_1\upsilon_{x,1}^2 + (m_1 + m_2)m_2\upsilon_{x,2}^2 + \cdots] =$
$V + \frac{1}{2}(m_1\upsilon_{x,1}^2 + m_2\upsilon_{x,2}^2 + m_1\upsilon_{y,1}^2 + m_2\upsilon_{y,2}^2 + m_1\upsilon_{z,1}^2 + m_2\upsilon_{z,2}^2) =$
$V + \frac{1}{2}m_1\upsilon_1^2 + \frac{1}{2}m_2\upsilon_2^2 = V + m_1^2\upsilon_1^2/2m_1 + m_2^2\upsilon_2^2/2m_2 = p_1^2/2m_1 + p_2^2/2m_2 + V$.
(The dots indicate similar terms in y and z.)

17.56 (a) $\mu = m_1m_2/(m_1 + m_2) = [(12.0\ \text{g/mol})/N_A][(16.0\ \text{g/mol})/N_A]/$ $[(28.0\ \text{g/mol})/N_A] = (6.86\ \text{g/mol})/N_A = 1.14 \times 10^{-23}$ g.

(b) $I = \mu d^2 = (1.14 \times 10^{-26}\ \text{kg})(1.13 \times 10^{-10}\ \text{m})^2 = 1.45 \times 10^{-46}\ \text{kg m}^2$.

(c) $E_{rot} = J(J+1)\hbar^2/2I$. $\hbar^2/2I = (6.626 \times 10^{-34}\text{ J s})^2/8\pi^2(1.45 \times 10^{-46}\text{ kg m}^2) = 3.83 \times 10^{-23}$ J. For $J = 0, 1, 2, 3$, we have $E_{rot} = 0, 7.66 \times 10^{-23}$ J, 23.0×10^{-23} J, 46.0×10^{-23} J, respectively. The levels are $(2J+1)$-fold degenerate, so the degeneracies are 1, 3, 5, 7.

(d) For $J = 0$ to 1, $\Delta E = 7.66 \times 10^{-23}\text{ J} - 0 = 7.66 \times 10^{-23}\text{ J} = h\nu = (6.626 \times 10^{-34}\text{ J s})\nu$ and $\nu = 1.16 \times 10^{11}\text{ s}^{-1}$. For $J = 1$ to 2, $\Delta E = (23.0 - 7.66)10^{-23}\text{ J} = h\nu$ and $\nu = 2.32 \times 10^{11}\text{ s}^{-1}$.

17.57 (a) Let $Nx(a-x)$ be the normalized function. So $\int_0^a [Nx(a-x)]^* Nx(a-x)\,dx = 1$ and $|N| = 1/[\int_0^a x^2(a-x)^2\,dx]^{1/2}$. From Example 17.8, $\int_0^a x^2(a-x)^2\,dx = a^5/30$, so $|N| = (30/a^5)^{1/2}$.

(b) $\langle x^2\rangle \cong (30/a^5)\int_0^a x(a-x)x^2x(a-x)\,dx = (30/a^5)\int_0^a (a^2x^4 - 2ax^5 + x^6)\,dx = (30/a^5)(a^7/5 - a^7/3 + a^7/7) = 30a^2/105 = 2a^2/7 = 0.2857a^2$. The true value is found by setting $n = 1$ in Prob. 17.42*c* to give $\langle x^2\rangle = a^2(1/3 - 1/2\pi^2) = 0.2827a^2$. The error is 1.1%.

17.58 The value of k that minimizes the variational integral W satisfies $\partial W/\partial k = 0$. We have $\partial W/\partial k = 0 = (\hbar^2/ma^2)[(8k+1)/(2k-1) - 2(4k^2+k)/(2k-1)^2] = (\hbar^2/ma^2)(8k^2 - 8k - 1)/(2k-1)^2$ and $8k^2 - 8k - 1 = 0$. The solutions are $k = 1.112372$ and -0.112372. The negative value of k makes $\phi = \infty$ at $x = 0$ and so is rejected. For $k = 1.112372$, $W = (\hbar^2/ma^2)(4k^2+k)/(2k-1) = 4.94949\hbar^2/ma^2 = 4.94949h^2/4\pi^2ma^2 = 0.125372h^2/ma^2$ compared with the true value $h^2/8ma^2 = 0.125h^2/ma^2$. The percent error is only 0.30%.

17.59 (a) $\int_0^a \phi^*\phi\,dx = \int_0^a x^2(a-x)^2x^2(a-x)^2\,dx = \int_0^a (a^4x^4 - 4a^3x^5 + 6a^2x^6 - 4ax^7 + x^8)\,dx = a^9/5 - 2a^9/3 + 6a^9/7 - a^9/2 + a^9/9 = a^9/630$. We have $\hat{H}\phi = (\hbar^2/2m)(d^2/dx^2)(x^2a^2 - 2ax^3 + x^4) = -(\hbar^2/2m)(2a^2 - 12ax + 12x^2)$. So $\int_0^a \phi^*\hat{H}\phi\,dx = -(\hbar^2/m)\int_0^a x^2(a-x)^2(a^2 - 6ax + 6x^2)\,dx = -(\hbar^2/m)\int_0^a (a^4x^2 - 8a^3x^3 + 19a^2x^4 - 18ax^5 + 6x^6)\,dx = -(\hbar^2/m)(a^7/3 - 2a^7 + 19a^7/5 - 3a^7 + 6a^7/7) = \hbar^2a^7/105m = h^2a^7/420\pi^2m$. Then $\int \phi^*\hat{H}\phi\,dx/\int \phi^*\phi\,dx = (h^2a^7/420\pi^2m)(630/a^9) = (3/2\pi^2)(h^2/ma^2) =$

$0.152h^2/ma^2 \approx E_{gs}$. The true E_{gs} is $h^2/8ma^2 = 0.125h^2/ma^2$. The error is 22%.

(b) It is discontinuous at $x = a$, since $\phi = 0$ outside the box.

17.60 $\hat{H} = \hat{H}^0 + \hat{H}'$, where $\hat{H}^0$ is the particle-in-a-box Hamiltonian operator and $\hat{H}' = kx$ for $0 \le x \le a$. We have $E_n^{(1)} = \int \psi_n^* \hat{H}'\psi_n \, d\tau =$ $(2/a) \int_0^a kx \sin^2(n\pi x/a)\, dx =$ $(2k/a)[\,\frac{1}{4}x^2 - \frac{1}{4}(ax/n\pi)\sin(2n\pi x/a) - (a^2/8n^2\pi^2)\cos(2n\pi x/a)]_0^a = \frac{1}{2}ak$, where $\sin 2n\pi = 0$ and $\cos 2n\pi = 1$ were used. $E^{(0)} = n^2h^2/8ma^2$, so $E^{(0)} + E^{(1)} =$ $n^2h^2/8ma^2 + \frac{1}{2}ak$.

17.61 **(a)** T; **(b)** F; **(c)** T; **(d)** F; **(e)** This question is defective. What was intended was to ask whether $\sum_m b_m c_m \delta_{mn} = b_n c_n$ (Eq. A) is true. Because the δ_{mn} factor makes all terms zero except the one with $m = n$, Eq. A is true.

17.62 We are given that $\hat{B}$ and $\hat{C}$ are Hermitian operators, so from (17.92) we have $\int f^* \hat{B}g\, d\tau = \int g(\hat{B}f)^*\, d\tau$ (1) and $\int f^* \hat{C}g\, d\tau = \int g(\hat{C}f)^*\, d\tau$ (2). To prove that $\hat{B}+\hat{C}$ is Hermitian, we shall prove that it satisfies (17.92): $\int f^*(\hat{B}+\hat{C})g\, d\tau = \int g[(\hat{B}+\hat{C})f]^*\, d\tau$ (A). The left side of equation (A) is $\int f^*(\hat{B}+\hat{C})g\, d\tau = \int f^*(\hat{B}g+\hat{C}g)\, d\tau = \int f^*\hat{B}g\, d\tau + \int f^*\hat{C}g\, d\tau =$ $\int g(\hat{B}f)^*\, d\tau + \int g(\hat{C}f)^*\, d\tau = \int g[(\hat{B}f)^* + (\hat{C}f)^*]\, d\tau = \int g[(\hat{B}f)+(\hat{C}f)]^*\, d\tau =$ $\int g[(\hat{B}+\hat{C})f]^*\, d\tau$, which completes the proof of Eq. (A). In the proof, we used the definition (17.51) of the sum of operators, the integral identity (1.53), the given equations (1) and (2), and the identity $(z_1 + z_2)^* = z_1^* + z_2^*$, which is easily proved by writing z_1 and z_2 as $a_1 + ib_1$ and $a_2 + ib_2$, respectively, where the a's and b's are real.

17.63 With $\Psi = f + cg$, Eq. (17.91) becomes $\int (f+cg)^* \hat{M}(f+cg)\, d\tau =$ $\int (f+cg)[\hat{M}(f+cg)]^*\, d\tau$. Using the identity $(z_1 + z_2)^* = z_1^* + z_2^*$ (which is easily proved by writing z_1 and z_2 as $a_1 + ib_1$ and $a_2 + ib_2$, respectively, where the a's and b's are real) and using the result of Prob. 17.19 and the fact that $\hat{M}$

is a linear operator, we get

$$\int f^* \hat{M}f\, d\tau + c^* \int g^* \hat{M}f\, d\tau + c \int f^* \hat{M}g\, d\tau + c^* c \int g^* \hat{M}g\, d\tau =$$

$$\int f(\hat{M}f)^*\, d\tau + c \int g(\hat{M}f)^*\, d\tau + c^* \int f(\hat{M}g)^*\, d\tau + cc^* \int g(\hat{M}g)^*\, d\tau$$

Equation (17.91) with Ψ replaced by either f or by g shows that the first integral on the left side of this equation equals the first integral on the right side, and that the last integral on the left side equals the last integral on the right side. Therefore we are left with

$$c^* \int g^* \hat{M}f\, d\tau + c \int f^* \hat{M}g\, d\tau = c \int g(\hat{M}f)^*\, d\tau + c^* \int f(\hat{M}g)^*\, d\tau$$

Putting $c = 1$ we get

$$\int g^* \hat{M}f\, d\tau + \int f^* \hat{M}g\, d\tau = \int g(\hat{M}f)^*\, d\tau + \int f(\hat{M}g)^*\, d\tau$$

Putting $c = i$ and then dividing by i, we get (since $i^* = -i$)

$$-\int g^* \hat{M}f\, d\tau + \int f^* \hat{M}g\, d\tau = \int g(\hat{M}f)^*\, d\tau - \int f(\hat{M}g)^*\, d\tau$$

Adding the last two equations and dividing by 2, we get $\int f^* \hat{M}g\, d\tau = \int g(\hat{M}f)^*\, d\tau$.

17.64 **(a)** $\int f^* \hat{x}g\, d\tau = \int_{-\infty}^{\infty} f^* xg\, dx = \int_{-\infty}^{\infty} gx^* f^*\, dx = \int_{-\infty}^{\infty} g(xf)^*\, dx$, where $(ab)^* = a^*b^*$ [the equation after (17.19)] and the fact that x is real ($x = x^*$) were used.

(b) $\int f^* \hat{p}_x g\, d\tau = \int_{-\infty}^{\infty} f^*(\hbar/i)(\partial g/\partial x)\, dx$. Let $u = f^*$ and $d\upsilon = (\partial g/\partial x)\, dx$. Then $\upsilon = g$ and the integration-by-parts formula $\int u\, d\upsilon = u\upsilon - \int \upsilon\, du$ gives

$\int f^* \hat{p}_x g\, d\tau = (\hbar/i) f^* g \,|_{-\infty}^{\infty} - (\hbar/i)\int_{-\infty}^{\infty} g(\partial f^*/\partial x)\, dx$. Equation (17.92) requires that f and g be well-behaved, which includes the requirement of quadratic integrability. In order to be quadratically integrable, the functions f and g must go to zero as x goes to $\pm\infty$. So $\int f^* \hat{p}_x g\, d\tau = \int_{-\infty}^{\infty} g[(\hbar/i)\partial f/\partial x]^*\, dx = \int_{-\infty}^{\infty} g(\hat{p}_x f)^*\, dx$, which completes the proof.

17.65 We must prove that $\hat{M}(c_1 f_1 + c_2 f_2) = b(c_1 f_1 + c_2 f_2)$. Using the linearity equations given after Example 17.5, we have $\hat{M}(c_1 f_1 + c_2 f_2) = \hat{M}(c_1 f_1) + \hat{M}(c_2 f_2) = c_1 \hat{M}f_1 + c_2 \hat{M}f_2 = c_1 b f_1 + c_2 b f_2 = b(c_1 f_1 + c_2 f_2)$, where the given eigenvalue equations for $\hat{M}$ were used.

17.66 $\int g_1^* g_2\, d\tau = \int f_1^*(f_2 + kf_1)\, d\tau = \int f_1^* f_2\, d\tau + k\int f_1^* f_1\, d\tau =$
$\int f_1^* f_2\, d\tau - [\int f_1^* f_2\, d\tau / \int f_1^* f_1\, d\tau] \int f_1^* f_1\, d\tau = 0.$

17.67 $F \equiv x^2(1-x)$, $G \equiv \sum_{n=1}^{m} c_n \psi_n$, where $\psi_n = 2^{1/2}\sin(n\pi x)$. From Eq. (17.98), $c_n = \int \psi_n^* F\, d\tau = 2^{1/2}\int_0^1 (x^2 - x^3)\sin(n\pi x)\, dx$. A table of integrals (or use of the website integrals.wolfram.com or a calculator that can do symbolic integration) gives $\int x^2 \sin kx\, dx = k^{-3}(2 - k^2x^2)\cos kx + 2k^{-2}x\sin kx$ and $\int x^3 \sin kx\, dx = k^{-3}(6x - k^2x^3)\cos kx + k^{-4}(3k^2x^2 - 6)\sin kx$. We get $c_n = -2^{1/2}(n\pi)^{-3}[4(-1)^n + 2]$, where $\sin n\pi = 0$ and $\cos n\pi = (-1)^n$ were used. So $G = \sum_{n=1}^{m}(-2)(n\pi)^{-3}[4(-1)^n + 2]\sin n\pi x$. We set up a spreadsheet with x values going from 0 to 1 in steps of 0.02 in column A, the values of F at these points in column B, and the values of the first, second,..., fifth terms in the series G in columns C, D, E, F, and G. In column H we sum the first three terms of the series and in column I we sum the first 5 terms of the series. The data in columns B, H, and I are graphed versus x on the same plot. The five-term sum gives a more accurate representation of F than the three-term function. For example, some values are

x	0.1	0.2	0.4	0.6	0.8	0.9
F	0.0090	0.0320	0.0960	0.144	0.128	0.081
3 terms	0.0153	0.0344	0.0914	0.148	0.126	0.072
5 terms	0.0106	0.0308	0.0972	0.143	0.130	0.079

17.68 **(a)** The series is a sum of terms and the complex conjugate of a sum is the sum of the complex conjugates. This if $z_1 = x_1 + iy_1$ and $z_2 = x_2 + iy_2$, where x_1, y_1, x_2, and y_2 are real, then $(z_1 + z_2)^* = (x_1 + iy_1 + x_2 + iy_2)^* = [x_1 + x_2 + i(y_1 + y_2)]^* = x_1 + x_2 - i(y_1 + y_2) = x_1 - iy_1 + x_2 - iy_2 = z_1^* + z_2^*$. So $\phi^* = \sum_k (c_k\psi_k)^*$. Then the use of the result of Prob. 17.19 gives $\phi^* = \sum_k c_k^*\psi_k^* = \sum_j c_j^*\psi_j^*$, where the last equality holds because the summation index is a dummy variable (Sec. 1.8).

(b) $1 = \int \phi^*\phi\, d\tau = \int \sum_j c_j^*\psi_j^* \sum_k c_k\psi_k\, d\tau = \sum_j\sum_k \int c_j^*\psi_j^* c_k\psi_k\, d\tau =$
$\sum_j\sum_k c_j^* c_k \int \psi_j^*\psi_k\, d\tau = \sum_j\sum_k c_j^* c_k \delta_{jk}$, since the integral of a sum is

the sum of the integrals, and we used the orthonormality of the wave functions. When the sum over k is performed, the δ_{jk} factor makes all terms equal to zero except for the one term where $k = j$. Thus the sum over k equals $c_j^* c_j = |c_j|^2$ and $1 = \sum_j |c_j|^2 = \sum_k |c_k|^2$.

(c) $W = \int \phi^* \hat{H} \phi \, d\tau = \int \sum_j c_j^* \psi_j^* \hat{H} \sum_k c_k \psi_k \, d\tau = \int \sum_j c_j^* \psi_j^* \sum_k c_k \hat{H} \psi_k \, d\tau$

where we used the linearity of the operator $\hat{H}$. Now $\hat{H}\psi_k = E_k \psi_k$, so $W = \sum_j \sum_k c_j^* c_k E_k \int \psi_j^* \psi_k \, d\tau = \sum_j \sum_k c_j^* c_k E_k \delta_{jk} = \sum_j |c_j|^2 E_j = \sum_k |c_k|^2 E_k$.

(d) Since $|c_k|^2 E_k \geq |c_k|^2 E_{gs}$, each term in the sum $W = \sum_k |c_k|^2 E_k$ is equal to or greater than the corresponding term in the sum $\sum_k |c_k|^2 E_{gs}$, the first sum must be greater than or equal to the second sum, and so $W \geq \sum_k |c_k|^2 E_{gs}$. But part (b) gives $\sum_k |c_k|^2 = 1$, so $W \geq E_{gs}$.

17.69 **(a)** Since $|\psi|^2\, dx$ is a probability and probabilities are dimensionless, $|\psi|^2$ has units of length^{-1} and ψ has units of length$^{-1/2}$. The SI units of ψ are m$^{-1/2}$ for a one-particle one-dimensional system.

(b) $|\psi|^2\, dx\, dy\, dz$ is dimensionless and ψ has units of length$^{-3/2}$.

(c) $|\psi|^2\, dx_1\, dy_1\, dz_1\, dx_2\, dy_2\, dz_2$ is dimensionless and ψ has units of length^{-3}.

17.70 The blackbody function (17.2) depends on the combinations of constants h/c^2 and h/k. In 1900, c was known reasonably accurately, so by fitting the observed blackbody curves Planck obtained values for both h and k. Use of $R = N_A k$ then gave N_A. Use of $F = N_A e$ then gave e.

17.71 **(a)** $E = n^2h^2/8ma^2$ and doubling a multiplies E by ¼.

(b) $E = J(J+1)\hbar^2/2I = J(J+1)\hbar^2/2\mu d^2$ and doubling d multiplies E by ¼.

(c) $E = h\nu = h(1/2\pi)(k/m)^{1/2}$ and doubling m multiplies E by $1/\sqrt{2}$.

17.72 **(a)** T. **(b)** T. The future state is predicted by integrating the time-dependent Schrödinger equation. **(c)** T. **(d)** T. **(e)** F. "Sum" must be replaced by "product" to make the statement true. **(f)** T. **(g)** F. **(h)** F. **(i)** F.

CHAPTER 18

Atomic Structure

CHAPTER OUTLINE

Chapter 17 introduced some of the main ideas of quantum chemistry and looked at the solutions of the time-independent Schrödinger equation for the particle in a box, the rigid two-particle rotor, and the harmonic oscillator. We now use quantum mechanics to discuss the electronic structure of atoms (Chapter 18) and molecules (Chapter 19).

18.1 UNITS

The forces acting between the particles in atoms and molecules are electrical (Prob. 18.69). To formulate the Hamiltonian operator of an atom or molecule, we need the expression for the potential energy of interaction between two charged particles. Equation (13.12) gives the electric potential ϕ at a distance r from charge Q_1 as $\phi = Q_1/4\pi\varepsilon_0 r$. Equation (13.10) gives the potential energy V of interaction of a second charge Q_2 with this electric potential as $V = \phi Q_2 = (Q_1/4\pi\varepsilon_0 r)Q_2$. The potential energy of interaction between two charges separated by a distance r is therefore

$$V = \frac{1}{4\pi\varepsilon_0}\frac{Q_1 Q_2}{r} \qquad \textbf{(18.1)*}$$

Equation (18.1) is in SI units, with Q_1 and Q_2 in coulombs (C), r in meters, and V in joules. The values of the constants ε_0 and $1/4\pi\varepsilon_0$ are listed in Eq. (13.2) and in the fundamental-constants table inside the back cover. The $1/4\pi\varepsilon_0$ factor in (18.1) is a bit of a pain, but if you learn the value of the speed of light c, you can avoid looking up $1/4\pi\varepsilon_0$ by using the relation (Prob. 18.3)

$$\frac{1}{4\pi\varepsilon_0} = 10^{-7}c^2 \text{ N s}^2/\text{C}^2 \qquad (18.2)$$

Atomic and molecular energies are very small. A convenient unit for these energies is the **electronvolt** (eV), defined as the kinetic energy acquired by an electron accelerated through a potential difference of one volt. From (13.10), the change in potential energy for this process is $-e(1\text{ V})$, where $-e$ is the electron's charge. This loss in potential energy is matched by a gain of kinetic energy equal to $e(1\text{ V})$. Substitution of (18.4) for e gives $1\text{ eV} = (1.6022 \times 10^{-19}\text{ C})(1\text{ V})$; the use of $1\text{ V} = 1\text{ J/C}$ [Eq. (13.8)] gives

$$1\text{ eV} = 1.6022 \times 10^{-19}\text{ J} \qquad (18.3)$$

18.2 HISTORICAL BACKGROUND

In a low-pressure gas-discharge tube, bombardment of the negative electrode (the cathode) by positive ions causes the cathode to emit what nineteenth-century physicists called *cathode rays*. In 1897, J. J. Thomson measured the deflection of cathode rays in simultaneously applied electric and magnetic fields of known strengths. His experiment allowed calculation of the charge-to-mass ratio Q/m of the cathode-ray particles. Thomson found that Q/m was independent of the metal used for the cathode, and his experiments mark the discovery of the electron. [G. P. Thomson, who was one of the first people to observe diffraction effects with electrons (Sec. 17.4), was J. J.'s son. It has been said that J. J. Thomson got the Nobel Prize for proving the electron to be a particle, and G. P. Thomson got the Nobel Prize for proving the electron to be a wave.]

Let the symbol e denote the charge on the proton. The electron charge is then $-e$. Thomson found $e/m_e = 1.7 \times 10^8$ C/g, where m_e is the electron's mass.

The first accurate measurement of the electron's charge was made by R. A. Millikan and Harvey Fletcher in the period 1909–1913 (see H. Fletcher, *Physics Today,* June 1982, p. 43). They observed the motion of charged oil drops in oppositely directed electric and gravitational fields and found that all observed values of the charge Q of a drop satisfied $|Q| = ne$, where n was a small integer and was clearly the number of extra or missing electrons on the charged oil drop. The smallest observed difference between values of $|Q|$ could then be taken as the magnitude of the charge of the electron. The currently accepted value of the proton charge is

$$e = 1.6022 \times 10^{-19} \text{ C} \tag{18.4}$$

From the values of e and the Faraday constant F, an accurate value of the Avogadro constant N_A can be obtained. Equation (13.13) gives $N_A = F/e = (96485 \text{ C mol}^{-1})/(1.6022 \times 10^{-19} \text{ C}) = 6.022 \times 10^{23} \text{ mol}^{-1}$.

From the values of e and e/m_e, the electron mass m_e can be found. The modern value is 9.1094×10^{-28} g. The ^{1}H atom mass is $(1.0078 \text{ g})/(6.022 \times 10^{23}) = 1.6735 \times 10^{-24}$ g. This is 1837 times m_e, and so a proton is 1836 times as heavy as an electron. Nearly all the mass of an atom is in its nucleus.

The existence of the atomic nucleus was demonstrated by the 1909–1911 experiments of Rutherford, Geiger, and Marsden, who allowed a beam of alpha particles (He^{2+} nuclei) to fall on a very thin gold foil. Although most of the alpha particles passed nearly straight through the foil, a few were deflected through large angles. Since the very light electrons of the gold atoms cannot significantly deflect the alpha particles (in a collision between a truck and a bicycle, it is the bicycle that is deflected), one need consider only the force between the alpha particle and the positive charge of a gold atom. This force is given by Coulomb's law (13.1). To get a force large enough to produce the observed large deflections, Rutherford found that r in (13.1) had to be in the range 10^{-12} to 10^{-13} cm, which is much less than the known radius of an atom (10^{-8} cm). Rutherford therefore concluded in 1911 that the positive charge of an atom was not distributed throughout the atom but was concentrated in a tiny central region, the nucleus.

In 1913, Bohr proposed his theory of the hydrogen atom (Sec. 17.3). By the early 1920s, physicists realized that the Bohr theory was not correct.

In January 1926, Erwin Schrödinger formulated the Schrödinger equation. He solved the time-independent Schrödinger equation for the hydrogen atom in his first paper on quantum mechanics, obtaining energy levels in agreement with the observed spectrum. In 1929, Hylleraas used the quantum-mechanical variational method (Sec. 17.15) to obtain a ground-state energy for helium in accurate agreement with experiment.

18.3 THE HYDROGEN ATOM

The simplest atom is hydrogen. The Schrödinger equation can be solved exactly for the H atom but not for atoms with more than one electron. Ideas developed in treating the H atom provide a basis for dealing with many-electron atoms.

The hydrogen atom is a two-particle system in which a nucleus and an electron interact according to Coulomb's law. Instead of dealing only with the H atom, we shall consider the slightly more general problem of the **hydrogenlike atom;** this is an atom with one electron and Z protons in the nucleus. The values $Z = 1, 2, 3, \ldots$ give the species H, He^+, Li^{2+}, ... With the nuclear charge Q_1 set equal to Ze and the electron charge Q_2 set equal to $-e$, Eq. (18.1) gives the potential energy as $V = -Ze^2/4\pi\varepsilon_0 r$, where r is the distance between the electron and the nucleus.

The potential-energy function depends only on the relative coordinates of the two particles, and so the conclusions of Sec. 17.13 apply. The total energy E_{tot} of the atom is the sum of its translational energy and the energy of internal motion of the electron relative to the proton. The translational energy levels can be taken as the particle-in-a-box levels (17.47). The box is the container holding the gas of H atoms. We now focus on the energy E of internal motion. The Hamiltonian H for the internal motion is given by the terms in the first pair of brackets in (17.78), and the corresponding Hamiltonian operator $\hat{H}$ for the internal motion is

$$\hat{H} = -\frac{\hbar^2}{2\mu}\left(\frac{\partial^2}{\partial x^2} + \frac{\partial^2}{\partial y^2} + \frac{\partial^2}{\partial z^2}\right) - \frac{Ze^2}{4\pi\varepsilon_0 r} \tag{18.5}$$

where x, y, and z are the coordinates of the electron relative to the nucleus and $r = (x^2 + y^2 + z^2)^{1/2}$. The reduced mass μ is $\mu = m_1 m_2/(m_1 + m_2)$ [Eq. (17.79)], where m_1 and m_2 are the nuclear and electron masses. For a hydrogen atom, $m_{\text{nucleus}} = m_{\text{proton}} = 1836.15m_e$ and

$$\mu_{\text{H}} = \frac{1836.15m_e^2}{1837.15m_e} = 0.999456m_e \tag{18.6}$$

which differs only slightly from m_e.

The H-atom Schrödinger equation $\hat{H}\psi = E\psi$ is difficult to solve in cartesian coordinates but is relatively easy to solve in spherical coordinates. The spherical coordinates r, θ, and ϕ of the electron relative to the nucleus are defined in Fig. 18.1. (Math books usually interchange θ and ϕ.) The projection of r on the z axis is $r \cos \theta$, and its projection on the xy plane is $r \sin \theta$. The relation between cartesian and spherical coordinates is therefore

$$x = r \sin \theta \cos \phi, \qquad y = r \sin \theta \sin \phi, \qquad z = r \cos \theta \tag{18.7}$$

Note that $x^2 + y^2 + z^2 = r^2$. The ranges of the coordinates are

$$0 \le r \le \infty, \quad 0 \le \theta \le \pi, \quad 0 \le \phi \le 2\pi \tag{18.8)*}$$

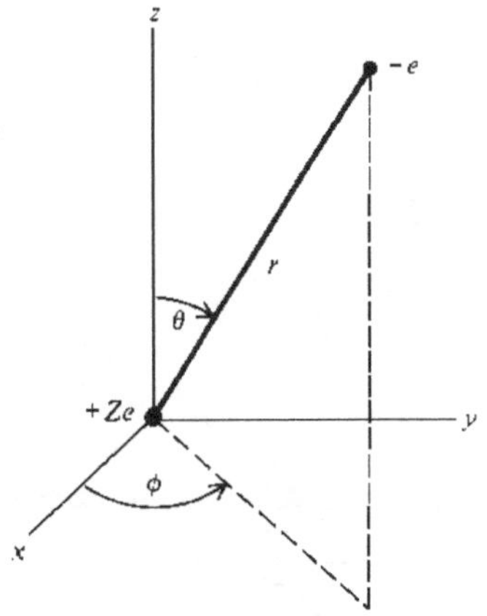

Figure 18.1

Coordinates of the electron relative to the nucleus in a hydrogenlike atom.

To solve the H-atom Schrödinger equation, one transforms the partial derivatives in the Hamiltonian operator (18.5) to derivatives with respect to r, θ, and ϕ and then uses the separation-of-variables procedure (Sec. 17.11). The details (which can be found in quantum chemistry texts) are omitted, and only an outline of the solution process will be given. The H-atom Schrödinger equation in spherical coordinates is found to be separable when the substitution

$$\psi = R(r)\Theta(\theta)\Phi(\phi)$$

is made, where R, Θ, and Φ are functions of r, θ, and ϕ, respectively. One obtains three separate differential equations, one for each coordinate.

The differential equation for $\Phi(\phi)$ is found to have solutions of the form $\Phi(\phi) = Ae^{im\phi}$, where $i = \sqrt{-1}$, A is an integration constant whose value is chosen to normalize Φ, and m (not to be confused with a mass) is a constant introduced in the process of separating the ϕ differential equation (recall the introduction of the separation constants E_x, E_y, and E_z in solving the problem of the particle in a three-dimensional box in Sec. 17.9). Since addition of 2π to the coordinate ϕ brings us back to the same point in space, the requirement that the wave function be single-valued (Sec. 17.7) means that we must have $\Phi(\phi) = \Phi(\phi + 2\pi)$. One finds (Prob. 18.18) that this equation is satisfied only if m is an integer (positive, negative, or zero).

The solution to the differential equation for $\Theta(\theta)$ is a complicated function of θ that involves a separation constant l and also the integer m that occurs in the ϕ equation. The $\Theta(\theta)$ solutions are not quadratically integrable except for values of l that satisfy $l = |m|$, $|m| + 1$, $|m| + 2, \ldots$, where $|m|$ is the absolute value of the integer m. Thus l is an integer with minimum possible value 0, since this is the minimum possible value of $|m|$. The condition $l \geq |m|$ means that m ranges from $-l$ to $+l$ in steps of 1.

The differential equation for $R(r)$ for the H atom contains the energy E of internal motion as a parameter and also contains the quantum number l. The choice of zero level of energy is arbitrary. The potential energy $V = -Ze^2/4\pi\varepsilon_0 r$ in (18.5) takes the zero level to correspond to infinite separation of the electron and nucleus, which is an ionized atom. If the internal energy E is less than zero, the electron is bound to the nucleus. If the internal energy is positive, the electron has enough energy to escape the attraction of the nucleus and is free. One finds that for negative E, the function $R(r)$ is not quadratically integrable except for values of E that satisfy $E = -Z^2e^4\mu/(4\pi\varepsilon_0)^2 2n^2\hbar^2$, where n is an integer such that $n \geq l + 1$. Since the minimum l is zero, the minimum n is 1. Also, l cannot exceed $n - 1$. Further, one finds that all positive values of E are allowed. When the electron is free, its energy is continuous rather than quantized.

In summary, the hydrogenlike-atom wave functions have the form

$$\psi = R_{nl}(r)\Theta_{lm}(\theta)\Phi_m(\phi) \tag{18.9}$$

where the radial function $R_{nl}(r)$ is a function of r whose form depends on the quantum numbers n and l, the theta factor depends on l and m, and the phi factor is

$$\Phi_m(\phi) = (2\pi)^{-1/2}e^{im\phi}, \qquad i \equiv \sqrt{-1} \tag{18.10}$$

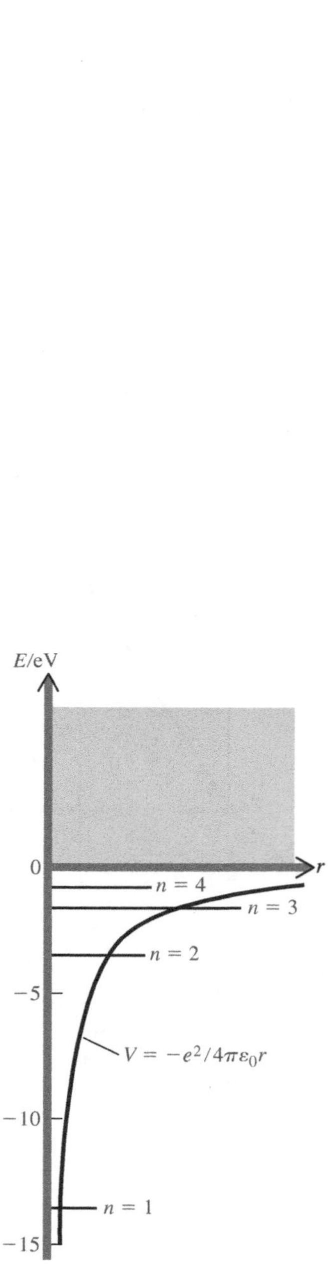

Figure 18.2

Energy levels and potential-energy function of the hydrogen atom. The shading indicates that all positive energies are allowed.

Since there are three variables, the solutions involve three quantum numbers: the **principal quantum number** n, the **angular-momentum quantum number** l, and the **magnetic quantum number** m (often symbolized by m_l). For ψ to be well behaved, the quantum numbers are restricted to the values

$$n = 1, 2, 3, \ldots \qquad \textbf{(18.11)*}$$

$$l = 0, 1, 2, \ldots, n - 1 \qquad \textbf{(18.12)*}$$

$$m = -l, -l + 1, \ldots, l - 1, l \qquad \textbf{(18.13)*}$$

For example, for $n = 2$, l can be 0 or 1. For $l = 0$, m is 0. For $l = 1$, m can be -1, 0, or 1. The allowed bound-state energy levels are

$$E = -\frac{Z^2}{n^2}\frac{e^2}{(4\pi\varepsilon_0)2a} \qquad \text{where } a \equiv \frac{\hbar^2(4\pi\varepsilon_0)}{\mu e^2} \tag{18.14}$$

where $n = 1, 2, 3, \ldots$. Also, all values $E \geq 0$ are allowed, corresponding to an ionized atom. Figure 18.2 shows some of the allowed energy levels and the potential-energy function.

The following letter code is often used to specify the l value of an electron:

l value	0	1	2	3	4	5
Code letter	s	p	d	f	g	h

(18.15)*

The value of n is given as a prefix to the l code letter, and the m value is added as a subscript. Thus, $2s$ denotes the $n = 2$, $l = 0$ state; $2p_{-1}$ denotes the $n = 2$, $l = 1$, $m = -1$ state.

The hydrogenlike-atom energy levels (18.14) depend only on n, but the wave functions (18.9) depend on all three quantum numbers n, l, and m. Therefore, there is degeneracy. For example, the $n = 2$ H-atom level is fourfold degenerate (spin considerations omitted), the wave functions (states) $2s$, $2p_1$, $2p_0$, and $2p_{-1}$ all having the same energy.

The defined quantity a in (18.14) has the dimensions of length. For a hydrogen atom, substitution of numerical values gives (Prob. 18.13) $a = 0.5295$ Å.

If the reduced mass μ in the definition of a is replaced by the electron mass m_e, we get the **Bohr radius** a_0:

$$a_0 = \hbar^2(4\pi\varepsilon_0)/m_e e^2 = 0.5292 \text{ Å} \tag{18.16}$$

a_0 was the radius of the $n = 1$ circle in the Bohr theory.

EXAMPLE 18.1 Ground-state energy of H

Calculate the ground-state hydrogen-atom energy E_{gs}. Also, express E_{gs} in electronvolts.

Setting $n = 1$ and $Z = 1$ in (18.14), we get

$$E_{gs} = -\frac{e^2}{(4\pi\varepsilon_0)2a} = -\frac{(1.6022 \times 10^{-19}\text{ C})^2}{4\pi(8.854 \times 10^{-12}\text{ C}^2\text{ J}^{-1}\text{ m}^{-1})2(0.5295 \times 10^{-10}\text{ m})}$$

$$= -2.179 \times 10^{-18}\text{ J}$$

The use of the conversion factor (18.3) gives for an H atom

$$E_{gs} = -e^2/(4\pi\varepsilon_0)(2a) = -13.60 \text{ eV} \tag{18.17}$$

Exercise

Find the wavelength of the longest-wavelength absorption line for a gas of ground-state hydrogen atoms. (*Answer*: 121.6 nm.)

$|E_{gs}|$ is the minimum energy needed to remove the electron from an H atom and is the **ionization energy** of H. The **ionization potential** of H is 13.60 V.

From (18.17), the energy levels (18.14) can be written as

$$E = -(Z^2/n^2)(13.60 \text{ eV}) \qquad \text{H-like atom} \tag{18.18*}$$

Although the reduced mass μ in (18.14) differs for different hydrogenlike species (H, He^+, Li^{2+}, . . .), the differences are very slight and have been ignored in (18.18).

Quantum chemists often use a system called **atomic units,** in which energies are reported in **hartrees** and distances in **bohrs.** These quantities are defined as

$$1 \text{ bohr} \equiv a_0 = 0.52918 \text{ Å}, \qquad 1 \text{ hartree} \equiv e^2/(4\pi\varepsilon_0)a_0 = 27.211 \text{ eV}$$

The ground-state energy (18.17) of H would be $-\frac{1}{2}$ hartree if a were approximated by a_0.

The first few $R_{nl}(r)$ and $\Theta_{lm}(\theta)$ factors in the wave functions (18.9) are

$$R_{1s} = 2(Z/a)^{3/2}e^{-Zr/a} \tag{18.19}$$

$$R_{2s} = 2^{-1/2}(Z/a)^{3/2}(1 - Zr/2a)e^{-Zr/2a} \tag{18.20}$$

$$R_{2p} = (24)^{-1/2}(Z/a)^{5/2}re^{-Zr/2a} \tag{18.21}$$

$$\Theta_{s_0} = 1/\sqrt{2}, \qquad \Theta_{p_0} = \tfrac{1}{2}\sqrt{6}\cos\theta, \qquad \Theta_{p_1} = \Theta_{p_{-1}} = \tfrac{1}{2}\sqrt{3}\sin\theta \tag{18.22}$$

where the code (18.15) was used for l. The general form of R_{nl} is

$$R_{nl}(r) = r^l e^{-Zr/na}(b_0 + b_1 r + b_2 r^2 + \cdots + b_{n-l-1} r^{n-l-1})$$

where $b_0, b_1, \ldots$ are certain constants whose values depend on n and l. As n increases, $e^{-Zr/na}$ dies off more slowly as r increases, so the average radius $\langle r \rangle$ of the atom increases as n increases. For the ground state, one finds (Prob. 18.15) $\langle r \rangle = 3a/2Z$, which is 0.79 Å for H. In (18.19) and (18.10), e is the base of natural logarithms, and not the proton charge.

Figure 18.3 shows some plots of $R_{nl}(r)$. The radial factor in ψ has $n - l - 1$ nodes (not counting the node at the origin for $l \neq 0$).

For s states ($l = 0$), Eqs. (18.10) and (18.22) give the angular factor in ψ as $1/\sqrt{4\pi}$, which is independent of θ and ϕ. For s states, ψ depends only on r and is therefore said to be **spherically symmetric.** For $l \neq 0$, the angular factor is not constant, and ψ is not spherically symmetric. Note from Fig. 18.3 that R_{nl}, and hence ψ, is nonzero at the nucleus ($r = 0$) for s states.

The ground-state wave function is found by multiplying R_{1s} in (18.19) by the s-state angular factor $(4\pi)^{-1/2}$ to give

$$\psi_{1s} = \pi^{-1/2}(Z/a)^{3/2} e^{-Zr/a} \tag{18.23}$$

Wave Functions of a Degenerate Energy Level

To deal with the $2p$ wave functions of the H atom, we need to use a quantum-mechanical theorem about wave functions of a degenerate level. By a **linear combination** of the functions $g_1, g_2, \ldots, g_k$, one means a function of the form $c_1 g_1 + c_2 g_2 + \cdots + c_k g_k$, where the c's are constants. Any linear combination of two or more stationary-state wave functions that belong to the same degenerate energy level is an eigenfunction of the Hamiltonian operator with the same energy value as that of the degenerate level. In other words, if $\hat{H}\psi_1 = E_1\psi_1$ and $\hat{H}\psi_2 = E_1\psi_2$, then $\hat{H}(c_1\psi_1 + c_2\psi_2) = E_1(c_1\psi_1 + c_2\psi_2)$. The linear combination $c_1\psi_1 + c_2\psi_2$ (when multiplied by a normalization constant) is therefore also a valid wave function, meaning that

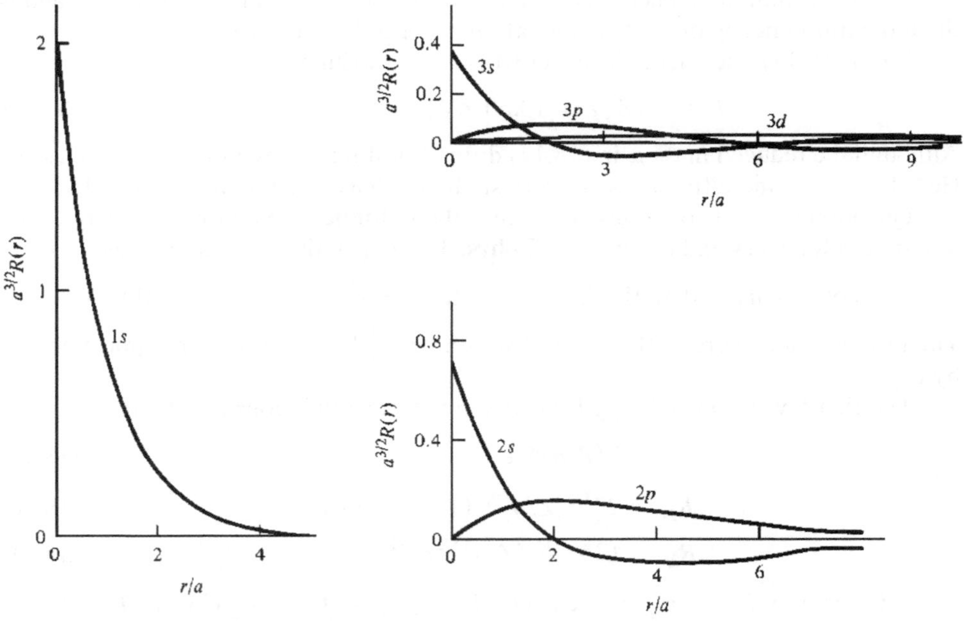

Figure 18.3

Radial factors in some hydrogen-atom wave functions.

it is an eigenfunction of $\hat{H}$ and therefore a solution of the Schrödinger equation. The proof follows from the fact that $\hat{H}$ is a linear operator (Sec. 17.11).

$$\hat{H}(c_1\psi_1 + c_2\psi_2) = \hat{H}(c_1\psi_1) + \hat{H}(c_2\psi_2) = c_1\hat{H}\psi_1 + c_2\hat{H}\psi_2$$
$$= c_1E_1\psi_1 + c_2E_1\psi_2 = E_1(c_1\psi_1 + c_2\psi_2)$$

Note that this theorem does not apply to wave functions belonging to two different energy levels. If $\hat{H}\psi_5 = E_5\psi_5$ and $\hat{H}\psi_6 = E_6\psi_6$ with $E_5 \neq E_6$, then $c_1\psi_5 + c_2\psi_6$ is not an eigenfunction of $\hat{H}$.

Real Wave Functions

The Φ factor (18.10) in the H-atom wave function (18.9) contains i and so is complex. Chemists often find it convenient to work with real wave functions instead. To get real functions, we use the theorem just stated.

From (18.9), (18.10), and (18.21), the complex $2p$ functions are

$$2p_{+1} = be^{-Zr/2a}r \sin\theta\, e^{i\phi}, \qquad 2p_{-1} = be^{-Zr/2a}r \sin\theta\, e^{-i\phi}$$
$$2p_0 = ce^{-Zr/2a}r \cos\theta$$

where $b \equiv (1/8\pi^{1/2})(Z/a)^{5/2}$ and $c \equiv \pi^{-1/2}(Z/2a)^{5/2}$. The $2p_0$ function is real as it stands. Equation (18.7) gives $r \cos\theta = z$, so the $2p_0$ function is also written as

$$2p_z \equiv 2p_0 = cze^{-Zr/2a}$$

(Don't confuse the nuclear charge Z with the z spatial coordinate.) The $2p_{+1}$ and $2p_{-1}$ functions are each eigenfunctions of $\hat{H}$ with the same energy eigenvalue, so we can take any linear combination of them and have a valid wave function.

As a preliminary, we note that

$$e^{i\phi} = \cos\phi + i \sin\phi \qquad \textbf{(18.24)*}$$

and $e^{-i\phi} = (e^{i\phi})^* = \cos\phi - i\sin\phi$. For a proof of (18.24), see Prob. 18.16.

We define the linear combinations $2p_x$ and $2p_y$ as

$$2p_x \equiv (2p_1 + 2p_{-1})/\sqrt{2}, \qquad 2p_y \equiv (2p_1 - 2p_{-1})/i\sqrt{2} \qquad (18.25)$$

The $1/\sqrt{2}$ factors normalize these functions. Using (18.24) and its complex conjugate, we find (Prob. 18.17) that

$$2p_x = cxe^{-Zr/2a}, \qquad 2p_y = cye^{-Zr/2a} \qquad (18.26)$$

where $c \equiv \pi^{-1/2}(Z/2a)^{5/2}$. The $2p_x$ and $2p_y$ functions have the same n and l values as the $2p_1$ and $2p_{-1}$ functions (namely, $n = 2$ and $l = 1$) but do not have a definite value of m. Similar linear combinations give real wave functions for higher H-atom states. The real functions (which have directional properties) are more suitable than the complex functions for use in treating the bonding of atoms to form molecules. Table 18.1 lists the $n = 1$ and $n = 2$ real hydrogenlike functions.

Orbitals

An **orbital** is a one-electron spatial wave function. Since a hydrogenlike atom has one electron, all the hydrogenlike wave functions are orbitals. The use of (one-electron) orbitals in many-electron atoms is considered later in this chapter.

The **shape** of an orbital is defined as a surface of constant probability density that encloses some large fraction (say 90%) of the probability of finding the electron. The probability density is $|\psi|^2$. When $|\psi|^2$ is constant, so is $|\psi|$. Hence $|\psi|$ *is constant on the surface of an orbital.*

TABLE 18.1

Real Hydrogenlike Wave Functions for $n = 1$ and $n = 2$

$1s = \pi^{-1/2}(Z/a)^{3/2}e^{-Zr/a}$

$2s = \frac{1}{4}(2\pi)^{-1/2}(Z/a)^{3/2}(2 - Zr/a)e^{-Zr/2a}$

$2p_x = \frac{1}{4}(2\pi)^{-1/2}(Z/a)^{5/2}re^{-Zr/2a}\sin\theta\cos\phi$

$2p_y = \frac{1}{4}(2\pi)^{-1/2}(Z/a)^{5/2}re^{-Zr/2a}\sin\theta\sin\phi$

$2p_z = \frac{1}{4}(2\pi)^{-1/2}(Z/a)^{5/2}re^{-Zr/2a}\cos\theta$

For an *s* orbital, ψ depends only on r, and $|\psi|$ is constant on the surface of a sphere with center at the nucleus. An *s* orbital has a spherical shape.

The volume element in spherical coordinates (see any calculus text) is

$$d\tau = r^2 \sin\theta \, dr \, d\theta \, d\phi \qquad \textbf{(18.27)*}$$

This is the volume of an infinitesimal solid for which the spherical coordinates lie in the ranges r to $r + dr$, θ to $\theta + d\theta$, and ϕ to $\phi + d\phi$.

EXAMPLE 18.2 1*s* orbital radius

Find the radius of the 1*s* orbital in H using the 90 percent probability definition.

The probability that a particle will be in a given region is found by integrating the probability density $|\psi|^2$ over the volume of the region. The region being considered here is a sphere of radius r_{1s}. For this region, θ and ϕ go over their full ranges 0 to π and 0 to 2π, respectively, and r goes from 0 to r_{1s}. Also, $\psi_{1s} = \pi^{-1/2}(Z/a)^{3/2}e^{-Zr/a}$ (Table 18.1). Using (18.27) for $d\tau$, we have as the probability that the electron is within distance r_{1s} from the nucleus:

$$0.90 = \int_0^{2\pi}\int_0^{\pi}\int_0^{r_{1s}} \pi^{-1}\left(\frac{Z}{a}\right)^3 e^{-2Zr/a}r^2 \sin\theta \, dr \, d\theta \, d\phi$$

$$0.90 = \frac{Z^3}{\pi a^3}\int_0^{2\pi} d\phi \int_0^{\pi} \sin\theta \, d\theta \int_0^{r_{1s}} e^{-2Zr/a}r^2 \, dr$$

where the integral identity of Prob. 17.14 was used. One next evaluates the integrals and uses trial and error or the Excel Solver or a calculator with equation-solving capability to find the value of r_{1s} that satisfies this equation with $Z = 1$. The remaining work is left as an exercise. One finds $r_{1s} = 1.4$ Å for H.

Exercise

Evaluate the integrals in this example (use a table of integrals for the r integral) and show that the result for $Z = 1$ is $e^{-2w}(2w^2 + 2w + 1) - 0.1 = 0$, where $w \equiv r_{1s}/a$. Solve this equation to show that $r_{1s} = 1.41$ Å.

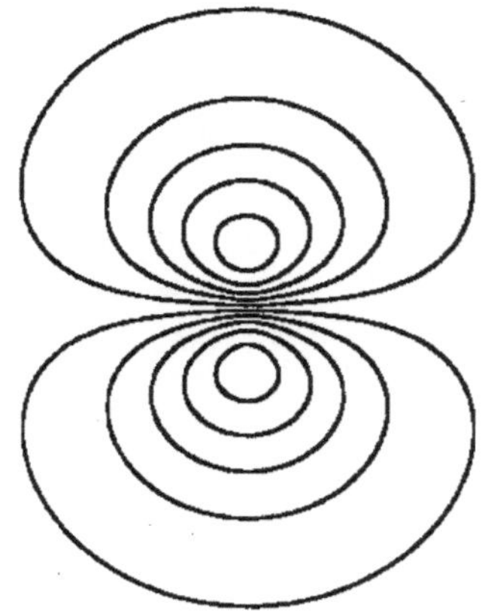

Figure 18.4

Contours of the $2p_z$ orbital in the *yz* plane. The *z* axis is vertical.

Consider the shapes of the real 2*p* orbitals. The $2p_z$ orbital is $2p_z = cze^{-Zr/2a}$, where c is a constant. The $2p_z$ function is zero in the *xy* plane (where $z = 0$), is positive above this nodal plane (where z is positive), and is negative below this plane. A detailed investigation (Prob. 18.70) gives the curves shown in Fig. 18.4 as the contours of constant $|\psi_{2p_z}|$ in the yz plane. The curves shown are for $|\psi/\psi_{\max}| = 0.9$ (the two innermost ovals), 0.7, 0.5, 0.3, and 0.1, where $\psi_{\max}$ is the maximum value of ψ_{2p_z}. The three-dimensional shape of the $2p_z$ orbital is obtained by rotating a cross section around the z axis. This gives two distorted ellipsoids, one above and one below the xy plane. The ellipsoids do not touch each other. This is obvious from the fact that ψ has opposite signs on each ellipsoid. The absolute value $|\psi|$ is the same on each ellipsoid of the $2p_z$ orbital. The $2p_x$, $2p_y$, and $2p_z$ orbitals have the same shape but different orientations in space. The two distorted ellipsoids are located on the x axis for the $2p_x$ orbital, on the y axis for the $2p_y$ orbital, and on the z axis for the $2p_z$ orbital.

The $2p_z$ wave function is a function of the three spatial coordinates: $\psi_{2p_z} = \psi_{2p_z}(x, y, z)$. Just as two dimensions are needed to graph a function of one variable, it would require four dimensions to graph $\psi_{2p_z}(x, y, z)$. Figure 18.5 shows a three-dimensional graph of $\psi_{2p_z}(0, y, z)$. In this graph, the value of ψ_{2p_z} at each point in the yz plane is given by the height of the graph above this plane. Note the resemblance to Fig. 17.13 for the $n_x = 1$, $n_y = 2$ particle-in-a-two-dimensional-box state.

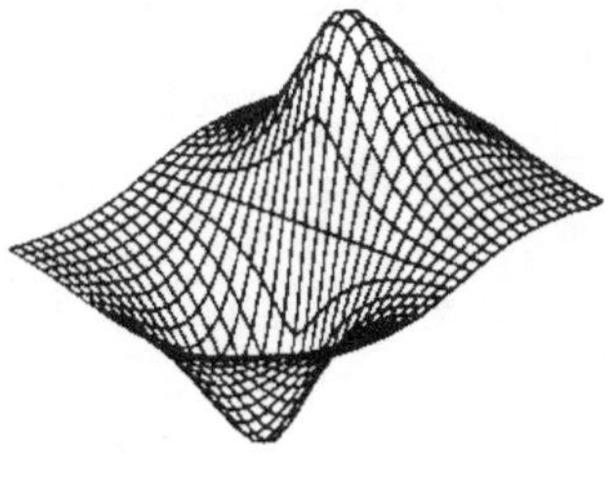

Figure 18.5

Three-dimensional graph of values of ψ_{2p_z} in the yz plane.

Figure 18.6 shows some hydrogen-atom orbital shapes. The plus and minus signs in Fig. 18.6 give the algebraic signs of ψ and have nothing to do with electric charge. The $3p_z$ orbital has a spherical node (shown by the dashed line in Fig. 18.6). The $3d_{z^2}$ orbital has two nodal cones (dashed lines). The other four $3d$ orbitals have the same shape as one another but different orientations; each of these orbitals has two nodal planes separating the four lobes.

Probability Density

The electron probability density for the hydrogenlike-atom ground state (Table 18.1) is $|\psi_{1s}|^2 = (Z^3/\pi a^3)e^{-2Zr/a}$. The $1s$ probability density is a maximum at the nucleus ($r = 0$). Figure 18.7 is a schematic indication of this, the density of the dots indicating the relative probability densities in various regions. Since $|\psi_{1s}|^2$ is nonzero everywhere, the

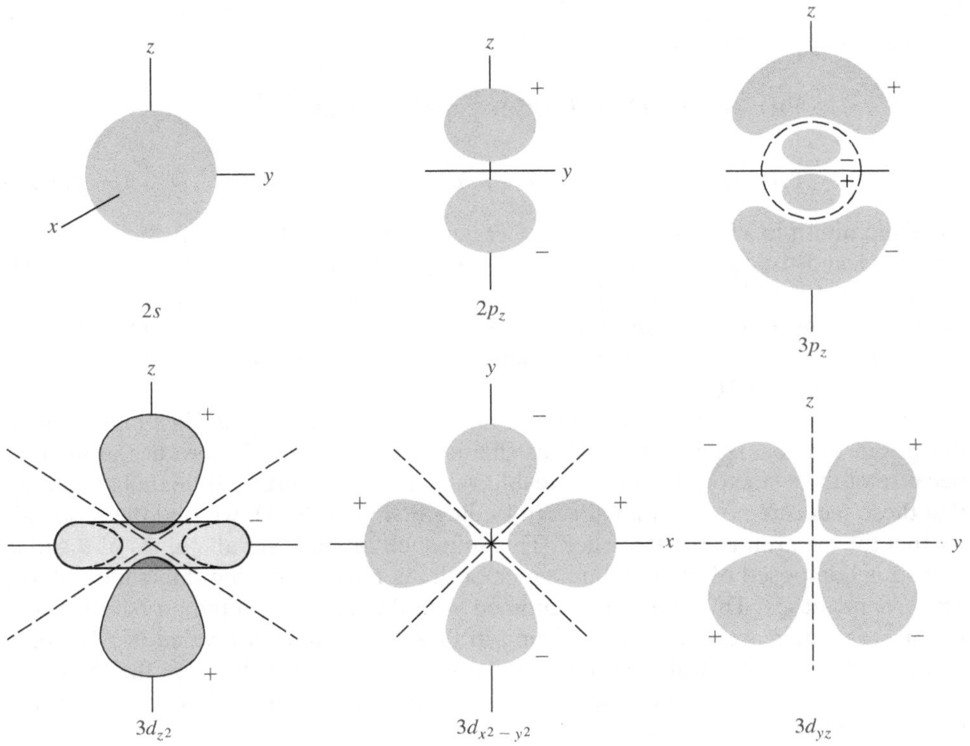

Figure 18.6

Shapes of some hydrogen-atom orbitals. (Not drawn to scale.) Note the different orientation of the axes in the $3d_{x^2-y^2}$ figure compared with the others. Not shown are the $3d_{xy}$ and $3d_{xz}$ orbitals; these have their lobes between the x and y axes and between the x and z axes, respectively.

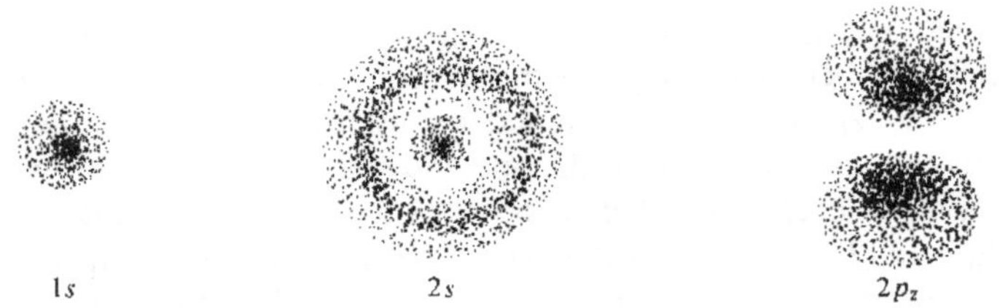

Figure 18.7

Probability densities in three hydrogen-atom states. (Not drawn to scale.)

electron can be found at any location in the atom (in contrast to the Bohr theory, where it had to be at a fixed r). Figure 18.7 also indicates the variation in probability density for the $2s$ and $2p_z$ states. Note the nodal sphere in the $2s$ function.

Radial Distribution Function

Suppose we want the probability $\Pr(r \rightarrow r + dr)$ that the electron–nucleus distance is between r and $r + dr$. This is the probability of finding the electron in a thin spherical shell whose center is at the nucleus and whose inner and outer radii are r and $r + dr$. For an s orbital, ψ is independent of θ and ϕ and so is essentially constant in the thin shell. Hence, the desired probability is found by multiplying $|\psi_s|^2$ (the probability per unit volume) by the volume of the thin shell. This volume is $\frac{4}{3}\pi(r + dr)^3 - \frac{4}{3}\pi r^3 = 4\pi r^2\, dr$, where the terms in $(dr)^2$ and $(dr)^3$ are negligible. Therefore, for an s state $\Pr(r \rightarrow r + dr) = 4\pi r^2 |\psi_s|^2\, dr$.

For a non-s state, ψ depends on the angles, so $|\psi|^2$ is not constant in the thin shell. Let us divide the shell into tiny volume elements such that the spherical coordinates range from r to $r + dr$, from θ to $\theta + d\theta$, and from ϕ to $\phi + d\phi$ in each tiny element. The volume $d\tau$ of each such element is given by (18.27), and the probability that the electron is in an element is $|\psi|^2\, d\tau = |\psi|^2 r^2 \sin\theta\, dr\, d\theta\, d\phi$. To find $\Pr(r \rightarrow r + dr)$, we must sum these infinitesimal probabilities over the thin shell. Since the shell goes over the full range of θ and ϕ, the desired sum is the definite integral over the angles. Hence, $\Pr(r \rightarrow r + dr) = \int_0^{2\pi} \int_0^{\pi} |\psi|^2 r^2 \sin\theta\, dr\, d\theta\, d\phi$. The use of $\psi = R\Theta\Phi$ [Eq. (18.9)] and a result similar to that in Prob. 17.14 gives

$$\Pr(r \rightarrow r + dr) = |R|^2 r^2\, dr \int_0^{\pi} |\Theta|^2 \sin\theta\, d\theta \int_0^{2\pi} |\Phi|^2\, d\phi$$

$$\Pr(r \rightarrow r + dr) = [R_{nl}(r)]^2 r^2\, dr \qquad (18.28)$$

since the multiplicative constants in the Θ and Φ functions have been chosen to normalize Θ and Φ; $\int_0^{\pi} |\Theta|^2 \sin\theta\, d\theta = 1$ and $\int_0^{2\pi} |\Phi|^2\, d\phi = 1$ (Prob. 18.26). Equation (18.28) holds for both s and non-s states. The function $[R(r)]^2 r^2$ in (18.28) is the **radial distribution function** and is plotted in Fig. 18.8 for several states. For the ground state, the radial distribution function is a maximum at $r = a/Z$ (Prob. 18.24), which is 0.53 Å for H.

For the hydrogen-atom ground state, the probability density $|\psi|^2$ is a maximum at the origin (nucleus), but the radial distribution function $R^2 r^2$ is zero at the nucleus because of the r^2 factor; the most probable value of r is 0.53 Å. A little thought shows that these facts are not contradictory. In finding $\Pr(r \rightarrow r + dr)$, we find the probability that the electron is in a thin shell. This thin shell ranges over all values of θ and ϕ and so is composed of many volume elements. As r increases, the thin-shell volume $4\pi r^2\, dr$ increases. This increase, combined with the decrease in the probability density $|\psi|^2$ as r increases, gives a maximum in $\Pr(r \rightarrow r + dr)$ for a value of r between 0 and ∞. The radial distribution function is zero at the nucleus because the thin-shell volume $4\pi r^2\, dr$ is zero here. (Note the resemblance to the discussion of the distribution function for speeds in a gas; Sec. 14.4.)

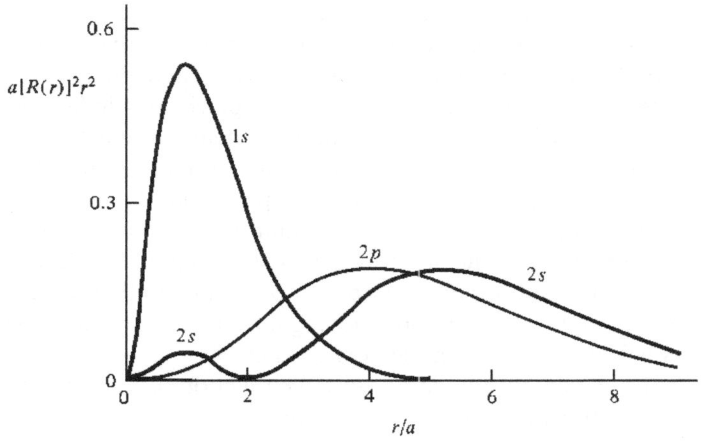

Figure 18.8

Radial distribution functions for some hydrogen-atom states.

Average Values

To find the average value of any property M of a stationary-state hydrogen atom, one uses $\langle M\rangle = \int \psi^* \hat{M}\, \psi\, d\tau$, Eq. (17.63).

EXAMPLE 18.3 Finding $\langle r\rangle$

Find the average value of the electron–nucleus separation in a hydrogenlike atom in the $2p_z$ state.

We have $\langle r\rangle = \int \psi^* \hat{r} \psi\, d\tau$. The $2p_z$ wave function is given in Table 18.1 and is real, so $\psi^* = \psi$. The operator $\hat{r}$ is multiplication by r. Thus, $\psi^* \hat{r} \psi = \psi^2 r$. The volume element is $d\tau = r^2 \sin\theta\, dr\, d\theta\, d\phi$ [Eq. (18.27)], and (18.8) gives the coordinate limits. Therefore

$$\langle r\rangle = \int \psi^* \hat{r} \psi\, d\tau$$

$$\langle r\rangle = \frac{1}{16(2\pi)}\left(\frac{Z}{a}\right)^5 \int_0^{2\pi}\int_0^{\pi}\int_0^{\infty} r^2 e^{-Zr/a} \cos^2\theta\, (r) r^2 \sin\theta\, dr\, d\theta\, d\phi$$

$$= \frac{1}{32\pi}\left(\frac{Z}{a}\right)^5 \int_0^{2\pi} d\phi \int_0^{\pi} \cos^2\theta \sin\theta\, d\theta \int_0^{\infty} r^5 e^{-Zr/a}\, dr$$

where the integral identity of Prob. 17.14 was used. A table of definite integrals gives $\int_0^\infty x^n e^{-bx}\, dx = n!/b^{n+1}$ for $b > 0$ and n a positive integer. Evaluation of the integrals (Prob. 18.23) gives $\langle r\rangle = 5a/Z$, where a is defined by (18.14) and equals 0.53 Å. As a partial check, note that $5a/Z$ has units of length.

Exercise

Find $\langle z^2\rangle$ for the $2p_z$ H-atom state. (*Answer:* $18a^2 = 5.05$ Å^2.)

18.4 ANGULAR MOMENTUM

The H-atom quantum numbers l and m are related to the angular momentum of the electron. The (linear) momentum $\mathbf{p}$ of a particle of mass m and velocity $\mathbf{v}$ is defined classically by $\mathbf{p} \equiv m\mathbf{v}$. Don't confuse the mass m with the m quantum number. Let $\mathbf{r}$ be the vector from the origin of a coordinate system to the particle. The particle's **angular momentum L** with respect to the coordinate origin is defined classically as a vector of length $rp \sin \beta$ (where β is the angle between **r** and **p**) and direction perpendicular to both **r** and **p**; see Fig. 18.9. More concisely, $\mathbf{L} \equiv \mathbf{r} \times \mathbf{p}$, where $\times$ indicates the vector cross-product.

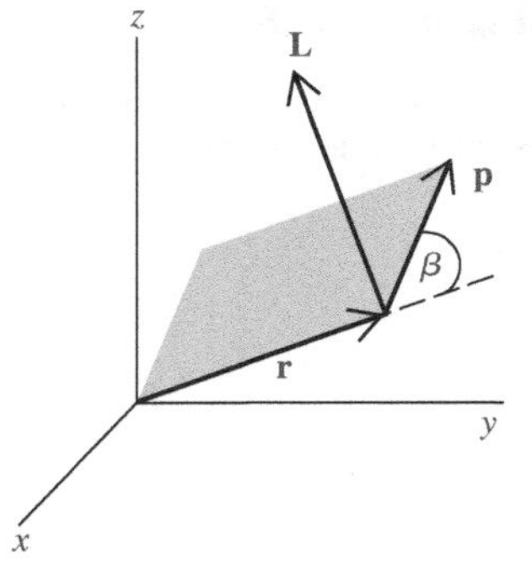

Figure 18.9

The angular-momentum vector **L** of a particle is perpendicular to the vectors **r** and **p** and has magnitude $rp \sin \beta$.

To deal with angular momentum in quantum mechanics, one uses the quantum-mechanical operators for the components of the **L** vector. We shall omit the quantum-mechanical treatment (see *Levine*, chaps. 5 and 6) and simply state the results. There are two kinds of angular momentum in quantum mechanics. **Orbital angular momentum** is the quantum-mechanical analog of the classical quantity **L** and is due to the motion of a particle through space. In addition to orbital angular momentum, many particles have an intrinsic angular momentum called *spin* angular momentum; this will be discussed in the next section.

The H-atom stationary-state wave functions $\psi_{nlm} = R_{nl}(r)\Theta_{lm}(\theta)\Phi_m(\phi)$ [Eq. (18.9)] are eigenfunctions of the energy operator $\hat{H}$ with eigenvalues given by Eq. (18.14); $\hat{H}\psi_{nlm} = E_n\psi_{nlm}$ [Eq. (17.61)], where $E_n = -(Z^2/n^2)(e^2/8\pi\varepsilon_0 a)$. This means that a measurement of the energy of an H atom in the state ψ_{nlm} must give the result E_n (see Sec. 17.11). One can show that the H-atom functions ψ_{nlm} are also eigenfunctions of the angular-momentum operators $\hat{L}^2$ and $\hat{L}_z$, where $\hat{L}^2$ is the operator for the square of the magnitude of the electron's orbital angular momentum **L** with respect to the nucleus, and $\hat{L}_z$ is the operator for the z component of **L.** The eigenvalues are $l(l + 1)\hbar^2$ for $\hat{L}^2$ and $m\hbar$ for $\hat{L}_z$:

$$\hat{L}^2\psi_{nlm} = l(l + 1)\hbar^2\psi_{nlm}, \qquad \hat{L}_z\psi_{nlm} = m\hbar\psi_{nlm} \tag{18.29}$$

where the quantum numbers l and m are given by (18.12) and (18.13). These eigenvalue equations mean that the magnitude $|\mathbf{L}|$ and the z component L_z of the electron's orbital angular momentum in the H-atom state ψ_{nlm} are

$$|\mathbf{L}| = \sqrt{l(l + 1)}\hbar, \qquad L_z = m\hbar \qquad \textbf{(18.30)*}$$

For s states ($l = 0$), the electronic orbital angular momentum is zero (a result hard to understand classically). For p states ($l = 1$), the magnitude of **L** is $\sqrt{2}\hbar$, and L_z can be $\hbar$, 0, or $-\hbar$. The possible orientations of **L** for $l = 1$ and $m = 1$, 0, and -1 are shown in Fig. 18.10. The quantum number l specifies the magnitude $|\mathbf{L}|$ of **L,** and m specifies the z component L_z of **L.** When $|\mathbf{L}|$ and L_z are specified in a quantum-mechanical system, it turns out that L_x and L_y cannot be specified, so **L** can lie anywhere on the surface of a cone about the z axis. For $m = 0$, the cone becomes a circle in the xy plane. Each of the three **L** vectors in Fig. 18.10 has length $|\mathbf{L}| = 2^{1/2}\,\hbar$; the z component of the $m = 1$ vector has length $m\hbar = \hbar$.

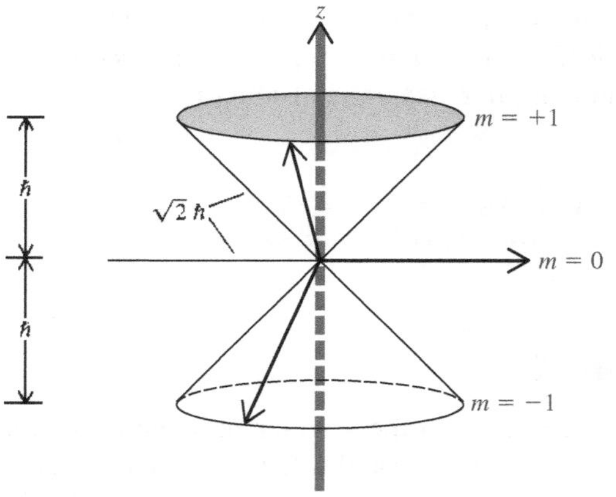

Figure 18.10

Allowed spatial orientations of the electronic orbital-angular-momentum vector **L** for $l = 1$ and $m = -1$, 0, and 1.

When an external magnetic field is applied to a hydrogen atom, the energies of the states depend on the m quantum number as well as on n.

The H-atom quantum numbers l and m are analogous to the two-particle-rigid-rotor quantum numbers J and M_J (Sec. 17.14). The functions Θ and Φ in the two-particle-rigid-rotor wave functions ψ_{rot} are the same functions as Θ and Φ in the H-atom wave functions (18.9).

The H-atom wave functions ψ_{nlm} are simultaneously eigenfunctions of the H-atom Hamiltonian operator $\hat{H}$ and of the angular-momentum operators $\hat{L}^2$ and $\hat{L}_z$ [Eq. (18.29)]. A theorem of quantum mechanics shows this is possible because the operators $\hat{H}$, $\hat{L}^2$, and $\hat{L}_z$ all **commute** with one another, meaning that the commutators (Example 17.4) $[\hat{H}, \hat{L}^2]$, $[\hat{H}, \hat{L}_z]$, and $[\hat{L}^2, \hat{L}_z]$ are all equal to zero. However, one finds that $[\hat{L}_z, \hat{L}_x] \neq 0$ and $[\hat{L}_z, \hat{L}_y] \neq 0$. Hence the ψ_{nlm} functions are not eigenfunctions of $\hat{L}_x$ and $\hat{L}_y$, and the quantities L_x and L_y cannot be specified for the states ψ_{nlm}. The $l = 0$ states are an exception. When $l = 0$, the orbital-angular-momentum magnitude $|\mathbf{L}|$ in (18.30) is zero and every component L_x, L_y, and L_z has the definite value of zero.

18.5 ELECTRON SPIN

The Schrödinger equation is a nonrelativistic equation and fails to account for certain relativistic phenomena. In 1928, the British physicist P. A. M. Dirac discovered the correct relativistic quantum-mechanical equation for a one-electron system. Dirac's relativistic equation predicts the existence of electron spin. Electron spin was first proposed by Uhlenbeck and Goudsmit in 1925 to explain certain observations in atomic spectra. In the nonrelativistic Schrödinger version of quantum mechanics that we are using, the existence of electron spin must be added to the theory as an additional postulate.

What is spin? **Spin** is an intrinsic (built-in) angular momentum possessed by elementary particles. This intrinsic angular momentum is in addition to the orbital angular momentum (Sec. 18.4) the particle has as a result of its motion through space. In a crude way, one can think of this intrinsic (or spin) angular momentum as being due to the particle's spinning about its own axis, but this picture should not be considered to represent reality. Spin is a nonclassical effect.

Quantum mechanics shows that the magnitude of the orbital angular momentum $\mathbf{L}$ of any particle can take on only the values $[l(l + 1)]^{1/2}\hbar$, where $l = 0, 1, 2, \ldots$; the z component L_z can take on only the values $m\hbar$, where $m = -l, \ldots, +l$. We mentioned this for the electron in the H atom, Eq. (18.30).

Let $\mathbf{S}$ be the spin-angular-momentum vector of an elementary particle. By analogy to orbital angular momentum [Eqs. (18.13) and (18.30)], we postulate that the magnitude of $\mathbf{S}$ is

$$|\mathbf{S}| = [s(s + 1)]^{1/2}\hbar \qquad \textbf{(18.31)*}$$

and that S_z, the component of the spin angular momentum along the z axis, can take on only the values

$$S_z = m_s\hbar \qquad \text{where } m_s = -s, -s + 1, \ldots, s - 1, s \qquad \textbf{(18.32)*}$$

The spin-angular-momentum quantum numbers s and m_s are analogous to the orbital-angular-momentum quantum numbers l and m, respectively. The analogy is not complete, since one finds that a given kind of elementary particle can have only one value for s and this value may be half-integral ($\frac{1}{2}, \frac{3}{2}, \ldots$) as well as integral (0, 1, ...). Experiment shows that electrons, protons, and neutrons all have $s = \frac{1}{2}$. Therefore, $m_s = -\frac{1}{2}$ or $+\frac{1}{2}$ for these particles.

$$s = \tfrac{1}{2}, \qquad m_s = +\tfrac{1}{2}, -\tfrac{1}{2} \qquad \text{for an electron} \qquad \textbf{(18.33)*}$$

With $s = \frac{1}{2}$, the magnitude of the electron spin-angular-momentum vector is $|\mathbf{S}| = [s(s+1)]^{1/2}\hbar = (3/4)^{1/2}\hbar$ and the possible values of S_z are $\frac{1}{2}\hbar$ and $-\frac{1}{2}\hbar$. Figure 18.11 shows the orientations of **S** for these two spin states. Chemists often use the symbols ↑ and ↓ to indicate the $m_s = +\frac{1}{2}$ and $m_s = -\frac{1}{2}$ states, respectively.

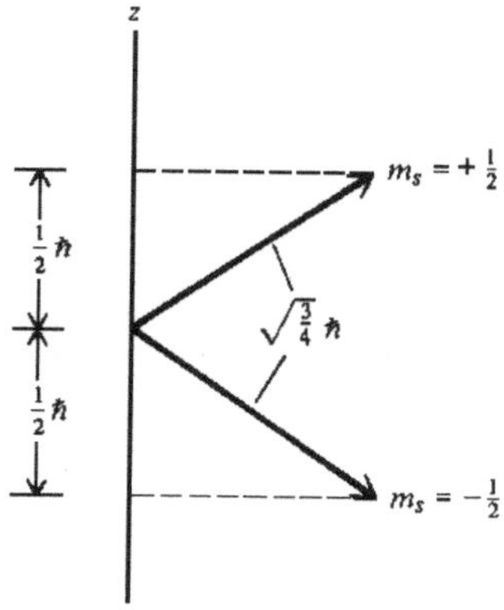

Figure 18.11

Orientations of the electron spin vector **S** with respect to the z axis. For $m_s = +\frac{1}{2}$, the vector **S** must lie on the surface of a cone about the z axis; similarly for $m_s = -\frac{1}{2}$.

Photons have $s = 1$. However, because photons are relativistic entities traveling at speed c, it turns out that they don't obey (18.32). Instead, photons can have only $m_s = +1$ or $m_s = -1$. These two m_s values correspond to left- and right-circularly polarized light.

The wave function is supposed to describe the state of the system as fully as possible. An electron has two possible spin states, namely, $m_s = +\frac{1}{2}$ and $m_s = -\frac{1}{2}$, and the wave function should indicate which spin state the electron is in. We therefore postulate the existence of two spin functions α and β that indicate the electron's spin state: α means that m_s is $+\frac{1}{2}$; β means that m_s is $-\frac{1}{2}$. The spin functions α and β can be considered to be functions of some hypothetical internal coordinate ω (omega) of the electron: $\alpha = \alpha(\omega)$ and $\beta = \beta(\omega)$. Since nothing is known of the internal structure of an electron (or even whether it has an internal structure), ω is purely hypothetical.

Since the spin function α has $m_s = \frac{1}{2}$ and β has $m_s = -\frac{1}{2}$ and both have $s = \frac{1}{2}$, by analogy to (18.29), we write

$$\hat{S}^2\alpha = \tfrac{3}{4}\hbar^2\alpha, \qquad \hat{S}^2\beta = \tfrac{3}{4}\hbar^2\beta, \qquad \hat{S}_z\alpha = \tfrac{1}{2}\hbar\alpha, \qquad \hat{S}_z\beta = -\tfrac{1}{2}\hbar\beta \tag{18.34}$$

where these equations are purely symbolic, in that we have not specified forms for the spin functions α and β or for the operators $\hat{S}^2$ and $\hat{S}_z$.

For a one-electron system, the spatial wave function $\psi(x, y, z)$ is multiplied by either α or β to form the complete wave function including spin. To a very good approximation, the spin has no effect on the energy of a one-electron system. For the hydrogen atom, the electron spin simply doubles the degeneracy of each level. For the H-atom ground level, there are two possible wave functions, $1s\alpha$ and $1s\beta$, where $1s = \pi^{-1/2}(Z/a)^{3/2}e^{-Zr/a}$. A one-electron wave function like $1s\alpha$ or $1s\beta$ that includes both spatial and spin functions is called a **spin-orbital.**

With inclusion of electron spin in the wave function, ψ of an n-electron system becomes a function of $4n$ variables: $3n$ spatial coordinates and n **spin variables** or **spin coordinates.** The normalization condition (17.17) must be modified to include an integration or summation over the spin variables as well as an integration over the spatial coordinates. If one uses the hypothetical spin coordinate ω, one integrates over ω. A common alternative is to take the spin quantum number m_s of each electron as being the spin variable of that electron. In this case, one sums over the two possible m_s values of each electron in the normalization equation. Such sums or integrals over the spin variables are, like (18.34), purely symbolic.

18.6 THE HELIUM ATOM AND THE SPIN–STATISTICS THEOREM

The Helium Atom

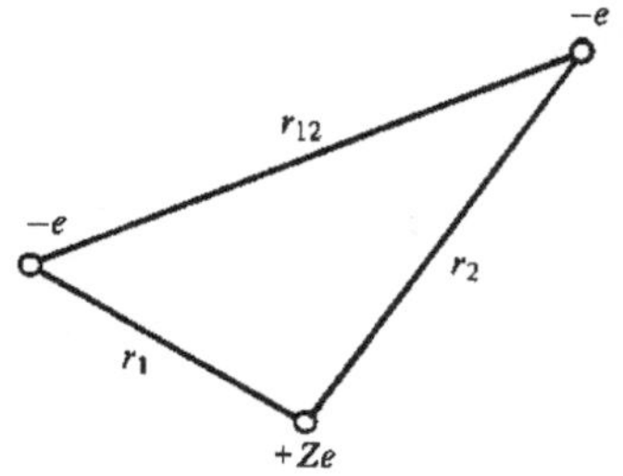

Figure 18.12

Interparticle distances in the heliumlike atom.

The helium atom consists of two electrons and a nucleus (Fig. 18.12). Separation of the translational energy of the atom as a whole from the internal motion is more complicated than for a two-particle problem and won't be gone into here. We shall just assume that it is possible to separate the translational motion from the internal motions.

The Hamiltonian operator for the internal motions in a heliumlike atom is

$$\hat{H} = -\frac{\hbar^2}{2m_e}\nabla_1^2 - \frac{\hbar^2}{2m_e}\nabla_2^2 - \frac{Ze^2}{4\pi\varepsilon_0 r_1} - \frac{Ze^2}{4\pi\varepsilon_0 r_2} + \frac{e^2}{4\pi\varepsilon_0 r_{12}} \tag{18.35}$$

The first term is the operator for the kinetic energy of electron 1. In this term, $\nabla_1^2 \equiv \partial^2/\partial x_1^2 + \partial^2/\partial y_1^2 + \partial^2/\partial z_1^2$ (where x_1, y_1, z_1 are the coordinates of electron 1, the origin being taken at the nucleus) and m_e is the electron mass. It would be more accurate to replace m_e by the reduced mass μ, but μ differs almost negligibly from m_e for He and heavier atoms. The second term is the operator for the kinetic energy of electron 2, and $\nabla_2^2 \equiv \partial^2/\partial x_2^2 + \partial^2/\partial y_2^2 + \partial^2/\partial z_2^2$. The third term is the potential energy of interaction between electron 1 and the nucleus and is obtained by putting $Q_1 = -e$ and $Q_2 = Ze$ in $V = Q_1Q_2/4\pi\varepsilon_0 r$ [Eq. (18.1)]. For helium, the atomic number Z is 2. In the third term, r_1 is the distance between electron 1 and the nucleus: $r_1^2 = x_1^2 + y_1^2 + z_1^2$. The fourth term is the potential energy of interaction between electron 2 and the nucleus. The last term is the potential energy of interaction between electrons 1 and 2 separated by distance r_{12} and is found by putting $Q_1 = Q_2 = -e$ in $V = Q_1Q_2/4\pi\varepsilon_0 r$. There is no term for kinetic energy of the nucleus, because we are considering only the internal motion of the electrons relative to the nucleus.

The Schrödinger equation is $\hat{H}\psi = E\psi$, where ψ is a function of the spatial coordinates of the electrons relative to the nucleus: $\psi = \psi(x_1, y_1, z_1, x_2, y_2, z_2)$, or $\psi = \psi(r_1, \theta_1, \phi_1, r_2, \theta_2, \phi_2)$ if spherical coordinates are used. Electron spin is being ignored for now and will be taken care of later.

Because of the interelectronic repulsion term $e^2/4\pi\varepsilon_0 r_{12}$, the helium-atom Schrödinger equation can't be solved exactly. As a crude approximation, we can ignore the $e^2/4\pi\varepsilon_0 r_{12}$ term. The Hamiltonian (18.35) then has the approximate form $\hat{H}_{\text{approx}} = \hat{H}_1 + \hat{H}_2$, where $\hat{H}_1 \equiv -(\hbar^2/2m_e)\nabla_1^2 - Ze^2/4\pi\varepsilon_0 r_1$ is a hydrogenlike Hamiltonian for electron 1 and $\hat{H}_2 \equiv -(\hbar^2/2m_e)\nabla_2^2 - Ze^2/4\pi\varepsilon_0 r_2$ is a hydrogenlike Hamiltonian for electron 2. Since $\hat{H}_{\text{approx}}$ is the sum of Hamiltonians for two noninteracting particles, the approximate energy is the sum of energies of each particle and the approximate wave function is the product of wave functions for each particle [Eqs. (17.68) to (17.70)]:

$$E \approx E_1 + E_2 \quad \text{and} \quad \psi \approx \psi_1(r_1, \theta_1, \phi_1)\psi_2(r_2, \theta_2, \phi_2) \tag{18.36}$$

where $\hat{H}_1\psi_1 = E_1\psi_1$ and $\hat{H}_2\psi_2 = E_2\psi_2$. Since $\hat{H}_1$ and $\hat{H}_2$ are hydrogenlike Hamiltonians, E_1 and E_2 are hydrogenlike energies and ψ_1 and ψ_2 are hydrogenlike wave functions (orbitals).

Let us check the accuracy of this approximation. Equations (18.14) and (18.18) give $E_1 = -(Z^2/n_1^2)(e^2/8\pi\varepsilon_0 a) = -(Z^2/n_1^2)(13.6 \text{ eV})$, where n_1 is the principal quantum number of electron 1 and a has been replaced by the Bohr radius a_0, since the reduced mass μ was replaced by the electron mass in (18.35). A similar equation holds for E_2. For the helium-atom ground state, the principal quantum numbers of the electrons are $n_1 = 1$ and $n_2 = 1$; also, $Z = 2$. Hence,

$$E \approx E_1 + E_2 = -4(13.6 \text{ eV}) - 4(13.6 \text{ eV}) = -108.8 \text{ eV}$$

The experimental first and second ionization energies of He are 24.6 eV and 54.4 eV, so the true ground-state energy is -79.0 eV. (The first and second ionization energies are the energy changes for the processes $\text{He} \rightarrow \text{He}^+ + e^-$ and $\text{He}^+ \rightarrow \text{He}^{2+} + e^-$, respectively.) The approximate result -108.8 eV is grossly in error, as might be expected from the fact that the $e^2/4\pi\varepsilon_0 r_{12}$ term we ignored is not small.

The approximate ground-state wave function is given by Eq. (17.68) and Table 18.1 as

$$\psi \approx (Z/a_0)^{3/2}\pi^{-1/2}e^{-Zr_1/a_0} \cdot (Z/a_0)^{3/2}\pi^{-1/2}e^{-Zr_2/a_0} \tag{18.37}$$

with $Z = 2$. We shall abbreviate (18.37) as

$$\psi \approx 1s(1)1s(2) \tag{18.38}$$

where $1s(1)$ indicates that electron 1 is in a $1s$ hydrogenlike orbital (one-electron spatial wave function). We have the familiar He ground-state configuration $1s^2$.

Two-Electron Spin Functions

To be fully correct, electron spin must be included in the wave function. One's first impulse might be to write down the following four spin functions for two-electron systems:

$$\alpha(1)\alpha(2), \qquad \beta(1)\beta(2), \qquad \alpha(1)\beta(2), \qquad \beta(1)\alpha(2) \tag{18.39}$$

where the notation $\beta(1)\alpha(2)$ means electron 1 has its spin quantum number m_{s1} equal to $-\frac{1}{2}$ and electron 2 has $m_{s2} = +\frac{1}{2}$. However, the last two functions in (18.39) are not valid spin functions because they distinguish between the electrons. Electrons are identical to one another, and there is no way of experimentally determining which electron has $m_s = +\frac{1}{2}$ and which has $m_s = -\frac{1}{2}$. In classical mechanics, we can distinguish two identical particles from each other by following their paths. However, the Heisenberg uncertainty principle makes it impossible to follow the path of a particle in quantum mechanics. Therefore, the wave function must not distinguish between the electrons. Thus, the fourth spin function in (18.39), which says that electron 1 has spin β and electron 2 has spin α, cannot be used. Instead of the third and fourth spin functions in (18.39), it turns out (see below for the justification) that one must use the functions $2^{-1/2}[\alpha(1)\beta(2) - \beta(1)\alpha(2)]$ and $2^{-1/2}[\alpha(1)\beta(2) + \beta(1)\alpha(2)]$. For each of these functions, electron 1 has both spin α and spin β, and so does electron 2. The $2^{-1/2}$ in these functions is a normalization constant.

The proper two-electron spin functions are therefore

$$\alpha(1)\alpha(2), \qquad \beta(1)\beta(2), \qquad 2^{-1/2}[\alpha(1)\beta(2) + \beta(1)\alpha(2)] \tag{18.40*}$$

$$2^{-1/2}[\alpha(1)\beta(2) - \beta(1)\alpha(2)] \tag{18.41*}$$

The three spin functions in (18.40) are unchanged when electrons 1 and 2 are interchanged. For example, interchanging the electrons in the third function gives $2^{-1/2}[\alpha(2)\beta(1) + \beta(2)\alpha(1)]$, which equals the original function. These three spin functions are said to be **symmetric** with respect to electron interchange. The spin function (18.41) is multiplied by -1 when the electrons are interchanged, since interchange gives

$$2^{-1/2}[\alpha(2)\beta(1) - \beta(2)\alpha(1)] = -2^{-1/2}[\alpha(1)\beta(2) - \beta(1)\alpha(2)]$$

The function (18.41) is **antisymmetric,** meaning that interchange of the coordinates of two particles multiplies the function by -1.

The Spin–Statistics Theorem

Since two identical particles cannot be distinguished from each other in quantum mechanics, interchange of two identical particles in the wave function must leave all physically observable properties unchanged. In particular, the probability density $|\psi|^2$ must be unchanged. We therefore expect that ψ itself would be multiplied by either $+1$ or -1 by such an interchange or relabeling. It turns out that only one of these possibilities occurs, depending on the spin of the particles. A particle whose spin quantum number s is an integer ($s = 0$ or 1 or 2 or . . .) is said to have **integral spin,** whereas a particle with $s = \frac{1}{2}$ or $\frac{3}{2}$ or $\frac{5}{2}$ or . . . has **half-integral spin.** Experimental evidence shows the validity of the following statement:

The complete wave function (including both spatial and spin coordinates) of a system of identical particles with half-integral spin must be antisymmetric with respect to interchange of all the coordinates (spatial and spin) of any two particles. For a system of identical particles with integral spin, the complete wave function must be symmetric with respect to such interchange.

This fact is called the **spin–statistics theorem.** (The word statistics is used because the symmetry or antisymmetry requirement of the wave function of identical particles leads to different results for integral-spin particles versus half-integral-spin particles as to how many particles can occupy a given state and this affects the statistical mechanics of systems of such particles; see Secs. 18.8 and 21.5, and Prob. 21.22.) In 1940, Pauli proved the spin–statistics theorem using relativistic quantum field theory. In the nonrelativistic quantum mechanics that we are using, the spin–statistics theorem must be regarded as an additional postulate.

Particles requiring antisymmetric wave functions and having half-integral spin are called **fermions** (after the Italian–American physicist Enrico Fermi). Particles requiring symmetric wave functions and having integral spin are called **bosons** (after the Indian physicist S. N. Bose). Electrons have $s = \frac{1}{2}$ and are fermions.

We are now ready to include spin in the ground-state He wave function. The approximate ground-state spatial function $1s(1)1s(2)$ of (18.38) is symmetric with respect to electron interchange, since $1s(2)1s(1) = 1s(1)1s(2)$. Since electrons have $s = \frac{1}{2}$, the spin–statistics theorem demands that the complete wave function be antisymmetric. To get an antisymmetric ψ, we must multiply $1s(1)1s(2)$ by the antisymmetric spin function (18.41). Use of the symmetric spin functions in (18.40) would give a symmetric wave function, which is forbidden for fermions. With inclusion of spin, the approximate ground-state He wave function becomes

$$\psi \approx 1s(1)1s(2) \cdot 2^{-1/2}[\alpha(1)\beta(2) - \beta(1)\alpha(2)] \tag{18.42}$$

Interchange of the electrons multiplies ψ by -1, so (18.42) is antisymmetric. Note that the two electrons in the $1s$ orbital have opposite spins.

The wave function (18.42) can be written as the determinant

$$\psi \approx \frac{1}{\sqrt{2}} \begin{vmatrix} 1s(1)\alpha(1) & 1s(1)\beta(1) \\ 1s(2)\alpha(2) & 1s(2)\beta(2) \end{vmatrix} \tag{18.43}$$

A second-order determinant is defined by

$$\begin{vmatrix} a & b \\ c & d \end{vmatrix} \equiv ad - bc \qquad \textbf{(18.44)*}$$

The use of (18.44) in (18.43) gives (18.42).

The justification for replacing the third and fourth spin functions in (18.39) by the linear combinations in (18.40) and (18.41) is that the latter two functions are the only normalized linear combinations of $\alpha(1)\beta(2)$ and $\beta(1)\alpha(2)$ that are either symmetric or antisymmetric with respect to electron interchange and that therefore do not distinguish between the electrons.

Improved Ground-State Wave Functions for Helium

For one- and two-electron systems, the wave function is a product of a spatial factor and a spin factor. The atomic Hamiltonian (to a very good approximation) contains no terms involving spin. Because of these facts, the spin part of the wave function need not be explicitly included in calculating the energy of one- and two-electron systems and will be omitted in the calculations in this section.

We saw above that ignoring the $e^2/4\pi\varepsilon_0 r_{12}$ term in $\hat{H}$ and taking E as the sum of two hydrogenlike energies gave a 38% error in the ground-state He energy. To improve on this dismal result, we can use the variation method. The most obvious choice of variational function is the $1s(1)1s(2)$ function of (18.37) and (18.38), which is a normalized product of hydrogenlike $1s$ orbitals. The variational integral in (17.84) is then $W = \int 1s(1)1s(2)\hat{H}1s(1)1s(2)\, d\tau$, where $\hat{H}$ is the true Hamiltonian (18.35) and

$d\tau = d\tau_1\, d\tau_2$, with $d\tau_1 = r_1^2 \sin\theta_1\, dr_1\, d\theta_1\, d\phi_1$. Since $e^2/4\pi\varepsilon_0 r_{12}$ is part of $\hat{H}$, the effect of the interelectronic repulsion will be included in an average way, rather than being ignored as it was in (18.36). Evaluation of the variational integral is complicated and is omitted here. The result is $W = -74.8$ eV, which is reasonably close to the true ground-state energy -79.0 eV.

A further improvement is to use a variational function having the same form as (18.37) and (18.38) but with the nuclear charge Z replaced by a variational parameter ζ (zeta). We then vary ζ to minimize the variational integral $W = \int \phi^* \hat{H} \phi\, d\tau$, where the normalized variation function ϕ is $(\zeta/a_0)^3 \pi^{-1} e^{-\zeta r_1/a_0} e^{-\zeta r_2/a_0}$. Substitution of the He Hamiltonian (18.35) with $Z = 2$ and evaluation of the integrals leads to

$$W = (\zeta^2 - 27\zeta/8)e^2/4\pi\varepsilon_0 a$$

(See *Levine,* sec. 9.4, for the details.) The minimization condition $\partial W/\partial\zeta = 0$ then gives $0 = (2\zeta - 27/8)e^2/4\pi\varepsilon_0 a$ and the optimum value of ζ is $27/16 = 1.6875$. This value of ζ gives $W = -2.848(e^2/4\pi\varepsilon_0 a) = -2.848(2 \times 13.6 \text{ eV}) = -77.5$ eV, where (18.17) was used. This result is only 2% above the true ground-state energy -79.0 eV.

The parameter ζ is called an *orbital exponent.* The fact that ζ is less than the atomic number $Z = 2$ can be attributed to the **shielding** or **screening** of one electron from the nucleus by the other electron. When electron 1 is between electron 2 and the nucleus, the repulsion between electrons 1 and 2 subtracts from the attraction between electron 2 and the nucleus. Thus, ζ can be viewed as the "effective" nuclear charge for the $1s$ electrons. Since both electrons are in the same orbital, the screening effect is not great and ζ is only 0.31 less than Z.

By using complicated variational functions, workers have obtained agreement to 1 part in 2 million between the theoretical and the experimental ionization energies of ground-state He. [C. L. Pekeris, *Phys. Rev.*, **115,** 1216 (1959); C. Schwartz, *Phys. Rev.*, **128,** 1146 (1962).]

Excited-State Wave Functions for Helium

We saw that the approximation of ignoring the interelectronic repulsion in the Hamiltonian gives the helium wave functions as products of two hydrogenlike functions [Eq. (18.36)]. The hydrogenlike $2s$ and $2p$ orbitals have the same energy, and we might expect the approximate spatial wave functions for the lowest excited energy level of He to be $1s(1)2s(2)$, $1s(2)2s(1)$, $1s(1)2p_x(2)$, $1s(2)2p_x(1)$, $1s(1)2p_y(2)$, $1s(2)2p_y(1)$, $1s(1)2p_z(2)$, and $1s(2)2p_z(1)$, where $1s(2)2p_z(1)$ is a function with electron 2 in the $1s$ orbital and electron 1 in the $2p_z$ orbital. Actually, these functions are incorrect, in that they distinguish between the electrons. As we did above with the spin functions, we must take linear combinations to give functions that don't distinguish between the electrons. Analogous to the linear combinations in (18.40) and (18.41), the correct normalized approximate spatial functions are

$$2^{-1/2}[1s(1)2s(2) + 1s(2)2s(1)] \qquad (18.45)$$

$$2^{-1/2}[1s(1)2s(2) - 1s(2)2s(1)] \qquad (18.46)$$

$$2^{-1/2}[1s(1)2p_x(2) + 1s(2)2p_x(1)] \quad \text{etc.} \qquad (18.47)$$

$$2^{-1/2}[1s(1)2p_x(2) - 1s(2)2p_x(1)] \quad \text{etc.} \qquad (18.48)$$

where each "etc." indicates two similar functions with $2p_x$ replaced by $2p_y$ or $2p_z$.

If ψ_k is the true wave function of state k of a system, then $\hat{H}\psi_k = E_k\psi_k$, where E_k is the energy of state k. We therefore have $\int \psi_k^* \hat{H}\psi_k\, d\tau = \int \psi_k^* E_k \psi_k\, d\tau = E_k \int \psi_k^* \psi_k\, d\tau = E_k$, since ψ_k is normalized. This result suggests that if we have an approximate wave

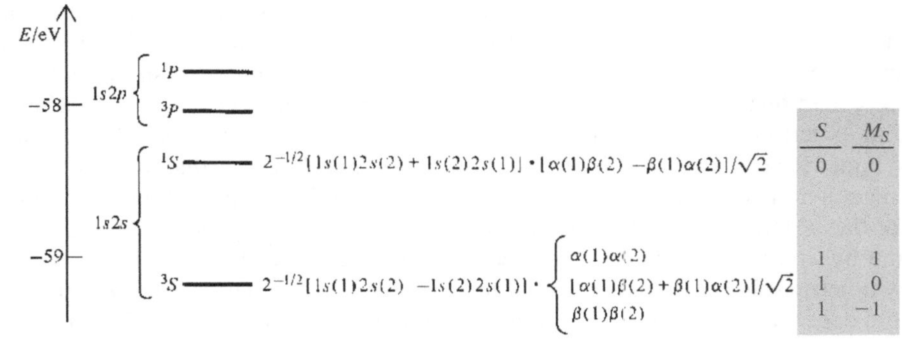

Figure 18.13

Energies of the terms arising from the helium-atom $1s2s$ and $1s2p$ electron configurations.

function $\psi_{k,\text{approx}}$ for state k, an approximate energy can be obtained by replacing ψ_k with $\psi_{k,\text{approx}}$ in the integral:

$$E_k \approx \int \psi^*_{k,\text{approx}} \hat{H} \psi^*_{k,\text{approx}} \, d\tau \tag{18.49}$$

where $\hat{H}$ is the true Hamiltonian, including the interelectronic repulsion term(s).

Use of the eight approximate functions (18.45) to (18.48) in Eq. (18.49) then gives approximate energies for these states. Because of the difference in signs, the two states (18.45) and (18.46) that arise from the $1s2s$ configuration will clearly have different energies. The state (18.46) turns out to be lower in energy. The approximate wave functions are real, and the contribution of the $e^2/4\pi\varepsilon_0 r_{12}$ interelectronic repulsion term in $\hat{H}$ to the integral in Eq. (18.49) is $\int \psi^2_{k,\text{approx}} e^2/4\pi\varepsilon_0 r_{12}\, d\tau$. One finds that the integral $\int (18.45)^2 e^2/4\pi\varepsilon_0 r_{12}\, d\tau$ differs in value from $\int (18.47)^2 e^2/4\pi\varepsilon_0 r_{12}\, d\tau$, where "(18.45)" and "(18.47)" stand for the functions in Eqs. (18.45) and (18.47). Hence, the $1s2p$ state in (18.47) differs in energy from the corresponding $1s2s$ state in (18.45). Likewise, the states (18.46) and (18.48) differ in energy from each other. Since the $2p_x$, $2p_y$, and $2p_z$ orbitals have the same shape, changing $2p_x$ to $2p_y$ or $2p_z$ in (18.47) or (18.48) doesn't affect the energy.

Thus, the states of the $1s2s$ configuration give two different energies and the states of the $1s2p$ configuration give two different energies, for a total of four different energies (Fig. 18.13). The $1s2s$ states turn out to lie lower in energy than the $1s2p$ states. Although the $2s$ and $2p$ orbitals have the same energy in one-electron (hydrogenlike) atoms, the interelectronic repulsions in atoms with two or more electrons remove the $2s$-$2p$ degeneracy. The reason the $2s$ orbital lies below the $2p$ orbital can be seen from Figs. 18.8 and 18.3. The $2s$ orbital has more probability density near the nucleus than the $2p$ orbital. Thus, a $2s$ electron is more likely than a $2p$ electron to penetrate within the probability density of the $1s$ electron. When it penetrates, it is no longer shielded from the nucleus and feels the full nuclear charge and its energy is thereby lowered. Similar penetration effects remove the l degeneracy in higher orbitals. For example, $3s$ lies lower than $3p$, which lies lower than $3d$ in atoms with more than one electron.

What about electron spin? The function (18.45) is symmetric with respect to electron interchange and so must be combined with the antisymmetric two-electron spin function (18.41) to give an overall ψ that is antisymmetric: $\psi \approx (18.45) \times (18.41)$. The function (18.46) is antisymmetric and so must be combined with one of the symmetric spin functions in (18.40). Because there are three symmetric spin functions, inclusion of spin in (18.46) gives three different wave functions, each having the same spatial factor. Since the spin factor doesn't affect the energy, there is a threefold spin degeneracy associated with the function (18.46). Similar considerations hold for the $1s2p$ states.

Figure 18.13 shows the energies and some of the approximate wave functions for the states arising from the $1s2s$ and $1s2p$ configurations. The labels ^3S, ^1S, ^3P, and ^1P and the S and M_S values are explained shortly. The atomic energies shown in Fig. 18.13 are called **terms,** rather than energy levels, for a reason to be explained in Sec. 18.7. The ^3S term of the $1s2s$ configuration is threefold degenerate, because of the three symmetric spin functions. The ^1S term is onefold degenerate (that is, nondegenerate), since there is only one wave function for this term. The approximate wave functions for the ^3P term are obtained from those of the ^3S term by replacing $2s$ by $2p_x$, by $2p_y$, and by $2p_z$. Each of these replacements gives three wave functions (due to the three symmetric spin functions), so the ^3P term is ninefold degenerate. The ^1P term is threefold degenerate, since three functions are obtained by replacement of $2s$ in the ^1S function with $2p_x$, with $2p_y$, and with $2p_z$.

Note that the helium wave functions (18.42) and (18.45) to (18.48) are only approximations, since at best they take account of the interelectronic repulsion in only an average way. Thus, to say that the helium ground state has the electron configuration $1s^2$ is only approximately true. The use of orbitals (one-electron wave functions) in many-electron atoms is only an approximation.

18.7 TOTAL ORBITAL AND SPIN ANGULAR MOMENTA

The magnitude of the orbital angular momentum of the electron in a one-electron atom is given by (18.30) as $[l(l+1)]^{1/2}\hbar$, where $l = 0, 1, 2, \ldots$. For an atom with more than one electron, the orbital angular momentum vectors $\mathbf{L}_i$ of the individual electrons add to give a **total electronic orbital angular momentum L,** given by $\mathbf{L} = \Sigma_i \mathbf{L}_i$. Quantum mechanics shows that the magnitude of **L** is given by

$$|\mathbf{L}| = [L(L+1)]^{1/2}\hbar, \qquad \text{where } L = 0, 1, 2, \ldots \tag{18.50}$$

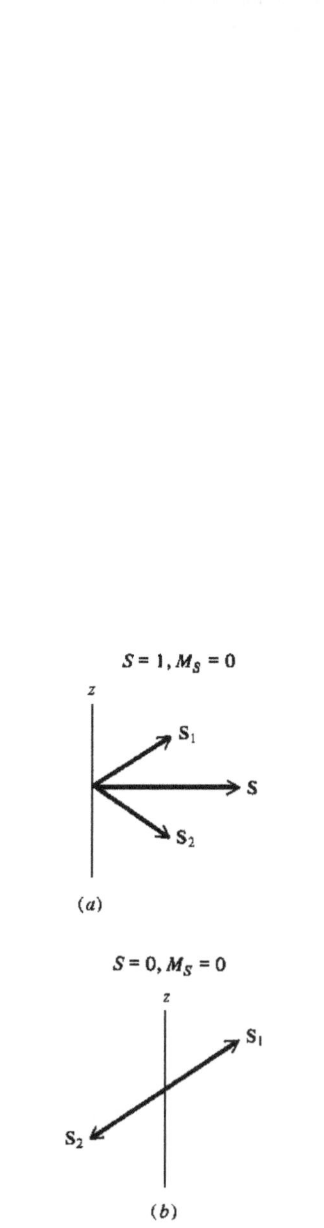

Figure 18.14

Spin orientations corresponding to the spin functions (*a*) $2^{-1/2}[\alpha(1)\beta(2) + \beta(1)\alpha(2)]$ and (*b*) $2^{-1/2}[\alpha(1)\beta(2) - \beta(1)\alpha(2)]$. **S** is the total electronic spin angular momentum.

The value of the total electronic orbital-angular-momentum quantum number L is indicated by a code letter similar to (18.15), except that capital letters are used:

L value	0	1	2	3	4	5
Code letter	S	P	D	F	G	H

For the $1s2s$ configuration of the He atom, both electrons have $l = 0$. Hence, the total orbital angular momentum is zero, and the code letter S is used for each of the two terms that arise from the $1s2s$ configuration (Fig. 18.13). For the $1s2p$ configuration, one electron has $l = 0$ and one has $l = 1$. Hence, the total-orbital-angular-momentum quantum number L equals 1, and the code letter P is used.

The **total electronic spin angular momentum S** of an atom (or molecule) is the vector sum of the spin angular momenta of the individual electrons: $\mathbf{S} = \Sigma_i \mathbf{S}_i$. The magnitude of **S** has the possible values $[S(S+1)]^{1/2}\hbar$, where the total electronic spin quantum number S can be $0, \frac{1}{2}, 1, \frac{3}{2}, \ldots$. (Don't confuse the spin quantum number S with the orbital-angular-momentum code letter S.) The component of **S** along the z axis has the possible values $M_S\hbar$, where $M_S = -S, -S+1, \ldots, S-1, S$.

For a two-electron system such as He, each electron has spin quantum number $s = \frac{1}{2}$, and the total spin quantum number S can be 0 or 1, depending on whether the two electron spin vectors point in opposite directions or in approximately the same direction. (See Prob. 18.46.) For $S = 1$, the total spin angular momentum is $[S(S+1)]^{1/2}\hbar = (1 \cdot 2)^{1/2}\hbar = 1.414\hbar$. The spin angular momentum of each electron is $(\frac{1}{2} \cdot \frac{3}{2})^{1/2}\hbar = 0.866\hbar$. The algebraic sum of the spin angular momenta of the individual electrons is $0.866\hbar + 0.866\hbar = 1.732\hbar$, which is greater than the magnitude of the total spin angular momentum. Hence, the two spin-angular-momentum vectors of the electrons cannot be exactly parallel; see, for example, Fig. 18.14*a*.

For spin quantum number $S = 0$, M_S must be zero, and there is only one possible spin state. This spin state corresponds to the antisymmetric spin function (18.41).

For $S = 1$, M_S can be -1, 0, or $+1$. The $M_S = -1$ spin state arises when each electron has $m_s = -\frac{1}{2}$ and so corresponds to the symmetric spin function $\beta(1)\beta(2)$ in (18.40). The $M_S = +1$ spin state corresponds to the function $\alpha(1)\alpha(2)$. For the $M_S = 0$ state, one electron must have $m_s = +\frac{1}{2}$ and the other $m_s = -\frac{1}{2}$. This is the function $2^{-1/2}[\alpha(1)\beta(2) + \beta(1)\alpha(2)]$ in (18.40). Although the z components of the two electron spins are in opposite directions, the two spin vectors can still add to give a total electronic spin with $S = 1$, as shown in Fig. 18.14*a*. The three symmetric spin functions in (18.40) thus correspond to $S = 1$.

The quantity $2S + 1$ (where S is the total spin quantum number) is called the **spin multiplicity** of an atomic term and is written as a left superscript to the code letter for L. The lowest term in Fig. 18.13 has spin wave functions that correspond to total spin quantum number $S = 1$. Hence, $2S + 1$ equals 3 for this term, and the term is designated ^3S (read as "triplet ess"). The second-lowest term in Fig. 18.13 has the $S = 0$ spin function and so has $2S + 1 = 1$. This is a ^1S ("singlet ess") term. The S and M_S total spin quantum numbers are listed in Fig. 18.13 for each state (wave function) of the ^1S and ^3S terms.

Note that the triplet term of the $1s2s$ configuration lies lower than the singlet term. The same is true for the terms of the $1s2p$ configuration. This illustrates **Hund's rule:** *For a set of terms arising from a given electron configuration, the lowest-lying term is generally the one with the maximum spin multiplicity.* There are several exceptions to Hund's rule. The theoretical basis for Hund's rule is discussed in R. L. Snow and J. L. Bills, *J. Chem. Educ.,* **51,** 585 (1974); I. Shim and J. P. Dahl, *Theor. Chim. Acta,* **48,** 165 (1978); J. W. Warner and R. S. Berry, *Nature*, **313,** 160 (1985).

The maximum spin multiplicity is produced by having the maximum number of electrons with parallel spins. Two electrons are said to have **parallel** spins when their spin-angular-momentum vectors point in approximately the same direction, as, for example, in Fig. 18.14*a*. Two electrons have **antiparallel** spins when their spin vectors point in opposite directions to give a net spin angular momentum of zero, as in Fig. 18.14*b*. The ^3S and ^1S terms of the $1s2s$ configuration can be represented by the diagrams

$$^3S: \quad \underset{1s}{\underline{\uparrow}} \; \underset{2s}{\underline{\uparrow}} \qquad\qquad ^1S: \quad \underset{1s}{\underline{\uparrow}} \; \underset{2s}{\underline{\downarrow}}$$

where the spins are parallel in ^3S and antiparallel in ^1S.

Electrons in a filled subshell (for example, the electrons in $2p^6$) have all their spins paired and contribute zero to the total electronic spin angular momentum. For each electron in a closed subshell with a positive value for the m quantum number, there is an electron with the corresponding negative value of m. (For example, in $2p^6$, there are two electrons with $m = +1$ and two with $m = -1$.) Therefore, electrons in a filled subshell contribute zero to the total electronic orbital angular momentum. Hence, electrons in closed subshells can be ignored when finding the possible values of the quantum numbers L and S for the total orbital and spin angular momenta. For example, the $1s^22s^22p^63s3p$ electron configuration of Mg gives rise to the same terms as the $1s2p$ configuration of He, namely, ^3P and ^1P.

The He-atom Hamiltonian (18.35) is not quite complete in that it omits a term called the *spin–orbit interaction* arising from the interaction between the spin and orbital motions of the electrons. The spin–orbit interaction is very small (except in heavy atoms), but it partly removes the degeneracy of a term, splitting an atomic term into a number of closely spaced **energy levels.** (See Prob. 18.47.) For example, the ^3P term in Fig. 18.13 is split into three closely spaced levels; the other three terms are each slightly shifted in energy by the spin–orbit interaction but are not split. Because of this spin–orbit splitting, the energies shown in Fig. 18.13 do not quite correspond to the actual pattern of atomic energy levels, and the energies in this figure are therefore called terms rather than energy levels.

An atomic term corresponds to definite values of the total orbital angular-momentum quantum number L and the total spin angular-momentum quantum number S. The L value is indicated by a code letter (S, P, D, . . .), and the S value is indicated by writing the value of $2S + 1$ as a left superscript to the L code letter.

For further discussion on the addition of angular momenta and on the way the terms arising from a given atomic electron configuration are found, see Probs. 18.46 and 18.47.

18.8 MANY-ELECTRON ATOMS AND THE PERIODIC TABLE

Lithium and the Pauli Exclusion Principle

As we did with helium, we can omit the interelectronic repulsion terms $(e^2/4\pi\varepsilon_0)\cdot(1/r_{12} + 1/r_{13} + 1/r_{23})$ from the Li-atom Hamiltonian to give an approximate Hamiltonian that is the sum of three hydrogenlike Hamiltonians. The approximate wave function is then the product of hydrogenlike (one-electron) wave functions. For the ground state, we might expect the approximate wave function $1s(1)1s(2)1s(3)$. However, we have not taken account of electron spin or the spin–statistics theorem (Sec. 18.6). Electrons, which have $s = \frac{1}{2}$ and are fermions, require a complete wave function that is antisymmetric. Therefore, the symmetric spatial function $1s(1)1s(2)1s(3)$ must be multiplied by an antisymmetric three-electron spin function. One finds, however, that it is impossible to write an antisymmetric spin function for three electrons. With three or more electrons, the antisymmetry requirement cannot be satisfied by writing a wave function that is the product of separate spatial and spin factors.

The clue to constructing an antisymmetric wave function for three or more electrons lies in Eq. (18.43), which shows that the ground-state wave function of helium can be written as a determinant. The reason a determinant gives an antisymmetric wave function follows from the theorem: *Interchange of two rows of a determinant changes the sign of the determinant.* (For a proof, see *Sokolnikoff and Redheffer,* app. A.) Interchange of rows 1 and 2 of the determinant in (18.43) amounts to interchanging the electrons. Thus a determinantal ψ is multiplied by -1 by such an interchange and therefore satisfies the antisymmetry requirement.

Let f, g, and h be three spin-orbitals. (Recall that a spin-orbital is the product of a spatial orbital and a spin factor; Sec. 18.5.) We can get an antisymmetric three-electron wave function by writing the following determinant (called a **Slater determinant**)

$$\frac{1}{\sqrt{6}}\begin{vmatrix} f(1) & g(1) & h(1) \\ f(2) & g(2) & h(2) \\ f(3) & g(3) & h(3) \end{vmatrix} \tag{18.51}$$

The $1/\sqrt{6}$ is a normalization constant; there are six terms in the expansion of this determinant.

A third-order determinant is defined by

$$\begin{vmatrix} a & b & c \\ d & e & f \\ g & h & i \end{vmatrix} \equiv a\begin{vmatrix} e & f \\ h & i \end{vmatrix} - b\begin{vmatrix} d & f \\ g & i \end{vmatrix} + c\begin{vmatrix} d & e \\ g & h \end{vmatrix}$$

$$= aei - ahf - bdi + bgf + cdh - cge \tag{18.52}$$

where (18.44) was used. The second-order determinant that multiplies a in the expansion is found by striking out the row and the column that contains a in the third-order determinant; similarly for the multipliers of $-b$ and c. The reader can verify that interchange of two rows multiplies the determinant's value by -1.

To get an antisymmetric approximate wave function for Li, we use (18.51). Let us try to put all three electrons into the $1s$ orbital by taking the spin-orbitals to be $f = 1s\alpha$, $g = 1s\beta$, and $h = 1s\alpha$. The determinant (18.51) becomes

$$\frac{1}{\sqrt{6}}\begin{vmatrix} 1s(1)\alpha(1) & 1s(1)\beta(1) & 1s(1)\alpha(1) \\ 1s(2)\alpha(2) & 1s(2)\beta(2) & 1s(2)\alpha(2) \\ 1s(3)\alpha(3) & 1s(3)\beta(3) & 1s(3)\alpha(3) \end{vmatrix} \tag{18.53}$$

Expansion of this determinant using (18.52) shows it to equal zero. This can be seen without multiplying out the determinant by using the following theorem (*Sokolnikoff and Redheffer,* app. A): *If two columns of a determinant are identical, the determinant equals zero.* The first and third columns of (18.53) are identical, and so (18.53) vanishes. If any two of the spin-orbitals f, g, and h in (18.51) are the same, two columns of the determinant are the same and the determinant vanishes. Of course, zero is ruled out as a possible wave function, since there would then be no probability of finding the electrons.

The requirement that the electronic wave function be antisymmetric thus leads to the conclusion:

No more than one electron can occupy a given spin-orbital.

This is the **Pauli exclusion principle,** first stated by Pauli in 1925. An orbital (or one-electron spatial wave function) is defined by giving its three quantum numbers (n, l, m in an atom). A spin-orbital is defined by giving the three quantum numbers of the orbital and the m_s quantum number ($+\frac{1}{2}$ for spin function α, $-\frac{1}{2}$ for β). Thus, in an atom, the exclusion principle requires that no two electrons have the same values for all four quantum numbers n, l, m, and m_s.

"There is no one fact in the physical world which has a greater impact on the way things *are,* than the Pauli Exclusion Principle. To this great Principle we credit the very existence of the hierarchy of matter, both nuclear and atomic, as ordered in Mendelejev's Periodic Table of the chemical elements, which makes possible all of nuclear and atomic physics, chemistry, biology, and the macroscopic world that we see." (I. Duck and E. C. G. Sudarshan, *Pauli and the Spin-Statistics Theorem,* World Scientific, 1997, p. 21.)

Some physicists have speculated that small violations of the Pauli exclusion principle might occur. To test this, physicists passed a large current through a copper strip and searched for x-rays that would occur if an electron in the current dropped into the $1s$ orbital of a Cu atom to give a $1s^3$ atom. No such x-rays were found, and the experiment showed that the probability that a new electron introduced into copper would violate the Pauli principle is less than 5×10^{-28} [S. Bartalucci et al., *Phys. Lett. B,* **641,** 18 (2006)].

The antisymmetry requirement holds for any system of identical fermions (Sec. 18.6), so *in a system of identical fermions each spin-orbital can hold no more than one fermion.* In contrast, ψ is symmetric for bosons, so *there is no limit to the number of bosons that can occupy a given spin-orbital.*

Coming back to the Li ground state, we can put two electrons with opposite spins in the $1s$ orbital ($f = 1s\alpha$, $g = 1s\beta$), but to avoid violating the exclusion principle, the third electron must go in the $2s$ orbital ($h = 2s\alpha$ or $2s\beta$). The approximate Li ground-state wave function is therefore

$$\psi \approx \frac{1}{\sqrt{6}}\begin{vmatrix} 1s(1)\alpha(1) & 1s(1)\beta(1) & 2s(1)\alpha(1) \\ 1s(2)\alpha(2) & 1s(2)\beta(2) & 2s(2)\alpha(2) \\ 1s(3)\alpha(3) & 1s(3)\beta(3) & 2s(3)\alpha(3) \end{vmatrix} \tag{18.54}$$

When (18.54) is multiplied out, it becomes a sum of six terms, each containing a spatial and a spin factor, so ψ cannot be written as a single spatial factor times a single spin factor. Because the $2s$ electron could have been given spin β, the ground state is doubly degenerate. The elements in each row of the Slater determinant (18.54) involve the same electron. The elements in each column involve the same spin-orbital.

Since all the electrons are s electrons with $l = 0$, the total-orbital-angular-momentum quantum number L is 0. The $1s$ electrons have antiparallel spins, so the total electronic spin of the atom is due to the $2s$ electron and the total-electronic-spin quantum number S is $\frac{1}{2}$. The spin multiplicity $2S + 1$ is 2, and the ground term of Li is designated ^2S.

A variational treatment using (18.54) would replace Z in the $1s$ function in Table 18.1 by a parameter ζ_1 and Z in the $2s$ function by a parameter ζ_2. These parameters are "effective" atomic numbers that allow for electron screening. One finds the optimum values to be $\zeta_1 = 2.69$ and $\zeta_2 = 1.78$. As expected, the $2s$ electron is much better screened from the $Z = 3$ nucleus than the $1s$ electrons. The calculated variational energy turns out to be -201.2 eV, compared with the true ground-state energy -203.5 eV.

The Periodic Table

A qualitative and semiquantitative understanding of atomic structure can be obtained from the orbital approximation. As we did with He and Li, we write an approximate wave function that assigns the electrons to hydrogenlike spin-orbitals. In each orbital, the nuclear charge is replaced by a variational parameter that represents an effective nuclear charge Z_{eff} and allows for electron screening. To satisfy the antisymmetry requirement, the wave function is written as a Slater determinant. For some atomic states, the wave function must be written as a linear combination of a few Slater determinants, but we won't worry about this complication.

Since an electron has two possible spin states (α or β), the exclusion principle requires that no more than two electrons occupy the same orbital in an atom or molecule. Two electrons in the same orbital must have antiparallel spins, and such electrons are said to be **paired.** A set of orbitals with the same n value and the same l value constitutes a **subshell.** The lowest few subshells are $1s$, $2s$, $2p$, $3s$, ... An s subshell has $l = 0$ and $m = 0$ and hence can hold at most two electrons without violating the exclusion principle. A p subshell has $l = 1$ and the three possible m values $-1, 0, +1$; hence, a p subshell has a capacity of 6 electrons; d and f subshells hold a maximum of 10 and 14 electrons, respectively.

The hydrogenlike energy formula (18.18) can be modified to approximate crudely the energy ε of a given atomic orbital as

$$\varepsilon \approx -(Z_{\text{eff}}^2/n^2)(13.6\text{ eV}) \qquad (18.55)$$

where n is the principal quantum number and the effective nuclear charge Z_{eff} differs for different subshells in the same atom. We write $Z_{\text{eff}} = Z - s$, where Z is the atomic number and the **screening constant** s for a given subshell is the sum of contributions from other electrons in the atom.

Figure 18.15 shows orbital energies for neutral atoms calculated using an approximate method. The scales in this figure are logarithmic. Note that the energy of an orbital depends strongly on the atomic number, decreasing with increasing Z, as would be expected from Eq. (18.55). Note the square-root sign on the vertical axis. Because Z_{eff} for the $1s$ electrons in Ne ($Z = 10$) is almost 10 times as large as Z_{eff} for the $1s$ electron in H, the $1s$ orbital energy ε_{1s} for Ne is roughly 100 times ε_{1s} for H. (Of course, ε_{1s} is negative.) As mentioned earlier, the l degeneracy that exists for $Z = 1$ is removed in many-electron atoms. For most values of Z, the $3s$ and $3p$ orbitals are much closer together than the $3p$ and $3d$ orbitals, and we get the familiar stable octet

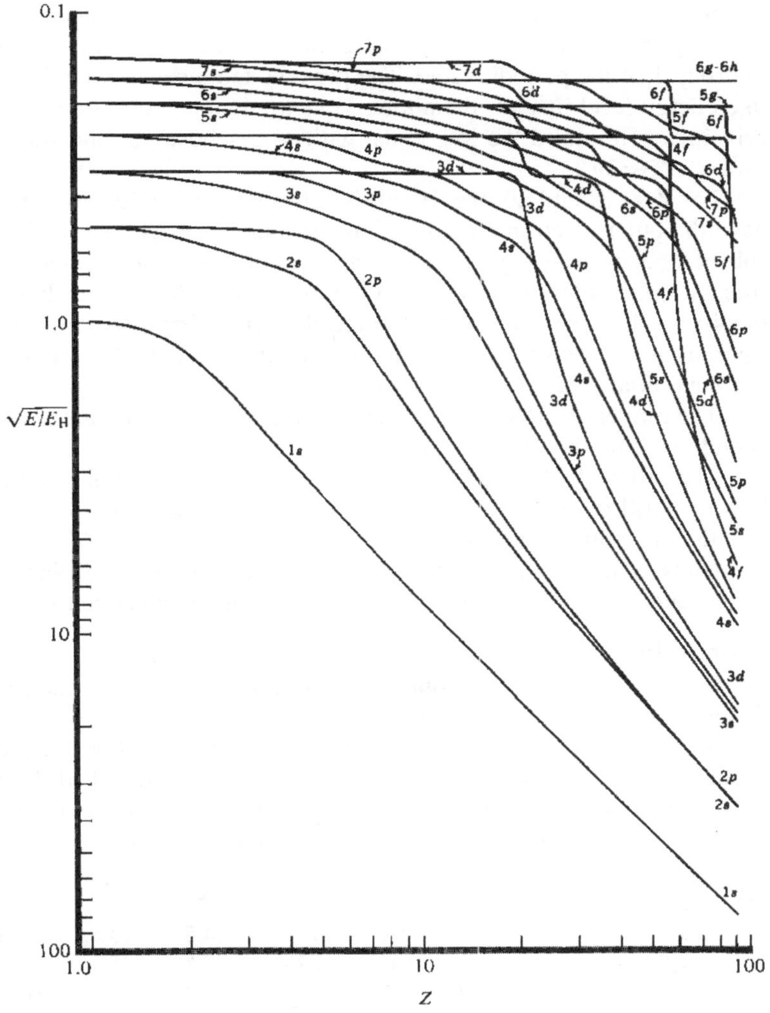

Figure 18.15

Approximate orbital energies versus atomic number Z in neutral atoms. $E_H = -13.6$ eV, the ground-state H-atom energy. [Redrawn by M. Kasha from R. Latter, *Phys. Rev.*, **99,** 510 (1955).]

of outer electrons (ns^2np^6). For Z between 7 and 21, the $4s$ orbital lies below the $3d$ (s orbitals are more penetrating than d orbitals), but for $Z > 21$, the $3d$ lies lower.

We saw that the ground-state configuration of Li is $1s^22s$, where the superscripts give the numbers of electrons in each subshell and a superscript of 1 is understood for the $2s$ subshell. We expect Li to readily lose one electron (the $2s$ electron) to form the Li^+ ion, and this is the observed chemical behavior.

The ground-state configurations of Be and B are $1s^22s^2$ and $1s^22s^22p$, respectively. For C, the ground-state configuration is $1s^22s^22p^2$. A given electron configuration can give rise to more than one atomic term. For example, the He $1s2s$ configuration produces the two terms ^3S and ^1S (Fig. 18.13). Figuring out the terms arising from the $1s^22s^22p^2$ configuration is complicated and is omitted. Hund's rule tells us that the lowest-lying term will have the two $2p$ spins parallel:

$$\underset{1s}{\underline{\uparrow\downarrow}} \quad \underset{2s}{\underline{\uparrow\downarrow}} \quad \underbrace{\underline{\uparrow} \;\; \underline{\uparrow} \;\; \underline{}}_{2p}$$

Putting the two $2p$ electrons in different orbitals minimizes the electrostatic repulsion between them. The $2p$ subshell is filled at $_{10}Ne$, whose electron configuration is $1s^22s^22p^6$. Like helium, neon does not form chemical compounds.

Sodium has the ground-state configuration $1s^22s^22p^63s$, and its chemical and physical properties resemble those of Li (ground-state configuration $1s^22s$), its predecessor in group 1 of the periodic table. The periodic table is a consequence of the hydrogenlike energy-level pattern, the allowed electronic quantum numbers, and the exclusion principle. The third period ends with Ar, whose ground-state configuration is $1s^22s^22p^63s^23p^6$.

For $Z = 19$ and 20, the 4*s* subshell lies below the 3*d* (Fig. 18.15) and K and Ca have the outer electron configurations 4*s* and $4s^2$, respectively. With $Z = 21$, the 3*d* subshell begins to fill, giving the first series of *transition elements*. The 3*d* subshell is filled at $_{30}$Zn, outer electron configuration $3d^{10}4s^2$ (3*d* now lies lower than 4*s*). Filling the 4*p* subshell then completes the fourth period. For discussion of the electron configurations of transition-metal atoms and ions, see L. G. Vanquickenborne et al., *J. Chem. Educ.,* **71,** 469 (1994); *Levine,* sec. 11.2.

The rare earths (lanthanides) and actinides in the sixth and seventh periods correspond to filling the 4*f* and 5*f* subshells.

The order of filling of subshells in the periodic table is given by the $n + l$ rule: Subshells fill in order of increasing $n + l$ values; for subshells with equal $n + l$ values, the one with the lower *n* fills first.

Niels Bohr rationalized the periodic table in terms of filling the atomic energy levels, and the familiar long form of the periodic table is due to him.

Atomic Properties

The **first, second, third, . . . ionization energies** of atom A are the energies required for the processes $A \rightarrow A^+ + e^-$, $A^+ \rightarrow A^{2+} + e^-$, $A^{2+} \rightarrow A^{3+} + e^-$, . . . , where A, A^+, etc., are isolated atoms or ions in their ground states. Ionization energies are traditionally expressed in eV. The corresponding numbers in volts are called the **ionization potentials.** Some first, second, and third ionization energies in eV are (C. E. Moore, Ionization Potentials and Ionization Limits, *Nat. Bur. Stand. U.S. Publ. NSRDS-NBS 34,* 1970):

H	He	Li	Be	B	C	N	O	F	Ne	Na
13.6	24.6	5.4	9.3	8.3	11.3	14.5	13.6	17.4	21.6	5.1
	54.4	75.6	18.2	25.2	24.4	29.6	35.1	35.0	41.0	47.3
		122.5	153.9	37.9	47.9	47.4	54.9	62.7	63.4	71.6

Note the low value for removal of the 2*s* electron from Li and the high value for removal of a 1*s* electron from Li^+. Ionization energies clearly show the "shell" structure of atoms.

The first ionization energy decreases going down a group in the periodic table because the increase in quantum number *n* of the valence electron increases the average distance of the electron from the nucleus, making it easier to remove. The first ionization energy generally increases going across a period (Fig. 18.16). As we go across a period, the nuclear charge increases, but the electrons being added have the same or a similar value of *n* and so don't screen one another very effectively; the effective nuclear charge $Z_{eff} = Z - s$ in (18.55) increases across a period, since *Z* is increasing faster than *s*, and the valence electrons become more tightly bound. Metals have lower ionization energies than nonmetals.

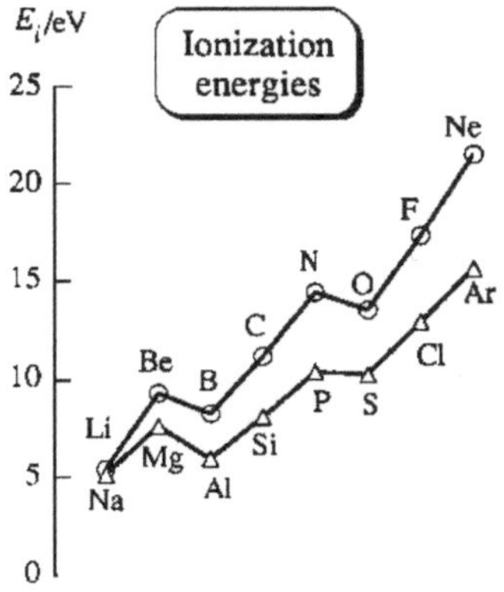

Figure 18.16

Ionization energies of second- and third-period elements.

The **electron affinity** of atom A is the energy released in the process $A + e^- \rightarrow A^-$. Some values in eV are [T. Andersen et al., *J. Phys. Chem. Ref. Data,* **28,** 1511 (1999)]:

H	He	Li	Be	B	C	N	O	F	Ne	Na
0.8	<0	0.6	<0	0.3	1.3	−0.1	1.5	3.4	<0	0.5

Note the opposite convention in the definitions of ionization energy and electron affinity. The ionization energy is ΔE accompanying loss of an electron. The electron affinity is $-\Delta E$ accompanying gain of an electron.

The motion of an electric charge produces a magnetic field. The orbital angular momentum of atomic electrons with $l \neq 0$ therefore produces a magnetic field. The "spinning" of an electron about its own axis is a motion of electric charge and also produces a magnetic field. Because of the existence of electron spin, an electron acts like a tiny magnet. (Magnetic interactions between electrons are much smaller than the electrical forces and can be neglected in the Hamiltonian except in very precise calculations.) The magnetic fields of electrons with opposite spins cancel each other. It follows that an atom in a state with $L \neq 0$ and/or $S \neq 0$ produces a magnetic field and is said to be **paramagnetic.** In a magnetized piece of iron, the majority of electron spins are aligned in the same direction to produce the observed magnetic field.

The **radius** of an atom is not a well-defined quantity, as is obvious from Figs. 18.7 and 18.8. From observed bond lengths in molecules and interatomic distances in crystals, various kinds of atomic radii can be deduced. (See Secs. 19.1 and 23.6.) Atomic radii decrease going across a given period because of the increase in Z_{eff} and increase going down a given group because of the increase in n.

The energies of excited states of most atoms of the periodic table have been determined from atomic spectral data and are tabulated in C. E. Moore, Atomic Energy Levels, *Nat. Bur. Stand. U.S. Circ.* 467, vols. I, II, and III, 1949, 1952, and 1958; C. E. Moore, Atomic Energy Levels, *Nat. Bur. Stand. Publ. NSRDS-NBS 35,* vols. I, II, and III, 1971. For online atomic-energy-level data, see physics.nist.gov/PhysRefData/ASD/index.html.

Figure 18.17 shows some of the term energies (as determined from emission and absorption spectra) of the Na atom, whose ground-state electron configuration is $1s^22s^22p^63s$. The zero level of energy has been taken at the ground state. This choice differs from the convention used in Fig. 18.2 and Eq. (18.5), where the zero level of energy corresponds to all charges being infinitely far from one another. (Each ^2P, ^2D, ^2F, . . . term in Na is split by spin–orbit interaction into two closely spaced energy levels, which are not shown.)

Each excited term in Fig. 18.17 arises from an electron configuration with the $3s$ valence electron excited to a higher orbital, which is written preceding the term symbol. Terms that correspond to electron configurations with an inner-shell Na electron excited have energies higher than the first ionization energy of Na (the dashed line), and such terms are not readily observed spectroscopically. Because Na has only one valence electron, each electron configuration with filled inner shells gives rise to only one term, and the Na term-energy diagram is simple. Note from Fig. 18.17 that the $4s$ ^2S term of Na lies below the $3d$ ^2D term, as might be expected from Fig. 18.15.

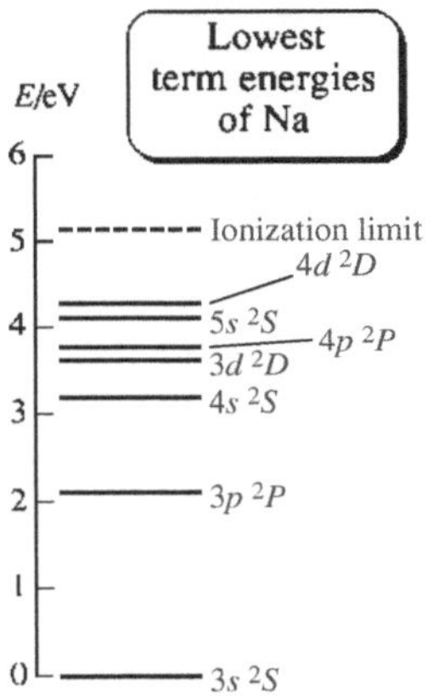

Figure 18.17

Some term energies of Na. The orbital of the excited electron is given preceding each term symbol.

18.9 HARTREE–FOCK AND CONFIGURATION-INTERACTION WAVE FUNCTIONS

Hartree–Fock Wave Functions

Wave functions like (18.42) for He and (18.54) for Li are approximations. How can these approximate wave functions be improved? One way is by not restricting the one-electron spatial functions to hydrogenlike functions. Instead, for the He ground state, we take as a trial variation function

$$\phi(1)\phi(2) \cdot 2^{-1/2}[\alpha(1)\beta(2) - \beta(1)\alpha(2)] \qquad (18.56)$$

and we look for the function ϕ that minimizes the variational integral, where ϕ need not be a hydrogenlike $1s$ orbital but can have any form. For the Li ground state, we use a function like (18.54) but with the $1s$ and $2s$ functions replaced by unknown functions f and g, and we look for those functions f and g that minimize the variational integral. These variation functions are still antisymmetrized products of one-electron spin-orbitals, so the functions ϕ, f, and g are still atomic orbitals.

In the period 1927–1930, the English physicist Hartree and the Russian physicist Fock developed a systematic procedure for finding the best possible forms for the orbitals. A variational wave function that is an antisymmetrized product of the best possible orbitals is called a **Hartree–Fock wave function.** For each state of a given system, there is a single Hartree–Fock wave function. Hartree and Fock showed that the Hartree–Fock orbitals ϕ_i satisfy the equation

$$\hat{F}\phi_i = \varepsilon_i \phi_i \tag{18.57}$$

where the Hartree–Fock operator $\hat{F}$ is a complicated operator whose form is discussed later in this section. Each of the spatial orbitals ϕ_i is a function of the three spatial coordinates; ε_i is the energy of orbital i.

The Hartree–Fock operator $\hat{F}$ in (18.57) is peculiar in that its form depends on what the eigenfunctions ϕ_i are. Since the orbitals ϕ_i are not known before the Hartree–Fock equations (18.57) are solved, the operator $\hat{F}$ is initially unknown. To solve (18.57), one starts with an initial guess for the orbitals ϕ_i, which allows one to calculate an initial guess for $\hat{F}$. One uses this initial estimate of $\hat{F}$ to solve (18.57) for an improved set of orbitals, and then uses these orbitals to calculate an improved $\hat{F}$, which is then used to solve for further improved orbitals, etc. The process is continued until no further significant change in the orbitals occurs from one iteration to the next.

Each spatial orbital in (18.57) is a function of the three spatial coordinates of a single electron. Likewise, the Hartree–Fock operator $\hat{F}$ contains the coordinates of a single electron and derivatives with respect to those coordinates. If we use the number 1 to label the electron in (18.57), we can write (18.57) as $\hat{F}(1)\phi_i(1) = \varepsilon_i\phi_i(1)$. The operator $\hat{F}(1)$ for an atom is the sum of the following terms: (*a*) The kinetic-energy operator $-(\hbar^2/2m_e)\nabla_1^2$ for electron 1. (*b*) The potential energy $-Ze^2/4\pi\varepsilon_0 r_1$ of attraction between electron 1 and the nucleus. (*c*) The potential energy of repulsion between electron 1 and a hypothetical continuous spatial distribution of negative charge whose charge density (charge per unit volume) is calculated by imagining that each of the other electrons in the atom is smeared out into a charge cloud whose charge density at each point is $-e|\phi_j|^2$, where ϕ_j is the orbital occupied by the hypothetical smeared-out electron. One adds the charge densities of the smeared-out electrons to get a charge cloud whose interaction with electron 1 is calculated. Part c of $\hat{F}(1)$ is called the *Coulomb operator.* (*d*) An *exchange operator* that involves the occupied orbitals and is present so as to make the overall wave function antisymmetric with respect to exchange of electrons. Parts c and d of $\hat{F}$ depend on the occupied orbitals, which are unknown at the start of the calculation.

Originally, Hartree–Fock orbitals were calculated numerically, and the results expressed as a table of values of ϕ_i at various points in space. In 1951, Roothaan showed that the most convenient way to express Hartree–Fock orbitals is as linear combinations of a set of functions called **basis functions.** A set of functions is said to be a **complete set** if *every* well-behaved function can be written as a linear combination of the members of the complete set. If the functions $g_1, g_2, g_3, \ldots$ form a complete set, then every well-behaved function f that depends on the same variables as the g functions can be expressed as

$$f = \sum_k c_k g_k \tag{18.58}$$

where the coefficients c_k are constants that depend on what the function f is. (More details on complete sets are given in Sec. 17.16.) It generally requires an infinite number of functions $g_1, g_2, \ldots$ to have a complete set (but not every infinite set of functions is complete). The basis functions g_k used to express the Hartree–Fock orbitals ϕ_i must be a complete set. We have

$$\phi_i = \sum_k b_k g_k \tag{18.59}$$

An orbital ϕ_i is specified by stating what the set of basis functions g_k is and giving the coefficients b_k. Roothaan showed how to calculate the b_k's that give the best possible orbitals. A different set of coefficients in (18.59) is used to express each of the orbitals.

A complete set of basis functions commonly used in atomic Hartree–Fock calculations is the set of **Slater-type orbitals** (STOs). An STO has the form $G_n(r)\Theta_{lm}(\theta)\Phi_m(\phi)$, where $\Theta_{lm}(\theta)$ and $\Phi_m(\phi)$ are the same functions as in the hydrogenlike orbitals (18.9). The radial factor has the form $G_n(r) = Nr^{n-1}e^{-\zeta r/a_0}$, where N is a normalization constant, n is the principal quantum number, and ζ is a variational parameter (the *orbital exponent*). The function $G_n(r)$ differs from a hydrogenlike radial factor in containing r^{n-1} in place of r^l times a polynomial in r. It has been shown that the set of STOs with n, l, and m given by (18.11) to (18.13) and with all possible positive values of ζ forms a complete set. Although, in principle, one needs an infinite number of basis functions to express a Hartree–Fock orbital, in practice, each atomic Hartree–Fock orbital can be very accurately approximated using only a few well-chosen STOs.

For example, for the helium ground state, Clementi expressed the Hartree–Fock orbital ϕ of the electrons as a linear combination of five 1s STOs that differ in their values of ζ. Thus $\phi = \Sigma_{k=1}^{5} b_k g_k$, where the coefficients b_k are found by solving the Hartree–Fock equation (18.57) and each g_k function is a 1s STO with the form $g_k = N_k \exp(-\zeta_k r/a_0)$. Each N_k is a normalization constant, and each g_k function has a fixed value of ζ_k. (The ζ_k values used were 1.417, 2.377, 4.396, 6.527, and 7.943, and the values found for b_k were 0.768, 0.223, 0.041, -0.010, and 0.002, respectively.) The Hartree–Fock helium ground-state wave function is then (18.56) with the function ϕ given by the five-term sum just discussed.

To solve (18.57) for the Hartree–Fock orbitals of an atom or molecule with many electrons requires a huge amount of computation, and it wasn't until the advent of high-speed computers in the 1960s that such calculations became practicable. Hartree–Fock wave functions have been computed for the ground states and certain excited states of the first 54 atoms of the periodic table. Hartree–Fock wave functions play a key role in the quantum chemistry of molecules (Chapter 19).

Although a Hartree–Fock wave function is an improvement on one that uses hydrogenlike orbitals, it is still only an approximation to the true wave function. The Hartree–Fock wave function assigns each electron pair to its own orbital. The forms of these orbitals are computed to take interelectronic repulsions into account in an average way. However, electrons are not actually smeared out into a static distribution of charge but interact with one another instantaneously. An orbital wave function cannot account for these instantaneous interactions, so the true wave function cannot be expressed as an antisymmetrized product of orbitals.

For helium, the use of a hydrogenlike 1s orbital with a variable orbital exponent gives a ground-state energy of -77.5 eV (Sec. 18.6) compared with the true value -79.0 eV. The Hartree–Fock wave function for the helium ground state gives an energy of -77.9 eV, which is still in error by 1.1 eV. The energy error of the Hartree–Fock wave function is called the **correlation energy,** since it results from the fact that the Hartree–Fock wave function neglects the instantaneous correlations in the motions of the electrons. Electrons repel one another and correlate their motions to avoid being close together; this phenomenon is called **electron correlation.**

Configuration Interaction

A method used to improve a Hartree–Fock wave function is configuration interaction. When a Hartree–Fock ground-state wave function of an atom or molecule is calculated, one also obtains expressions for unoccupied excited-state orbitals. It is possible to show that the set of functions obtained by making all possible assignments of electrons to the available orbitals is a complete set. Hence, the true wave function ψ of the ground state can be expressed as

$$\psi = \sum_j a_j \psi_{\text{orb},j} \tag{18.60}$$

where the $\psi_{\text{orb},j}$'s are approximate orbital wave functions that differ in the assignment of electrons to orbitals. Each $\psi_{\text{orb},j}$ is a Slater determinant of spin-orbitals. The functions $\psi_{\text{orb},j}$ are called *configuration functions* (or *configurations*). One uses a variational procedure to find the values of the coefficients a_j that minimize the variational integral. This type of calculation is called **configuration interaction** (CI).

For the helium ground state, the term with the largest coefficient in the CI wave function will be a Slater determinant with both electrons in orbitals resembling 1*s* orbitals, but Slater determinants with electrons in 2*s*-like and higher orbitals will also contribute. A CI wave function for the He ground state has the form $\psi = a_1\psi(1s^2) + a_2\psi(1s2s) + a_3\psi(1s3s) + a_4\psi(2s^2) + a_5\psi(2p^2) + a_6\psi(2s3s) + a_7\psi(3s^2) + a_8\psi(2p3p) + a_9\psi(3d^2) + \cdots$, where the a's are numerical coefficients and $\psi(1s2s)$ indicates a Slater determinant with one electron in a 1*s*-like orbital and one in a 2*s*-like orbital. (This last statement is inaccurate; see Prob. 18.62.)

CI computer calculations are extremely time-consuming, since it often requires a linear combination of thousands or even millions of configuration functions to give an accurate representation of ψ.

18.10 SUMMARY

The potential energy of interaction between two charges separated by distance r is $V = Q_1Q_2/4\pi\varepsilon_0 r$.

The Schrödinger equation for the H-atom internal motion is separable in spherical coordinates r, θ, and ϕ. The stationary-state hydrogenlike-atom wave functions and bound-state energies are $\psi = R_{nl}(r)\Theta_{lm}(\theta)\Phi_m(\phi)$ and $E = -(Z^2/n^2)(e^2/8\pi\varepsilon_0 a)$, where $a \equiv \hbar^2 4\pi\varepsilon_0/\mu e^2$ (μ is the reduced mass) and the quantum numbers take the values $n = 1, 2, 3, \ldots$; $l = 0, 1, 2, \ldots, n-1$; $m = -l, \ldots, +l$. The letters $s, p, d, f, \ldots$ indicate $l = 0, 1, 2, 3, \ldots$, respectively. The magnitude of the electron's orbital angular momentum with respect to the nucleus is $|\mathbf{L}| = [l(l+1)]^{1/2}\hbar$, and L_z equals $m\hbar$. An orbital is a one-electron spatial wave function. The shape of an orbital is defined as a surface of constant $|\psi|$ that encloses some large fraction of the probability density. Figure 18.6 shows some H-atom orbital shapes.

The average value of any function f of r for a hydrogen-atom stationary state is given by $\langle f(r)\rangle = \int |\psi|^2 f(r)\, d\tau$. Use of $\psi = R\Theta\Phi$, $d\tau = r^2 \sin\theta\, dr\, d\theta\, d\phi$, and the limits $0 \le r \le \infty$, $0 \le \theta \le \pi$, $0 \le \phi \le 2\pi$ gives

$$\langle f(r)\rangle = \int_0^\infty f(r)|R(r)|^2 r^2\, dr \int_0^\pi |\Theta(\theta)|^2 \sin\theta\, d\theta \int_0^{2\pi} |\Phi(\phi)|^2\, d\phi$$

Electrons and other elementary particles have a built-in angular momentum (spin angular momentum) **S** of magnitude $[s(s+1)]^{1/2}\hbar$, where $s = \frac{1}{2}$ for an electron. The z component of **S** is $m_s\hbar$, where $m_s = \pm\frac{1}{2}$ for an electron. The symbols α and β denote spin functions with $m_s = +\frac{1}{2}$ and $m_s = -\frac{1}{2}$, respectively. A product of one-electron spatial and spin functions is called a spin-orbital. The complete wave function of a system of identical particles with half-integral spin must be antisymmetric with respect to interchange of all coordinates (spatial and spin) of any two particles, meaning that such an interchange multiplies the wave function by -1. The wave function of a system of identical integral-spin particles must be symmetric.

There are three symmetric two-electron spin functions [Eq. (18.40)] and one antisymmetric one [Eq. (18.41)]. An approximate ground-state wave function for He is $1s(1)1s(2)$ times (18.41).

In a many-electron atom, the total electronic orbital and spin angular momenta are $[L(L+1)]^{1/2}\hbar$ and $[S(S+1)]^{1/2}\hbar$, respectively, where the quantum number L can be 0, 1, 2, . . . and the quantum number S can be 0, $\frac{1}{2}$, 1, $\frac{3}{2}$, ... The L value is indicated using the letter code $S, P, D, F, \ldots$, and the value of $2S + 1$ (the spin multiplicity) is written as a left superscript to the L code letter. Atomic states that correspond to the same electron configuration and have the same L and S values belong to the same atomic term. Usually, the lowest-energy term of a given electron configuration is the term with the largest S value (Hund's rule).

An approximate (antisymmetric) wave function for a many-electron atom can be written as a Slater determinant of spin-orbitals. In such an approximate wave function, no more than one electron can occupy a given spin-orbital (the Pauli exclusion principle). The variation of atomic-orbital energy with atomic number is given in Fig. 18.15. The periodic table, ionization energies, and electron affinities were discussed.

The best possible (that is, lowest-energy) wave function that assigns each electron to a single spin-orbital is called the Hartree–Fock wave function. Hartree–Fock orbitals are expressed as linear combinations of basis functions. The Hartree–Fock wave function is still an approximation to the true wave function. In a configuration-interaction (CI) calculation, the wave function is written as a linear combination of the Hartree–Fock wave function and functions in which some of the electrons occupy excited orbitals. A CI wave function can approach the true wave function if enough configuration functions are included.

FURTHER READING

Hanna, chap. 6; *Karplus and Porter*, chaps. 3, 4; *Levine*, chaps. 10, 11; *Lowe and Peterson*, chaps. 4, 5; *McQuarrie* (1983), chap. 8; *Ratner and Schatz*, chaps. 6–9; *Atkins and Friedman*, chap. 7.

PROBLEMS

Section 18.1

18.1 True or false? (*a*) Doubling the distance between two charges multiplies the force between them by one-half. (*b*) Doubling the distance between two charges multiplies the potential energy of their interaction by one-half. (*c*) One joule is many orders of magnitude larger than one electronvolt.

18.2 Use the relation $V = Q_1Q_2/4\pi\varepsilon_0 r$ to deduce the SI units of $4\pi\varepsilon_0$.

18.3 Maxwell's electromagnetic theory of light (Sec. 20.1) shows that the speed of light in vacuum is $c = (\mu_0\varepsilon_0)^{-1/2}$, where ε_0 occurs in the proportionality constant in Coulomb's law and μ_0 occurs in the proportionality constant in Ampère's law for the magnetic field produced by an electric current. μ_0 is arbitrarily assigned the value $\mu_0 \equiv 4\pi \times 10^{-7}$ N s^2/C^2 in the SI system. Use $c = (\mu_0\varepsilon_0)^{-1/2}$ to verify Eq. (18.2) for $1/4\pi\varepsilon_0$.

18.4 (*a*) Calculate the electrostatic potential energy of two electrons separated by 3.0 Å in vacuum. Express your answer in joules and in electronvolts. (*b*) Calculate the electrostatic potential energy in eV of a system of two electrons and a proton in vacuum if the electrons are separated by 3.0 Å and the electron–proton distances are 4.0 and 5.0 Å.

Section 18.2

18.5 What fraction of the volume of an atom of radius 10^{-8} cm is occupied by its nucleus if the nuclear radius is 10^{-12} cm?

Section 18.3

18.6 True or false? (*a*) The photon emitted in an $n = 3$ to $n = 2$ transition in the H atom has a lower frequency than the photon for an $n = 2$ to $n = 1$ H-atom transition. (*b*) The ground-state energy of He^+ is about 4 times the ground-state energy of H. (*c*) ψ is zero at the nucleus for all H-atom stationary states. (*d*) For the ground state of the H atom, $|\psi|^2$ is a maximum at the nucleus. (*e*) The most probable value of the electron–nucleus distance in a ground-state H atom is zero. (*f*) The smallest allowed value of the atomic quantum number n is 0. (*g*) For H-atom stationary states with $l = 0$, ψ is independent of θ and ϕ. (*h*) For the H-atom ground state, the electron is confined to move on the surface of a sphere centered around the nucleus. (*i*) For the H-atom ground state, the electron is confined to move within a sphere of fixed radius.

18.7 Match each of the spherical coordinates r, θ, and ϕ with each of the following descriptions and give the range of each coordinate. (*a*) Angle between the positive z axis and the radius vector. (*b*) Distance to the origin. (*c*) Angle between the positive x axis and the projection of the radius vector in the xy plane.

18.8 True or false? For the hydrogen atom, (*a*) the allowed energy levels are $E = -(13.60 \text{ eV})/n^2$ and $E \geq 0$; (*b*) any photon with energy $E_{\text{photon}} \geq 13.60$ eV can ionize a hydrogen atom in the $n = 1$ state; (*c*) Any photon with $E_{\text{photon}} \geq 0.75(13.60 \text{ eV})$ can cause a hydrogen atom to go from the $n = 1$ state to the $n = 2$ state.

18.9 Give the allowed values of (*a*) l for $n = 5$ and (*b*) m if $l = 5$.

18.10 Omitting spin considerations, give the degeneracy of the hydrogenlike energy level with (*a*) $n = 1$; (*b*) $n = 2$; (*c*) $n = 3$.

18.11 Calculate the ionization potential in V of (*a*) He^+; (*b*) Li^{2+}.

18.12 Calculate the wavelength of the photon emitted when an electron goes from the $n = 3$ to $n = 2$ level of a hydrogen atom.

18.13 Calculate a in Eq. (18.14).

18.14 Positronium is a species consisting of an electron bound to a positron. Calculate its ionization potential. A positron has the same mass as an electron and the same charge as a proton.

18.15 Show that $\langle r \rangle = 3a/2Z$ for a ground-state hydrogenlike atom. Use a table of integrals.

18.16 Use the Taylor-series expansions about $\phi = 0$ for $e^{i\phi}$, $\sin \phi$, and $\cos \phi$ to verify that $e^{i\phi} = \cos \phi + i \sin \phi$.

18.17 Verify Eq. (18.26) for $2p_x$ and $2p_y$.

18.18 (*a*) Let $z_1 = a_1 + ib_1$ and $z_2 = a_2 + ib_2$, where $i = \sqrt{-1}$ and the a's and b's are real. If $z_1 = z_2$, what must be true about the a's and b's? (*b*) Verify that the requirement that $\Phi(\phi) = \Phi(\phi + 2\pi)$ leads to the requirement that m in (18.10) be an integer.

18.19 Draw a rough graph (not to scale) of the value of ψ_{2p_z} along the z axis versus z. Then do the same for $|\psi_{2p_z}|^2$.

18.20 (*a*) Find r_{2s} for H using the 90% probability definition. (*b*) Find r_{2s} for H using a 95% probability definition.

18.21 Verify that the 1s wave function in Table 18.1 is an eigenfunction of the hydrogenlike Hamiltonian operator. (Use the chain rule to find the partial derivatives.)

18.22 Show that the average potential energy $\langle V \rangle$ for a ground-state hydrogenlike atom is $-Z^2e^2/4\pi\varepsilon_0 a$.

18.23 (*a*) Complete Example 18.3 in Sec. 18.3 and find $\langle r \rangle$ for a hydrogen atom in the $2p_z$ state. (*b*) Without doing any calculations, give the value of $\langle r \rangle$ for a $2p_x$ H atom. (*c*) Verify your answer to (*b*) by evaluating the appropriate triple integral.

18.24 Show that the maximum in the radial distribution function of a ground-state hydrogenlike atom is at a/Z.

18.25 For a hydrogen atom in a 1s state, calculate the probability that the electron is between 0 and 2.00 Å from the nucleus.

18.26 Verify that $\int_0^{2\pi} |\Phi|^2 \, d\phi = 1$, where Φ is given by (18.10).

Section 18.4

18.27 True or false for the classical-mechanical angular momentum **L**? (*a*) **L** of a particle depends on which point is chosen as the origin. (*b*) For a particle vibrating back and forth on a straight line through the origin, **L** is zero. (*c*) For a particle revolving around the origin on a circle, **L** is nonzero.

18.28 Calculate the angles the three angular-momentum vectors make with the z axis in Fig. 18.10.

18.29 (*a*) From the definition of angular momentum in Sec. 18.4, show that for a classical particle of mass m moving on a circle of radius r, the magnitude of the angular momentum with respect to the circle's center is mvr. (*b*) What is the direction of the **L** vector for this system?

18.30 Give the magnitude of the ground-state orbital angular momentum of the electron in a hydrogen atom according to (*a*) quantum mechanics; (*b*) the Bohr theory.

18.31 Calculate the magnitude of the orbital angular momentum of a 3p electron in a hydrogenlike atom.

Section 18.5

18.32 Calculate in SI units the magnitude of the spin angular momentum of an electron.

18.33 Calculate the angles between the spin vectors and the z axis in Fig. 18.11.

18.34 State what physical property is associated with each of the following quantum numbers in a one-electron atom and give the value of this physical property in terms of the quantum number. (*a*) l; (*b*) m; (*c*) s; (*d*) m_s.

18.35 For a particle with $s = 3/2$: (*a*) sketch the possible orientations of the **S** vector with the z axis; (*b*) calculate the smallest possible angle between **S** and the z axis.

Section 18.6

18.36 True or false? (*a*) The spatial factor in the ground-state wave function of He is antisymmetric. (*b*) All two-electron spin functions are antisymmetric. (*c*) The wave function of every system of identical particles must be antisymmetric with respect to exchange of all coordinates of any two particles. (*d*) Interchange of electrons 1 and 2 in the He-atom Hamiltonian (18.35) does not change this Hamiltonian.

18.37 State whether each of these functions is symmetric, antisymmetric, or neither: (*a*) $f(1)g(2)$; (*b*) $g(1)g(2)$; (*c*) $f(1)g(2) - g(1)f(2)$; (*d*) $r_1^2 - 2r_1r_2 + r_2^2$; (*e*) $(r_1 - r_2)e^{-br_{12}}$, where r_{12} is the distance between particles 1 and 2.

18.38 If $g(1, 2)$ is a symmetric function for particles 1 and 2, and $h(1, 2)$ and $k(1, 2)$ are each antisymmetric functions, state whether each of the following functions is symmetric, antisymmetric, or neither. (*a*) $g(1, 2)h(1, 2)$; (*b*) $h(1, 2)k(1, 2)$; (*c*) $h(1, 2) + k(1, 2)$; (*d*) $g(1, 2) + k(1, 2)$.

18.39 A professor does a variational calculation on the ground state of He and finds that the variational integral equals -86.7 eV. Explain why it is certain that the professor made an error.

Section 18.7

18.40 Give the term symbol for the term arising from each of the following H-atom electron configurations: (*a*) $1s$; (*b*) $3p$; (*c*) $3d$.

18.41 Give the values of L and S for a ^4F term.

18.42 State what physical property is associated with each of the following quantum numbers in a many-electron atom and give the value of this property in terms of the quantum number: (*a*) L; (*b*) S; (*c*) M_S.

18.43 For a ^3D term, give the value of (*a*) the total electronic orbital angular momentum; (*b*) the total electronic spin angular momentum.

18.44 Give the terms arising from each of the following electron configurations of K: (*a*) $1s^22s^22p^63s^23p^63d$; (*b*) $1s^22s^22p^63s^23p^64p$.

18.45 Draw a sketch like Fig. 18.14 that shows the orientations of $\mathbf{S}_1$, $\mathbf{S}_2$, and **S** for the spin function $\alpha(1)\alpha(2)$. (*Hint:* Begin by finding the angles between the z axis and each of $\mathbf{S}_1$, $\mathbf{S}_2$, and **S**.)

18.46 Consider two angular momenta $\mathbf{M}_1$ and $\mathbf{M}_2$ (these can be orbital or spin angular momenta) whose magnitudes are $[j_1(j_1 + 1)]^{1/2}\hbar$ and $[j_2(j_2 + 1)]^{1/2}\hbar$, respectively. Let $\mathbf{M}_1$ and $\mathbf{M}_2$ combine with each other to give a total angular momentum **M**, which is the vector sum of $\mathbf{M}_1$ and $\mathbf{M}_2$; $\mathbf{M} = \mathbf{M}_1 + \mathbf{M}_2$. The magnitude of **M** can be shown to be $[J(J + 1)]^{1/2}\hbar$, where the quantum number J has the possible values (*Levine*, sec. 11.4)

$$j_1 + j_2,\ j_1 + j_2 - 1,\ j_1 + j_2 - 2, \ldots, |j_1 - j_2|$$

For example, when the spins of two electrons with spin quantum numbers $s_1 = \frac{1}{2}$ and $s_2 = \frac{1}{2}$ combine to give a total electronic spin, the possible values of the total spin quantum number are $\frac{1}{2} + \frac{1}{2} = 1$ and $|\frac{1}{2} - \frac{1}{2}| = 0$. (*a*) For terms arising from the electron configuration $1s^22s^22p^63s^23p3d$, give the possible values of the total electronic orbital-angular-momentum quantum number L (electrons in filled subshells contribute zero to the total orbital angular momentum and to the total spin angular momentum and so can be ignored) and give the possible values of the total electronic spin quantum number S. (*b*) Pair each possible value of L with each possible value of S to give the terms that arise from the . . . $3p3d$ electron configuration. (*Note:* For an electron configuration like $1s^22s^22p^2$ that has two or more electrons in a partly filled subshell, the Pauli exclusion principle restricts the possible terms, and special techniques must be used to find the terms in this case. See *Levine*, sec. 11.5.)

18.47 When spin–orbit interaction splits an atomic term into energy levels, each energy level can be characterized by a total electronic angular momentum **J** that is the vector sum of the total electronic orbital and spin angular momenta: $\mathbf{J} = \mathbf{L} + \mathbf{S}$. The magnitude of **J** is $[J(J + 1)]^{1/2}\hbar$, where the possible values of the quantum number J are given by the angular-momentum addition rule in Prob. 18.46 as

$$J = L + S,\ L + S - 1,\ L + S - 2, \ldots |L - S|$$

Each level is indicated by writing its J value as a subscript on the term symbol. For example, the Na electron configuration $1s^22s^22p^63p$ gives rise to the term ^2P with $L = 1$ and $S = \frac{1}{2}$. With $L = 1$ and $S = \frac{1}{2}$, the possible J values are $1 + \frac{1}{2} = \frac{3}{2}$ and $|1 - \frac{1}{2}| = \frac{1}{2}$. Therefore, a ^2P term has two energy levels, $^2P_{3/2}$ and $^2P_{1/2}$. Give the levels that arise from each of the following terms: (*a*) ^2S; (*b*) ^4P; (*c*) ^5F; (*d*) ^3D.

Section 18.8

18.48 Write down the Hamiltonian operator for the internal motion in Li.

18.49 For a system of two electrons in a one-dimensional box, write down the approximate wave functions (interelectronic repulsion ignored) including spin for states that have one electron with $n = 1$ and one electron with $n = 2$. Which of these states has (have) the lowest energy?

18.50 Write down an approximate wave function for the Be ground state.

18.51 Which of the first 10 elements in the periodic table have paramagnetic ground states?

18.52 Calculate the eighteenth ionization potential of Ar.

18.53 Use the ionization-potential data in Sec. 18.8 to calculate Z_{eff} for the $2s$ electrons in (*a*) Li; (*b*) Be.

18.54 (*a*) Suppose the electron had spin quantum number $s = \frac{3}{2}$. What would be the ground-state configurations of atoms with 3, 9, and 17 electrons? (*b*) Suppose the electron had $s = 1$. What would be the ground-state configurations of atoms with 3, 9, and 17 electrons?

18.55 For each pair, state which would have the higher first ionization potential (refer to a periodic table): (*a*) Na, K; (*b*) K, Ca; (*c*) Cl, Br; (*d*) Br, Kr.

18.56 Use Fig. 18.15 to calculate Z_{eff} for the 1*s*, 2*s*, and 2*p* electrons in Ne.

18.57 True or false? (*a*) The 2*s* orbital energy in K is lower than the 1*s* orbital energy in H. (*b*) Interchange of two rows of a determinant multiplies the determinant's value by -1.

18.58 Of the elements with $Z \leq 10$, which one has the largest number of unpaired electrons in its ground state?

18.59 Consider the systems (*a*) $Na^+ + 2e^-$, (*b*) $Na + e^-$, (*c*) Na^-, where in each system the Na atom or ion and the electron(s) are at infinite separation from one another. Use data in Sec. 18.8 to decide which system has the lowest energy and which has the highest energy.

18.60 Which species in each of the following pairs has the larger atomic radius: (*a*) Ca, Sr; (*b*) F, Ne; (*c*) Ar, K; (*d*) C, O; (*e*) Cl^-, Ar?

General

18.61 Write out the explicit form for the five-term ground-state helium-atom Hartree–Fock orbital given in Sec. 18.9. The only nonnumerical constants in your expression should be a_0 and π.

18.62 Let D_1, D_2, D_3, and D_4 be two-row Slater determinants that contain the following spin-orbitals: $1s\alpha$ and $2s\alpha$ in D_1; $1s\alpha$ and $2s\beta$ in D_2; $1s\beta$ and $2s\alpha$ in D_3; $1s\beta$ and $2s\beta$ in D_4. Consider the four helium-atom approximate wave functions given in Fig. 18.13 for states of the 1*s*2*s* configuration. Show that two of these wave functions are each equal to one of the determinants D_1, D_2, D_3, D_4 but that the other two wave functions must each be expressed as a linear combination of two of these determinants. Thus, an orbital wave function for a state with partly filled orbitals must sometimes be expressed as a linear combination of more than one Slater determinant. [In the CI wave function (18.60), each $\psi_{orb,j}$ should have the same *S* and M_S spin quantum numbers as the wave function ψ. Thus, for a CI wave function for the He ground state, the $\psi_{orb,j}$ that corresponds to the 1*s*2*s* configuration must have the spin function (18.41) (Fig. 18.13). As shown in this problem, this $\psi_{orb,j}$ is a linear combination of two Slater determinants.]

18.63 Derive the formula for the volume of a sphere by integrating the spherical coordinate volume element (18.27) over the sphere's volume.

18.64 For each of the following pairs, state which quantity (if any) is larger: (*a*) the ground-state energy of H or He^+; (*b*) the ionization energy of K or K^+; (*c*) the wavelength of the longest-wavelength electronic absorption of ground-state H or He^+; (*d*) the ionization energy of Cl^- or the electron affinity of Cl?

18.65 Give an example of a quantum-mechanical system for which the spacing between energy levels: (*a*) increases as *E* increases; (*b*) remains the same as *E* increases; (*c*) decreases as *E* increases.

18.66 For the ground state of the hydrogen atom, find the probability the electron is in a tiny spherical region of radius 1.0×10^{-3} Å if this sphere is centered at a point that is (*a*) at the origin (nucleus); (*b*) a distance 0.50 Å from the nucleus; (*c*) a distance 5.0 Å from the nucleus. Consider the tiny sphere to be infinitesimal.

18.67 For the ground state of the hydrogen atom, find the probability that the distance between the electron and the proton lies in each of the following ranges (treat each range as infinitesimal): (*a*) 0.100 and 0.101 Å; (*b*) 0.500 and 0.501 Å; (*c*) 1.000 and 1.001 Å; (*d*) 5.000 and 5.001 Å.

18.68 For each of the following systems, give the expression for $d\tau$ in the equation $\int |\psi|^2\, d\tau = 1$ and give the limits on each coordinate: (*a*) one-dimensional harmonic oscillator; (*b*) particle in a three-dimensional rectangular box with edges *a*, *b*, and *c*; (*c*) the hydrogen atom internal motion using spherical coordinates.

18.69 Is there a gravitational attraction between the electron and the proton in the H atom? If there is, why is this not taken into account in the Hamiltonian? Do a calculation to support your answer.

18.70 (*a*) Show that the maximum value of ψ_{2p_z} for $Z = 1$ is $\psi_{max} = 1/(2a)^{3/2}\pi^{1/2}e$. (*b*) Write a computer program that will vary z/a from 0.01 to 10 in steps of 0.01 and for each value of z/a will calculate values of y/a for which $|\psi_{2p_z}/\psi_{max}|$ is equal to a certain constant *k*, where the value of *k* is input at the start of the program. Note that for some values of z/a, there are no values of y/a that satisfy the condition. Be careful that spurious values of y/a are eliminated. (The output of this program can be used as input to a graphing program to graph contours of the $2p_z$ orbital.)

18.71 *Physicist trivia question.* Name the physicist referred to in each of the following descriptions. All names appear in Chapter 18. Two of these physicists have elements named after them. (*a*) This experimental physicist (rated the 10th greatest physicist of all times in a 1999 poll) was weak in mathematics. Norman Ramsey took a course given by him in the 1930s and found that when this physicist tried to derive in class the formula for Rutherford scattering of alpha particles, "he got completely fouled up in the math, and he finally ended up telling us to go home and work it out for ourselves." Later, Ramsey came to recognize the great physical insight this physicist had and Ramsey concluded that "an ability to make a formal mathematical derivation was not the criterion of being a good physicist." (*b*) He was friends with the Swiss psychoanalyst Carl Jung and contributed a chapter to a book written by Jung. Jung published analyses of many of the dreams of this physicist; the number 4 often occurred in these dreams. (*c*) He was noted for his sarcastic comments about other physicists' work. His first wife (a cabaret entertainer) left him after less than a year of marriage and after the divorce married the chemist Paul Goldfinger. This physicist then became deeply depressed and sought psychological treatment. He remarked that "Had she taken a bullfighter, I would have understood, but a chemist . . . ” (*d*) He was one of the few twentieth-century physicists who did outstanding work in both experiments and theory. In the mid-1930s, he and coworkers bombarded many elements with neutrons and produced radioactive products. He found that uranium irradiated with neutrons gave products whose atomic numbers did not lie in the range 86 to 92 and concluded that he had produced new elements with atomic numbers of 93 and 94, which he called ausenium and hasperium, respectively. He received a Nobel Prize in physics “for his demonstration of the existence of new radioactive elements produced by neutron irradiation, and for his related discovery of nuclear reactions brought about by slow neutrons.” In fact, he had not prepared elements with $Z > 92$. One month after he received his Nobel Prize, Hahn and Strassmann published work showing that neutron irradiation of uranium gave barium as one product. Meitner and Frisch used Bohr’s liquid-drop model of the nucleus to interpret the Hahn–Strassmann results as the fission of a uranium nucleus to produce two lighter nuclei. On December 2, 1942, the first human-produced self-sustaining nuclear-fission chain reaction was achieved on a squash court at the University of Chicago in a uranium pile constructed under the direction of the subject of this question. The success of the experiment was reported in a coded telephone conversation with the words “The Italian navigator has just landed in the New World.”

18.72 True or false? (*a*) In this chapter, e stands for the charge on an electron. (*b*) The exact helium-atom ground-state wave function is a product of wave functions for each electron. (*c*) The wave function of every system of fermions must be antisymmetric with respect to interchange of all coordinates of any two particles. (*d*) The spin quantum number s of an electron has the possible values $\pm\frac{1}{2}$. (*e*) The shape of a $2p_z$ orbital is two tangent spheres. (*f*) All states belonging to the same electron configuration of a given atom must have the same energy. (*g*) Every solution of the time-independent Schrödinger equation is a possible stationary state. (*h*) The ground-state wave function of a lithium atom cannot be expressed as a spatial factor times a spin factor. (*i*) Every linear combination of two solutions of the time-independent Schrödinger equation is a solution of this equation.

Chapter 18

18.1 **(a)** F; **(b)** T; **(c)** T.

18.2 The equation shows that $4\pi\varepsilon_0$ has the same units as Q_1Q_2/rV, which are $C^2/(m\ J) = C^2\ N^{-1}\ m^{-2}$, since 1 J = 1 N m.

18.3 $c^2 = 1/\mu_0\varepsilon_0$, so $1/4\pi\varepsilon_0 = \mu_0 c^2/4\pi = (4\pi \times 10^{-7}\ N\ s^2/C^2)c^2/4\pi = 10^{-7}c^2\ N\ s^2\ C^{-2}$.

18.4 $V = Q_1Q_2/4\pi\varepsilon_0 r$.

(a) $V = (1.602 \times 10^{-19}\ C)^2/4\pi(8.854 \times 10^{-12}\ C^2/N\text{-}m^2)(3.0 \times 10^{-10}\ m) = 7.7 \times 10^{-19}\ J = (7.7 \times 10^{-19}\ J)\ (1\ eV/1.602 \times 10^{-19}\ J) = 4.8\ eV$.

(b) Let the electrons be numbered 1 and 2, let the proton be p and let e denote the proton charge. Then $V = V_{12} + V_{1p} + V_{2p} = (1/4\pi\varepsilon_0) \times [(-e)^2/(3.0 \times 10^{-10}\ m) - e^2/(4.0 \times 10^{-10}\ m) - e^2/(5.0 \times 10^{-10}\ m)] = [(1.602 \times 10^{-19}\ C)^2/4\pi(8.854 \times 10^{-12}\ C^2/N\text{-}m^2)](-1.167 \times 10^9\ m^{-1}) = -2.7 \times 10^{-19}\ J = (-2.7 \times 10^{-19}\ J)(1\ eV/1.602 \times 10^{-19}\ J) = -1.7\ eV$.

18.5 $V = 4\pi r^3/3$, so $V_{nuc}/V_{atom} = r^3_{nuc}/r^3_{atom} = (10^{-12}\ cm)^3/(10^{-8}\ cm)^3 = 1 \times 10^{-12}$.

18.6 **(a)** T; **(b)** T; **(c)** F; $\psi_{1s} \neq 0$ at $r = 0$. **(d)** T; **(e)** F; **(f)** F; **(g)** T; **(h)** F; **(i)** F.

18.7 **(a)** θ; **(b)** r; **(c)** ϕ.

18.8 **(a)** T; **(b)** T; **(c)** F.

18.9 **(a)** 0, 1, 2, 3, 4.

(b) −5, −4, −3, −2, −1, 0, 1, 2, 3, 4, 5.

18.10 **(a)** The only $n = 1$ state is $1s_0$, so the degeneracy is 1 (i.e., nondegenerate), if spin is not considered.

(b) The states $2s$, $2p_1$, $2p_0$, $2p_{-1}$ have the same energy, so the degeneracy is 4.

(c) The states $3s$, $3p_1$, $3p_0$, $3p_{-1}$, $3d_2$, $3d_1$, $3d_0$, $3d_{-1}$, $3d_{-2}$ have the same energy and the degeneracy is 9. (The general formula is n^2.)

18.11 These are hydrogenlike species, so $E = -(Z^2/n^2)(13.60 \text{ eV})$. The ionization energy IE is $-E$ for the ground state, $n = 1$.

(a) IE $= 2^2(13.60 \text{ eV}) = 54.4$ eV and the ionization potential IP is 54.4 V.

(b) IE $= 3^2(13.60 \text{ eV}) = 122.4$ eV and IP = 122.4 V.

18.12 $|\Delta E| = -(13.60 \text{ eV})(1.602 \times 10^{-19} \text{ J/1 eV})(1/3^2 - 1/2^2) = 3.026 \times 10^{-19} \text{ J} = h\nu$ and $\nu = (3.026 \times 10^{-19} \text{ J})/(6.626 \times 10^{-34} \text{ J s}) = 4.567 \times 10^{14}\text{/s}$.
$\lambda = c/\nu = (2.9979 \times 10^8 \text{ m/s})/(4.567 \times 10^{14}\text{/s}) = 6.564 \times 10^{-7} \text{ m} = 656.4$ nm.

18.13 For a hydrogen atom, $\mu = m_e m_p/(m_e + m_p) =$

$$\frac{(9.1095 \times 10^{-31} \text{ kg})(1.67265 \times 10^{-27} \text{ kg})}{(9.1095 \times 10^{-31} \text{ kg} + 1.67265 \times 10^{-27} \text{ kg})} = 9.1045 \times 10^{-31} \text{ kg}$$

We have $a = 4\pi\varepsilon_0 \hbar^2/\mu e^2 = 4\pi(8.854 \times 10^{-12} \text{ C}^2/\text{N-m}^2)(6.6262 \times 10^{-34} \text{ J} \cdot \text{s})^2/$ $4\pi^2(9.1045 \times 10^{-31} \text{ kg})(1.6022 \times 10^{-19} \text{ C})^2 = 5.295 \times 10^{-11} \text{ m} = 0.5295$ Å.

18.14 This is a hydrogenlike species, so its energy levels are given by Eq. (18.14) as $E = -(Z^2/n^2)[e^2/(4\pi\varepsilon_0)2a] = -(Z^2/n^2)[\mu e^4/2(4\pi\varepsilon_0)^2\hbar^2]$. Let m_e be the electron mass; the positron has mass m_e. So $\mu_{\text{positronium}} = m_e^2/(m_e + m_e) = m_e/2$, as compared with $\mu \approx m_e$ for an H atom. Since E is proportional to μ, each positronium energy level is half the corresponding H-atom energy. The positronium ionization potential is thus $\frac{1}{2}(13.6 \text{ V}) = 6.8$ V.

18.15 Using Table 18.1 and Eq. (18.27), $\langle r \rangle = \int \psi^* r \psi \, d\tau = \pi^{-1}(Z/a)^3 \times$ $\int_0^{2\pi}\int_0^{\pi}\int_0^{\infty} e^{-Zr/a} r e^{-Zr/a} r^2 \sin\theta \, dr \, d\theta \, d\phi = (Z^3/\pi a^3)\int_0^{\infty} r^3 e^{-2Zr/a} \, dr \int_0^{\pi} \sin\theta \, d\theta \int_0^{2\pi} d\phi$.
A table of integrals gives $\int z^3 e^{bz} \, dz = e^{bz}(z^3/b - 3z^2/b^2 + 6z/b^3 - 6/b^4)$, so $\int_0^{\infty} r^3 e^{-2Zr/a} \, dr = 6/(-2Z/a)^4 = 3a^4/8Z^4$, since $e^{-2Zr/a}$ vanishes at $r = \infty$. (This result also follows from the definite integral $\int_0^{\infty} z^n e^{-bz} \, dz = n!/b^{n+1}$ found in most tables.) Then $\langle r \rangle = (Z^3/\pi a^3)(3a^4/8Z^4)(2)(2\pi) = 3a/2Z$.

18.16 $e^{i\phi} = 1 + i\phi + (i\phi)^2/2! + (i\phi)^3/3! + (i\phi)^4/4! + (i\phi)^5/5! + \cdots =$
$1 + i\phi - \phi^2/2! - i\phi^3/3! + \phi^4/4! + i\phi^5/5! + \cdots$.
Also, $\cos\phi + i\sin\phi = (1 - \phi^2/2! + \phi^4/4! - \cdots) + i(\phi - \phi^3/3! + \phi^5/5! - \cdots) =$
$1 + i\phi - \phi^2/2! - i\phi^3/3! + \phi^4/4! + i\phi^5/5! + \cdots = e^{i\phi}$.

18.17 $2p_x = 2^{-1/2}(2p_1 + 2p_{-1}) = 2^{-1/2}(1/8\pi^{1/2})(Z/a)^{5/2}e^{-Zr/2a}r\sin\theta\,(e^{i\phi} + e^{-i\phi})$. We have $e^{i\phi} + e^{-i\phi} = \cos\phi + i\sin\phi + \cos(-\phi) + i\sin(-\phi) = \cos\phi + i\sin\phi + \cos\phi - i\sin\phi = 2\cos\phi$, so $2p_x = (2^{1/2}/2^3\pi^{1/2})(Z/a)^{5/2}e^{-Zr/2a}r\sin\theta\cos\phi =$ $\pi^{-1/2}(Z/2a)^{5/2}xe^{-Zr/2a}$, where Eq. (18.7) was used. Also, $2p_y = (2p_1 - 2p_{-1})/i\sqrt{2} =$ $(2^{-1/2}/i)(1/8\pi^{1/2})(Z/a)^{5/2}e^{-Zr/2a}\,r\sin\theta\,(e^{i\phi} - e^{-i\phi})$. We have $e^{i\phi} - e^{-i\phi} = \cos\phi + i\sin\phi - (\cos\phi - i\sin\phi) = 2i\sin\phi$, so $2p_y =$ $(2^{1/2}/2^3\pi^{1/2})(Z/a)^{5/2}e^{-Zr/2a}\,r\sin\theta\sin\phi = \pi^{-1/2}(Z/2a)^{5/2}ye^{-Zr/2a}$.

18.18 **(a)** $a_1 + ib_1 = a_2 + ib_2$. We must have $a_1 = a_2$ and $b_1 = b_2$.

(b) $(2\pi)^{-1/2}e^{im\phi} = (2\pi)^{-1/2}e^{im(\phi + 2\pi)} = (2\pi)^{-1/2}e^{im\phi}e^{2\pi mi}$, so $1 = e^{2\pi mi} =$ $\cos(2\pi m) + i\sin(2\pi m)$, where (18.24) was used. Equating the real parts and the imaginary parts of this last equation [as shown in part (a)], we have $\cos(2\pi m) = 1$ and $\sin(2\pi m) = 0$. The cosine function equals 1 only for angles of $0, \pm 2\pi, \pm 4\pi, \pm 6\pi, \ldots$ and the sine vanishes at these angles. Therefore $2\pi m = 0, \pm 2\pi, \pm 4\pi, \ldots$ and $m = 0, \pm 1, \pm 2, \ldots$.

18.19 ψ_{2p_z} has the form bze^{-dr}; b and d are constants. Along the z axis, $x = 0 = y$ and $r = (x^2 + y^2 + z^2)^{1/2} = (z^2)^{1/2} = |z|$. Thus $\psi_{2p_z} = bze^{-d|z|}$ along the z axis. Near $z = 0$, $e^{-d|z|} \approx 1$ and $\psi_{2p_z} \approx bz$ (a straight line through the origin). For large values of $|z|$, the exponential causes ψ to fall to zero. Also, $|\psi_{2p_z}|^2 = |b|^2z^2e^{-2d|z|}$ along the z axis. $|\psi|^2$ is parabolic near the origin and is positive for negative values of z. The graphs are:

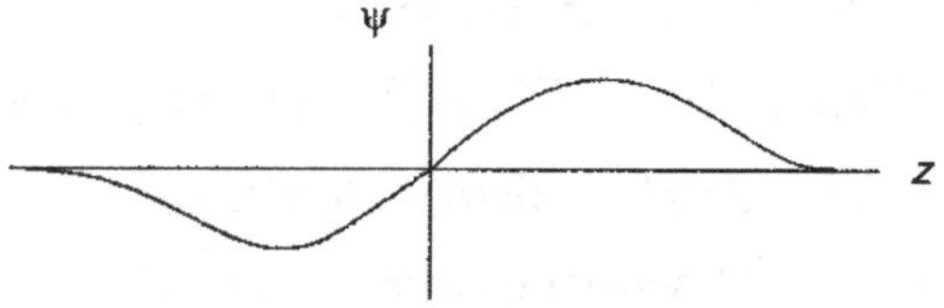

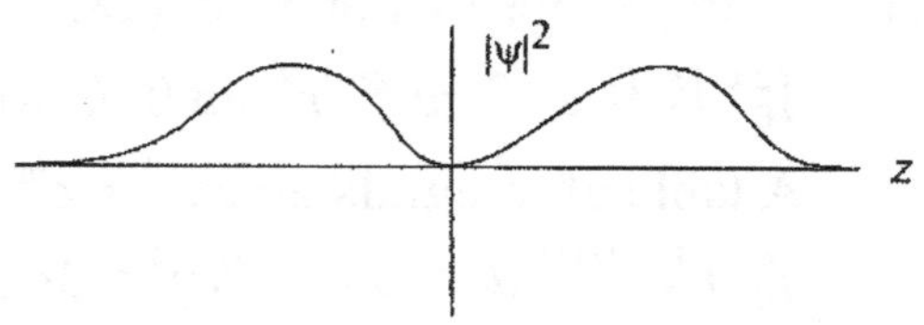

18.20 **(a)** We have $0.9 = (1/32\pi a^3)\int_0^{r_{2s}}(2 - r/a)^2r^2e^{-r/a}\,dr\int_0^{\pi}\sin\theta\,d\theta\int_0^{2\pi}d\phi =$ $e^{-r_{2s}/a}[-r_{2s}^2/2a^2 - r_{2s}/a - 1 - r_{2s}^4/8a^4] + 1$, where a table of integrals was used. Let $\upsilon \equiv r_{2s}/a$. We must solve $e^{-\upsilon}(\upsilon^2/2 + \upsilon + 1 + \upsilon^4/8) = 0.1$. Trial and error (or a spreadsheet Solver) gives $\upsilon = 9.125$ and $r_{2s} = 9.125a =$ 4.83 Å.

(b) $0.95 = e^{-r_{2s}/a}[-r_{2s}^2/2a^2 - r_{2s}/a - 1 - r_{2s}^4/8a^4] + 1$ and $e^{-\upsilon}(\upsilon^2/2 + \upsilon + 1 + \upsilon^4/8) = 0.05$. We find $\upsilon = 10.283$ and $r_{2s} = 10.283a =$ 5.44 Å.

18.21 $\partial e^{-Zr/a}/\partial x = (\partial e^{-Zr/a}/\partial r)(\partial r/\partial x)$. We have $r = (x^2 + y^2 + z^2)^{1/2}$ and $\partial r/\partial x = \frac{1}{2}(x^2 + y^2 + z^2)^{-1/2}(2x) = x/r$. So $\partial e^{-Zr/a}/\partial x = -(Zx/ra)e^{-Zr/a}$. $\partial^2 e^{-Zr/a}/\partial x^2 = (\partial/\partial x)(-Zxe^{-Zr/a}/ra) = -Ze^{-Zr/a}/ra - x[(\partial/\partial r)(Ze^{-Zr/a}/ra)](\partial r/\partial x) = -Ze^{-Zr/a}/ra + (x^2Z^2/r^2a^2)e^{-Zr/a} + (Zx^2/ar^3)e^{-Zr/a}$. Similar equations hold for $\partial^2/\partial y^2$ and $\partial^2/\partial z^2$ of $e^{-Zr/a}$. Then $(\partial^2/\partial x^2 + \partial^2/\partial y^2 + \partial^2/\partial z^2)e^{-Zr/a} = -3Ze^{-Zr/a}/ra + [(x^2 + y^2 + z^2)Z^2/r^2a^2]e^{-Zr/a} + [Z(x^2 + y^2 + z^2)/ar^3]e^{-Zr/a} = -3Ze^{-Zr/a}/ra + (Z^2/a^2)e^{-Zr/a} + (Z/ra)e^{-Zr/a} = -2Ze^{-Zr/a}/ra + (Z^2/a^2)e^{-Zr/a} = -2\mu e^2Ze^{-Zr/a}/4\pi\varepsilon_0 r\hbar^2 + (Z^2/a^2)e^{-Zr/a}$, where (18.14) was used. The Hamiltonian operator is (18.5) and $\hat{H}\psi_{1s} = -(\hbar^2/2\mu)(\partial^2/\partial x^2 + \partial^2/\partial y^2 + \partial^2/\partial z^2)[\pi^{-1/2}(Z/a)^{3/2}e^{-Zr/a}] - (Ze^2/4\pi\varepsilon_0 r)[\pi^{-1/2}(Z/a)^{3/2}e^{-Zr/a}] = -(Ze^2/4\pi\varepsilon_0 r)\pi^{-1/2}(Z/a)^{3/2}e^{-Zr/a} - \pi^{-1/2}(Z/a)^{3/2}(\hbar^2/2\mu)[-2\mu e^2Ze^{-Zr/a}/4\pi\varepsilon_0 r\hbar^2 + (Z^2/a^2)e^{-Zr/a}] = -(\hbar^2Z^2/2\mu a^2)\pi^{-1/2}(Z/a)^{3/2}e^{-Zr/a} = -[Z^2e^2/2(4\pi\varepsilon_0)a][\pi^{-1/2}(Z/a)^{3/2}e^{-Zr/a}] = E_{1s}\psi_{1s}$.

18.22 $\langle V\rangle = \langle -Ze^2/4\pi\varepsilon_0 r\rangle = -(Ze^2/4\pi\varepsilon_0)\int \psi_{1s}^* r^{-1}\psi_{1s}\, d\tau = -(Z^4e^2/4\pi\varepsilon_0 a^3\pi)\int_0^{2\pi}\int_0^{\pi}\int_0^{\infty} e^{-Zr/a}r^{-1}e^{-Zr/a}r^2 \sin\theta\, dr\, d\theta\, d\phi = -(Z^4e^2/4\pi\varepsilon_0 a^3\pi)\int_0^{\infty} re^{-2Zr/a}\, dr \int_0^{\pi}\sin\theta\, d\theta \int_0^{2\pi} d\phi$. Using either the definite integral $\int_0^{\infty} r^n e^{-br}\, dr = n!/b^{n+1}$ or the indefinite integral $\int re^{-br}\, dr = -e^{-br}(r/b + 1/b^2)$, we get $\int_0^{\infty} re^{-2Zr/a}\, dr = a^2/4Z^2$. Then $\langle V\rangle = -(Z^4e^2/4\pi\varepsilon_0 a^3\pi)(a^2/4Z^2)2(2\pi) = -Z^2e^2/4\pi\varepsilon_0 a$ for the ground state.

18.23 **(a)** We have: $\langle r\rangle = \int \psi_{2p_z}^* r\psi_{2p_z}\, d\tau = (Z^5/32\pi a^5)\int_0^{2\pi}\int_0^{\pi}\int_0^{\infty} re^{-Zr/2a}\cos\theta\, r\, re^{-Zr/2a}\cos\theta\, r^2\sin\theta\, dr\, d\theta\, d\phi = (Z^5/32\pi a^5)\int_0^{\infty} r^5e^{-Zr/a}\, dr\int_0^{2\pi} d\phi\int_0^{\pi}\cos^2\theta\sin\theta\, d\theta$. A table of definite integrals gives $\int_0^{\infty} r^ne^{-br}\, dr = n!/b^{n+1}$ for $b > 0$ and n a positive integer. So $\int_0^{\infty} r^5e^{-Zr/a}\, dr = 5!a^6/Z^6$. Let $t = \cos\theta$; then $dt = -\sin\theta\, d\theta$ and

$\int_0^\pi \cos^2\theta \sin\theta \, d\theta = -\int_1^{-1} t^2 \, dt = 2/3$. So
$\langle r \rangle = (Z^5/32\pi a^5)(120a^6/Z^6)(2/3)(2\pi) = 5a/Z$.

(b) The $2p_z$ and $2p_x$ orbitals have the same shape and the same size and differ only in spatial orientation. Since r does not depend on spatial orientation, $\langle r \rangle$ must be the same for the $2p_x$ and $2p_z$ states.

(c) $\langle r \rangle_{2p_x} = \int \psi_{2p_x}^* r \psi_{2p_x} \, d\tau = (Z^5/32\pi a^5) \int_0^\infty r^5 e^{-Zr/a} \, dr \int_0^{2\pi} \cos^2\phi \, d\phi \times \int_0^\pi \sin^3\theta \, d\theta$. The r integral was found in (a). A table of integrals gives $\int \cos^2\phi \, d\phi = \frac{1}{2}\phi + \frac{1}{4}\sin 2\phi$ and $\int \sin^3\theta \, d\theta = -\frac{2}{3}\cos\theta - \frac{1}{3}\cos\theta \sin^2\theta$ and we find $\langle r \rangle_{2p_x} = (Z^5/32\pi a^5)(120a^6/Z^6)(\pi)(4/3) = 5a/Z$.

18.24 Equations (18.28) and (18.19) give the ground-state H-atom radial distribution function as $R_{1s}^2 r^2 = 4(Z/a)^3 r^2 e^{-2Zr/a}$. The maximum is found by setting the derivative equal to zero: $0 = 4(Z/a)^3[2re^{-2Zr/a} - (2Zr^2/a)e^{-2Zr/a}]$ and $r = a/Z$. (The root $r = 0$ is a minimum.)

18.25 Let $c = 2.00$ Å. To find the desired probability, we integrate $\psi^*\psi \, d\tau$ over the volume of a sphere of radius c. The angles go over their full ranges and r goes from 0 to c. Table 18.1 and Eq. (18.27) give the probability as $(1/\pi a^3) \times \int_0^c \int_0^\pi \int_0^{2\pi} e^{-2r/a} r^2 \sin\theta \, dr \, d\theta \, d\phi = (1/\pi a^3) \int_0^c r^2 e^{-2r/a} \, dr \int_0^\pi \sin\theta \, d\theta \int_0^{2\pi} d\phi$. The radial integral has the same form as the radial integral in Prob. 18.20 except that r_{2s} is replaced by c. So $\int_0^c r^2 e^{-2r/a} \, dr = -e^{-2c/a}(\frac{1}{2}ac^2 + \frac{1}{2}a^2c + \frac{1}{4}a^3) + \frac{1}{4}a^3$. The θ and ϕ integrals are given in Prob. 18.20, and the desired probability is $-e^{-2c/a}(2c^2/a^2 + 2c/a + 1) + 1$. We have $c/a = (2.00\text{ Å})/(0.5295\text{ Å}) = 3.77_7$, and the probability is $1 - e^{-2(3.777)}[2(3.777)^2 + 2(3.777) + 1] = 0.981$.

18.26 $\int_0^{2\pi} |\Phi|^2 \, d\phi = \int_0^{2\pi} \Phi^*\Phi \, d\phi = (1/2\pi) \int_0^{2\pi} e^{-im\phi} e^{im\phi} \, d\phi = (1/2\pi) \int_0^{2\pi} d\phi = 1$.

18.27 **(a)** T; **(b)** T; **(c)** T.

18.28 Let θ be the angle between the positive z axis and an angular-momentum vector. For $m = +1$ in Fig. 18.10, $\cos\theta = L_z/|\mathbf{L}| = \hbar/\sqrt{2}\hbar = 1/\sqrt{2} = 0.7071$ and $\theta = 45°$. For $m = 0$, $\theta = 90°$. For $m = -1$, $\theta = 180° - 45° = 135°$.

18.29 **(a)** $L = rp \sin \beta$, where β is the angle between **r** and **p**. For circular motion, the velocity vector **v** is perpendicular to the radius, and so is **p** $= m$**v**; thus $\beta = 90°$ and $\sin \beta = 1$. Since $p = mv$, we get $L = mvr$.

(b) The **L** vector is perpendicular to both **r** and **p** and **r** and **p** lie in the plane of the circular motion. Hence **L** is perpendicular to the plane of the circular orbit.

18.30 **(a)** From Sec. 18.4, $|\mathbf{L}| = [l(l+1)]^{1/2}\hbar = 0$, since $l = 0$ for the 1*s* state.

(b) From Sec. 17.3, Bohr had $|\mathbf{L}| = mvr = nh/2\pi = h/2\pi$ for the ground state. The Bohr theory had the wrong value of $|\mathbf{L}|$.

18.31 $|\mathbf{L}| = [l(l+1)]^{1/2}\hbar = [1(2)]^{1/2}\hbar = \sqrt{2}\hbar = \sqrt{2}(6.626\times10^{-34}\text{ J s})/2\pi =$ 1.491×10^{-34} J s.

18.32 $|\mathbf{S}| = [s(s+1)]^{1/2}\hbar = [0.5(1.5)]^{1/2}(6.626 \times 10^{-34}\text{ J s})/2\pi = 9.13 \times 10^{-35}$ J s.

18.33 Let θ be the angle between the *z* axis and a spin vector. For $m_s = +\frac{1}{2}$, we have $\cos\theta = \frac{1}{2}\hbar/\sqrt{3/4}\hbar = 1/\sqrt{3} = 0.57735$ and $\theta = 54.7°$. For $m_s = -\frac{1}{2}$, $\theta = 180° - 54.7° = 125.3°$.

18.34 **(a)** Electronic orbital angular momentum; $|\mathbf{L}| = [l(l+1)]^{1/2}\hbar$.

(b) *z* component of electronic orbital angular momentum; $L_z = m\hbar$.

(c) Electronic spin angular momentum; $|\mathbf{S}| = [s(s+1)]^{1/2}\hbar$.

(d) *z* component of electronic spin angular momentum; $S_z = m_s\hbar$.

18.35 **(a)** For $s = 3/2$, Eq. (18.32) gives $m_s = 3/2, 1/2, -1/2$, and $-3/2$. The possible *z* components of the spin are $m_s\hbar$. The length of the spin vector is $\sqrt{s(s+1)}\hbar = \frac{1}{2}\sqrt{15}\hbar$. The possible orientations are

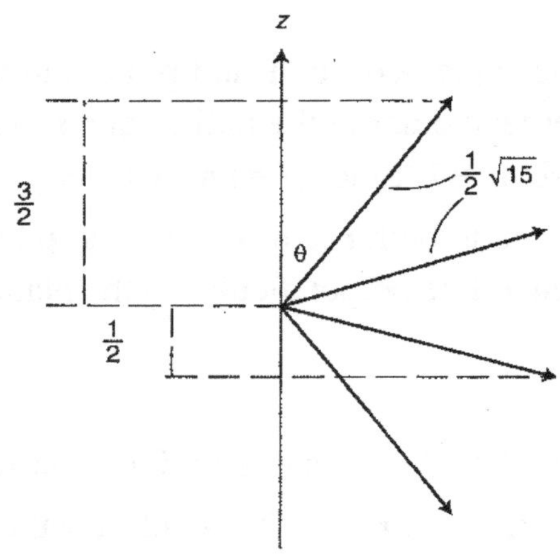

(b) $\cos\theta = 1.5\hbar/\frac{1}{2}\sqrt{15}\hbar =$ 0.7746 and $\theta = 39.2°$.

18.36 **(a)** F; **(b)** F; **(c)** F; **(d)** T.

18.37 **(a)** Neither, since $f(2)g(1) \neq \pm f(1)g(2)$.
(b) Symmetric, since $g(2)g(1) = g(1)g(2)$.
(c) Antisymmetric, since $f(2)g(1) - g(2)f(1) = -[f(1)g(2) - g(1)f(2)]$.
(d) Symmetric. **(e)** Antisymmetric.

18.38 **(a)** Antisymmetric. **(b)** Symmetric. **(c)** Antisymmetric, since interchange of 1 and 2 multiplies h by –1 and multiplies k by – 1. **(d)** Neither.

18.39 The true ground-state energy of He is –79.0 eV (Sec. 18.6). The variational value –86.7 eV is less than the true E_{gs}; this violates the variation theorem (17.86), so there must be an error in the calculation.

18.40 There is one electron, so $S = s = \frac{1}{2}$ and $2s + 1 = 2$; also, $L = l$.
(a) ^2S. **(b)** ^2P. **(c)** ^2D.

18.41 $2S + 1 = 4$, so $S = 3/2$. The code letter F means $L = 3$.

18.42 **(a)** Total electronic orbital angular momentum; $|\mathbf{L}| = [L(L+1)]^{1/2}\hbar$.
(b) Total electronic spin angular momentum; $|\mathbf{S}| = [S(S+1)]^{1/2}\hbar$.
(c) z component of total electronic spin angular momentum; $S_z = M_S\hbar$.

18.43 **(a)** D means $L = 2$, so $|\mathbf{L}| = [L(L+1)]^{1/2}\hbar = 6^{1/2}\hbar$.

(b) $2S + 1 = 3$ and $S = 1$, so $|\mathbf{S}| = [S(S+1)]^{1/2}\hbar = 2^{1/2}\hbar$.

18.44 **(a)** Electrons in filled subshells can be ignored. The $3d$ electron has $s = ½$ and $l = 2$. With only one electron outside filled subshells, there is only one term, namely ^2D, with $L = 2$ and $S = ½$.

(b) ^2P.

18.45 Let θ be the angle between the z axis and a spin vector. For spin function $\alpha(1)$ and vector $\mathbf{S}_1$, we have $\cos\theta = \frac{1}{2}\hbar/(3/4)^{1/2}\hbar = 1/3^{1/2} = 0.57735$ and $\theta = 54.7°$ (as in Prob. 18.35). Likewise, $\theta = 54.7°$ for $\mathbf{S}_2$. For $\alpha(1)\alpha(2)$, the total spin vector $\mathbf{S}$ has magnitude $2^{1/2}\hbar$ and z component $\hbar$, so $\cos\theta = \hbar/2^{1/2}\hbar = 1/2^{1/2} = 0.7071$ and $\theta = 45°$ for $\mathbf{S}$. $\mathbf{S}$ lies closer to the z axis than do $\mathbf{S}_1$ and $\mathbf{S}_2$. $\mathbf{S}_1$ and $\mathbf{S}_2$ lie on the surface of a cone making angle 54.7° with the z axis and $\mathbf{S}$ lies within this cone:

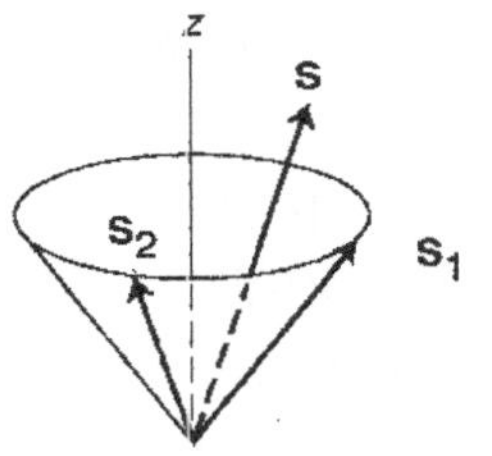

18.46 **(a)** $l = 1$ for a $3p$ electron and $l = 2$ for a $3d$ electron. The maximum and minimum L values are $2 + 1$ and $|2 - 1|$ and the possible L values are 3, 2, and 1. A $3p$ electron has $s = \frac{1}{2}$ and so does a $3d$ electron, so the maximum and minimum S values are $\frac{1}{2} + \frac{1}{2}$ and $|\frac{1}{2} - \frac{1}{2}|$; the possible S values are 1 and 0. (Electrons in filled subshells were ignored.)

(b) Combining each S value with each L value, we have as the terms: ^3F, ^3D, ^3P, ^1F, ^1D, and ^1P, where $2S + 1 = 3$ and 1 for $S = 1$ and 0, and P, D, F denote $L = 1, 2, 3$.

18.47 **(a)** $L = 0$ and $S = \frac{1}{2}$, so $J = \frac{1}{2}$ and the only level is $^2S_{1/2}$.

(b) $L = 1$ and $S = 3/2$. So $J = 5/2, 3/2, 1/2$, and the levels are $^4P_{5/2}$, $^4P_{3/2}$, $^4P_{1/2}$.

(c) $L = 3$ and $S = 2$, so the levels are $^5F_5$, $^5F_4$, $^5F_3$, $^5F_2$, $^5F_1$.

(d) $L = 2$ and $S = 1$; the levels are $^3D_3$, $^3D_2$, $^3D_1$.

18.48 Let the electrons be numbered 1, 2, and 3. The nuclear charge is $3e$. As was done with He in Eq. (18.35), we use the electron mass m_e in the Hamiltonian. Then $\hat{H} = -(\hbar^2/2m_e)\nabla_1^2 - (\hbar^2/2m_e)\nabla_2^2 - (\hbar^2/2m_e)\nabla_3^2 - 3e^2/4\pi\varepsilon_0 r_1 - 3e^2/4\pi\varepsilon_0 r_2 - 3e^2/4\pi\varepsilon_0 r_3 + e^2/4\pi\varepsilon_0 r_{12} + e^2/4\pi\varepsilon_0 r_{13} + e^2/4\pi\varepsilon_0 r_{23}$, where r_1 is the distance between electron 1 and the nucleus.

18.49 Let f and g denote the $n = 1$ and $n = 2$ spatial functions, i.e., $f = (2/a)^{1/2} \sin(\pi x/a)$ and $g = (2/a)^{1/2} \sin(2\pi x/a)$. With interelectronic repulsion ignored, the spatial wave function is a product of one-electron spatial functions. Analogous to Eqs. (18.45) and (18.46), we form the linear combinations $2^{-1/2}[f(1)g(2) + f(2)g(1)]$ and $2^{-1/2}[f(1)g(2) - f(2)g(1)]$ that don't distinguish between the electrons. To satisfy the Pauli principle, the symmetric spatial function must be combined with the antisymmetric two-electron spin function (18.41) and the antisymmetric spatial function must be combined with one of the symmetric spin functions. The approximate wave functions are therefore

$2^{-1/2}[f(1)g(2) + f(2)g(1)]2^{-1/2}[\alpha(1)\beta(2) - \beta(1)\alpha(2)]$
$2^{-1/2}[f(1)g(2) - f(2)g(1)]\alpha(1)\alpha(2)$
$2^{-1/2}[f(1)g(2) - f(2)g(1)]\beta(1)\beta(2)$
$2^{-1/2}[f(1)g(2) - f(2)g(1)]2^{-1/2}[\alpha(1)\beta(2) + \beta(1)\alpha(2)]$

The first wave function has $S = 0$. The second, third and fourth have $S = 1$ and have the same energy as one another (since they have the same spatial factor). According to Hund's rule, the $S = 1$ functions lie lower.

18.50 The ground-state configuration is $1s^2 2s^2$. To make the approximate wave function antisymmetric, we use a Slater determinant. Analogous to Eq. (18.54), we have

$$\psi_{gs} \approx N \begin{vmatrix} 1s(1)\alpha(1) & 1s(1)\beta(1) & 2s(1)\alpha(1) & 2s(1)\beta(1) \\ 1s(2)\alpha(2) & 1s(2)\beta(2) & 2s(2)\alpha(2) & 2s(2)\beta(2) \\ 1s(3)\alpha(3) & 1s(3)\beta(3) & 2s(3)\alpha(3) & 2s(3)\beta(3) \\ 1s(4)\alpha(4) & 1s(4)\beta(4) & 2s(4)\alpha(4) & 2s(4)\beta(4) \end{vmatrix}$$

where N is a normalization constant (equal to $1/\sqrt{24}$).

18.51 H ($1s$), Li ($1s^2 2s$), B ($1s^2 2s^2 2p$), C ($1s^2 2s^2 2p^2$), N ($1s^2 2s^2 2p^3$), O ($1s^2 2s^2 2p^4$), and F ($1s^2 2s^2 2p^5$) all have one or more unpaired electrons and so have $S \neq 0$ and have paramagnetic ground states. He ($1s^2$), Be ($1s^2 2s^2$), and Ne ($1s^2 2s^2 2p^6$) have all electrons paired, have $S = 0$ and $L = 0$ and do not have paramagnetic ground states. (Ne has two $2p$ electrons with $m = +1$, two $2p$ electrons with $m = 0$, and two $2p$ electrons with $m = -1$, and so has total orbital angular-momentum quantum number $L = 0$.)

18.52 We want the energy needed for $_{18}Ar^{17+} \rightarrow Ar^{18+}$. The ion Ar^{17+} has one electron and so is a hydrogenlike species. From Eq. (18.18) with $n = 1$, the ionization potential is $(18)^2(13.6\text{ V}) = 4406\text{ V}$.

18.53 $\varepsilon = -(Z_{\text{eff}}^2/n^2)(13.6\text{ eV})$.

(a) In Li, the first ionization potential is for removal of a $2s$ electron, so $5.4\text{ eV} = (Z_{\text{eff}}^2/2^2)(13.6\text{ eV})$ and $Z_{\text{eff}} = 1.26$.

(b) $9.3\text{ eV} = (Z_{\text{eff}}^2/2^2)(13.66\text{ eV})$ and $Z_{\text{eff}} = 1.65$. (The increase over Li is due to the poor screening of one $2s$ electron by the other.)

18.54 **(a)** If $s = 3/2$, the m_s values are $3/2, 1/2, -1/2, -3/2$. For $s = 3/2$, the electrons are still fermions and the Pauli exclusion principle still holds. The four values of m_s mean that 4 electrons (instead of 2) can go in each orbital. The $1s$, $2s$, and $2p$ subshells would therefore hold 4, 4, and 12 electrons (double their capacities for $s = ½$). The ground-state configurations are $1s^3$, $1s^4 2s^4 2p$, and $1s^4 2s^4 2p^9$.

(b) For $s = 1$, the electrons would be bosons and there would be no restriction on the number of electrons in a spin-orbital. The ground-state configurations would therefore be $1s^3$, $1s^9$, and $1s^{17}$.

18.55 **(a)** The outer electron in K is further from the nucleus, so Na has the higher ionization potential.

(b) The ineffective screening of one $4s$ electron by the other makes Z_{eff} greater in Ca than in K, so Ca has the higher ionization potential.

(c) Cl.

(d) Kr.

18.56 For $Z = 10$, the figure gives $\sqrt{\varepsilon/\varepsilon_H} = 8, 2.2,$ and 1.6. Since $\varepsilon_H = -13.6$ eV, we get $\varepsilon_{1s} \approx -870$ eV, $\varepsilon_{2s} \approx -66$ eV, and $\varepsilon_{2p} \approx -35$ eV. Substitution in $\varepsilon = -(Z_{\text{eff}}^2/n^2)(13.6 \text{ eV})$ gives $Z_{\text{eff},1s} \approx 8$, $Z_{\text{eff},2s} \approx 4.4$, $Z_{\text{eff},2p} \approx 3.2$.

18.57 **(a)** T; **(b)** T.

18.58 Nitrogen, with 3 unpaired electrons.

18.59 Ionization energy data in Sec. 18.8 show that $\Delta E = 5.1$ eV for $\text{Na} \rightarrow \text{Na}^+ + \text{e}^-$, so $E(\text{Na}^+ + \text{e}^-) > E(\text{Na})$ and $E(\text{Na}^+ + 2\text{e}^-) > E(\text{Na} + \text{e}^-)$. Electron affinity data give $\Delta E = -0.5$ eV for $\text{Na} + \text{e}^- \rightarrow \text{Na}^-$, so $E(\text{Na}^-) < E(\text{Na} + \text{e}^-)$. The lowest-energy (most stable) system is Na^-; the highest-energy system is $\text{Na}^+ + 2\text{e}^-$.

18.60 **(a)** Sr; **(b)** F; **(c)** K; **(d)** C; **(e)** Cl^-. Cl^- and Ar are isoelectronic and the higher Z in Ar means a smaller size.

18.61 $\phi = \sum_{k=1}^{5} b_k g_k$, where $g_k = N_k \exp(-\zeta_k r/a_0)$. Replacement of Z by ζ_k in the $1s$ orbital in Table 18.1 gives $N_k = \pi^{-1/2}(\zeta_k/a_0)^{3/2}$. So $\phi = b_1 g_1 + b_2 g_2 + \cdots + b_5 g_5 = 0.768\pi^{-1/2}(1.417/a_0)^{3/2}\exp(-1.417r/a_0) + 0.233\pi^{-1/2}(2.377/a_0)^{3/2}\exp(-2.377r/a_0) + 0.041\pi^{-1/2}(4.396/a_0)^{3/2}\exp(-4.396r/a_0) - 0.010\pi^{-1/2}(6.527/a_0)^{3/2}\exp(-6.527r/a_0) + 0.002\pi^{-1/2}(7.943/a_0)^{3/2}\exp(-7.943r/a_0)$.

18.62 $D_1 = 2^{-1/2}\begin{vmatrix} 1s(1)\alpha(1) & 2s(1)\alpha(1) \\ 1s(2)\alpha(2) & 2s(2)\alpha(2) \end{vmatrix} =$

$2^{-1/2}[1s(1)\alpha(1)2s(2)\alpha(2) - 1s(2)\alpha(2)2s(1)\alpha(1)] =$
$2^{-1/2}[1s(1)2s(2) - 1s(2)2s(1)]\alpha(1)\alpha(2)$, which is the $S = 1$, $M_S = 1$ function in Fig. 18.13. Replacement of α by β in the preceding equations shows that D_4 equals the $S = 1$, $M_S = -1$ function.

$D_2 = 2^{-1/2}\begin{vmatrix} 1s(1)\alpha(1) & 2s(1)\beta(1) \\ 1s(2)\alpha(2) & 2s(2)\beta(2) \end{vmatrix} =$

$2^{-1/2}[1s(1)2s(2)\alpha(1)\beta(2) - 2s(1)1s(2)\beta(1)\alpha(2)]$. Interchange of α and β in D_2 gives $D_3 = 2^{-1/2}[1s(1)2s(2)\beta(1)\alpha(2) - 2s(1)1s(2)\alpha(1)\beta(2)]$. We have $2^{-1/2}(D_2 + D_3) = 2^{-1/2}\{1s(1)2s(2)2^{-1/2}[\alpha(1)\beta(2) + \beta(1)\alpha(2)] - 2s(1)1s(2)2^{-1/2} \times$

$[\beta(1)\alpha(2) + \alpha(1)\beta(2)]\}$, which is the $S = 1$, $M_s = 0$ function. Similarly, $2^{-1/2}(D_2 - D_3)$ is found to be the $S = 0$, $M_S = 0$ function.

18.63 $\int_0^{2\pi} \int_0^{\pi} \int_0^{a} r^2 \sin\theta \, dr \, d\theta \, d\phi = \int_0^{2\pi} d\phi \int_0^{\pi} \sin\theta \, d\theta \int_0^{a} r^2 \, dr = 2\pi(2)(a^3/3) = \frac{4}{3}\pi a^3$.

18.64 **(a)** E is proportional to $-Z^2$, so $E_H > E_{He^+}$.

(b) The ionization energy of K^+.

(c) The energy-level spacing for these one-electron species is proportional to Z^2, so ν is proportional to Z^2. Thus $\nu_{He^+} > \nu_H$ and $\lambda_H > \lambda_{He^+}$.

(d) These quantities are equal.

18.65 **(a)** Particle in a box; rigid rotor.

(b) Harmonic oscillator.

(c) Hydrogenlike atom.

18.66 **(a)** $d\tau = \frac{4}{3}\pi r^3 = \frac{4}{3}\pi(0.0010 \text{ Å})^3 = 4.19 \times 10^{-9} \text{ Å}^3$. $|\psi|^2 = (1/\pi a^3)e^{-2r/a} = [\pi(0.5295 \text{ Å})^3]^{-1} = 2.14 \text{ Å}^{-3}$. $|\psi|^2 \, d\tau = 9.0 \times 10^{-9}$.

(b) $|\psi|^2 = (1/\pi a^3)e^{-2(0.50 \text{ Å}/a)} = 0.324$. $d\tau = 4.19 \times 10^{-9} \text{ Å}^3$. $|\psi|^2 \, d\tau = 1.4 \times 10^{-9}$.

(c) $|\psi|^2 = 1.35 \times 10^{-8}$ and $|\psi|^2 \, d\tau = 5.7 \times 10^{-17}$.

18.67 From (18.28) and (18.19), this probability is $\Pr = |R_{1s}(r)|^2 r^2 \, dr = (4/a^3)e^{-2r/a} r^2 \, dr$.

(a) $\Pr = 4(0.5295 \text{ Å})^{-3} e^{-2(0.100/0.5295)}(0.100 \text{ Å})^2(0.001 \text{ Å}) = 0.00018_5$.

(b) $4(0.5295 \text{ Å})^{-3}\exp[-2(0.500/0.5295)](0.500 \text{ Å})^2(0.001 \text{ Å}) = 0.00102$.

(c) 0.00062.

(d) 4.2×10^{-9}. (See also Fig. 18.8.)

18.68 **(a)** $d\tau = dx$ and $-\infty \le x \le \infty$.

$0 \le x \le a,\ 0 \le y \le b,\ 0 \le z \le c.$

(c) $d\tau = r^2 \sin\theta\, dr\, d\theta\, d\phi$. $0 \le r \le \infty,\ 0 \le \theta \le \pi,\ 0 \le \phi \le 2\pi$.

18.69 Yes. The gravitational force is far smaller than the electrostatic force and so can be neglected. $|F_{\text{grav}}|/|F_{\text{el}}| = (Gm_e m_p/r^2)/(e^2/4\pi\varepsilon_0 r^2) = 4\pi\varepsilon_0 G m_e m_p/e^2 =$ $4\pi(8.85 \times 10^{-12}\ \text{C}^2/\text{N-m}^2)(6.67 \times 10^{-11}\ \text{m}^3/\text{s}^2\text{-kg})(9.1 \times 10^{-31}\ \text{kg}) \times$ $(1.67 \times 10^{-27}\ \text{kg})/(1.6 \times 10^{-19}\ \text{C})^2 = 4 \times 10^{-40}$.

18.70 **(a)** $\psi_{2p_z} = \frac{1}{4}(2\pi)^{-1/2}a^{-5/2}re^{-r/2a}\cos\theta$. The maximum value of $\cos\theta$ occurs at $\theta = 0$, where $\cos\theta = 1$. Setting $\cos\theta = 1$ in ψ and then taking $\partial\psi/\partial r = 0$, we get $\psi_{2p_z}/\partial r = 0 = \frac{1}{4}(2\pi)^{-1/2}a^{-5/2}[e^{-r/2a} - (1/2a)re^{-r/2a}]$. Solving for r, we get $r = 2a$. Setting $r = 2a$ and $\cos\theta = 1$ in ψ_{2p_z}, we get $\psi_{\text{max}} =$ $\frac{1}{4}(2\pi)^{-1/2}a^{-5/2}(2a)e^{-1} = \pi^{-1/2}(2a)^{-3/2}e^{-1}$.

(b) This problem is incompletely stated, in that it is intended that the calculations be done for points in the *yz* plane, where $x = 0$. In the *yz* plane, $|\psi_{2p_z}/\psi_{\text{max}}| = \frac{1}{2}(|z|/a)e\ \exp\{-\frac{1}{2}[(y/a)^2 + (z/a)^2]^{1/2}\} = k$.
A BASIC program is

```
15 INPUT "PSI/PSIMAX";K
25 PRINT "PSI/PSIMAX=";K
35 FOR Z = 0.01 TO 10 STEP 0.01
45 W=LOG(Z/2)+1–LOG(K)
55 IF W<Z/2 THEN 85
65 Y=SQR(4*W*W*–Z*Z)
75 PRINT "Z/A=";Z;" Y1/A=";
   Y;" Y2/A=";–Y
85 NEXT Z
95 STOP
```

18.71 As a hint, elements were named for the physicists in (a) and (d).

18.72 **(a)** F; **(b)** F. **(c)** F; this is true only for *identical* fermions. **(d)** F. **(e)** F. **(f)** F; **(g)** F; only well-behaved solutions are possible stationary states. **(h)** T. **(i)** F; this is true only if the states have the same energy.

CHAPTER 19 Molecular Electronic Structure

CHAPTER OUTLINE

A full and correct treatment of molecules must be based on quantum mechanics. Indeed, the stability of a covalent bond cannot be understood without quantum mechanics. Because of the mathematical difficulties involved in the application of quantum mechanics to molecules, chemists developed a variety of empirical concepts to describe bonding. Section 19.1 discusses some of these concepts. Section 19.2 describes how the molecular Schrödinger equation is separated into Schrödinger equations for electronic motion and for nuclear motion. The one-electron molecule H_2^+ is discussed in Sec. 19.3 to develop some ideas about electron orbitals in molecules. A major approximation method used in describing molecular electronic structure is the molecular-orbital method, developed in Secs. 19.4 to 19.6. Section 19.8 shows how molecular properties are calculated from electronic wave functions. Section 19.9 discusses some of the remarkable advances in calculation of molecular electronic structure made in recent years. The currently most widely used method for calculating molecular properties, density-functional theory, is presented in Sec. 19.10. Section 19.11 discusses semiempirical methods of calculation, which can treat large molecules. Section 19.12 gives details on how electronic-structure calculations are done. Section 19.13 presents the molecular-mechanics method, a nonquantum-mechanical method that can be applied to very large molecules.

19.1 CHEMICAL BONDS

Bond Radii

The length of a bond in a molecule is the distance between the nuclei of the two atoms forming the bond. Spectroscopic and diffraction methods (Chapters 20 and 23) enable bond lengths to be measured accurately. Bond lengths range from 0.74 Å in H_2 to 4.65 Å in Cs_2 and are usually in the range 1–2 Å for bonds between elements in the first, second, and third periods. The length of a given kind of bond is found to be approximately constant from molecule to molecule. For example, the carbon–carbon single-bond length in most nonconjugated molecules lies in the range 1.53 to 1.54 Å. Moreover, one finds that the bond length d_{AB} between atoms A and B is *approximately* equal to $\frac{1}{2}(d_{AA} + d_{BB})$, where d_{AA} and d_{BB} are the typical A—A and B—B bond lengths. For example, let A and B be Cl and C. The bond length in Cl_2 is 1.99 Å, and $\frac{1}{2}(d_{AA} + d_{BB}) = \frac{1}{2}(1.99 + 1.54)$ Å $= 1.76$ Å, in good agreement with the observed bond length 1.76_6 Å in CCl_4.

One can therefore take $\frac{1}{2}d_{AA}$ as the **bond radius** (or **covalent radius**) r_A for atom A and use a table of bond radii to estimate the bond length d_{AB} as $r_A + r_B$. Double and triple bonds are shorter than the corresponding single bonds, and so different bond radii are used for single, double, and triple bonds. Some bond radii (due mainly to Pauling) in angstroms (Å) are (1 Å ≡ 10^{-8} cm ≡ 10^{-10} m):

	H	C	N	O	F	P	S	Cl	Br	I
Single	0.30	0.77	0.70	0.66	0.64	1.10	1.04	0.99	1.14	1.33
Double		0.67	0.60	0.56		1.00	0.94			
Triple		0.60	0.55							

The bond length 0.74 Å in H_2 indicates $r_A = 0.37$ Å for H, but the listed value 0.30 Å works better in predicting bond lengths between H and other elements.

When atoms A and B differ substantially in electronegativity, the observed bond length is often shorter than $r_A + r_B$.

The carbon–carbon bond length in benzene is 1.40 Å. This lies between the carbon–carbon single-bond length 1.54 Å and double-bond length 1.34 Å, which indicates that the benzene bonds are intermediate between single and double bonds.

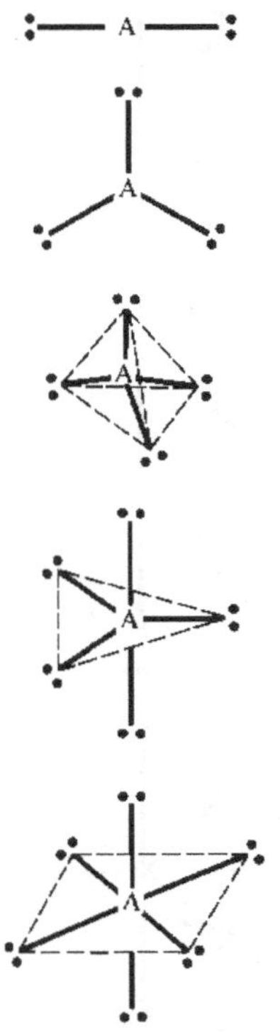

Figure 19.1

Arrangements of valence electron pairs around a central atom.

Bond Angles

The VSEPR (valence-shell electron-pair repulsion) method estimates bond angles at an atom A by counting the number of valence electron pairs that surround atom A in the molecule's Lewis electron-dot formula. The valence pairs around A are arranged in space to minimize electrostatic repulsions between pairs. The VSEPR arrangements for various numbers of pairs are (Fig. 19.1):

Number of pairs	2	3	4	5	6
Arrangement	linear	trigonal planar	tetrahedral	trigonal bipyramidal	octahedral
Angles	180°	120°	109.5°	90°, 120°	90°

For five valence electron pairs, lone pairs are placed in the equatorial position(s) (Fig. 19.2). A double bond or a triple bond is counted as one pair for the purposes of the VSEPR method. Lone pairs are more spread out in space than bonding pairs, so the lone pairs push the bonding pairs on an atom together a bit, making the bond angle(s) at that atom a bit less than the values listed in the table just given. For example, the bond angle in H_2O (which has two bonding pairs and two lone pairs on O) is 104.5° instead of 109.5°, and the bond angles in ClF_3 (Fig. 19.2) are a bit less than 90°. In $H_2C{=}CH_2$, the double bond exerts greater repulsions than the single bonds and so the HCH angle is 117°, somewhat less than the 120° trigonal-planar angle.

Dihedral Angles

Except for small molecules, specification of bond lengths and bond angles does not completely specify a molecule's geometry. One must also specify the dihedral angles of rotation about single bonds, which differ for different conformations. For example, Figs. 19.33 and 20.24 show different conformations of butane and glycine. A molecular bond angle is defined by specifying three atoms in a particular order. A dihedral angle is defined by specifying four atoms in a particular order. Figure 19.3 is a Newman projection of the molecule BrIFCC(O)H; one carbon atom is directly behind the other, and the O atom on the rear carbon is directly behind the I atom on the front carbon. The dihedral angle D(BrCCO) is 120° because a clockwise rotation of 120° about the C—C single bond is needed to make the CBr bond eclipse the CO bond. The usual convention is to take the molecular dihedral angle D(ABFG) (also called the **torsion angle**) to be in the range $-180° < D(\text{ABFG}) \le 180°$, with a negative angle meaning that a counterclockwise rotation of the front AB bond is needed to make it eclipse the rear FG bond. The D(FCCO) dihedral angle in Fig. 19.3 is $-120°$, since a counterclockwise rotation of 120° is needed to make the CF bond eclipse the CO bond. The D(ICCO) angle is 0° and D(ICCH) = 180°.

Some empirical rules for predicting dihedral angles in organic compounds without rings are [J. A. Pople and M. Gordon, *J. Am. Chem. Soc.*, **89,** 4253 (1967)]

1. If atoms A and B both have tetrahedral bond angles, the conformation about the A—B bond is usually staggered. For example, in CH_3CH_3 and CH_3NH_2, the bonds are staggered.
2. If atom A has tetrahedral bond angles and atom B has trigonal (120°) bond angles, then (*a*) one of the non-B atoms bonded to atom A lies in the plane defined by atom B and the atoms bonded to it; (*b*) in the lowest-energy conformation, one of the single bonds to A eclipses the double bond to B. For example, in CH_3CHO, one of the methyl CH bonds eclipses the CO double bond.
3. If atoms A and B both have trigonal bond angles, all the atoms bonded to A and to B lie in the same plane. For example, ethene, CH_2CH_2, is planar.

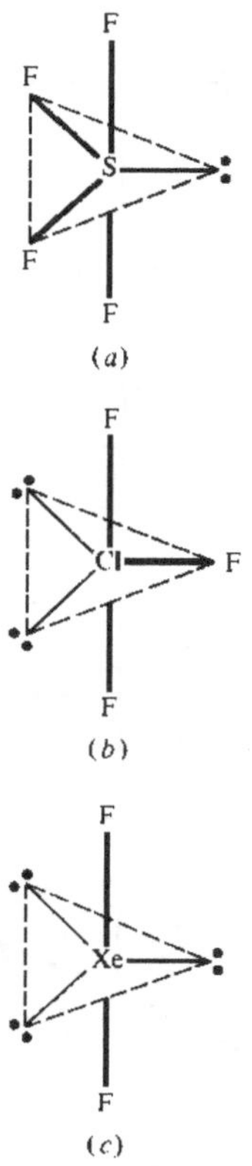

Figure 19.2

Some molecules with five valence electron pairs around the central atom.

Bond Energies

Section 5.10 explained how experimental ΔH°_{298} values for gas-phase atomization processes can be used to give average bond energies (which can be used to estimate ΔH°_{298} for gas-phase reactions). Table 19.1 lists some average bond energies. The values listed for H—H, F—F, Cl—Cl, O=O, and N≡N are ΔH°_{298} for dissociation of the appropriate gas-phase diatomic molecule. Double and triple bonds are stronger than single bonds. The N—N, O—O, and N—O single bonds are quite weak.

The fact that the bond energy of a carbon–carbon double bond is less than twice the energy of a carbon–carbon single bond makes vinyl addition polymerizations possible. The reaction $RCH_2CH_2\cdot + CH_2{=}CH_2 \rightarrow RCH_2CH_2CH_2CH_2\cdot$ has ΔS° negative, since two molecules are replaced by one, and has ΔH° negative, since one C=C bond is replaced by two C—C bonds.

Tabulated bond energies are on a per-mole basis. To convert to a per-molecule basis, we divide by the Avogadro constant. One kJ/mol corresponds to (1 kJ/mol)/N_A = 1.66054×10^{-21} J per molecule. Since 1 eV = 1.60218×10^{-19} J [Eq. (18.3)], 1 kJ/mol corresponds to 0.010364 eV per molecule. Thus

$$1 \text{ eV/molecule corresponds to } 96.485 \text{ kJ/mol } (23.061 \text{ kcal/mol}) \qquad (19.1)$$

Bond Moments

The **electric dipole moment** $\boldsymbol{\mu}$ of a charge distribution is defined by Eq. (13.82). Molecular dipole moments can be found by microwave spectroscopy (Sec. 20.7) or by dielectric-constant measurements (Sec. 13.14). From (13.82), the SI unit of μ is the

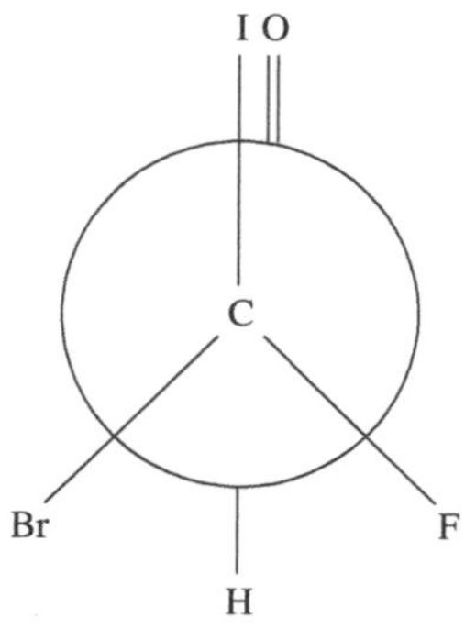

Figure 19.3

Newman projection of one conformation of the CFBrICHO molecule. The O and H atoms are bonded to the rear carbon.

TABLE 19.1

Average Bond Energies in kJ/mol[a]

C—H	C—C	C—O	C—N	C—S	C—F	C—Cl	C—Br	C—I	F—F
415	344	350	292	259	441	328	276	240	158
N—H	O—H	S—H	S—S	N—O	O—O	N—N	N—Cl	H—H	Cl—Cl
391	463	368	266	175	143	159	200	436	243
C=C	C=O	C=N	N=N	O=O	C≡C	C≡N	N≡N		
615	725	615	418	498	812	890	946		

[a]Data from L. Pauling, *General Chemistry,* 3d ed., Freeman, 1970, p. 913.

coulomb-meter (C m). Molecular dipole moments are usually quoted in units of **debyes** (D), where

$$1\ \text{D} \equiv 3.335641 \times 10^{-30}\ \text{C m} \tag{19.2}$$

For example μ of HCl is 1.07 D = 3.57×10^{-30} C m.

The dipole moment of a molecule can be roughly estimated by taking the *vector* sum of assigned **bond dipole moments** for the bonds. Some bond moments in debyes are:

H—O	H—N	H—C	C—Cl	C—Br	C—O	C=O	C—N	C≡N
1.5	1.3	0.4	1.5	1.4	0.8	2.5	0.5	3.5

where the first-listed atom is the positive end of the bond moment. The value for H—C is an assumed one, and the other moments involving C depend on the magnitude and sign of this assumed value. The above table uses the traditionally assumed polarity H^+—C^-.

The H—O and H—N bond moments are calculated from the observed dipole moments of H_2O and NH_3 without explicitly considering the contributions of the lone pairs to the dipole moment. Their contributions are absorbed into the values calculated for the OH and NH moments. For example, the observed μ for H_2O is 1.85 D, and the bond angle is 104.5°; Fig. 19.4 gives $2\mu_{OH} \cos 52.2° = 1.85$ D, and the O—H bond moment is $\mu_{OH} = 1.5$ D. The other moments listed are calculated from the experimental dipole moments and geometries of CH_3Cl, CH_3Br, CH_3OH, $(CH_3)_2CO$, $(CH_3)_3N$, and CH_3CN using the assumed CH bond moment and the OH and NH moments.

A shortcut in bond-moment calculations is to note that the vector sum of the three CH bond moments of a tetrahedral CH_3 group equals the moment of one CH bond. This follows from the zero dipole moment of methane (HCH_3).

Electronegativity

The **electronegativity** x of an element is a measure of the ability of an atom of that element to attract the electrons in a bond. The degree of polarity of an A—B bond is related to the difference in the electronegativities of the atoms forming the bond.

Many electronegativity scales have been proposed [J. Mullay, *Structure and Bonding,* **66,** 1 (1987); L. C. Allen, *Acc. Chem. Res.,* **23,** 175 (1990)]. The best-known is the Pauling scale, based on bond energies. Pauling observed that the A—B average bond energy generally exceeds the mean of the A—A and B—B average bond energies by an amount that increases with increasing polarity of the A—B bond. The Pauling electronegativity scale defines the electronegativity difference between elements A and B as

$$|x_A - x_B| \equiv 0.102(\Delta_{AB}/\text{kJ mol}^{-1})^{1/2} \tag{19.3}$$

where $\Delta_{AB} \equiv E(\text{A—B}) - \frac{1}{2}[E(\text{A—A}) + E(\text{B—B})]$ and where the E's are average single-bond energies. The electronegativity of H is arbitrarily set at 2.2.

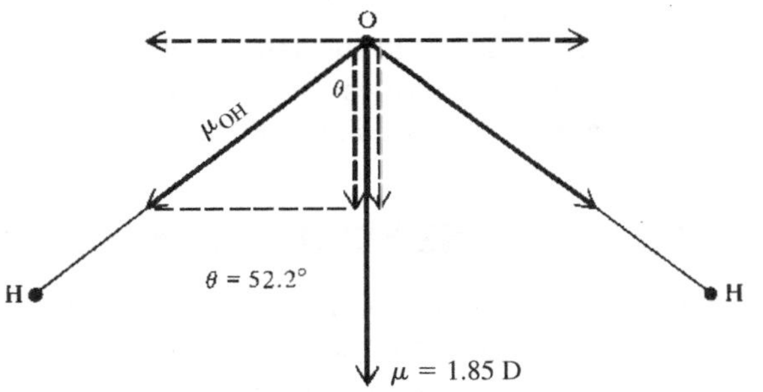

Figure 19.4

Calculation of the OH bond moment in H_2O. The dashed vectors are the bond-moment components along $\boldsymbol{\mu}$ and perpendicular to $\boldsymbol{\mu}$.

TABLE 19.2

Some Pauling Electronegativities[a]

H 2.2	Li 1.0	Be 1.6	B 2.0	C 2.5	N 3.0	O 3.4	F 4.0
	Na 0.9	Mg 1.3	Al 1.6	Si 1.9	P 2.2	S 2.6	Cl 3.2
	K 0.8	Ca 1.0	Ga 1.8	Ge 2.0	As 2.2	Se 2.6	Br 3.0
	Rb 0.8	Sr 0.9	In 1.8	Sn 2.0	Sb 2.1	Te	I 2.7

[a]Data from A. L. Allred, *J. Inorg. Nucl. Chem.*, **17,** 215 (1961).

The exothermicity of the combustion of hydrocarbons can be explained in terms of electronegativities. The large electronegativity differences between C and O and between O and H lead to highly polar bonds in the products CO_2 and H_2O, whereas the C—H, C—C, and O=O bonds in the reactants have low or no polarity. Therefore the total bond energy of the products is substantially greater than that of the reactants and the reaction is very exothermic.

The Allred–Rochow scale [A. L. Allred and E. G. Rochow, *J. Inorg. Nucl. Chem.*, **5,** 264, 269 (1958)] defines the electronegativity x_A of element A as

$$x_A \equiv 0.359 Z_{eff}/(r_A/\text{Å})^2 + 0.744 \tag{19.4}$$

where r_A is the bond radius of A and Z_{eff} is the effective nuclear charge [Eq. (18.55)] that would act on an electron added to the valence shell of a neutral A atom. The quantity $Z_{eff}e^2/4\pi\varepsilon_0 r_A^2$ is the average force exerted by atom A on an added electron. The constants 0.359 and 0.744 were chosen to make the scale as consistent as possible with the Pauling scale.

The Allen scale [L. C. Allen, *J. Am. Chem. Soc.*, **111,** 9003 (1989); J. B. Mann et al., *J. Am. Chem. Soc.*, **122,** 2780, 5132 (2000)] takes the electronegativity x of an atom as proportional to the average ionization energy $\langle E_{i,val}\rangle$ of the valence-shell electrons of the ground-state free atom: $x = 0.169\langle E_{i,val}\rangle/\text{eV}$.

The Nagle scale [J. K. Nagle, *J. Am. Chem. Soc.*, **112,** 4741 (1990)] defines the electronegativity x in terms of the polarizability α (Sec. 13.14) of the atom: $x = 1.66[n(4\pi\varepsilon_0\ \text{Å}^3/\alpha)]^{1/3} + 0.37$, where n is the number of valence electrons of the atom and 1 Å $= 10^{-10}$ m. Nagle assumed $n = 2$ (the valence s electrons) for each transition element.

Some electronegativities on the Pauling scale are given in Table 19.2. Electronegativities tend to decrease going down a group in the periodic table (because of the increasing distance of the valence electrons from the nucleus) and to increase going across a period (mainly because of the increasing Z_{eff} resulting from the lesser screening by electrons added to the same shell). Although electronegativity is an imprecise concept, electronegativities on various scales generally agree well. Defects of the Pauling scale are discussed in L. R. Murphy et al., *J. Phys. Chem. A*, **104,** 5867 (2000).

19.2 THE BORN–OPPENHEIMER APPROXIMATION

All molecular properties are, in principle, calculable by solving the Schrödinger equation for the molecule. Because of the great mathematical difficulties involved in solving the molecular Schrödinger equation, one must make approximations. Until about

1960, the level of approximations was such that the calculations gave only qualitative and not quantitative information. Since then, the use of computers has made molecular wave-function calculations accurate enough to give reliable quantitative information in many cases.

The Hamiltonian operator for a molecule is

$$\hat{H} = \hat{K}_N + \hat{K}_e + \hat{V}_{NN} + \hat{V}_{Ne} + \hat{V}_{ee} \tag{19.5}$$

where $\hat{K}_N$ and $\hat{K}_e$ are the kinetic-energy operators for the nuclei and the electrons, respectively, $\hat{V}_{NN}$ is the potential energy of repulsions between the nuclei, $\hat{V}_{Ne}$ is the potential energy of attractions between the electrons and the nuclei, and $\hat{V}_{ee}$ is the potential energy of repulsions between the electrons.

The Born–Oppenheimer Approximation

The molecular Schrödinger equation $\hat{H}\psi = E\psi$ is extremely complicated, and it would be almost hopeless to attempt an exact solution, even for small molecules. Fortunately, the fact that nuclei are much heavier than electrons allows the use of a very accurate approximation that greatly simplifies things. In 1927, Max Born and J. Robert Oppenheimer showed that it is an excellent approximation to treat the electronic and nuclear motions separately. The mathematics of the Born–Oppenheimer approximation is complicated, and so we shall give only a qualitative physical discussion.

Because of their much greater masses, the nuclei move far more slowly than the electrons, and the electrons carry out many "cycles" of motion in the time it takes the nuclei to move a short distance. The electrons see the heavy, slow-moving nuclei as almost stationary point charges, whereas the nuclei see the fast-moving electrons as essentially a three-dimensional distribution of charge.

One therefore assumes a fixed configuration of the nuclei, and for this configuration one solves an electronic Schrödinger equation to find the molecular electronic energy and wave function. This process is repeated for many different fixed nuclear configurations to give the electronic energy as a function of the positions of the nuclei. The nuclear configuration that corresponds to the minimum value of the electronic energy is the equilibrium geometry of the molecule. Having found how the electronic energy varies as a function of the nuclear configuration, one then uses this electronic energy function as the potential-energy function in a Schrödinger equation for the nuclear motion, thereby obtaining the molecular vibrational and rotational energy levels for a given electronic state.

The electronic Schrödinger equation is formulated for a fixed set of locations for the nuclei. Therefore, the nuclear kinetic-energy operator $\hat{K}_N$ in (19.5) is omitted from the Hamiltonian, and the **electronic Hamiltonian** $\hat{H}_e$ and **electronic Schrödinger equation** are

$$\hat{H}_e = \hat{K}_e + \hat{V}_{Ne} + \hat{V}_{ee} + \hat{V}_{NN} \tag{19.6}$$

$$\hat{H}_e\psi_e = E_e\psi_e \tag{19.7}$$

E_e is the **electronic energy,** including the energy V_{NN} of nuclear repulsion. Note that V_{NN} in (19.6) is a constant, since the nuclei are held fixed. The electronic wave function ψ_e is a function of the $3n$ spatial and n spin coordinates (Sec. 18.5) of the n electrons of the molecule. The electronic energy E_e contains potential and kinetic energy of the electrons and potential energy of the nuclei.

Consider a diatomic (two-atom) molecule with nuclei A and B with atomic numbers Z_A and Z_B. The spatial configuration of the nuclei is specified by the distance R between the two nuclei. The potential-energy operator $\hat{V}_{Ne}$ depends on R as a parameter, as does the internuclear repulsion $\hat{V}_{NN}$, which equals $Z_AZ_Be^2/4\pi\varepsilon_0R$ [Eq. (18.1)]. (A **parameter** is a quantity that is constant for one set of circumstances but may vary for other circumstances.) Hence, at each value of R, we get a different electronic wave function and energy. These quantities depend on R as a parameter and vary continuously as R varies. We therefore have $\psi_e = \psi_e(q_1, \ldots, q_n; R)$ and $E_e = E_e(R)$, where q_n stands for the spatial coordinates and spin coordinate of electron n. For a polyatomic molecule, ψ_e and E_e will depend parametrically on the locations of all the nuclei:

$$\psi_e = \psi_e(q_1, \ldots, q_n; Q_1, \ldots, Q_{\mathcal{N}}), \qquad E_e = E_e(Q_1, \ldots, Q_{\mathcal{N}}) \qquad (19.8)$$

where the Q's are the coordinates of the $\mathcal{N}$ nuclei.

Of course, a molecule has many different possible electronic states. For each such state, there is a different electronic wave function and energy, which vary as the nuclear configuration varies. Figure 19.5 shows $E_e(R)$ curves for the ground electronic state and some excited states of H_2. Since $E_e(R)$ is the potential-energy function for motion of the nuclei, a state with a minimum in the $E_e(R)$ curve is a bound state, with the atoms bonded to each other. For an electronic state with no minimum, $E_e(R)$ increases continually as R decreases. This means that the atoms repel each other as they come together, and this is not a bound state. The colliding atoms simply bounce off each other. The two lowest electronic states in Fig. 19.5 each dissociate to two ground-state ($1s$) hydrogen atoms. [Note from (18.18) that -27.2 eV is the energy of two $1s$ hydrogen atoms.] The ground electronic state of H_2 dissociates to (and arises from) $1s$ H atoms with opposite electronic spins, whereas the repulsive first excited electronic state arises from $1s$ H atoms with parallel electron spins.

The internuclear distance R_e at the minimum in the E_e curve for a bound electronic state is the **equilibrium bond length** for that state. (Because of molecular zero-point vibrations, R_e is not quite the same as the observed bond length.) As R goes to zero, E_e goes to infinity, because of the internuclear repulsion V_{NN}. As R goes to infinity, E_e goes to the sum of the energies of the separated atoms into which the molecule decomposes. The difference $E_e(\infty) - E_e(R_e)$ is the **equilibrium dissociation energy** D_e of the molecule (Fig. 19.5). Some D_e and R_e values (found by spectroscopy) for the ground electronic states of diatomic molecules are given in Table 19.3. Note the high D_e values of CO and N_2, which have triple bonds.

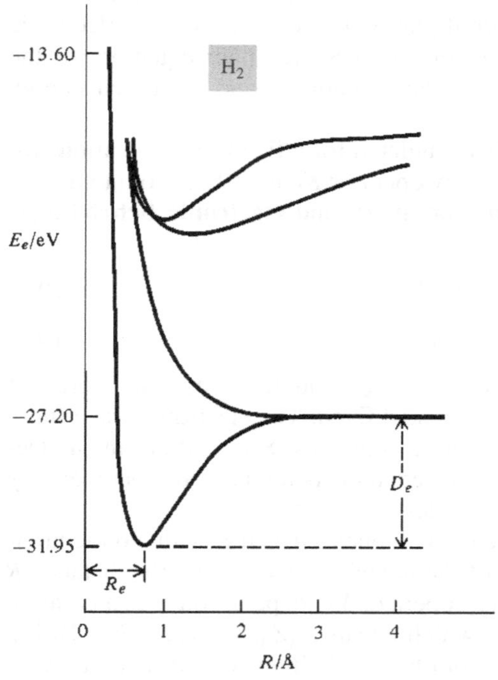

Figure 19.5

Potential-energy curves for the lowest few electronic states of H_2. R_e and D_e of the ground electronic state are shown.

TABLE 19.3

Diatomic-Molecule Ground-State D_e and R_e Values

	H_2^+	H_2	He_2^+	Li_2	C_2	N_2	O_2	F_2
D_e/eV	2.8	4.75	2.5	1.1	6.3	9.9	5.2	1.7
R_e/Å	1.06	0.74	1.1	2.7	1.24	1.10	1.21	1.41
	CH	CO	NaCl	OH	HCl	CaO	NaH	NaK
D_e/eV	3.6	11.2	4.3	4.6	4.6	4.8	2.0	0.6
R_e/Å	1.12	1.13	2.36	0.97	1.27	1.82	1.89	3.59

We now resume consideration of the Born–Oppenheimer approximation. Having solved the electronic Schrödinger equation (19.7) to obtain the electronic energy $E_e(Q_1, \ldots, Q_{\mathcal{N}})$ as a function of the nuclear coordinates, we use this as the potential-energy function in the Schrödinger equation for nuclear motion:

$$(\hat{K}_N + E_e)\psi_N \equiv \hat{H}_N\psi_N = E\psi_N \tag{19.9}$$

The Hamiltonian $\hat{H}_N$ for nuclear motion equals the nuclear kinetic-energy operator $\hat{K}_N$ plus the electronic energy function E_e, so E in (19.9) includes both electronic and nuclear energies and is the total energy of the molecule. The nuclear wave function ψ_N is a function of the $3\mathcal{N}$ spatial and $\mathcal{N}$ spin coordinates of the $\mathcal{N}$ nuclei.

E_e is the potential energy for nuclear vibration. As the relatively sluggish nuclei vibrate, the rapidly moving electrons almost instantaneously adjust their wave function ψ_e and energy E_e to follow the nuclear motion. The electrons act somewhat like springs connecting the nuclei. As the internuclear distances change, the energy stored in the "springs" (that is, in the electronic motions) changes.

The nuclear kinetic-energy operator $\hat{K}_N$ involves vibrational, rotational, and translational kinetic energies. (Rotational and translational motions do not change the electronic energy E_e.) We shall deal with nuclear vibrations and rotations in Chapter 20. The remainder of this chapter deals with the electronic wave function and energy.

The Born–Oppenheimer treatment shows that the complete molecular wave function ψ is to a very good approximation equal to the product of electronic and nuclear wave functions: $\psi = \psi_e\psi_N$.

In addition to making the Born–Oppenheimer approximation, one usually neglects relativistic effects in treating molecules. This is a very good approximation for molecules composed of light atoms, but it is not good for molecules containing heavy atoms. Inner-shell electrons in atoms of high atomic number move at very high speeds and are significantly affected by the relativistic increase of mass with speed. The valence electrons can undergo relativistic effects due to interactions with the inner-shell electrons and to the portion of the valence-electrons' probability density that deeply penetrates the inner-shell electrons. Relativistic effects have substantial influence on bond lengths and binding energies of molecules containing atoms of high atomic number (for example, Au). See P. Pyykkö, *Chem. Rev.*, **88,** 563 (1988).

Ionic and Covalent Bonding

A bound electronic state of a diatomic molecule has a minimum in its curve of electronic energy E_e versus internuclear distance R (Fig. 19.5). Why is E_e lower in the molecule than in the separated atoms?

An ionic molecule like NaCl is held together by the Coulombic attraction between the ions. Solid NaCl consists of an array of alternating Na^+ and Cl^- ions, and one cannot pick out individual NaCl molecules. However, gas-phase NaCl consists of individual ionic NaCl molecules. (In aqueous solution, hydration of the ions makes the separated hydrated ions more stable than Na^+Cl^- molecules.) Ionic molecules dissociate to neutral atoms in the gas phase. Consider, for example, NaCl. The ionization energy of Na is 5.14 eV, whereas the electron affinity of Cl is only 3.61 eV. Hence, isolated Na and Cl atoms are more stable than isolated Na^+ and Cl^- ions. Thus as the internuclear distance R increases, the bonding in NaCl shifts from ionic to covalent at very large R values.

EXAMPLE 19.1 D_e and μ of NaCl

Use the model of an NaCl molecule as consisting of nonoverlapping spherical Na^+ and Cl^- ions separated by the experimentally observed distance $R_e = 2.36$ Å (Table 19.3) to estimate the equilibrium dissociation energy D_e and the dipole moment of NaCl.

Equation (18.1) gives the potential energy of interaction between two charges as $V = Q_1Q_2/4\pi\varepsilon_0 r$. Therefore the energy needed to take the Na^+ and Cl^- ions from a 2.36-Å separation to an infinite separation (where $V = 0$) is $e^2/4\pi\varepsilon_0 R_e$. The use of (18.4) for e and of (18.3) gives

$$\frac{e^2}{4\pi\varepsilon_0 R_e} = \frac{(1.602 \times 10^{-19}\text{ C})^2}{4\pi(8.854 \times 10^{-12}\text{ C}^2/\text{N-m}^2)(2.36 \times 10^{-10}\text{ m})}$$
$$= 9.77 \times 10^{-19}\text{ J} = 6.10\text{ eV}$$

However, this is not the estimate of D_e, since (as already noted) NaCl dissociates to neutral atoms. Breaking the dissociation into two hypothetical steps, we have

$$\text{NaCl} \xrightarrow{(a)} \text{Na}^+ + \text{Cl}^- \xrightarrow{(b)} \text{Na} + \text{Cl}$$

where the two ions (and the two atoms) are at infinite separation from each other. We estimated the energy change for step (*a*) as 6.10 eV. Addition of an electron to Na^+ lowers the energy by the Na ionization energy 5.14 eV, and removal of an electron from Cl^- raises the energy by the Cl electron affinity 3.61 eV. Hence, the nonoverlapping-spherical-ion model gives the energy needed to dissociate NaCl into Na + Cl as

$$6.10\text{ eV} - 5.14\text{ eV} + 3.61\text{ eV} = 4.57\text{ eV}$$

which is only 7 percent away from the experimental value $D_e = 4.25$ eV. The error results from neglect of the repulsion between the slightly overlapping electron probability densities of the Na^+ and Cl^- ions, which makes the molecule less stable than calculated.

The dipole moment of a charge distribution is given by Eq. (13.82) as $\mu = \Sigma_i\, Q_i\mathbf{r}_i$. The charge on Na^+ equals the proton charge e. Taking the coordinate origin at the center of the Cl^- ion, we estimate μ as

$$\mu = eR_e = (1.602 \times 10^{-19}\text{ C})(2.36 \times 10^{-10}\text{ m}) = 3.78 \times 10^{-29}\text{ C m} = 11.3\text{ D}$$

where (19.2) was used. This value is not far from the experimental value 9.0 D. The error can be attributed to polarization of one ion by the other.

Exercise

The LiF molecule has $R_e = 1.56$ Å. Estimate D_e and μ of LiF. Use data in Chapter 18. (*Answers:* 7.2 eV and 7.5 D.)

The ionic bonding in NaCl can be contrasted with the nonpolar covalent bonding in H_2 and other homonuclear diatomic molecules. Here, the bonding electrons are shared equally. For a diatomic molecule formed from different nonmetals (for example, HCl, BrCl) or from different metals (for example, NaK), the bonding is polar covalent, the more electronegative atom having a greater share of the electrons and a partial negative charge. Bonds between metals with relatively high electronegativities and nonmetals are sometimes polar covalent, rather than ionic, as noted in Sec. 10.5.

The physical reason for the stability of a covalent bond is not a fully settled question. A somewhat oversimplified statement is that the stability is due to the decrease in the average potential energy of the electrons forming the bond. This decrease results from the greater electron–nuclear attractions in the molecule compared with those in the separated atoms. The electrons in the bond can feel the simultaneous attractions of two nuclei. This decrease in electronic potential energy outweighs the increases in interelectronic repulsions and internuclear repulsions that occur as the atoms come together.

19.3 THE HYDROGEN MOLECULE ION

The simplest molecule is H_2^+, which consists of two protons and one electron.

Adopting the Born–Oppenheimer approximation, we hold the nuclei at a fixed distance R and deal with the electronic Schrödinger equation $\hat{H}_e\psi_e = E_e\psi_e$ [Eq. (19.7)]. The electronic Hamiltonian including nuclear repulsion for H_2^+ is given by Eqs. (19.6), (18.1), and (17.58) as

$$\hat{H}_e = -\frac{\hbar^2}{2m_e}\nabla^2 - \frac{e^2}{4\pi\varepsilon_0 r_A} - \frac{e^2}{4\pi\varepsilon_0 r_B} + \frac{e^2}{4\pi\varepsilon_0 R} \qquad (19.10)$$

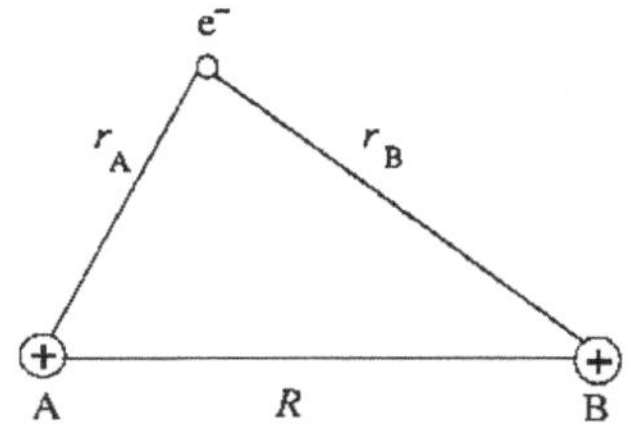

Figure 19.6
Interparticle distances in the H_2^+ molecule.

where r_A and r_B are the distances from the electron to nuclei A and B and R is the internuclear distance (Fig. 19.6). The first term on the right side of (19.10) is the operator for the kinetic energy of the electron. The second and third terms are the potential energies of attraction between the electron and the nuclei. The last term is the repulsion between the nuclei. Since H_2^+ has only one electron, there is no interelectronic repulsion. Figure 19.7 is a three-dimensional plot of values of $-e^2/4\pi\varepsilon_0 r_A - e^2/4\pi\varepsilon_0 r_B$ in a plane containing the nuclei.

The electronic Schrödinger equation $\hat{H}_e\psi_e = E_e\psi_e$ can be solved exactly for H_2^+, but the solutions are complicated. For our purposes, an approximate treatment will suffice. The lowest electronic state of H_2^+ will dissociate to a ground-state (1s) H atom and a proton as R goes to infinity. Suppose the electron in H_2^+ is close to nucleus A and rather far from nucleus B. The H_2^+ electronic wave function should then resemble a ground-state H-atom wave function for atom A; that is, ψ_e will be approximately given by the function (Table 18.1 in Sec. 18.3)

$$1s_A \equiv (1/a_0)^{3/2}\pi^{-1/2}e^{-r_A/a_0} \qquad (19.11)$$

where the Bohr radius a_0 is used since the nuclei are fixed. Similarly, when the electron is close to nucleus B, ψ_e can be roughly approximated by

$$1s_B \equiv (1/a_0)^{3/2}\pi^{-1/2}e^{-r_B/a_0}$$

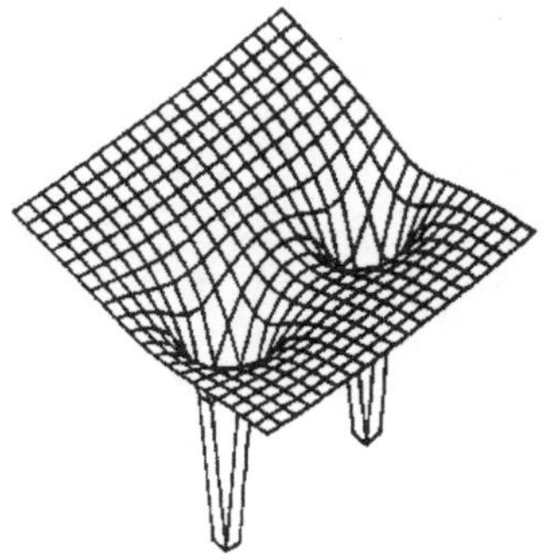

Figure 19.7
Three-dimensional plot of the potential energy of attraction between the electron and the nuclei of H_2^+ in a plane containing the nuclei.

This suggests as an approximate wave function for the H_2^+ ground electronic state:

$$\phi = c_A 1s_A + c_B 1s_B = a_0^{-3/2}\pi^{-1/2}(c_A e^{-r_A/a_0} + c_B e^{-r_B/a_0}) \qquad (19.12)$$

which is a linear combination of the $1s_A$ and $1s_B$ atomic orbitals. When the electron is very close to nucleus A, then r_A is much less than r_B and e^{-r_A/a_0} is much greater than e^{-r_B/a_0}. Hence, the $1s_A$ term in (19.12) dominates, and the wave function resembles that of an H atom at nucleus A, as it should. Similarly for the electron close to nucleus B. One multiplies the spatial function of (19.12) by a spin function (either α or β) to get the complete approximate wave function.

The function (19.12) can be regarded as a variation function and the constants c_A and c_B chosen to minimize the variational integral $W = \int \phi^* \hat{H} \phi \, d\tau / \int \phi^* \phi \, d\tau$. The function (19.12) is a linear combination of two functions, and, as noted in Sec. 17.15, the conditions $\partial W/\partial c_A = 0 = \partial W/\partial c_B$ will be satisfied by two sets of values of c_A and c_B. These sets will yield approximate wave functions and energies for the lowest two electronic states of H_2^+. We need not go through the details of evaluating W and setting $\partial W/\partial c_A = 0 = \partial W/\partial c_B$, since the fact that the nuclei are identical requires that the electron probability density be the same on each side of the molecule. Restricting ourselves to a real variation function, the electron probability density is $\phi^2 = c_A^2(1s_A)^2 + c_B^2(1s_B)^2 + 2c_Ac_B 1s_A 1s_B$. To have ϕ^2 be the same at corresponding points on each side of the molecule, we must have either $c_B = c_A$ or $c_B = -c_A$. For $c_B = c_A$, we have

$$\phi = c_A(1s_A + 1s_B), \qquad \phi^2 = c_A^2(1s_A^2 + 1s_B^2 + 2 \cdot 1s_A 1s_B) \tag{19.13}$$

For $c_B = -c_A$,

$$\phi' = c_A'(1s_A - 1s_B), \qquad \phi'^2 = c_A'^2(1s_A^2 + 1s_B^2 - 2 \cdot 1s_A 1s_B) \tag{19.14}$$

The constants c_A and c_A' are found by requiring that ϕ and ϕ' be normalized.

The normalization condition for the function in (19.13) is

$$1 = \int \phi^2 \, d\tau = c_A^2\left(\int 1s_A^2 \, d\tau + \int 1s_B^2 \, d\tau + 2\int 1s_A 1s_B \, d\tau\right)$$

The H-atom wave functions are normalized, so $\int 1s_A^2 \, d\tau = \int 1s_B^2 \, d\tau = 1$. Defining the **overlap integral** S as

$$S \equiv \int 1s_A 1s_B \, d\tau$$

we get $1 = c_A^2(2 + 2S)$ and $c_A = (2 + 2S)^{-1/2}$. Similarly, one finds $c_A' = (2 - 2S)^{-1/2}$. Hence the normalized approximate wave functions for the lowest two H_2^+ electronic states are

$$\phi = (2 + 2S)^{-1/2}(1s_A + 1s_B), \qquad \phi' = (2 - 2S)^{-1/2}(1s_A - 1s_B) \tag{19.15}$$

For completeness, each spatial function should be multiplied by a one-electron spin function, either α or β.

The value of the overlap integral $\int 1s_A 1s_B \, d\tau$ depends on how much the functions $1s_A$ and $1s_B$ overlap each other. Only regions of space where both $1s_A$ and $1s_B$ are of significant magnitude will contribute substantially to S. The main contribution to S therefore comes from the region between the nuclei. The value of S clearly depends on the internuclear distance R. For $R = 0$, we have $1s_A = 1s_B$ and $S = 1$. For $R = \infty$, the $1s_A$ and $1s_B$ atomic orbitals don't overlap, and $S = 0$. For R between 0 and ∞, S is between 0 and 1 and is easily evaluated from the expressions for $1s_A$ and $1s_B$.

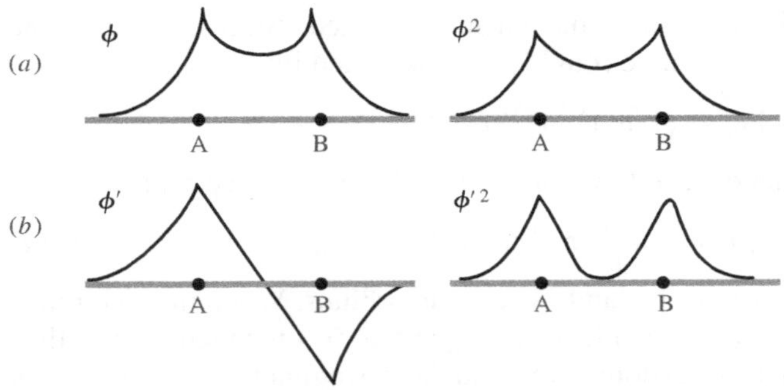

Figure 19.8

Graphs of (*a*) ground state and (*b*) first excited state H_2^+ approximate wave functions for points on the internuclear axis. (Not drawn to scale.)

The probability density ϕ^2 in (19.13) can be written as $c_A^2(1s_A^2 + 1s_B^2)$ plus $2c_A^2 1s_A 1s_B$. The $1s_A^2 + 1s_B^2$ part of ϕ^2 is proportional to the probability density due to two separate noninteracting $1s$ H atoms. The term $2c_A^2 1s_A 1s_B$ is large only in regions where both $1s_A$ and $1s_B$ are reasonably large. This term therefore increases the electron probability density in the region between the nuclei. This buildup of probability density between the nuclei (at the expense of regions outside the internuclear region) causes the electron to feel the attractions of both nuclei at once, thereby lowering its average potential energy and providing a stable covalent bond. The bonding is due to the overlap of the atomic orbitals $1s_A$ and $1s_B$.

Figure 19.8*a* graphs ϕ and ϕ^2 of (19.13) for points along the line joining the nuclei. The probability-density buildup between the nuclei is evident.

For the function ϕ' of (19.14), the term $-2c_A'^2 1s_A 1s_B$ decreases the electron probability density between the nuclei. At any point on a plane midway between the nuclei and perpendicular to the internuclear axis we have $r_A = r_B$ and $1s_A = 1s_B$. Hence $\phi' = 0 = \phi'^2$ on this plane, which is a *nodal plane* for the function ϕ'. Figure 19.8*b* shows ϕ' and ϕ'^2 for points along the internuclear axis.

The functions ϕ and ϕ' depend on the internuclear distance R, since r_A and r_B in $1s_A$ and $1s_B$ depend on R (see Fig. 19.6). The variational integral W is therefore a function of R. When W is evaluated for ϕ and ϕ', one finds that ϕ gives an electronic energy curve $W(R) \approx E_e(R)$ with a minimum; see the lower curve in Fig. 19.9. In contrast, the $W(R)$-versus-R curve for the H_2^+ function ϕ' in (19.15) has no minimum (Fig. 19.9), indicating an unbound electronic state. These facts are understandable from the preceding electron-probability-density discussion.

The true values of R_e and D_e for H_2^+ are 1.06 Å and 2.8 eV. The function (19.13) gives $R_e = 1.32$ Å and $D_e = 1.8$ eV, which is rather poor. Substantial improvement can be obtained if a variational parameter ζ is included in the exponentials, so that $1s_A$ and $1s_B$ become proportional to $e^{-\zeta r_A/a_0}$ and $e^{-\zeta r_B/a_0}$. One then finds $R_e = 1.07$ Å and $D_e = 2.35$ eV (Fig. 19.10). The parameter ζ depends on R and is found to equal 1.24 at R_e.

An orbital is a one-electron spatial wave function. H_2^+ has but one electron, and the approximate wave functions ϕ and ϕ' in (19.13) and (19.14) are approximations to the orbitals of the two lowest electronic states of H_2^+. An orbital for an atom is called an **atomic orbital** (AO). An orbital for a molecule is a **molecular orbital** (MO). Just as the wave function of a many-electron atom can be approximated by use of AOs, the wave function of a many-electron molecule can be approximated by use of MOs. Each MO can hold two electrons of opposite spin.

The situation is more complicated for molecules than for atoms, in that the number of nuclei varies from molecule to molecule. Whereas hydrogenlike orbitals with effective nuclear charges are useful for all many-electron atoms, the H_2^+-like orbitals with effective nuclear charges are directly applicable only to molecules with two identical nuclei, that is, **homonuclear** diatomic molecules. We shall later see, however, that since a molecule is held together by bonds and since (with some exceptions) each bond is between two atoms, we can construct an approximate molecular wave function using bond orbitals (and lone-pair and inner-shell orbitals), where the bond orbitals resemble diatomic-molecule MOs.

Let us consider further excited electronic states of H_2^+. We expect such states to dissociate to a proton and a $2s$ or $2p$ or $3s$ or . . . H atom. Therefore, analogous to the functions (19.13) and (19.14), we write as approximate wave functions (molecular orbitals) for excited H_2^+ states

$$N(2s_A + 2s_B), \quad N(2s_A - 2s_B), \quad N(2p_{xA} + 2p_{xB}), \quad N(2p_{xA} - 2p_{xB}), \quad \text{etc.} \quad (19.16)$$

where the normalization constant N differs for different states. Actually, because of the degeneracy of the $2s$ and $2p$ states in the H atom, we should expect extensive mixing together of $2s$ and $2p$ AOs in the H_2^+ MOs. Since we are mainly interested in H_2^+ MOs

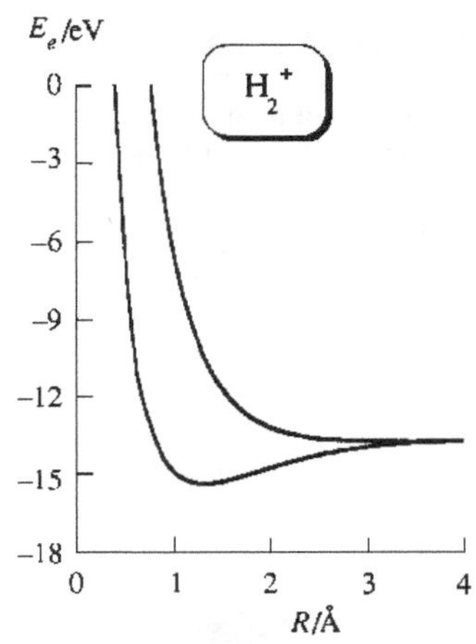

Figure 19.9

Electronic energy (including internuclear repulsion) versus R for the ground state and first excited state of H_2^+ as calculated from the approximate wave functions $N(1s_A + 1s_B)$ and $N'(1s_A - 1s_B)$ [Eq. (19.15)].

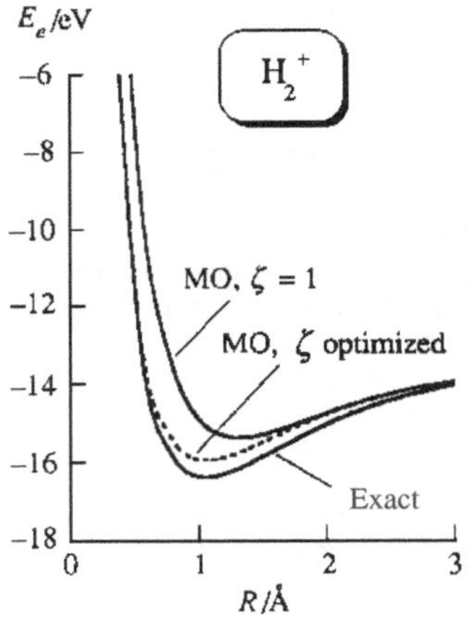

Figure 19.10

Electronic energy including internuclear repulsion for the H_2^+ ground electronic state. The curves are calculated from the exact wave function, from the LCAO MO function with optimized orbital exponent ζ, and from the LCAO MO function with $\zeta = 1$ (as in the lower curve in Fig. 19.9).

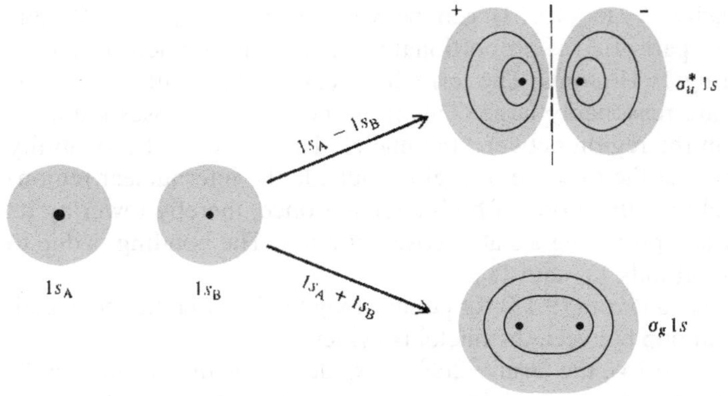

Figure 19.11

Formation of homonuclear diatomic MOs from $1s$ AOs. The dashed line indicates a nodal plane.

for use in many-electron molecules, and since the $2s$ and $2p$ levels are not degenerate in many-electron atoms, we shall ignore such mixing for now. A wave function like $2s_A + 2s_B$ expresses the fact that there is a 50-50 probability as to which nucleus the electron will go with when the molecule dissociates ($R \to \infty$).

The MOs in (19.15) and (19.16) are *l*inear *c*ombinations of *a*tomic *o*rbitals and so are called LCAO MOs. There is no necessity for MOs to be expressed as linear combinations of AOs, but this approximate form is a very convenient one. Let us see what these MOs look like.

Since the functions $1s_A$ and $1s_B$ are both positive in the internuclear region, the function $1s_A + 1s_B$ shows a buildup of probability density between the nuclei, whereas the linear combination $1s_A - 1s_B$ has a nodal plane between the nuclei. Figure 19.11 shows contours of constant probability density for the two MOs (19.13) and (19.14) formed from $1s$ AOs. The three-dimensional shape of these orbitals is obtained by rotating the contours about the line joining the nuclei. See also Fig. 19.12.

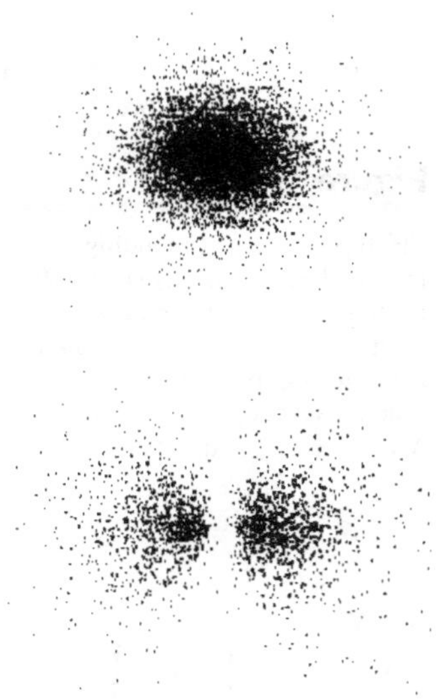

Figure 19.12

Electron probability density in a plane containing the nuclei for the ground and first excited states of H_2^+.

A word about terminology. The component of electronic orbital angular momentum along the internuclear (z) axis of H_2^+ can be shown to have the possible values $L_z = m\hbar$, where $m = 0, \pm 1, \pm 2, \ldots$. (Unlike the H atom, there is no l quantum number in H_2^+, since the magnitude of the total electronic orbital angular momentum is not fixed in H_2^+. This is because there is spherical symmetry in H but only axial symmetry in H_2^+.) The following code letters are used to indicate the $|m|$ value:

$\|m\|$	0	1	2	3	...
Letter	σ	π	δ	ϕ	...

(19.17)

These are the Greek equivalents of s, p, d, f.

The AOs $1s_A$ and $1s_B$ have zero electronic orbital angular momentum along the molecular axis, and so the two MOs formed from these AOs have $m = 0$ and from (19.17) are σ (sigma) MOs. We call these the $\sigma_g 1s$ MO and the $\sigma_u^* 1s$ MO. The $1s$ indicates that they originate from separated-atom $1s$ AOs. The star indicates the **antibonding** character of the $1s_A - 1s_B$ MO, associated with the nodal plane and the charge depletion between the nuclei.

The g subscript (from the German *gerade,* "even") means that the orbital has the same value at two points that are on diagonally opposite sides of the center of the molecule and equidistant from the center. The u subscript (*ungerade,* "odd") means that the values of the orbital differ by a factor -1 at two such points. [The point diagonally opposite (x, y, z) is at $(-x, -y, -z)$. An **even function** of x, y, and z is one for which $f(-x, -y, -z) = f(x, y, z)$. An **odd function** is one that satisfies $f(-x, -y, -z) = -f(x, y, z)$.]

The linear combinations $2s_A + 2s_B$ and $2s_A - 2s_B$ give the $\sigma_g 2s$ and $\sigma_u^* 2s$ MOs, whose shapes resemble those of the $\sigma_g 1s$ and $\sigma_u^* 1s$ MOs.

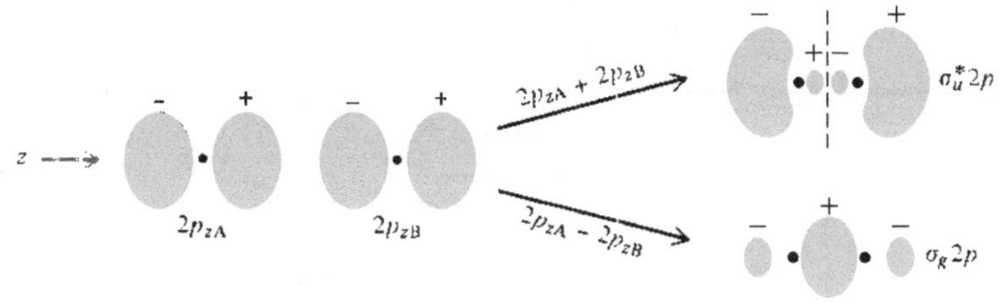

Figure 19.13

Formation of homonuclear diatomic MOs from $2p_z$ AOs.

Let the molecular axis be the z axis. Because of the opposite signs of the right lobe of $2p_{zA}$ and the left lobe of $2p_{zB}$ (Fig. 19.13), the linear combination $2p_{zA} + 2p_{zB}$ has a nodal plane midway between the nuclei, as indicated by the dashed line. The charge depletion between the nuclei makes this an antibonding MO. The linear combination $2p_{zA} - 2p_{zB}$ gives charge buildup between the nuclei and is a bonding MO. The $2p_z$ AO has atomic quantum number $m = 0$ (Sec. 18.3) and so has $L_z = 0$. The MOs formed from $2p_z$ AOs are therefore σ MOs, the $\sigma_g 2p$ and $\sigma_u^* 2p$ MOs. Their three-dimensional shapes are obtained by rotating the contours in Fig. 19.13 about the internuclear (z) axis.

Formation of homonuclear diatomic MOs from the $2p_x$ AOs is shown in Fig. 19.14. The p_x AO is a linear combination of $m = 1$ and $m = -1$ AOs [see Eq. (18.25)] and has $|m| = 1$. Therefore the MOs made from the $2p_x$ AOs have $|m| = 1$ and are π MOs [Eq. (19.17)]. The linear combination $N(2p_{xA} + 2p_{xB})$ has charge buildup in the internuclear regions above and below the z axis and is therefore bonding. This MO has opposite signs at the diagonally opposite points c and d in Fig. 19.14 and so is a u MO, the $\pi_u 2p_x$ MO. The linear combination $N(2p_{xA} - 2p_{xB})$ gives the antibonding $\pi_g^* 2p_x$ MO.

The σ MOs in Figs. 19.11 and 19.13 are symmetric about the internuclear axis; the orbital shapes are figures of rotation about the z axis. In contrast, the $\pi_u 2p_x$ and $\pi_g^* 2p_x$ MOs consist of blobs of probability density above and below the yz plane, which is a nodal plane for these MOs.

The linear combinations $2p_{yA} + 2p_{yB}$ and $2p_{yA} - 2p_{yB}$ give the $\pi_u 2p_y$ and $\pi_g^* 2p_y$ MOs. These MOs have the same shapes as the $\pi_u 2p_x$ and $\pi_g^* 2p_x$ MOs but are rotated by 90° about the internuclear axis compared with the $\pi 2p_x$ MOs. Since they have the same shapes, the $\pi_u 2p_x$ and $\pi_u 2p_y$ MOs have the same energy. Likewise, the $\pi_g^* 2p_x$ and $\pi_g^* 2p_y$ MOs have the same energy (see Fig. 19.15).

The σ MOs have no nodal planes containing the internuclear axis. (Some σ MOs have a nodal plane or planes perpendicular to the internuclear axis.) Each π MO has one nodal plane containing the internuclear axis. This is true provided one uses the real $2p$ AOs to form the MOs, as we have done. It turns out that δ MOs have two nodal planes containing the internuclear axis (see Fig. 19.29c). We shall later use the number of nodal planes to classify bond orbitals in polyatomic molecules.

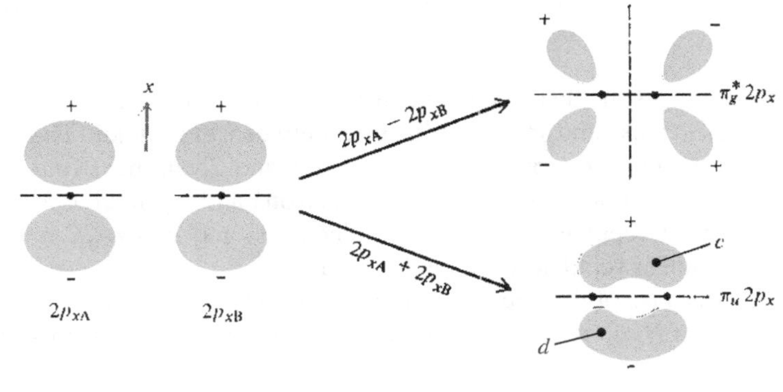

Figure 19.14

Formation of homonuclear diatomic MOs from $2p_x$ AOs.

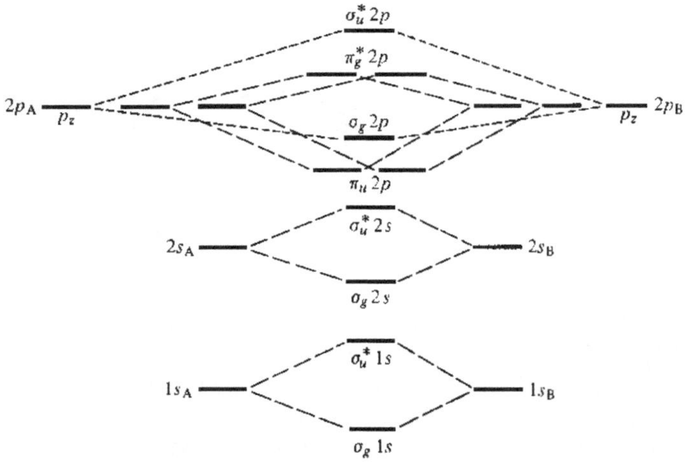

Figure 19.15

Lowest-lying homonuclear diatomic MOs. The dashed lines show which AOs contribute to each MO.

19.4 THE SIMPLE MO METHOD FOR DIATOMIC MOLECULES

MOs for Homonuclear Diatomic Molecules

Just as we constructed approximate wave functions for many-electron atoms by feeding electrons two at a time into hydrogenlike AOs, we shall construct approximate wave functions for many-electron homonuclear diatomic molecules by feeding electrons two at a time into H_2^+-like MOs. Figure 19.15 shows the lowest-lying H_2^+-like MOs (Sec. 19.3). Similar to AO energies (Fig. 18.15), the energies of these MOs vary from molecule to molecule. They also vary with varying internuclear distance in the same molecule. The energy order shown in the figure is the order in which the MOs are filled in going through the periodic table, as shown by spectroscopic observations. The AOs at the sides are connected by dashed lines to the MOs to which they contribute. Note that each pair of AOs leads to the formation of two MOs, a bonding MO with energy lower than that of the AOs and an antibonding MO with energy higher than that of the AOs.

The Hydrogen Molecule

H_2 consists of two protons (A and B) and two electrons (1 and 2); see Fig. 19.16. The electronic Hamiltonian (including nuclear repulsion) is [Eqs. (19.6), (18.1), and (17.58)]

$$\hat{H}_e = -\frac{\hbar^2}{2m_e}\nabla_1^2 - \frac{\hbar^2}{2m_e}\nabla_2^2 - \frac{e^2}{4\pi\varepsilon_0 r_{1A}} - \frac{e^2}{4\pi\varepsilon_0 r_{1B}} - \frac{e^2}{4\pi\varepsilon_0 r_{2A}} - \frac{e^2}{4\pi\varepsilon_0 r_{2B}} + \frac{e^2}{4\pi\varepsilon_0 r_{12}} + \frac{e^2}{4\pi\varepsilon_0 R} \tag{19.18}$$

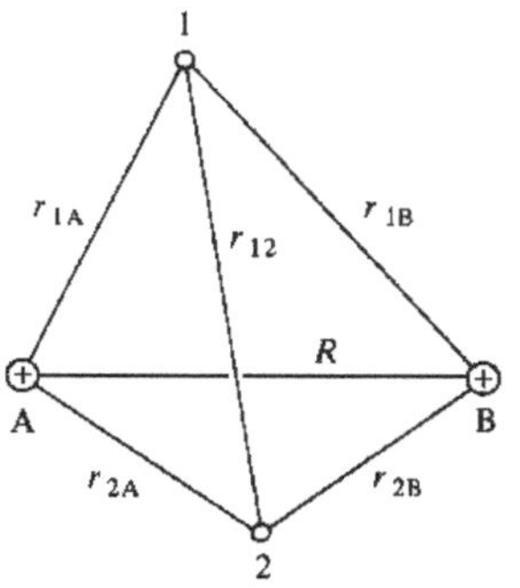

Figure 19.16

Interparticle distances in the H_2 molecule.

where r_{1A} is the distance between electron 1 and nucleus A, r_{12} is the distance between the electrons, and R is the distance between the nuclei. The first two terms are the kinetic-energy operators for electrons 1 and 2, the next four terms are the potential energy of attractions between the electrons and the nuclei, $e^2/4\pi\varepsilon_0 r_{12}$ is the potential energy of repulsion between the electrons, and $e^2/4\pi\varepsilon_0 R$ is the potential energy of internuclear repulsion. R is held fixed.

Because of the interelectronic repulsion term $e^2/4\pi\varepsilon_0 r_{12}$, the electronic Schrödinger equation $\hat{H}_e\psi_e = E_e\psi_e$ cannot be solved exactly for H_2. If this term is ignored, we get an approximate electronic Hamiltonian that is the sum of two H_2^+-like electronic Hamiltonians, one for electron 1 and one for electron 2. [This isn't quite true, because the internuclear repulsion $e^2/4\pi\varepsilon_0 R$ is the same in (19.10) and (19.18). However, $e^2/4\pi\varepsilon_0 R$ is a constant and therefore only shifts the energy by $e^2/4\pi\varepsilon_0 R$ but does not affect the wave functions; see Prob. 19.33.] The approximate electronic wave function for H_2 is then the product of two H_2^+-like electronic wave functions, one for each electron [Eq. (17.68)]. This is exactly analogous to approximating the He wave function by the product of two H-like wave functions in Sec. 18.6.

The function $(2 + 2S)^{-1/2}(1s_A + 1s_B)$ in Eq. (19.15) is an approximate wave function for the H_2^+ ground electronic state, and so the MO approximation to the H_2 ground-electronic-state spatial wave function is

$$\sigma_g 1s(1) \cdot \sigma_g 1s(2) = N[1s_A(1) + 1s_B(1)] \cdot [1s_A(2) + 1s_B(2)] \qquad (19.19)$$

where the normalization constant N is $(2 + 2S)^{-1}$. The numbers in parentheses refer to the electrons. For example, $1s_A(2)$ is proportional to e^{-r_{2A}/a_0}. The MO wave function (19.19) is analogous to the He ground-state AO wave function $1s(1)1s(2)$ in Eq. (18.38); the MO $\sigma_g 1s$ replaces the AO $1s$.

To take care of spin and to satisfy the requirement that the complete electronic wave function be antisymmetric (Sec. 18.6), the symmetric two-electron spatial function (19.19) must be multiplied by the antisymmetric spin function (18.41). The approximate MO ground-state wave function for H_2 is then

$$\sigma_g 1s(1)\sigma_g 1s(2)2^{-1/2}[\alpha(1)\beta(2) - \beta(1)\alpha(2)] = \frac{1}{\sqrt{2}}\begin{vmatrix} \sigma_g 1s(1)\alpha(1) & \sigma_g 1s(1)\beta(1) \\ \sigma_g 1s(2)\alpha(2) & \sigma_g 1s(2)\beta(2) \end{vmatrix} \qquad (19.20)$$

where we introduced the Slater determinant (Sec. 18.8). Just as the ground-state electron configuration of He is $1s^2$, the ground-state electron configuration of H_2 is $(\sigma_g 1s)^2$; compare (19.20) with (18.43).

We have put each electron in H_2 into an MO. This allows for interelectronic repulsion only in an average way, so the treatment is an approximate one. We are using the crudest possible version of the MO approximation.

Using the approximate wave function (19.20), one evaluates the variational integral W to get W as a function of R. Since evaluation of molecular quantum-mechanical integrals is complicated, we shall just quote the results. With inclusion of a variable orbital exponent, the function (19.20) and (19.19) gives a $W(R)$ curve (Fig. 19.17) with a minimum at $R = 0.73$ Å, which is close to the observed value $R_e = 0.74$ Å in H_2. The calculated D_e is 3.49 eV, which is far from the experimental value 4.75 eV. This is the main failing of the MO method; molecular dissociation energies are not accurately calculated.

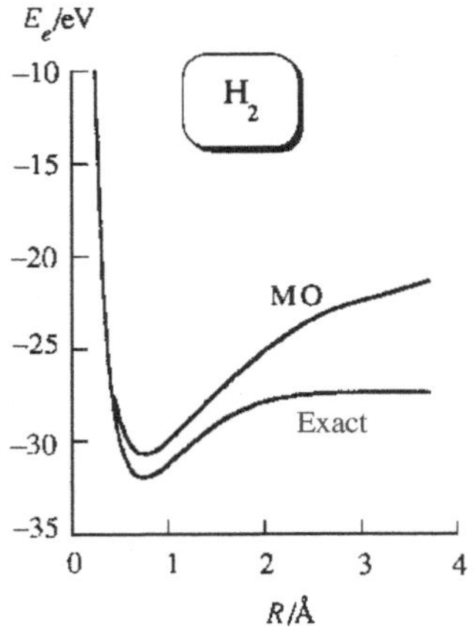

Figure 19.17

Ground-state electronic energy including internuclear repulsion for H_2 as calculated from the LCAO-MO wave function (19.19) with an optimized orbital exponent compared with the exact ground-state electronic energy curve. Note the incorrect behavior of the MO function as $R \to \infty$.

We approximated the $\sigma_g 1s$ MO in the H_2 ground-state approximate wave function (19.19) by the linear combination $N(1s_A + 1s_B)$. To improve the MO wave function, we can look for the best possible form for the $\sigma_g 1s$ MO, still writing the spatial wave function as the product of an orbital for each electron. The best possible MO wave function is the Hartree–Fock wave function (Secs. 18.9 and 19.5). Finding the Hartree–Fock wave function for H_2 is not too difficult. The H_2 Hartree–Fock wave function predicts $R_e = 0.73$ Å and $D_e = 3.64$ eV; D_e is still substantially in error. As noted in Sec. 18.9, the Hartree–Fock wave function is not the true wave function, because of neglect of electron correlation.

In the 1960s, Kolos and Wolniewicz used very complicated variational functions that go beyond the Hartree–Fock approximation. With the inclusion of relativistic corrections and corrections for deviations from the Born-Oppenheimer approximation, they calculated $D_0/hc = 36117.9\ \text{cm}^{-1}$ for H_2. (D_0 differs from D_e by the zero-point vibrational energy; see Chapter 20.) At the time the calculation was completed, the experimental D_0/hc was $36114 \pm 1\ \text{cm}^{-1}$, and the $4\ \text{cm}^{-1}$ discrepancy was a source of embarrassment to the theoreticians. Finally, reinvestigations of the spectrum of H_2 showed that the experimental result was in error and gave the new experimental value $36118.1 \pm 0.2\ \text{cm}^{-1}$, in excellent agreement with the value calculated from quantum mechanics.

What about excited electronic states for H_2? The lowest-lying excited H_2 MO is the σ_u^*1s MO. Just as the lowest excited electron configuration of He is $1s2s$, the lowest excited electron configuration of H_2 is $(\sigma_g 1s)(\sigma_u^*1s)$, with one electron in each of the MOs $\sigma_g 1s$ and σ_u^*1s. Like the He $1s2s$ configuration, the $(\sigma_g 1s)(\sigma_u^*1s)$ H_2 configuration gives rise to two terms, a singlet with total spin quantum number $S = 0$ and a triplet with total spin quantum number $S = 1$. In accord with Hund's rule, the triplet lies lower and is therefore the lowest excited electronic level of H_2. In analogy with (18.46), the triplet has the MO wave functions

$$2^{-1/2}[\sigma_g 1s(1)\sigma_u^*1s(2) - \sigma_g 1s(2)\sigma_u^*1s(1)] \times \text{spin function} \qquad (19.21)$$

where the spin function is one of the three symmetric spin functions (18.40). With one electron in a bonding orbital and one in an antibonding orbital, we expect no net bonding. This is borne out by experiment and by accurate theoretical calculations, which show the $E_e(R)$ curve to have no minimum (Fig. 19.5).

The H_2 levels (19.20) and (19.21) both dissociate into two H atoms in $1s$ states. The bonding level (19.20) has the electrons paired with opposite spins and a net spin of zero. The repulsive level (19.21) has the electrons unpaired with approximately parallel spins. Whether two approaching $1s$ H atoms attract or repel each other depends on whether their spins are antiparallel or parallel.

Other Homonuclear Diatomic Molecules

The simple MO treatment of He_2 places the four electrons into the two lowest available MOs to give the ground-state configuration $(\sigma_g 1s)^2(\sigma_u^*1s)^2$. The MO wave function is a Slater determinant with four rows and four columns. With two bonding and two antibonding electrons, we expect no net bonding and no stability for the ground electronic state. This is in agreement with experiment. When two ground-state He atoms approach each other, the electronic energy curve $E_e(R)$ resembles the second lowest curve of Fig. 19.5. Since $E_e(R)$ is the potential energy for nuclear motion, two $1s^2$ He atoms strongly repel each other. Actually, in addition to the strong, relatively short-range repulsion, there is a very weak attraction at relatively large values of R that produces a very slight minimum in the He-He potential-energy curve. This attraction is responsible for the liquefaction of He at very low temperature and produces an extremely weakly bound ground-state He_2 molecule ($D_0 = 10^{-7}$ eV) that has been detected at $T = 10^{-3}$ K (see Sec. 21.10). At ordinary temperatures, the He_2 concentration is negligible.

Similar to the repulsion between two $1s^2$ He atoms is the observed repulsion whenever two closed-shell atoms or molecules approach each other closely. This repulsion is important in chemical kinetics, since it is related to the activation energy of chemical reactions (see Sec. 22.2). Part of this repulsion is due to the Coulombic repulsion between electrons, but a major part of the repulsion is a consequence of the antisymmetry requirement (Sec. 18.6), as we now show. Let $\psi(q_1, q_2, q_3, \ldots)$ be the wave function for a system of electrons, where q_1 stands for the four coordinates (three spatial and one spin) of electron 1. Since ψ is antisymmetric, interchange of the coordinates of electrons 1 and 2 multiplies ψ by -1. Therefore, $\psi(q_2, q_1, q_3, \ldots) = -\psi(q_1, q_2, q_3, \ldots)$. Now suppose that electrons 1 and 2 have the same spin coordinate (both α or both β) and the same spatial coordinates. Then $q_1 = q_2$, and $\psi(q_1, q_1, q_3, \ldots) = -\psi(q_1, q_1, q_3, \ldots)$. Hence, $2\psi(q_1, q_1, q_3, \ldots) = 0$, and $\psi(q_1, q_1, q_3, \ldots) = 0$.

The vanishing of $\psi(q_1, q_1, q_3, \ldots)$ shows that there is zero probability for two electrons to have the same spatial and spin coordinates. Two electrons that have the same spin (both with $m_s = \frac{1}{2}$ or both with $m_s = -\frac{1}{2}$) have zero probability of being at the same point in space. Moreover, because ψ is a continuous function, the probability that two electrons with the same spin will approach each other closely must be very small. Electrons with the same spin tend to avoid each other and act as if they repelled each other over and above the Coulombic repulsion. This apparent extra repulsion of electrons with like spins is called the **Pauli repulsion.** The Pauli repulsion is not a real physical force. It is an apparent force that is a consequence of the antisymmetry requirement of the wave function.

When two $1s^2$ He atoms approach, the antisymmetry requirement causes an apparent Pauli repulsion between the spin-α electron on one atom and the spin-α electron on the other atom; likewise for the spin-β electrons. As the He atoms approach each other, there is a depletion of electron probability density in the region between the nuclei (and a corresponding buildup of probability density in regions outside the nuclei) and the atoms repel each other.

The ground-state electron configurations of Li_2, Be_2, etc., are formed by filling in the homonuclear diatomic MOs in Fig. 19.15 (see Prob. 19.30). For example, O_2 has 16 electrons, and Fig. 19.15 gives the ground-state configuration

$$(\sigma_g 1s)^2(\sigma_u^* 1s)^2(\sigma_g 2s)^2(\sigma_u^* 2s)^2(\pi_u 2p)^4(\sigma_g 2p)^2(\pi_g^* 2p)^2$$

Actually, spectroscopic evidence shows that in O_2 the $\sigma_g 2p$ MO lies slightly lower than the $\pi_u 2p$ MOs, so $\sigma_g 2p$ precedes $\pi_u 2p$ in the electron configuration. Figure 19.18 shows the distribution of the valence electrons in MOs in the O_2 ground state. In accord with Hund's rule of maximum multiplicity for the ground state, the two antibonding π electrons are placed in separate orbitals to allow a triplet ground state. This agrees with the observed paramagnetism of ground-state O_2. In O_2, there are four more bonding electrons than antibonding electrons, and so the MO theory predicts a double bond (composed of one σ bond and one π bond) for O_2. Note the higher D_e for O_2 compared with the single-bonded species F_2 and Li_2 (Table 19.3). The double bond makes R_e of O_2 less than R_e of Li_2. R_e of O_2 is greater than R_e of H_2 because of the presence of the inner-shell $1s$ electrons on the O atoms.

In O_2, the high nuclear charge draws the $1s$ orbitals on each atom in close to the nuclei, and there is virtually no overlap between these AOs. Therefore, the $\sigma_g 1s$ and $\sigma_u^* 1s$ MO energies in O_2 are each nearly the same as the $1s$ AO energy in an O atom. Inner-shell electrons play no real part in chemical bonding, other than to screen the valence electrons from the nuclei.

Figure 19.19 plots R_e, D_e, and the MO *bond order* (defined as half the difference between the number of bonding and antibonding electrons) for some second-row homonuclear diatomic molecules. The higher the bond order, the greater is D_e and the smaller is R_e.

Instead of the separated-atoms notation for homonuclear diatomic MOs, quantum chemists prefer a notation in which the lowest σ_g MO is called $1\sigma_g$, the next lowest σ_g MO is called $2\sigma_g$, etc. In this notation, the MOs in Fig. 19.15 are called (in order of increasing energy) $1\sigma_g$, $1\sigma_u$, $2\sigma_g$, $2\sigma_u$, $1\pi_u$, $3\sigma_g$, $1\pi_g$, $3\sigma_u$.

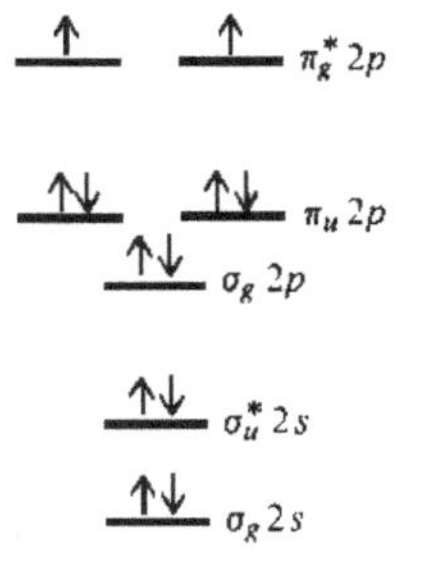

Figure 19.18

Occupied valence MOs in the O_2 ground electronic state. (Not to scale.)

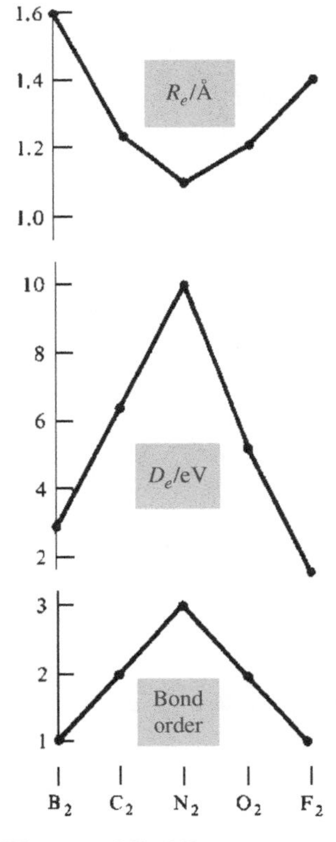

Figure 19.19

Correlation between bond orders and bond lengths and dissociation energies for some homonuclear diatomic molecules.

Heteronuclear Diatomic Molecules

The MO method feeds the electrons of a heteronuclear diatomic molecule into molecular orbitals. In the crudest approximation, each bonding MO is taken as a linear combination of two AOs, one from each atom. In constructing MOs, one uses the principle that *only AOs of reasonably similar energies contribute substantially to a given MO.*

As an example, consider HF. Figure 18.15 shows that the energy of a $2p$ AO in $_9$F is reasonably close to the $1s$ AO energy in H, but the $2s$ AO in F is substantially lower in energy than the $1s$ H AO. (The logarithmic scale makes the fluorine $2s$ level appear closer to the $2p$ level than it actually is.) The $2p$ AO in F lies somewhat lower than the $1s$ AO in H because the five $2p$ electrons in F screen one another rather poorly, giving a large Z_{eff} for the $2p$ electrons [Eq. (18.55)]; this large Z_{eff} makes F more electronegative than H [Eq. (19.4)].

Let the HF molecular axis be the z axis, and let F$2p$ and H$1s$ denote a $2p$ AO on F and a $1s$ AO on H. The F$2p_z$ AO has quantum number $m = 0$ and has no nodal plane containing the internuclear axis. The overlap of this AO with the H$1s$ AO, which also has $m = 0$ and no nodal plane containing the z axis, therefore gives rise to a σ MO (Fig. 19.20). We therefore form the linear combination $c_1\text{H}1s + c_2\text{F}2p_z$. Minimization of the variational integral will lead to two sets of values for c_1 and c_2, one set giving a bonding MO and the other an antibonding MO:

$$\sigma = c_1\text{H}1s + c_2\text{F}2p_z \quad \text{and} \quad \sigma^* = c_1'\text{H}1s - c_2'\text{F}2p_z \tag{19.22}$$

The σ MO in (19.22) has c_1 and c_2 both positive and is bonding because of the charge buildup between the nuclei. The antibonding σ^* MO in (19.22) has opposite signs for the coefficients of the AOs and so has charge depletion between the nuclei. This MO is unoccupied in the HF ground state. The g, u designation does not apply to heteronuclear diatomics.

In contrast to the F$2p_z$ AO, the F$2p_x$ and F$2p_y$ AOs have $|m| = 1$ and have one nodal plane containing the internuclear (z) axis. These AOs will therefore be used to form π MOs in HF. Since H has no valence-shell AOs with $|m| = 1$, the π MOs in HF will consist entirely of F AOs, and these MOs are $\pi_x = \text{F}2p_x$ and $\pi_y = \text{F}2p_y$.

The $1s$ and $2s$ AOs in F are too low in energy to take a substantial part in the bonding and therefore form nonbonding σ MOs in HF. Don't confuse a nonbonding MO with an antibonding MO. A nonbonding MO shows neither charge depletion nor charge buildup between the nuclei.

In the standard notation for heteronuclear diatomic molecules, the lowest σ MO is called the 1σ MO, the next lowest σ MO is the 2σ MO, etc. The lowest π energy level is called the 1π level, etc. In our crude approximation, the occupied MOs in hydrogen fluoride are

$$\begin{gathered} 1\sigma = \text{F}1s, \qquad 2\sigma = \text{F}2s, \qquad 3\sigma = c_1\text{H}1s + c_2\text{F}2p_z \\ 1\pi_x = \text{F}2p_x, \qquad 1\pi_y = \text{F}2p_y \end{gathered} \tag{19.23}$$

where $1\pi_x$ and $1\pi_y$ have the same energy. Since F is more electronegative than H, we expect $|c_2| > |c_1|$ in the 3σ MO; the electrons of the bond are more likely to be found close to F than to H.

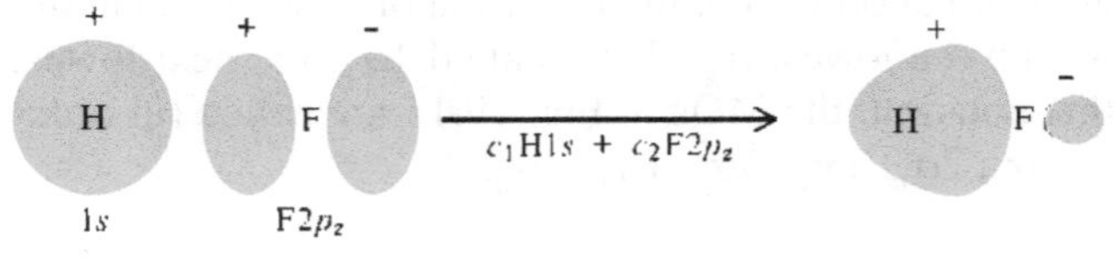

Figure 19.20

Formation of the bonding MO in HF.

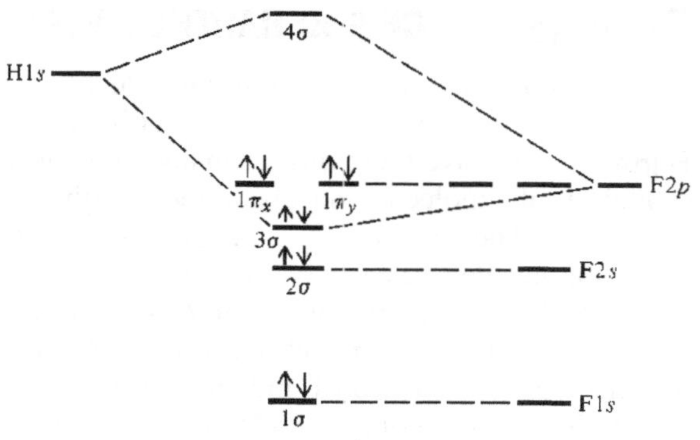

Figure 19.21

MO energies in HF. (Not to scale.)

Figure 19.21 shows the energy-level scheme for HF in the simple approximation (19.23). The 1π MOs are lone-pair AOs on F and have nearly the same energy as F2p AOs. The 2σ MO is also a lone-pair orbital.

An H atom is special, since it has no p valence orbitals. Consider a polar-covalent heteronuclear diatomic molecule AB, where both A and B are from the second or a higher period and hence have s and p valence levels. Let B be somewhat more electronegative than A. We draw Fig. 19.22 similar to Figs. 19.15 and 19.21 to show the formation of valence MOs from the ns and np valence AOs of A and the $n's$ and $n'p$ valence AOs of B; n and n' are the principal quantum numbers of the valence electrons and equal the periods of A and B in the periodic table. [It is assumed that B and A don't differ greatly in electronegativity. If B were much more electronegative than A (as, for example, in BF), then the valence p level of B might lie close to the valence s level of A, and the p_z AO of B would combine mainly with the s valence AO of A.] The MO shapes are similar to those in Figs. 19.11 to 19.14 for homonuclear diatomics, except that in each bonding MO the probability density is greater around the more electronegative element B than around A, and each bonding MO contour is therefore larger around B than A. In each antibonding MO, the probability density is larger around A, since more of the atom-B AO has been "used up" in forming the corresponding bonding MO.

To get the valence MO configuration of molecules like CN, NO, CO, or ClF, we feed the valence electrons into the MOs of Fig. 19.22. For example, CO has 10 valence electrons and has the configuration $(\sigma_s)^2(\sigma_s^*)^2(\pi)^4(\sigma_p)^2$. With six more bonding than antibonding electrons, the molecule has a triple bond (composed of one σ and two π bonds), in accord with the dot structure :C≡O:. The lowest two MOs in CO are the 1σ and 2σ MOs, formed from linear combinations of C1s and O1s AOs, and the complete MO configuration of CO in the standard notation is $1\sigma^2 2\sigma^2 3\sigma^2 4\sigma^2 1\pi^4 5\sigma^2$.

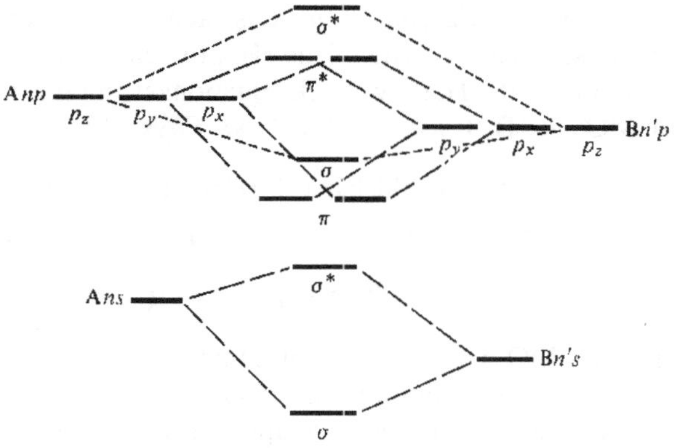

Figure 19.22

MOs formed from valence s and p AOs of atoms A and B with rather similar electronegativities.

19.5 SCF AND HARTREE–FOCK WAVE FUNCTIONS

The best possible wave function with electrons assigned to orbitals is the Hartree–Fock wave function. Starting in the 1960s, the use of electronic computers allowed Hartree–Fock wave functions for many molecules to be calculated. The Hartree–Fock orbitals ϕ_i of a molecule must be found by solving the Hartree–Fock equations (18.57): $\hat{F}\phi_i = \varepsilon_i\phi_i$. The terms in the Fock operator $\hat{F}$ for an atom were discussed after (18.57). $\hat{F}$ for a molecule is the same as $\hat{F}$ for an atom except that the electron–nucleus attraction $-Ze^2/4\pi\varepsilon_0 r_1$ in an atom (term *b*) is replaced by $-\Sigma_\alpha Z_\alpha e^2/4\pi\varepsilon_0 r_{1\alpha}$, which gives the potential energy of the attractions between electron 1 and all the nuclei; $r_{1\alpha}$ is the distance between electron 1 and nucleus α.

As is done for atoms, each Hartree–Fock MO is expressed as a linear combination of a set of functions called basis functions. If enough basis functions are included, one can get MOs that differ negligibly from the true Hartree–Fock MOs. Any functions can be used as basis functions, so long as they form a complete set (as defined in Sec. 18.9). Since molecules are made of bonded atoms, it is most convenient to use atomic orbitals as the basis functions. Each MO is then written as a linear combination of the basis-set AOs, and the coefficients of the AOs are found by solving the Hartree–Fock equations.

To have an accurate representation of an MO requires that the MO be expressed as a linear combination of a complete set of functions. This means that all the AOs of a given atom, whether occupied or unoccupied in the free atom, contribute to the MOs. To simplify the calculation, one frequently solves the Hartree–Fock equations using in the basis set only those AOs from each atom whose principal quantum number does not exceed the principal quantum number of the atom's valence electrons. Such a basis set limited to inner-shell and valence-shell AOs is called a **minimal basis set.** Use of a minimal basis set gives only an approximation to the Hartree–Fock MOs. Any wave function found by solving the Hartree–Fock equations is called a **self-consistent-field (SCF) wave function.** Only if the basis set is very large is an SCF wave function accurately equal to the Hartree–Fock wave function.

Which AOs contribute to a given MO is determined by the symmetry properties of the MO. For example, we saw at the end of Sec. 19.3 that MOs of a diatomic molecule can be classified as σ, π, δ, . . . according to whether they have 0, 1, 2, . . . nodal planes containing the internuclear axis. Only AOs that have 0 such nodal planes contribute to a σ MO; only AOs that have 1 such nodal plane contribute to a π MO; etc. In Sec. 19.4, we took each diatomic MO as a linear combination of only two AOs. This is the crudest approximation and does not give an accurate representation of MOs. In actuality, all σ AOs of the two atoms contribute to each σ MO; similarly for π MOs. (By a σ AO is meant one with no nodal plane containing the internuclear axis.)

Consider, for example, a minimal-basis-set calculation of HF. The valence electron in H has $n = 1$, so we use only the H1*s* AO. The valence electrons in F have $n = 2$, so we use the 1*s*, 2*s*, $2p_x$, $2p_y$, and $2p_z$ AOs of F. This gives a total of six basis functions. The H1*s*, F1*s*, F2*s*, and F$2p_z$ AOs each have 0 nodal planes containing the internuclear (*z*) axis, so each σ MO of the HF molecule is a linear combination of these four AOs. Solution of the Hartree–Fock equations using this minimal basis set gives the occupied σ MOs as [B. J. Ransil, *Rev. Mod. Phys.*, **32,** 245 (1960)]

$$
\begin{aligned}
1\sigma &= 1.000(\mathrm{F}1s) + 0.012(\mathrm{F}2s) + 0.002(\mathrm{F}2p_z) - 0.003(\mathrm{H}1s) \\
2\sigma &= -0.018(\mathrm{F}1s) + 0.914(\mathrm{F}2s) + 0.090(\mathrm{F}2p_z) + 0.154(\mathrm{H}1s) \qquad (19.24) \\
3\sigma &= -0.023(\mathrm{F}1s) - 0.411(\mathrm{F}2s) + 0.711(\mathrm{F}2p_z) + 0.516(\mathrm{H}1s)
\end{aligned}
$$

The 1σ MO has a significant contribution only from the F1*s* AO. The 2σ MO has a significant contribution only from the F2*s* AO. This is in accord with the simple arguments that led to the very approximate MOs in (19.23). The 3σ MO has its largest contributions from H1s and F2p_z but [unlike the crude approximation of (19.23)] also has a significant contribution from the F2s AO. The mixing together of two or more AOs on the same atom is called **hybridization.** [Each AO in (19.24) and in other equations in this section and the next is actually an approximate AO whose form is given by a single Slater-type orbital (Sec. 18.9).]

The F2p_x and F2p_y AOs each have one nodal plane containing the internuclear axis, and these AOs form the occupied π MOs of HF:

$$1\pi_x = \text{F}2p_x, \qquad 1\pi_y = \text{F}2p_y \tag{19.25}$$

The 1π level is doubly degenerate, so any two linear combinations of the orbitals in (19.25) could be used. (Recall the theorem about degenerate levels in Sec. 18.3.)

For the molecule F_2, a minimal-basis-set SCF calculation (Ransil, op. cit.) gives the MO we previously called the $\sigma_g 2p$ MO as $-0.005(1s_A + 1s_B) - 0.179(2s_A + 2s_B) + 0.648(2p_{zA} - 2p_{zB})$. This can be compared with the simple expression $N(2p_{zA} - 2p_{zB})$ used earlier. When a larger basis set is used, this F_2 MO is found to have small contributions also from $3s$, $3d\sigma$, and $4f\sigma$ AOs, where $3d\sigma$ and $4f\sigma$ signify AOs with no nodal planes containing the molecular axis.

To reach the true molecular wave function, one must go beyond the Hartree–Fock approximation. Methods that do this are discussed in Secs. 19.9 and 19.10.

19.6 THE MO TREATMENT OF POLYATOMIC MOLECULES

As with diatomic molecules, one expresses the MOs of a polyatomic molecule as linear combinations of basis functions. Most commonly, AOs of the atoms forming the molecule are used as the basis functions. To find the coefficients in the linear combinations, one solves the Hartree–Fock equations (18.57). Which AOs contribute to a given MO is determined by the symmetry of the molecule.

The BeH_2 Molecule

We shall apply the MO method to BeH_2. Since the valence shell of Be has $n = 2$, a minimal-basis-set calculation uses the Be1s, Be2s, Be2p_x, Be2p_y, Be2p_z AOs and the H_A1s and H_B1s AOs, where H_A and H_B are the two H atoms. The molecule has six electrons, and these will fill the lowest three MOs in the ground state.

Accurate theoretical calculations show that the equilibrium geometry is linear and symmetric (HBeH), and we shall assume this structure. Each MO of this linear molecule can be classified as σ, π, δ, . . . according to whether it has 0, 1, 2, . . . nodal planes containing the internuclear axis. Further, since the molecule has a center of symmetry at the Be nucleus, we can classify each MO as g or u (as we did with homonuclear diatomics), according to whether it has the same or opposite signs on diagonally opposite sides of the Be atom.

The Be1s AO has a much lower energy than all the other AOs in the basis set (Fig. 18.15), so the lowest MO will be nearly identical to the Be1s AO. The function Be1s has no nodal planes containing the internuclear axis and is a σ function; it also has g symmetry. We therefore write

$$1\sigma_g = \text{Be}1s \tag{19.26}$$

where the 1 indicates that this is the lowest σ_g MO.

The $2s$ and $2p$ valence AOs of Be and the $1s$ valence AOs of H_A and H_B have similar energies and will be combined to form the remaining occupied MOs. In forming these MOs one must take the symmetry of the molecule into account. An MO without either g or u symmetry could not be a solution of the BeH_2 Hartree–Fock equations. The proof of this requires group theory and is omitted.

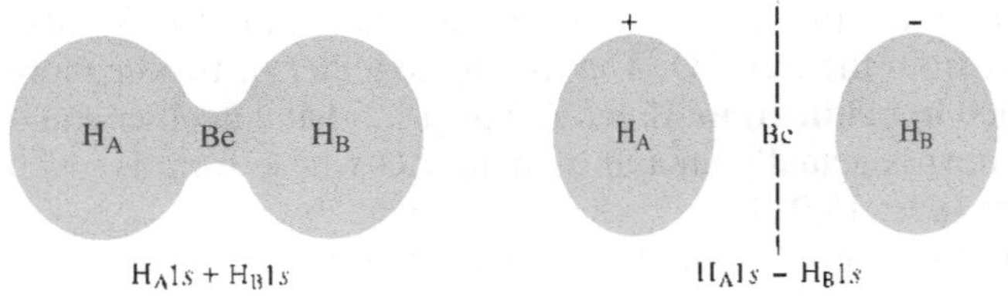

Figure 19.23

Linear combinations of H-atom 1*s* AOs in BeH_2 that have suitable symmetry.

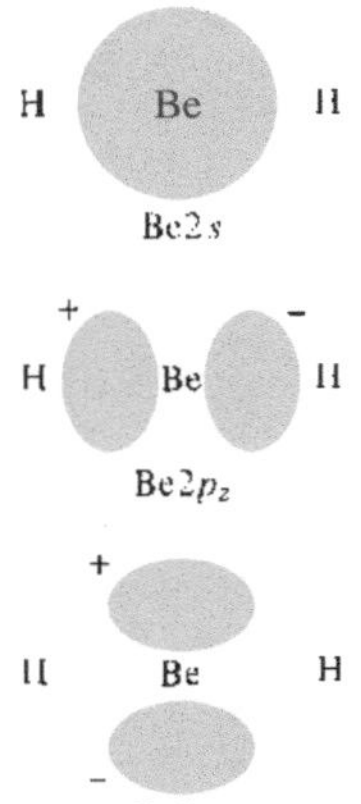

Figure 19.24

Be AOs in BeH_2.

For a BeH_2 MO to have *g* or *u* symmetry (that is, for the square of the MO to have the same value at corresponding diagonally opposite points on each side of the central Be atom) the squares of the coefficients of the H_A1s and H_B1s AOs must be equal in each BeH_2 MO. Just as the $1s_A$ and $1s_B$ AOs in a homonuclear diatomic molecule's MOs occur as the linear combinations $1s_A + 1s_B$ and $1s_A - 1s_B$, the H_A1s and H_B1s AOs in BeH_2 can occur only as the linear combinations $H_A1s + H_B1s$ and $H_A1s - H_B1s$ in the BeH_2 MOs that satisfy the Hartree–Fock equations. Both these linear combinations have no nodal plane containing the internuclear axis. Hence these linear combinations will contribute to σ MOs. The linear combination $H_A1s + H_B1s$ has equal values at points diagonally opposite the center of the molecule (Fig. 19.23) and so has σ_g symmetry. The linear combination $H_A1s - H_B1s$ has opposite signs at points diagonally opposite the molecular center and thus has σ_u symmetry.

What about the Be AOs? The Be2*s* AO has σ_g symmetry. Calling the internuclear axis the *z* axis, we see from Fig. 19.24 that the $Be2p_z$ AO has σ_u symmetry. The $Be2p_x$ and $Be2p_y$ AOs each have π_u symmetry.

The basis-set functions and their symmetries are thus

Be1*s*	Be2*s*	$Be2p_z$	$Be2p_x$	$Be2p_y$	$H_A1s + H_B1s$	$H_A1s - H_B1s$
σ_g	σ_g	σ_u	π_u	π_u	σ_g	σ_u

Combining functions that have σ_g symmetry and comparable energies, we form a σ_g MO as follows:

$$2\sigma_g = c_1\text{Be}2s + c_2(\text{H}_\text{A}1s + \text{H}_\text{B}1s) \qquad (19.27)$$

The 2 in $2\sigma_g$ indicates that this is the second lowest σ_g MO, the lowest being (19.26). The very-low-energy Be1*s* AO will make a very slight contribution to $2\sigma_g$, which we shall neglect. With c_1 and c_2 both positive, this MO has probability-density buildup between Be and H_A and between Be and H_B and is therefore bonding (Fig. 19.25).

Similarly, we form a bonding σ_u MO as (Fig. 19.25)

$$1\sigma_u = c_3\text{Be}2p_z + c_4(\text{H}_\text{A}1s - \text{H}_\text{B}1s) \qquad (19.28)$$

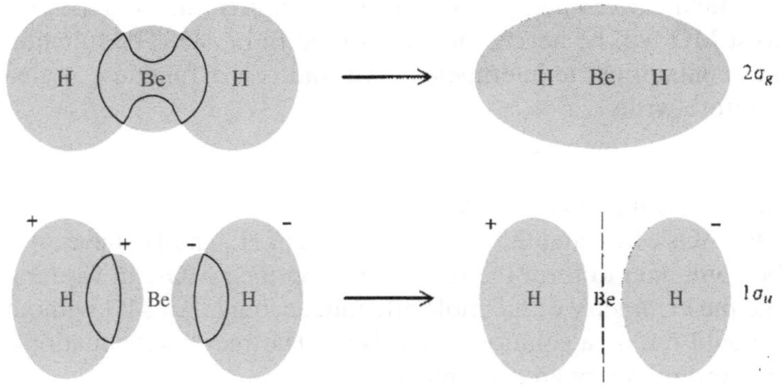

Figure 19.25

Formation of bonding MOs in BeH_2.

with c_3 and c_4 both positive. This bonding MO has its energy below the energies of the Be2p_z and H1s AOs from which it is formed.

The coefficients c_1, c_2, c_3, c_4 are found by solving the Hartree–Fock equations.

The Be2p_x and Be2p_y AOs form two π_u MOs:

$$1\pi_{u,x} = \text{Be}2p_x, \qquad 1\pi_{u,y} = \text{Be}2p_y \tag{19.29}$$

These two MOs have the same energy and constitute the doubly degenerate $1\pi_u$ energy level. The nonbonding π MOs of (19.29) have nearly the same energy as the Be2p_x and Be2p_y AOs and so lie above the bonding $2\sigma_g$ and $1\sigma_u$ MOs. The π MOs are therefore unoccupied in the ground state of this six-electron molecule.

A minimal-basis-set SCF calculation on BeH_2 [R. G. A. R. Maclagan and G. W. Schnuelle, *J. Chem. Phys.*, **55,** 5431 (1971)] gave the occupied MOs as

$$\begin{aligned} 1\sigma_g &= 1.00(\text{Be}1s) + 0.016(\text{Be}2s) - 0.002(\text{H}_\text{A}1s + \text{H}_\text{B}1s) \\ 2\sigma_g &= -0.09(\text{Be}1s) + 0.40(\text{Be}2s) + 0.45(\text{H}_\text{A}1s + \text{H}_\text{B}1s) \\ 1\sigma_u &= 0.44(\text{Be}2p_z) + 0.44(\text{H}_\text{A}1s - \text{H}_\text{B}1s) \end{aligned} \tag{19.30}$$

The $1\sigma_g$ MO is essentially a Be1s AO, as anticipated in (19.26). The $2\sigma_g$ and $1\sigma_u$ MOs have essentially the forms of (19.27) and (19.28).

There are also two antibonding MOs $3\sigma_g^*$ and $2\sigma_u^*$ formed from the same AOs as the two bonding MOs (19.27) and (19.28):

$$3\sigma_g^* = c_1'\text{Be}2s - c_2'(\text{H}_\text{A}1s + \text{H}_\text{B}1s), \quad 2\sigma_u^* = c_3'\text{Be}2p_z - c_4'(\text{H}_\text{A}1s - \text{H}_\text{B}1s) \tag{19.31}$$

Figure 19.26 sketches the AO and MO energies for BeH_2. Of course, this molecule has many higher unoccupied MOs that are not shown in the figure. These MOs are formed from higher AOs of Be and the H's. The BeH_2 ground-state configuration is $(1\sigma_g)^2(2\sigma_g)^2(1\sigma_u)^2$. There are four bonding electrons and hence two bonds. The $1\sigma_g$ electrons are nonbonding inner-shell electrons.

Note that a bonding MO has a lower energy than the AOs from which it is formed, an antibonding MO has a higher energy than the AOs from which it is formed, and a nonbonding MO has approximately the same energy as the AO or AOs from which it is formed.

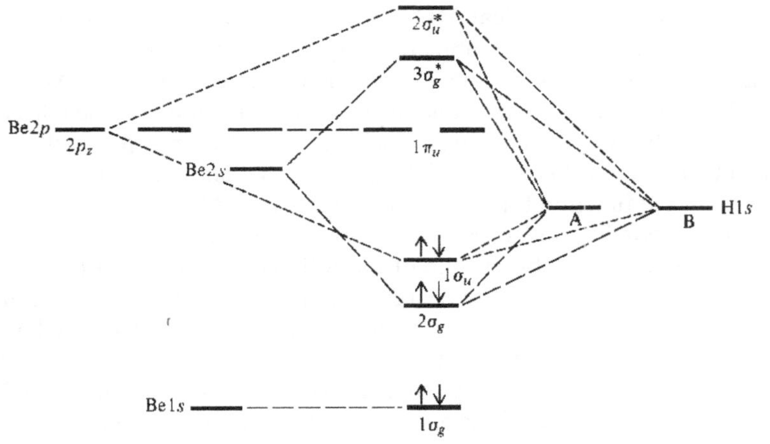

Figure 19.26

Energy-level scheme for BeH_2 MOs. (Not to scale.)

Localized MOs

The BeH_2 bonding MOs $2\sigma_g$ and $1\sigma_u$ in Fig. 19.25 are each delocalized over the entire molecule. The two electrons in the $2\sigma_g$ MO move over the entire molecule, as do the two in the $1\sigma_u$ MO. This is puzzling to a chemist, who likes to think in terms of individual bonds: H—Be—H or H:Be:H. The existence of bond energies, bond moments, and bond vibrational frequencies (Sec. 20.9) that are roughly the same for a given kind of bond in different molecules shows that there is much validity in the picture of individual bonds. How can we reconcile the existence of individual bonds with the delocalized MOs found by solving the Hartree–Fock equations?

Actually, we *can* use the MO method to arrive at a picture in accord with chemical experience, as we now show. The MO ground-state wave function for BeH_2 is a 6×6 Slater determinant (Sec. 18.8). The first two rows of this Slater determinant are

$$\begin{matrix} 1\sigma_g(1)\alpha(1) & 1\sigma_g(1)\beta(1) & 2\sigma_g(1)\alpha(1) & 2\sigma_g(1)\beta(1) & 1\sigma_u(1)\alpha(1) & 1\sigma_u(1)\beta(1) \\ 1\sigma_g(2)\alpha(2) & 1\sigma_g(2)\beta(2) & 2\sigma_g(2)\alpha(2) & 2\sigma_g(2)\beta(2) & 1\sigma_u(2)\alpha(2) & 1\sigma_u(2)\beta(2) \end{matrix}$$

The third row involves electron 3, etc. Each column has the same spin-orbital. Now it is a well-known theorem (*Sokolnikoff and Redheffer,* app. A) that addition of a constant times one column of a determinant to another column leaves the determinant unchanged in value. For example, if we add three times column 1 of the determinant in (18.44) to column 2, we get

$$\begin{vmatrix} a & b + 3a \\ c & d + 3c \end{vmatrix} = a(d + 3c) - c(b + 3a) = ad - bc = \begin{vmatrix} a & b \\ c & d \end{vmatrix}$$

Thus, if we like, we can add a multiple of one column of the Slater-determinant MO wave function to another column without changing the wave function. This addition will "mix" together different MOs, since each column is a different spin-orbital. We can therefore take linear combinations of MOs to form new MOs without changing the overall wave function. Of course, the new MOs should each be normalized and, for computational convenience, should also be orthogonal to one another.

The BeH_2 MOs $1\sigma_g$, $2\sigma_g$, and $1\sigma_u$ satisfy the Hartree–Fock equations (18.57) and have the symmetry of the molecule. Because they have the molecular symmetry, they are delocalized over the whole molecule. (More accurately, the $2\sigma_g$ and $1\sigma_u$ MOs are delocalized, but the inner-shell $1\sigma_g$ is localized on the central Be atom.) These delocalized MOs satisfying the Hartree–Fock equations and having the symmetry of the molecule are called the **canonical MOs.** The canonical MOs are unique (except for the possibility of taking linear combinations of degenerate MOs).

As just shown, we can take linear combinations of the canonical MOs to form a new set of MOs that will give the same overall wave function. The new MOs will not individually be solutions of the Hartree–Fock equations $\hat{F}\phi_i = \varepsilon_i\phi_i$, but the wave function formed from these MOs will have the same energy and the same total probability density as the wave function formed from the canonical MOs.

Of the many possible sets of MOs that can be formed, we want to find a set that will have each MO classifiable as one of the following: a **bonding** (b) orbital localized between two atoms and having charge buildup between the atoms, an **inner-shell** (i) orbital, or a **lone-pair** (l) orbital. We call such a set of MOs **localized MOs.** Each localized MO will not have the symmetry of the molecule, but the localized MOs will correspond closely to a chemist's picture of bonding. Since localized MOs are not eigenfunctions of the Hartree–Fock operator $\hat{F}$, in a certain sense each such MO does not correspond to a definite orbital energy. However, one can calculate an average energy of a localized MO by averaging over the orbital energies of the canonical MOs that form the localized MO.

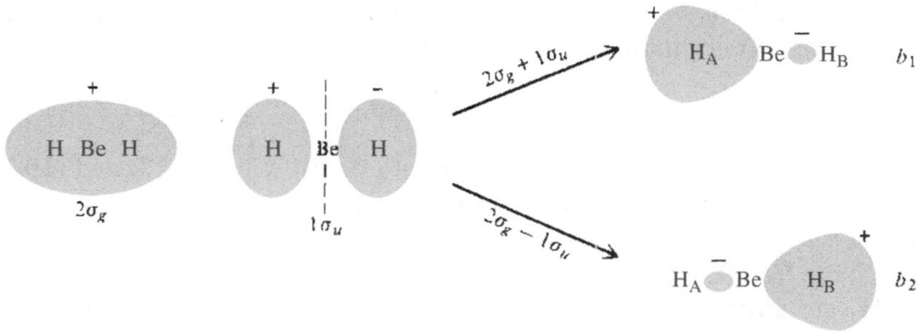

Figure 19.27

Formation of localized bonding MOs in BeH_2 from linear combinations of delocalized (canonical) MOs.

Consider BeH_2. The $1\sigma_g$ canonical MO is an inner-shell (i) AO on Be and can therefore be taken as one of the localized MOs: $i(\text{Be}) = 1\sigma_g = \text{Be}1s$. The $2\sigma_g$ and $1\sigma_u$ canonical MOs are delocalized. Figure 19.25 shows that the $1\sigma_u$ MO has opposite signs in the two halves of the molecule, whereas $2\sigma_g$ is essentially positive throughout the molecule. Hence, by taking linear combinations that are the sum and difference of these two canonical MOs, we get MOs that are each largely localized between only two atoms (Fig. 19.27). Thus, we take the localized bonding MOs b_1 and b_2 as

$$b_1 = 2^{-1/2}(2\sigma_g + 1\sigma_u), \qquad b_2 = 2^{-1/2}(2\sigma_g - 1\sigma_u) \tag{19.32}$$

where the $2^{-1/2}$ is a normalization constant. The b_1 localized MO corresponds to a bond between Be and H_A. The b_2 MO gives the Be—H_B bond.

Using these localized MOs, we write the BeH_2 MO wave function as a 6×6 Slater determinant whose first row is

$$i(1)\alpha(1) \quad i(1)\beta(1) \quad b_1(1)\alpha(1) \quad b_1(1)\beta(1) \quad b_2(1)\alpha(1) \quad b_2(1)\beta(1)$$

This localized-MO wave function is equal to the wave function that uses delocalized (canonical) MOs.

Equation (19.32) expresses the localized bonding MOs b_1 and b_2 in terms of the canonical MOs. Substitution of (19.27) and (19.28) into (19.32) gives

$$\begin{aligned} b_1 &= 2^{-1/2}[c_1\text{Be}2s + c_3\text{Be}2p_z + (c_2 + c_4)\text{H}_A1s + (c_2 - c_4)\text{H}_B1s] \\ b_2 &= 2^{-1/2}[c_1\text{Be}2s - c_3\text{Be}2p_z + (c_2 - c_4)\text{H}_A1s + (c_2 + c_4)\text{H}_B1s] \end{aligned} \tag{19.33}$$

As a rough approximation, we see from (19.27), (19.28), and (19.30) that $c_2 \approx c_4$ and $c_1 \approx c_3$. These approximations give

$$\begin{aligned} b_1 &\approx 2^{-1/2}[c_1(\text{Be}2s + \text{Be}2p_z) + 2c_2\text{H}_A1s] \\ b_2 &\approx 2^{-1/2}[c_1(\text{Be}2s - \text{Be}2p_z) + 2c_2\text{H}_B1s] \end{aligned} \tag{19.34}$$

The approximate MOs (19.34) are each fully localized between Be and one H atom, but the more accurate expressions (19.33) show that the Be—H_A bonding MO b_1 has a small contribution from the H_B1s AO and so is not fully localized between the two atoms forming the bond.

Note that [unlike the canonical MOs (19.27) and (19.28)] the b_1 and b_2 localized MOs each have the Be2s and Be2p_z AOs mixed together, or *hybridized*. **Hybridization** is the mixing of different AOs of the same atom. The precise degree of hybridization depends on the values of c_1 and c_3 in (19.33). In the approximation of

(19.34), the MOs b_1 and b_2 would each contain equal amounts of the Be2s and Be2p_z AOs. The two normalized linear combinations

$$2^{-1/2}(2s + 2p_z) \quad \text{and} \quad 2^{-1/2}(2s - 2p_z) \qquad (19.35)$$

are called ***sp* hybrid AOs.** Comparison of (19.30) with (19.27) and (19.28) gives $c_1 = 0.40$ and $c_3 = 0.44$, so c_1 and c_3 are not precisely equal, but are nearly equal. Thus, the Be AOs in the BeH_2 bonding MOs are not precisely *sp* hybrids but are nearly *sp* hybrids.

Note from Eq. (19.33) and Fig. 19.27 that the localized bonding MOs b_1 and b_2 in BeH_2 are **equivalent** to each other. By this we mean that b_1 and b_2 have the same shapes and are interchanged by a rotation that interchanges the two equivalent chemical bonds in BeH_2. If we rotate b_1 and b_2 180° about an axis through Be and perpendicular to the molecular axis (thereby interchanging H_A1s and H_B1s and changing Be2p_z to $-$Be2p_z), then b_1 is changed to b_2, and vice versa. Because of the symmetry of the molecule, we expect b_1 and b_2 to be equivalent orbitals. It is possible to show that the linear combinations in (19.32) are the only linear combinations of the $2\sigma_g$ and $1\sigma_u$ canonical MOs that meet the requirements of being normalized, equivalent, and orthogonal.

We arrived at the approximately *sp* hybrid Be AOs in the localized BeH_2 bonding MOs by first finding the delocalized canonical MOs and then transforming to localized MOs. A simpler (and more approximate) approach is often preferred by chemists for qualitative discussions of bonding. In this procedure, one omits consideration of the canonical MOs. Instead, one forms the required hybrid AOs on the free Be atom and then uses these hybrids to form localized bonding MOs with the H1s AOs. For BeH_2 with its 180° bond angle, we need two equivalent hybrid AOs on Be that point in opposite directions. The valence AOs of Be are 2s and 2p. Figure 19.28 shows that the linear combinations (19.35) give two equivalent hybridized AOs oriented 180° apart. It is possible to show that the *sp* hybrids (19.35) are the only linear combinations that give AOs at 180° that are equivalent, normalized, and orthogonal in the free atom. We then overlap each of these *sp* hybrid AOs with an H1s AO to form the two bonds. This gives the approximate localized bonding MOs of Eq. (19.34).

Although the *sp* hybrids (19.35) are the only linear combinations of 2s and 2p_z that give equivalent orbitals in the free Be atom, we must expect that in the BeH_2 molecule the interaction between the Be hybrids and the H atoms will alter the nature of these hybrids somewhat. What is really wanted is equivalent, normalized, orthogonal MOs in the BeH_2 molecule and not equivalent, normalized, orthogonal AOs in the Be atom. The equivalent MOs in BeH_2 are (19.33), and as noted above, these contain not precisely *sp* hybrids but only approximately *sp* hybrids. Another approximation involved in the use of *sp* Be hybrids to form the MOs (19.34) is neglect of the small contribution of the H_B1s AO to the bonding MO between Be and H_A. This is the term $(c_2 - c_4)H_B1s = 0.01(H_B1s)$ in (19.33).

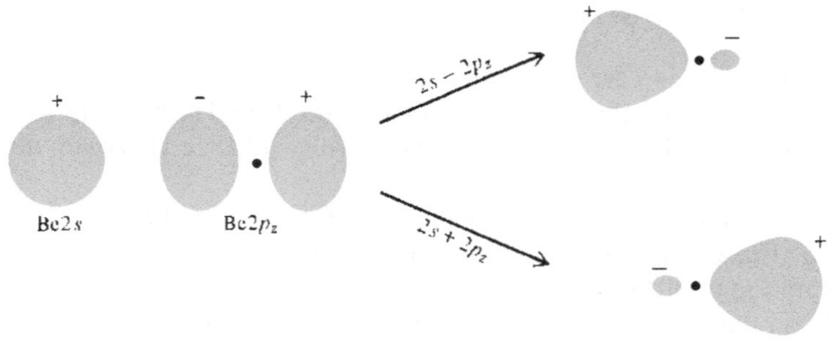

Figure 19.28

Formation of *sp* hybrid AOs in the Be atom.

Energy-Localized MOs

For BeH_2, the symmetry of the molecule enables one to determine what linear combination of canonical MOs to use to get localized bonding MOs. For less symmetric molecules, one cannot use symmetry, since the localized MOs need not be equivalent to one another. Several methods have been suggested for finding localized MOs from the canonical MOs. A widely accepted approach is that of Edmiston and Ruedenberg, who defined the **energy-localized MOs** as those orthogonal MOs that minimize the total of the Coulombic repulsions between the various pairs of localized MOs considered as charge distributions in space. This gives localized MOs that are separated as far as possible from one another.

In most cases, the energy-localized MOs agree with what one would expect from the Lewis dot formula. For example, for H_2O, the energy-localized MOs turn out to be one inner-shell MO, two bonding MOs, and two lone-pair MOs, in agreement with the dot formula H:Ö:H. The inner-shell MO is nearly identical to the O1*s* AO. One bonding energy-localized MO is largely localized in the O—H_A region, and the other is largely localized in the O—H_B region. The angle between the bonding localized MOs is 103°, which is nearly the same as the 104.5° experimental bond angle in water. The angle between the lone-pair localized orbitals is 114°. Each bonding localized MO is mainly a linear combination of 2*s* and 2*p* oxygen AOs and a 1*s* hydrogen AO. Each lone-pair MO is mainly a hybrid of 2*s* and 2*p* AOs on oxygen.

Sigma, Pi, and Delta Bonds

In most cases, each localized bonding MO of a molecule contains substantial contributions from AOs of only two atoms, the atoms forming the bond. In analogy with the classification used for diatomic molecules, each localized bonding MO of a polyatomic molecule is classified as σ, π, δ, . . . according to whether the MO has 0, 1, 2, . . . nodal planes containing the axis between the two bonded atoms. The BeH_2 MOs b_1 and b_2 in Fig. 19.27 are clearly σ MOs. One finds that a single bond between two atoms nearly always corresponds to a σ localized MO. Nearly always, a double bond between two atoms is composed of one σ localized MO and one π localized MO. Nearly always, a triple bond is composed of one σ-bond orbital and two π-bond orbitals. A quadruple bond is composed of one σ bond, two π bonds, and one δ bond.

A σ bond is formed by overlap of two AOs that have no nodal planes containing the bond axis. Figure 19.29*a* shows some kinds of AO overlap that produce σ localized bond MOs. Figure 19.29*b* shows some overlaps that lead to π bonds. Figure 19.29*c* shows formation of a δ bond.

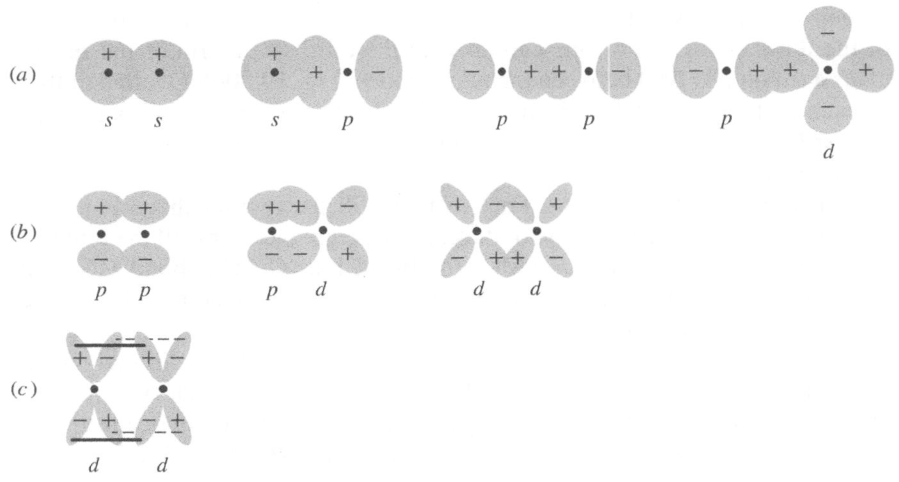

Figure 19.29

Overlap of AOs to form (*a*) σ bonds; (*b*) π bonds; (*c*) a δ bond. The lobes in (*c*) are in front of and behind the paper.

Chemists have known about σ and π bonds since the 1930s. In 1964, Cotton pointed out that the $Re_2Cl_8^{2-}$ ion has a quadruple bond between the two Re atoms, as shown by an abnormally short Re–Re bond distance. This bond is composed of one σ bond, two π bonds, and one δ bond, the δ bond being formed by overlap of two $d_{x^2-y^2}$ AOs, one on each Re atom. Several other transition-metal species have quadruple bonds. [F. A. Cotton, *Chem. Soc. Rev.*, **4,** 27 (1975); F. A. Cotton and R. A. Walton, *Multiple Bonds between Metal Atoms,* Wiley, 1982.]

Methane, Ethylene, and Acetylene

For CH_4, the canonical occupied MOs are found to consist of an MO that is a nearly pure C1*s* AO and four delocalized bonding MOs that each extend over much of the molecule. When the canonical MOs are transformed to energy-localized MOs, the localized MOs are found to consist of an inner-shell MO that is essentially a pure C1*s* AO and four localized bonding MOs, each bonding MO pointing toward one of the H atoms of the tetrahedral molecule. The localized bonding MO between C and atom H_A is [R. M. Pitzer, *J. Chem. Phys.*, **46,** 4871 (1967)]

$$0.02(C1s) + 0.292(C2s) + 0.277(C2p_x + C2p_y + C2p_z) + 0.57(H_A1s) - 0.07(H_B1s + H_C1s + H_D1s)$$

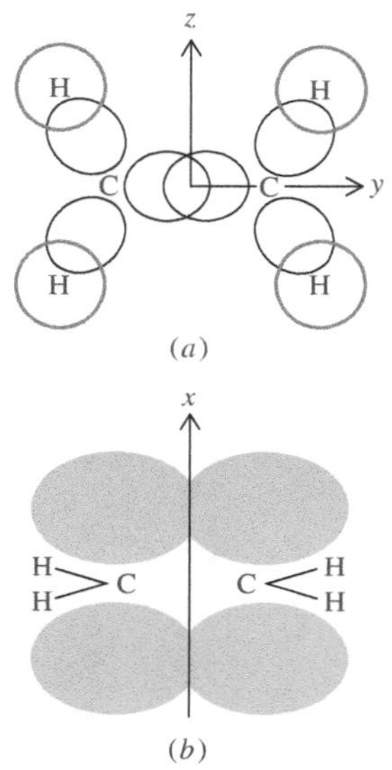

Figure 19.30

Bonding in ethylene: (*a*) σ bonds; (*b*) π bond.

The carbon 2*s* and 2*p* AOs make nearly equal contributions, and the hybridization on carbon is approximately sp^3. It would be exactly sp^3 if the coefficient of C2*s* equaled that of the C2*p* AOs. Atom H_A is in the positive octant of space, and the combination $C2p_x + C2p_y + C2p_z$ (which is proportional to $x + y + z$) has its maximum probability density along the line running through C and H_A and on both sides of the C nucleus. Addition of C2*s* to $C2p_x + C2p_y + C2p_z$ cancels most of the probability density on the side of C that is away from H_A and reinforces the probability density in the region between C and H_A. (This is the same thing that occurs in Fig. 19.28 for the BeH_2 *sp* hybrids.) Overlap of the C hybrid AO with the H_A1s AO then forms the bond. Each bonding localized MO has no nodal planes containing the axis between the bonded atoms and is a σ MO.

Consider ethylene ($H_2C{=}CH_2$). The molecule is planar, with the bond angles at each carbon close to 120°. A minimal basis set consists of four H1*s* AOs and two each of C1*s*, C2*s*, $C2p_x$, $C2p_y$, and $C2p_z$. Let the molecular plane be the *yz* plane. One way to form localized MOs for C_2H_4 is to use linear combinations of the C2*s*, $C2p_y$, and $C2p_z$ AOs at each carbon to form three sp^2 hybrid AOs at each carbon. These hybrids make 120° angles with one another. Overlap of two of the three sp^2 hybrids at each carbon with H1*s* AOs forms the C—H single bonds, and these are σ bonds. Overlap of the third sp^2 hybrid of one carbon with the third sp^2 hybrid of the second carbon gives a σ bonding MO between the two carbons (Fig. 19.30*a*). Overlap of the $2p_x$ AOs of each carbon gives a localized π bonding MO between the carbons (Fig. 19.30*b*). This π MO has a nodal plane coinciding with the molecular plane and containing the C—C axis.

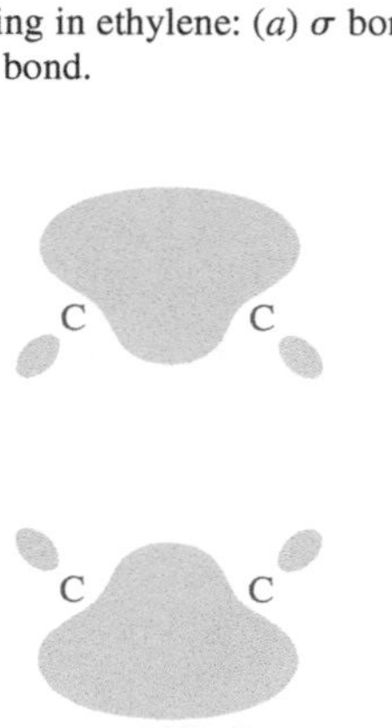

Figure 19.31

Equivalent "banana" bonds in ethylene. The view is the same as shown in Fig. 19.30*b*.

Ethylene has 16 electrons. Four of them fill the two localized inner-shell MOs, each of which is a 1*s* AO on one of the carbons; eight electrons fill the four C—H bond MOs; two fill the C—C σ-bond MO, and two fill the C—C π-bond MO. (Unlike that in diatomic molecules, the ethylene π-bond MO is nondegenerate.) In this picture, the carbon–carbon double bond consists of one σ bond and one π bond.

The above description of localized MOs for C_2H_4 is the traditional one. However, calculation of the energy-localized MOs in ethylene shows that two bent, equivalent "banana" bonds (Fig. 19.31) between the two carbons are more localized than the traditional σ-π description [U. Kaldor, *J. Chem. Phys.*, **46,** 1981 (1967)].

The traditional description of HC≡CH uses two *sp* hybrids on each carbon to overlap the H1*s* AOs and to form a σ bond between the carbons. The linear combinations $C_A 2p_x + C_B 2p_x$ and $C_A 2p_y + C_B 2p_y$ (where the *z* axis is the molecular axis) give two π bonds between C_A and C_B. In this picture, the triple bond consists of one σ bond and two π bonds. Again, actual calculation shows the energy-localized MOs to consist of three equivalent bent banana bonds.

Benzene

The carbons in benzene (C_6H_6) form a regular hexagon with 120° bond angles. We can use three sp^2 hybrid AOs at each carbon to form localized σ bonds with one hydrogen and with two adjacent carbons. This leaves a $2p_z$ AO at each carbon (where the *z* axis is perpendicular to the molecular plane). For benzene, two equivalent Lewis dot formulas can be written; the carbon–carbon single bonds and double bonds are interchanged in the two formulas. Moreover, $\Delta_f H°$ for benzene (Prob. 19.40) and the chemical behavior of benzene differ from what is expected of a species with localized double bonds. Hence, it would not be suitable to form three localized π MOs by pairwise interactions of the six $2p_z$ AOs. Instead, all six $2p_z$ AOs must be considered to interact with one another, and one uses delocalized (canonical) MOs to form the π bonds. The six $2p_z$ AOs give six delocalized π MOs, three bonding and three antibonding. These six MOs are linear combinations of the six $2p_z$ AOs; their forms are fully determined by the symmetry of benzene. (See Prob. 19.41 and *Levine,* sec. 17.2.)

Three of the four valence electrons of each carbon go into the three bonding MOs formed from the sp^2 hybrids, leaving one electron at each carbon to go into the π MOs. These remaining six electrons fill the three bonding π MOs. Each π MO has a nodal plane that coincides with the molecular plane (since each $2p_z$ AO has such a plane). This nodal plane is analogous to the nodal plane of a localized π bond joining two doubly bonded atoms (for example, as in ethylene), and so these benzene MOs are called π MOs.

A similar situation holds for other planar conjugated organic compounds. (A **conjugated** molecule has a framework consisting of alternating carbon–carbon single and double bonds.) One can form in-plane localized σ-bond MOs using sp^2 hybrids on each carbon, but one uses delocalized (canonical) MOs for the π MOs.

Three-Center Bonds

B_2H_6 has 12 valence electrons, which is not enough to allow one to write a Lewis dot structure with two electrons shared between each pair of bonded atoms. Calculation of the energy-localized MOs of B_2H_6 shows that two of the localized MOs each extend over two B atoms and one H atom to give two three-center bonds. [E. Switkes et al., *J. Chem. Phys.,* **51,** 2085 (1969).] The H atoms of the three-center bonds lie above and below the plane of the remaining six atoms and midway between the borons. Three-center bonds also occur in higher boron hydrides.

Multicenter bonding occurs in chemisorption. For example, experimental evidence shows that an ethylene molecule can bond to a metal's surface by forming two σ bonds with metal atoms (the stars in Fig. 19.32*a*) or by forming a π complex (Fig. 19.32*b*) in which electron probability density of the carbon–carbon π bond bonds the molecule to a metal atom as well as bonding the carbons to each other (*Bamford and Tipper,* vol. 20, pp. 22–23). Such a π complex is stabilized by donation of electron density from the C_2H_4 π electrons into vacant metal-atom orbitals and by donation of electron density from filled metal-atom orbitals into the vacant antibonding π MO in C_2H_4 (back bonding). π complexes occur in such organometallic compounds as dibenzenechromium, $C_6H_6CrC_6H_6$, ferrocene, $C_5H_5FeC_5H_5$, and the complex ion $[Pt(C_2H_4)Cl_3]^-$; see *DeKock and Gray,* sec. 6-4.

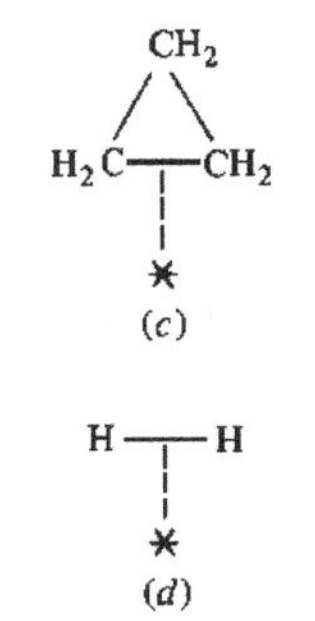

Figure 19.32

Species chemisorbed on the surface of a solid.

Cyclopropane can be weakly chemisorbed by forming a complex in which the electron probability density of a carbon–carbon bond bonds the molecule to a surface metal atom (*Bamford and Tipper,* vol. 20, p. 102), as shown in Fig. 19.32*c*. H_2 can be nondissociatively weakly chemisorbed to form the complex in Fig. 19.32*d*. In $[Cr(CO)_5(H_2)]$ and a few other coordination compounds, H_2 forms a three-center bond with the central metal atom; see G. Wilkinson (ed.), *Comprehensive Coordination Chemistry,* vol. 2, Pergamon, 1987, pp. 690–691.

Ligand-Field Theory

The application of MO theory to transition-metal complexes gives what is called *ligand-field theory*. See *DeKock and Gray,* sec. 6-7.

Canonical versus Localized MOs

For accurate quantitative calculations of molecular properties, one solves the Hartree–Fock equations (18.57) and obtains the canonical (delocalized) MOs. Since each canonical MO corresponds to a definite orbital energy, these MOs are also useful in discussing transitions to excited electronic states and ionization.

For qualitative discussion of bonding in the ground electronic state of a molecule, it is usually simplest to describe things in terms of localized bonding MOs (constructed from suitably hybridized AOs of pairs of bonded atoms), lone-pair MOs, and inner-shell MOs. One can usually get a reasonably good idea of the localized MOs without going through the difficult computations involved in first finding the canonical MOs and then using the Edmiston–Ruedenberg criterion to transform the canonical MOs to localized MOs. Localized MOs are approximately transferable from molecule to molecule; for example, the C—H localized MOs in CH_4 and C_2H_6 are very similar to each other. Canonical MOs are not transferable.

19.7 THE VALENCE-BOND METHOD

So far our discussion of molecular electronic structure has been based on the MO approximation. Historically, the first quantum-mechanical treatment of molecular bonding was the 1927 Heitler–London treatment of H_2. Their approach was extended by Slater and by Pauling to give the **valence-bond (VB) method.**

Heitler and London started with the idea that a ground-state H_2 molecule is formed from two $1s$ H atoms. If all interactions between the H atoms were ignored, the wave function for the system of two atoms would be the product of the separate wave functions of each atom. Hence, the first approximation to the H_2 spatial wave function is $1s_A(1)1s_B(2)$, where $1s_A(1) = \pi^{-1/2}a_0^{-3/2}e^{-r_{1A}/a_0}$. This product wave function is unsatisfactory, since it distinguishes between the identical electrons, saying that electron 1 is on nucleus A and electron 2 is on nucleus B. To take care of electron indistinguishability, we must write the approximation to the ground-state H_2 spatial wave function as the linear combination $N'[1s_A(1)1s_B(2) + 1s_A(2)1s_B(1)]$. This function is symmetric with respect to electron interchange and therefore requires the antisymmetric two-electron spin function (18.41). The ground-state H_2 Heitler–London VB wave function is then

$$N'[1s_A(1)1s_B(2) + 1s_A(2)1s_B(1)] \cdot 2^{-1/2}[\alpha(1)\beta(2) - \beta(1)\alpha(2)] \qquad (19.36)$$

Introducing a variable orbital exponent and using (19.36) in the variational integral, one finds a predicted D_e of 3.78 eV compared with the experimental value 4.75 eV and the Hartree–Fock value 3.64 eV.

The Heitler–London function (19.36) is a linear combination of two determinants:

$$N\begin{vmatrix} 1s_A(1)\alpha(1) & 1s_B(1)\beta(1) \\ 1s_A(2)\alpha(2) & 1s_B(2)\beta(2) \end{vmatrix} - N\begin{vmatrix} 1s_A(1)\beta(1) & 1s_B(1)\alpha(1) \\ 1s_A(2)\beta(2) & 1s_B(2)\alpha(2) \end{vmatrix} \tag{19.37}$$

The two determinants differ by giving different spins to the AOs $1s_A$ and $1s_B$ involved in the bonding.

When multiplied out, the MO spatial function (19.19) for H_2 equals

$$N[1s_A(1)1s_A(2) + 1s_B(1)1s_B(2) + 1s_A(1)1s_B(2) + 1s_B(1)1s_A(2)]$$

Because of the terms $1s_A(1)1s_A(2)$ and $1s_B(1)1s_B(2)$, the MO function gives a 50% probability that an H_2 molecule will dissociate into $H^- + H^+$, and a 50% probability for dissociation into H + H. In actuality, a ground-state H_2 molecule always dissociates to two neutral H atoms. This incorrect dissociation prediction is related to the poor dissociation energies predicted by the MO method. In contrast, the VB function (19.36) correctly predicts dissociation into H + H.

If, instead of the symmetric spatial function in (19.36), one uses the antisymmetric spatial function $N[1s_A(1)1s_B(2) - 1s_A(2)1s_B(1)]$ multiplied by one of the three symmetric spin functions in (18.40), one gets the VB functions for the first excited electronic level (a triplet level) of H_2. The minus sign produces charge depletion between the nuclei, and the atoms repel each other as they come together.

To apply the VB method to polyatomic molecules, one writes down all possible ways of pairing up the unpaired electrons of the atoms forming the molecule. Each way of pairing gives one of the **resonance structures** of the molecule. For each resonance structure, one writes down a function (called a *bond eigenfunction*) resembling (19.37), and the molecular wave function is taken as a linear combination of the bond eigenfunctions. The coefficients in the linear combination are found by minimizing the variational integral. Besides covalent pairing structures, one also includes ionic structures. For example, for H_2, the only covalent pairing structure is H—H, but one also has the ionic resonance structures $H^+ H^-$ and $H^- H^+$. These ionic structures correspond to the bond eigenfunctions $1s_A(1)1s_A(2)$ and $1s_B(1)1s_B(2)$. By symmetry, the two ionic structures contribute equally, so with inclusion of ionic structures, the VB spatial wave function for H_2 becomes

$$c_1[1s_A(1)1s_B(2) + 1s_B(1)1s_A(2)] + c_2[1s_A(1)1s_A(2) + 1s_B(1)1s_B(2)]$$

One says there is *ionic–covalent resonance*. One expects $c_2 \ll c_1$ for this nonpolar molecule.

In many cases, one uses hybrid atomic orbitals to form the bond eigenfunctions. For example, for the tetrahedral molecule CH_4, one combines four sp^3 hybrid AOs on carbon with the $1s$ AOs of the hydrogens.

For polyatomic molecules, the VB wave function is cumbersome. For example, CH_4 has four bonds, and the bond eigenfunction corresponding to the single most important resonance structure (the one with each H$1s$ AO paired with one of the carbon sp^3 hybrids) turns out to be a linear combination of $2^4 = 16$ determinants. Inclusion of other resonance structures further complicates the wave function.

The calculations of the VB method turn out to be more difficult than those of the MO method. The various MO approaches have overshadowed the VB method when it comes to actual computation of molecular wave functions and properties. However, the language of VB theory provides organic chemists with a simple qualitative tool for rationalizing many observed trends.

19.8 CALCULATION OF MOLECULAR PROPERTIES

This section considers the calculation of molecular properties from approximate molecular wave functions.

Molecular Geometry

The equilibrium geometry of a molecule is the spatial configuration of the nuclei for which the electronic energy (including nuclear repulsion) E_e in the electronic Schrödinger equation (19.7) is a minimum. To determine the equilibrium geometry theoretically, one calculates the molecular wave function and electronic energy for many different configurations of the nuclei, varying the bond distances, bond angles, and dihedral angles to find the minimum-energy configuration. A very efficient way to find the equilibrium geometry involves calculating the derivatives of the electronic energy with respect to each of the nuclear coordinates (this set of derivatives is called the **energy gradient**) for an initially guessed geometry. One then uses the values of these derivatives to change the nuclear coordinates to new values that are likely to be closer to the equilibrium geometry, and one then calculates the wave function, energy, and energy gradient at the new geometry. This process is repeated until the components of the energy gradient are all very close to zero, indicating that the energy minimum has been found.

One finds that even though the Hartree–Fock MO wave function differs significantly from the true wave function, it gives generally accurate bond distances and bond angles. Some examples of calculated Hartree–Fock (or approximate Hartree–Fock) geometries are (experimental values in parentheses):

H_2O: $r(\text{OH}) = 0.94$ Å (0.96 Å), HOH angle $= 106.1°$ (104.5°)

H_2CO: $r(\text{CH}) = 1.10$ Å (1.12 Å), $r(\text{CO}) = 1.22$ Å (1.21 Å)
HCH angle $= 114.8°$ (116.5°)

C_6H_6: $r(\text{CC}) = 1.39$ Å (1.40 Å), $r(\text{CH}) = 1.08$ Å (1.08 Å)

It has been found that to obtain an accurate geometry, one needs only an approximation to the Hartree–Fock wave function. Minimal-basis SCF wave functions (Sec. 19.5) usually give accurate geometries, but occasionally show large errors. A somewhat larger than minimal-basis-set SCF calculation is needed to obtain a reliable geometry.

Dipole Moments

The classical expression for the dipole moment of a charge distribution is $\boldsymbol{\mu} = \Sigma_i\, Q_i \mathbf{r}_i$ [Eq. (13.82)]. To calculate the dipole moment of a molecule from its equilibrium-geometry electronic wave function ψ, we use the right side of (13.82) as an operator in the average-value expression (17.63) to get

$$\boldsymbol{\mu} = \int \psi^* \sum_i Q_i \mathbf{r}_i \psi \, d\tau = \sum_i Q_i \int |\psi|^2 \mathbf{r}_i \, d\tau \qquad (19.38)$$

where the sum goes over all the electrons and nuclei, the integral is over the electronic coordinates, and Q_i is the charge on particle i. Evaluation of (19.38) once ψ is known is easy. The hard thing is to get a reasonably accurate approximation to ψ.

One finds that Hartree–Fock MO wave functions give generally accurate molecular dipole moments. Some values of the Hartree–Fock and the experimental dipole moments are:

	HCN	H_2O	LiH	NaCl	CO	NH_3
μ_{HF}/D	3.29	1.98	6.00	9.18	−0.11	1.66
μ_{exp}/D	2.98	1.85	5.88	9.00	+0.27	1.48

The experimental CO dipole moment has the carbon negative (Prob. 19.20); the calculated Hartree–Fock CO dipole moment is in the wrong direction. A calculation using a CI wave function gives the proper polarity for CO.

Minimal-basis-set SCF wave functions give only fairly accurate dipole moments, and larger than minimal basis sets are needed to get good accuracy.

Ionization Energies

The molecular ionization energy I is the energy needed to remove the most loosely held electron from the molecule. T. C. Koopmans (cowinner of the 1975 Nobel Prize in economics) proved in 1933 that the energy needed to remove an electron from an orbital of a closed-shell atom or molecule is well approximated by minus the Hartree–Fock orbital energy ε_i in (18.57). The molecular ionization energy can therefore be estimated by taking $-\varepsilon_i$ of the highest occupied MO. One finds pretty good agreement between Koopmans'-theorem Hartree–Fock ionization energies and experimental ionization energies. Some results are (experimental values in parentheses): 17.4 eV (15.6 eV) for N_2, 13.8 eV (12.6 eV) for H_2O, 9.1 eV (9.3 eV) for C_6H_6.

Dissociation Energies

To calculate D_e theoretically, one subtracts the calculated Hartree–Fock molecular energy at the equilibrium geometry from the Hartree–Fock energies of the separated atoms that form the molecule. Hartree–Fock wave functions give poor D_e values. Some results for binding energies are:

	H_2	BeO	N_2	CO	F_2	CO_2	H_2O	N_2O
$D_{e,HF}$/eV	3.64	2.0	5.3	7.9	−1.4	11.3	6.9	4.0
$D_{e,exp}$/eV	4.75	4.7	9.9	11.2	1.6	16.8	10.1	11.7

The Hartree–Fock wave functions predict the separated atoms F + F to be more stable than the F_2 molecule at R_e.

The **equilibrium dissociation energy** D_e is not directly observable because the nuclei vibrate about the equilibrium geometry. Recall that a one-dimensional harmonic oscillator has a ground-state zero-point vibrational energy of $\frac{1}{2}h\nu$ (Sec. 17.12). The molecule's energy in its ground vibrational level is higher than the equilibrium-geometry energy by the zero-point energy (ZPE) and so the observed **ground-state dissociation energy** D_0 is less than D_e, the difference being the ZPE. Figure 20.9 shows this for H_2.

The **atomization energy** of a species is the thermodynamic internal energy change for the atomization process of gas-phase molecules dissociating to gas-phase atoms. For example (5.45) is the atomization process for $CH_4(g)$. Most commonly, one deals with $\Delta_{at}U°$ at 0 K or at 298 K. Since $\Delta_{at}U°$ is a per-mole quantity, we have $\Delta_{at}U°_0 = N_A D_0$, where N_A is the Avogadro constant. One can calculate $\Delta_{at}U°_{298}$ from $\Delta_{at}U°_0$ using statistical mechanics (Sec. 21.8).

Rotational Barriers

The equilibrium conformation of ethane, $H_3C—CH_3$, has the hydrogens of one CH_3 group staggered with respect to the hydrogens of the other CH_3 group. The barrier to internal rotation about the single bond in ethane is quite low, being 0.13 eV, which corresponds to 3 kcal/mol. The barrier can be determined experimentally from thermodynamic data or the infrared spectrum. To calculate the rotational barrier B theoretically, one calculates wave functions and energies for the staggered and eclipsed geometries and takes the energy difference. SCF wave functions give pretty accurate rotational barriers, provided one uses a substantially larger than minimal basis set. Some results are:

	C_2H_6	CH_3CHO	CH_3OH	CH_3NH_2	CH_3SiH_3
B_{exp}/(kcal/mol)	2.9	1.2	1.1	2.0	1.7
B_{calc}/(kcal/mol)	3.2	1.1	1.4	2.4	1.4

Calculations indicate that the ethane rotational barrier is mainly due to steric repulsions in the eclipsed geometry [Y. Mo and J. Gao, *Acc. Chem. Res.,* **40,** 113 (2007)].

The Hartree–Fock method does well on barrier calculations because no bonds are broken or formed in going from the staggered to the eclipsed conformation and the correlation energy (which is the energy error in the Hartree–Fock method) is nearly the same for the two conformations. In contrast, when a molecule dissociates, bonds are broken and the correlation energy changes substantially. The Hartree–Fock wave function therefore cannot deal with dissociation.

Relative Energies of Isomers

Although energies of dissociation of molecules to atoms calculated by the Hartree–Fock method are very inaccurate, *relative* energies of isomeric molecules are generally predicted accurately by Hartree–Fock wave functions. Relative energies of isomers are calculated the same way rotational barriers are calculated. Minimal-basis-set SCF wave functions don't give accurate relative energies of isomers, and one must use a larger than minimal basis set to get good results. As an example, SCF calculations with a basis set substantially larger than minimal (but not large enough to give a near-Hartree–Fock wave function) gave the following ground-state electronic energies in kcal/mol of C_3H_4 isomers relative to that of propyne (experimental values in parentheses): allene, 1.7 (1.6); cyclopropene, 25.4 (21.9); they gave the following electronic energies of C_4H_6 isomers relative to 1,3-butadiene: 2-butyne, 6.7 (8.6); cyclobutene, 12.4 (11.2); methylenecyclopropane, 20.2 (21.7); bicyclobutane, 30.4 (25.6); 1-methylcyclopropene, 31.4 (32.1) [see M. C. Flanigan et al. in G. A. Segal (ed.), *Semiempirical Methods of Electronic Structure Calculation,* pt. B, Plenum, 1977, p. 1].

A major application of SCF energy and geometry calculations is to reaction intermediates, which often are too short-lived to have their structures determined by spectroscopy. For example, the relative energies and the geometries of carbonium ions containing up to eight carbon atoms have been calculated (*Hehre* et al., sec. 7.3).

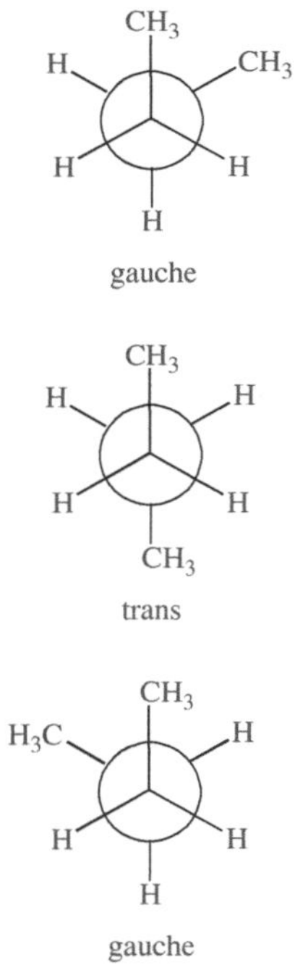

Figure 19.33

Conformers of butane, $CH_3CH_2CH_2CH_3$.

Relative Energies of Conformers

A **conformation** of a molecule is defined by specifying the dihedral angles of rotation about the single bonds. A conformation that corresponds to an energy minimum is called a **conformer.** Figure 19.33 shows the *gauche* and *trans* conformers of butane, $CH_3CH_2CH_2CH_3$, which occur at CCCC dihedral angles of about 65° (as shown by quantum-mechanical calculations) and 180°, respectively. Energy differences between conformers of a molecule are usually small (typically 0 to 2 kcal/mol for conformers that differ by rotation about one bond), and internal rotation barriers to interconversion of conformers are usually small. Therefore, different conformers of a molecule are usually not isolable, and a molecule with internal rotation about single bonds consists of a mixture of conformers whose relative amounts are determined by the Boltzmann distribution law (Secs. 14.9, 21.5, and 21.8). Minimal-basis-set SCF calculations are unreliable for predicting energy differences between conformers, and large-basis-set SCF calculations are needed to obtain reasonably reliable results here. Figure 19.34 plots the electronic energy of butane versus CCCC dihedral angle.

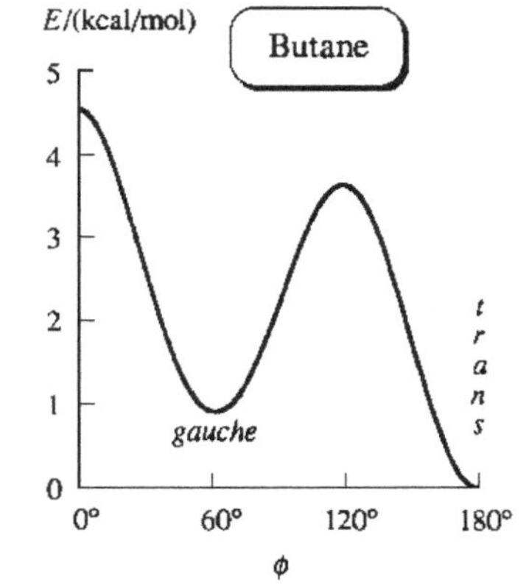

Figure 19.34

Electronic energy including nuclear repulsion of butane (Fig. 19.33) versus angle ϕ of internal rotation. This is the potential-energy function for torsional (twisting) vibration about the central C—C bond; it was found from vibrational transition frequencies in the Raman spectrum (Sec. 20.10) [D. A. C. Compton, S. Montero, and W. F. Murphy, *J. Phys. Chem.*, **84**, 3587 (1980)].

Electron Probability Density

Let $\rho(x, y, z)\,dx\,dy\,dz$ be the probability of finding an electron of a many-electron molecule in the box-shaped region located at x, y, z and having edges dx, dy, dz; by "an electron," we mean any electron, not a particular one. The electron probability density $\rho(x, y, z)$ can be calculated theoretically from the molecular electronic wave function ψ_e by integrating $|\psi_e|^2$ over the spin coordinates of all electrons and over the spatial coordinates of all but one electron and multiplying the result by the number of electrons in the molecule. One can find ρ experimentally by analyzing x-ray diffraction data of crystals (Chapter 23). Electron densities calculated from Hartree–Fock wave functions for small molecules agree well with experimentally determined densities; see P. Coppens and M. B. Hall (eds.), *Electron Distributions and the Chemical Bond*, Plenum, 1982, pp. 265, 331. ρ plays a key role in density-functional theory (Sec. 19.10), one of the most important methods for calculating molecular properties.

EXAMPLE 19.2 Particle probability density

Find the particle probability density function $\rho(x)$ for a system of two identical noninteracting particles each of which has spin $s = 0$ in a stationary state of a one-dimensional box of length a if the quantum numbers for the two particles are equal.

The particles do not have spin so the wave function is a function of only the spatial coordinates. Since the particles are noninteracting, the wave function is the product of a particle-in-a-box wave function for each particle. Particles with $s = 0$ are bosons and require a symmetric wave function. The normalized wave function is thus [similar to the He ground-state approximate wave function (18.38)]

$$\psi = f(1)f(2)$$

where f is a normalized particle-in-a-box wave function with quantum number n and the numbers in parentheses are the labels of the particles. Thus, $f(1) = (2/a)^{1/2}\sin(n\pi x_1/a)$ inside the box [Eq. (17.35)]. Integrating $|\psi|^2$ over the spatial coordinates of all particles except particle 1, and multiplying it by the number of particles, we have

$$\rho = 2[f(1)]^2 \int_0^a [f(2)]^2\,dx_2 = 2[f(1)]^2$$

since $f(2)$ is normalized. Dropping the unnecessary subscript 1, we have

$$\rho(x) = 2(2/a)\sin^2(n\pi x/a)$$

inside the box and $\rho = 0$ outside the box. The case where the two particles have different quantum numbers is considered in Prob. 19.45.

Exercise

Find $\int_{-\infty}^{\infty} \rho(x)\,dx$ for this problem. (*Answer:* 2.)

19.9 ACCURATE CALCULATION OF MOLECULAR ELECTRONIC WAVE FUNCTIONS AND PROPERTIES

The major sources of error in quantum-mechanical calculations of ground-state molecular properties are (*a*) inadequacy of the basis set; (*b*) neglect or incomplete treatment of electron correlation; (*c*) neglect of relativistic effects.

Relativistic effects strongly influence the properties of molecules containing very heavy atoms such as Au, Hg, and Pb, but for molecules composed of atoms with atomic number less than 54, relativistic effects can usually be neglected.

Basis Sets

Most quantum-mechanical calculations use a basis set to express the molecular orbitals [Eq. (18.59)]. As noted in Sec. 19.8, SCF calculations that use a minimal basis set (containing only inner- and valence-shell AOs) cannot be relied on to give accurate molecular properties. Calculations using the very large basis sets needed to come close to the Hartree–Fock wave function are feasible only for rather small molecules. For medium-size molecules, one must limit the basis-set size, and this is a major source of error in calculated properties.

Although the Slater-type orbitals (STOs) of Sec. 18.9 are often used as basis sets in atomic calculations, use of STOs as basis functions in polyatomic-molecule calculations produces integrals that are very time-consuming to evaluate on a computer. Therefore, most molecular quantum-mechanical calculations use Gaussian functions instead of STOs as the basis functions. A **Gaussian function** contains the factor $e^{-\zeta r^2}$ instead of the $e^{-\zeta r}$ factor in an STO, and molecular integrals with Gaussian basis functions are evaluated very rapidly on a computer. However, the factor $e^{-\zeta r^2}$ is not as accurate a representation of the actual behavior of an AO as the factor $e^{-\zeta r}$, so one must use a linear combination of a few Gaussian functions to represent an AO.

Many Gaussian basis sets have been devised for use in molecular calculations. Some of the most widely used are the basis sets devised by Pople and co-workers. These basis sets (listed in order of increasing size) include the STO-3G, 3-21G, 3-21G$^{(*)}$, 6-31G*, and 6-31G** sets, where the numbers and symbols are related to the number of basis functions on each atom. STO-3G is a minimal basis set and often gives unreliable results, so this set is essentially obsolete. The 6-31G** basis set, also called 6-31G(d, p), is not a particularly large set and many larger sets [such as 6-311+G(3df, 2p)] have been devised by adding functions to 6-31G**. The *correlation consistent* basis sets (such as cc-pVDZ, cc-pVTZ, cc-pVQZ, . . .) devised by Dunning and co-workers are also widely used. For details of these basis sets, see *Levine,* sec. 15.4.

Configuration Interaction

In order to calculate molecular properties to high accuracy, one must usually go beyond the Hartree–Fock (SCF) method and include electron correlation. One method to do this is configuration interaction (CI)—Sec. 18.9. The steps in a CI calculation are: (1) One chooses a set of basis functions $g_1, g_2, \ldots, g_i, \ldots$. (2) Each molecular orbital ϕ_j is written as a linear combination of the basis functions: $\phi_j = \Sigma_i c_i g_i$, where the expansion coefficients c_i are to be determined. This expression for the MOs is substituted into the Hartree–Fock equations $\hat{F}\phi_j = \varepsilon_j\phi_j$ [Eq. (18.57)], and the SCF iterative procedure [described in the paragraph after Eq. (18.57)] is carried out to solve for the coefficients c_i, which determine the MOs ϕ_j. Just as the simple linear variation function $c_A 1s_A + c_B 1s_B$ for H_2^+ gave two MOs, one occupied and one unoccupied in the ground state (Sec. 19.3), the SCF procedure will give expressions for many MOs, some occupied and some unoccupied. The unoccupied orbitals are called *virtual orbitals*. The ground-state SCF wave function is a Slater determinant with the lowest

MOs occupied. (3) The molecular wave function ψ is then expressed as a linear combination of the SCF wave function found in step 2 and functions with one or more electrons placed in virtual orbitals: $\psi = \sum_j a_j\psi_{\text{orb},j}$ [Eq. (18.60)], and the variation method is used to find the coefficients a_j that minimize the variational integral.

If the sum $\sum_j a_j\psi_{\text{orb},j}$ includes functions with all possible assignments of electrons to virtual orbitals (this is called *full* CI), and if the basis set $g_1, g_2, \ldots$ used to expand the MOs is a complete set, then one can prove that the exact wave function will be obtained. In practice, the basis set must be limited in size (and is therefore incomplete), and full CI takes far too much computer time to be feasible (except for small molecules and small basis sets).

Experience shows that CI calculations with small basis sets do not give highly accurate results for molecular properties, and basis sets at least as large as 6-31G* should be used in CI calculations.

Each of the functions $\psi_{\text{orb},j}$ in the CI wave function is classified as *singly excited, doubly excited, triply excited,* . . . according to whether one, two, three, . . . electrons are in virtual orbitals. Since full CI is impractical, one must limit the number of terms in the CI wave function. The most common kind of CI calculation is CISD, where the letters SD indicate that all possible singly and doubly excited configuration functions $\psi_{\text{orb},j}$ are included, but configuration functions with 3 or more electrons in virtual orbitals are omitted. For molecules with up to 10 or 20 electrons, CISD with an adequate-size basis set yields very accurate results. However, as the number of electrons increases, CISD becomes more and more inaccurate. A CISDTQ calculation (which includes all singly, doubly, triply, and quadruply excited configuration functions) with an adequate-size basis set will yield quite accurate results for molecules with up to 50 electrons. However, CISDTQ calculations with large basis sets require monumental amounts of computer time and are impractical to perform except for small molecules.

Møller–Plesset Perturbation Theory

Because of the inefficiency of CI calculations, quantum chemists have developed other methods to include electron correlation. One such method is Møller–Plesset (MP) perturbation theory. The Hartree–Fock wave function is an antisymmetrized product of spin-orbitals, where each MO ϕ_i is found by solving the Hartree–Fock equations $\hat{F}\phi_i = \varepsilon_i\phi_i$ [Eq. (18.57)]. MP perturbation theory writes the molecular electronic Hamiltonian $\hat{H}$ as the sum of an unperturbed Hamiltonian $\hat{H}^0$ and a perturbation $\hat{H}'$, where $\hat{H}^0$ is taken as the sum of the Hartree–Fock operators $\hat{F}$ for the electrons in the molecule. (See Sec. 17.15 for a general discussion of perturbation theory.) With this choice of $\hat{H}^0$, one finds that the unperturbed wave function $\psi^{(0)}$ is the Hartree–Fock wave function and the energy $E^{(0)} + E^{(1)}$ is the Hartree–Fock energy (see *Levine,* sec. 16.2 for details). To improve on the Hartree–Fock energy, one then calculates the higher-order energy corrections $E^{(2)}$, $E^{(3)}$, etc. MP calculations are designated MP2, MP3, MP4, according to whether the highest-order energy correction included is $E^{(2)}$, $E^{(3)}$, or $E^{(4)}$. MP2 calculations can be done much faster than CI calculations and are a common way to include correlation in calculations on ground-state molecules.

An MP calculation begins by choosing a basis set to express the MOs and then finds the SCF wave function corresponding to this basis set—steps 1 and 2 in the above CI discussion. The expressions for the MP energy corrections $E^{(2)}$, $E^{(3)}$, . . . involve integrals over the MOs (including the virtual orbitals), and these integrals can be related to integrals over the basis functions. Evaluation of these integrals and substitution in the relevant formulas gives $E^{(2)}$, $E^{(3)}$, etc. Most MP calculations are MP2 calculations. Almost never are calculations done beyond MP4. The energy gradient can be readily found in the MP method, and this allows the MP equilibrium geometry to be readily found.

The Coupled-Cluster Method

The coupled-cluster (CC) method is a third way to allow for electron correlation. Like the CI and MP methods, there are various levels of CC calculations: CCD, CCSD, CCSDT, . . . , where CCSDT stands for coupled-cluster singles, doubles, and triples. Because CCSDT calculations usually require prohibitive amounts of computer time, an approximation to CCSDT called CCSD(T) is more commonly used than CCSDT. CCSD(T) gives very accurate results but is limited to quite small molecules. For details, see *Levine,* sec. 16.3.

Density-Functional Theory

This highly popular method deserves its own section and is discussed in Sec. 19.10.

Applications

Quantum-chemistry calculations have advanced to the point where they can help answer many chemically significant questions.

For example, calculations using large basis sets and allowing for electron correlation by the MP, CISD, and CC methods have shown that the equilibrium structures of the vinyl cation and the ethyl cation are not the classical structures $CH_2{=}CH^+$ and $CH_3CH_2^+$, but are the bridged structures shown in Fig. 19.35, which have three-center bonds [B. Ruscic et al., *J. Chem. Phys.,* **91,** 114 (1989); C. Liang, T. P. Hamilton, and H. F. Schaefer, III, *J. Chem. Phys.,* **92,** 3653 (1990)]. The calculated energy differences between the classical and bridged structures are about 4 kcal/mol for $C_2H_3^+$ and 6 kcal/mol for $C_2H_5^+$. Spectroscopic observations confirm that the equilibrium structures are the bridged structures.

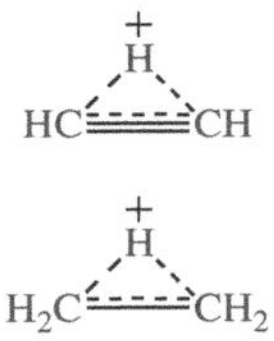

Figure 19.35

Equilibrium structures of the vinyl ($C_2H_3^+$) and ethyl ($C_2H_5^+$) cations.

Spectroscopic observations of 1,3-butadiene show the presence of two conformers. Two possible conformations are the *s-trans* and *s-cis* configurations:

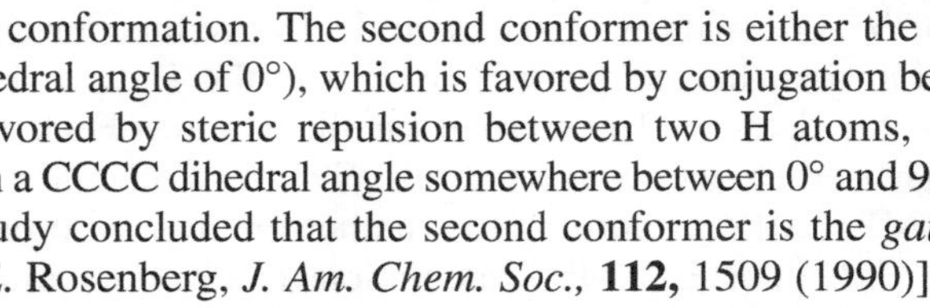

s-trans *s-cis*

Spectroscopic evidence shows that the *s-trans* form (CCCC dihedral angle of 180°) is the predominant conformation. The second conformer is either the planar *s-cis* form (with CCCC dihedral angle of 0°), which is favored by conjugation between the double bonds and disfavored by steric repulsion between two H atoms, or the nonplanar *gauche* form with a CCCC dihedral angle somewhere between 0° and 90°. One gas-phase spectroscopic study concluded that the second conformer is the *gauche* form [K. B. Wiberg and R. E. Rosenberg, *J. Am. Chem. Soc.,* **112,** 1509 (1990)], whereas a spectroscopic study of butadiene in a frozen Ar matrix concluded that in this environment the second conformer is the *s-cis* form [B. R. Arnold et al., *J. Am. Chem. Soc.,* **112,** 1808 (1990)]. Virtually all SCF, MP2, MP4, CISD, and CC calculations with adequate-size basis sets predict the second conformer to be the *gauche* form with a CCCC dihedral angle of 38° ± 10° and an energy that is 0.5 to 1 kcal/mol below that of the *s-cis* form, which the calculations show to be at an energy maximum (Fig. 19.36) [M. A. Murcko et al., *J. Phys. Chem.,* **100,** 16162 (1996)]; [A. Karpfen and V. Parasuk, *Mol. Phys.,* **102,** 819 (2004)].

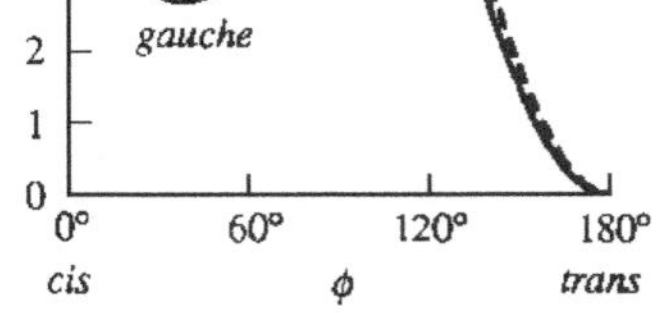

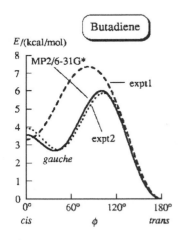

Figure 19.36

Potential energy for twisting about the central bond in 1,3-butadiene. The solid line is the result of an MP2/6-31G* calculation [H. Guo and M. Karplus, *J. Chem. Phys.,* **94,** 3679 (1991)]. The dashed and dotted lines are the results of two different possible interpretations of butadiene vibrational-spectroscopy data [J. R. Durig et al., *Can. J. Phys.,* **53,** 1832 (1975)].

An important question is the energy of the hydrogen bond between two H_2O molecules. Study of the temperature and pressure dependences of the thermal conductivity k of water vapor showed that the reaction $2H_2O(g) \rightarrow (H_2O)_2(g)$ has

$\Delta H^\circ_{373} = -3.6 \pm 0.5$ kcal/mol, which corresponds to an electronic energy change of $\Delta E_e = -5.0 \pm 0.7$ kcal/mol for nonvibrating, nonrotating molecules [E. Mas et al., *J. Chem. Phys.*, **113,** 6687 (2000)]. MP2 calculations with large correlation-consistent basis sets gave $\Delta E_e = -5.0 \pm 0.1$ kcal/mol [M. W. Feyereisen et al., *J. Phys. Chem.*, **100,** 2993 (1996)] and CCSD(T) calculations with large basis sets gave -5.02 ± 0.05 kcal/mol [W. Klopper et al., *Phys. Chem. Chem. Phys.*, **2,** 2227 (2000)]. Theoretical calculations have determined the dimerization energy with a much smaller uncertainty than the experimental result.

The Computational Chemistry Comparison and Benchmark Database (CCCBDB) tabulates the results of quantum-chemistry calculations on about 800 gas-phase molecules (srdata.nist.gov/cccbdb/).

19.10 DENSITY-FUNCTIONAL THEORY (DFT)

In the period 1995–2000, density-functional theory (DFT) showed a meteoric rise to popularity in quantum-chemistry calculations: "a substantial majority of the [quantum chemistry] papers published today are based on applications of density functional theory" [K. Raghavachari, *Theor. Chem. Acc.*, **103,** 361 (2000)].

The Hohenberg–Kohn Theorem

In DFT, one does not attempt to calculate the molecular wave function. Instead, one works with the electron probability density $\rho(x, y, z)$ (Sec. 19.8). DFT is based on a theorem proved in 1964 by Pierre Hohenberg and Walter Kohn that states that *the energy and all other properties of a ground-state molecule are uniquely determined by the ground-state electron probability density* $\rho(x, y, z)$. One therefore says that the ground-state electronic energy E_{gs} is a functional of ρ and one writes $E_{gs} = E_{gs}[\rho(x, y, z)]$ or simply

$$E_{gs} = E_{gs}[\rho] \tag{19.39}$$

where the square brackets denote a functional relation.

What is a functional? Recall that a *function* $y = f(x)$ is a rule that associates a number (the value of y) with each value of the independent variable x. For example, the function $y = 3x^2 - 2$ associates the value $y = 73$ with $x = 5$. A **functional** $z = G[f]$ is a rule that associates a number with each function f. For example, if $z = \int_0^1 f(x)\,dx$, then this functional associates the value $z = 3$ with the function $f(x) = 6x^2 + 1$. The variational integral $W = \int \phi^* \hat{H}\phi\,d\tau / \int \phi^*\phi\,d\tau$ (Sec. 17.15) associates a number with each well-behaved function ϕ and we can write $W = W[\phi]$.

The Kohn–Sham Method

Unfortunately, the functional $E_{gs}[\rho]$ in (19.39) is unknown, so the Hohenberg–Kohn theorem does not tell us *how* to calculate E_{gs} from ρ or how to find ρ without first finding the ground-state molecular electronic wave function. In 1965, Kohn and Sham devised a practical method to find ρ and to calculate E_{gs} from ρ. The Kohn–Sham equations do contain an unknown functional, but using a combination of physical insight and guesswork, physicists and chemists have developed approximations to this functional that allow accurate calculations of molecular properties, thereby turning DFT into a key method in quantum chemistry.

The Kohn–Sham (KS) method uses a fictitious **reference system** (usually denoted by the subscript s) that contains the same number of electrons (n) as the molecule we are dealing with, but that differs from the molecule in that (a) the electrons in the reference system do not exert forces on one another; (b) each electron i ($i = 1, 2, \ldots, n$) in the reference system experiences a potential energy $v_s(x_i, y_i, z_i)$, where v_s is the same function for each electron and is such as to make the electron probability density ρ_s in the reference system exactly equal to the ground-state electron probability density ρ in the real molecule: $\rho = \rho_s$. The actual form of v_s is not known. (In the real molecule, the electrons experience attractions to the nuclei, but these are not present in the reference system.)

Because the electrons in the reference system do not interact with one another, the Hamiltonian $\hat{H}_s$ of the reference system is the sum of Hamiltonians of the individual electrons:

$$\hat{H}_s = -\frac{\hbar^2}{2m_e}\sum_{i=1}^{n} \nabla_i^2 + \sum_{i=1}^{n} v_s(x_i, y_i, z_i) \equiv \sum_{i=1}^{n} \hat{h}_i^{\text{KS}}$$

$$\hat{h}_i^{\text{KS}} \equiv -\frac{\hbar^2}{2m_e}\nabla_i^2 + v_s(x_i, y_i, z_i) \tag{19.40}$$

$\hat{h}_i^{\text{KS}}$ is the one-electron Kohn–Sham Hamiltonian. Since the reference system s consists of noninteracting particles, its wave function, if spin and the antisymmetry requirement are neglected, is the product of one-electron spatial wave functions, each of which is an eigenfunction of $\hat{h}_i^{\text{KS}}$ [Eqs. (17.68) and (17.70)]. To allow for spin and the antisymmetry requirement, the ground-state wave function of the reference system is a Slater determinant (Sec. 18.8) of spin-orbitals, one for each electron. Each spin-orbital is the product of a spatial orbital θ_i^{KS} and a spin function (either α or β). The Kohn–Sham orbitals θ_i^{KS} are eigenfunctions of $\hat{h}_i^{\text{KS}}$:

$$\hat{h}_i^{\text{KS}}\theta_i^{\text{KS}} = \varepsilon_i^{\text{KS}}\theta_i^{\text{KS}} \tag{19.41}$$

where $\varepsilon_i^{\text{KS}}$ is the Kohn–Sham orbital energy of θ_i^{KS}. Each Kohn–Sham orbital holds two electrons of opposite spin.

One can prove that the probability density ρ for a wave function that is a Slater determinant (such as the wave function of the KS reference system) is the sum of the probability densities $|\theta_i^{\text{KS}}|^2$ of the individual orbitals. Also, by definition, the reference system has the same ρ as the molecule:

$$\rho = \rho_s = \sum_{i=1}^{n} |\theta_i^{\text{KS}}|^2 \tag{19.42}$$

The Kohn–Sham Energy Expression

We now need the expression for the molecule's ground-state electronic energy E_e. (The reference system and the molecule have the same probability density function but they do not have the same ground-state energy.) Kohn and Sham derived the following exact equation for E_e:

$$E_e = \langle K_{e,s}\rangle + \langle V_{Ne}\rangle + J + V_{NN} + E_{xc}[\rho] \tag{19.43}$$

We now discuss the meanings of the terms in (19.43). Complicated equations are given in small print at the end of this subsection, so they can be skimmed over, if desired.

$\langle K_{e,s}\rangle$ is the average electronic kinetic energy in the reference system. Its value can be calculated from the Kohn–Sham orbitals θ_i^{KS} of the reference system [see Eq. (19.46)].

$\langle V_{Ne}\rangle$ is the average potential energy of attractions between the electrons and the nuclei in the molecule. Its value can be calculated from the electron probability density $\rho(x, y, z)$ [see Eq. (19.47)]. $\rho(x, y, z)$ is the same for the molecule as for the reference system and is calculated from the Kohn–Sham orbitals θ_i^{KS} using Eq. (19.42).

J is the classical energy of electrical repulsion that arises between the infinitesimal charge elements of a hypothetical smeared-out electron charge cloud whose probability density is $\rho(x, y, z)$. J can be calculated from $\rho(x, y, z)$ [see Eq. (19.48)].

The internuclear repulsion energy V_{NN} is a constant that depends on the nuclear charges and the internuclear distances and is calculated from the molecular geometry at which the calculation is being done; one sums the potential energies (18.1) for each internuclear repulsion.

$E_{xc}[\rho]$ in (19.43), called the **exchange–correlation energy functional,** is a functional of ρ that is defined as

$$E_{xc}[\rho] \equiv \langle K_e \rangle - \langle K_{e,s} \rangle + \langle V_{ee} \rangle - J \qquad (19.44)$$

$E_{xc}[\rho]$ is the sum of two differences: (*a*) the difference $\langle K_e \rangle - \langle K_{e,s} \rangle$ between the average electronic kinetic energy in the molecule and in the reference system; and (*b*) the difference $\langle V_{ee} \rangle - J$ between the average potential energy of interelectronic repulsion $\langle V_{ee} \rangle$ in the molecule and the classical charge-cloud self-repulsion energy J. The values of $\langle K_e \rangle$ and $\langle K_{e,s} \rangle$ are expected to be similar to each other; also, the values of $\langle V_{ee} \rangle$ and J are similar. Hence, the two differences in (19.44) are relatively small quantities. These two differences are not zero because the requirement that the wave function be antisymmetric with respect to exchange produces *exchange* effects on the energy, and because the instantaneous correlations between the motions of the electrons produces *correlation* effects on the energy.

Equation (19.43) is an exact expression for the molecular electronic energy. However, no one knows what the true mathematical expression for the functional E_{xc} is. Hence, approximations to E_{xc} must be used. The term $E_{xc}[\rho]$ can be found if we know ρ and if we know what the functional E_{xc} is. Approximations to E_{xc} are discussed later in this section, and for now we shall assume that a good approximation to the functional E_{xc} is known.

Note that if (19.44) is substituted into (19.43), we get

$$E_e = \langle K_e \rangle + \langle V_{Ne} \rangle + \langle V_{ee} \rangle + V_{NN} \qquad (19.45)$$

an equation that follows from $\hat{H}_e = \hat{K}_e + \hat{V}_{Ne} + \hat{V}_{ee} + \hat{V}_{NN}$ [Eq. (19.6)] if we take the average value of each side of (19.6) (see Prob. 19.48).

We now give the explicit formulas for some of the terms in (19.43). $\langle K_{e,s} \rangle$ and $\langle V_{Ne} \rangle$ can be shown to be (see *Levine* chap. 16)

$$\langle K_{e,s} \rangle = -\frac{\hbar^2}{2m_e} \sum_{i=1}^{n} \int_{-\infty}^{\infty}\int_{-\infty}^{\infty}\int_{-\infty}^{\infty} \theta_i^{\mathrm{KS}}(1)^* \nabla_1^2 \theta_i^{\mathrm{KS}}(1)\, dx_1\, dy_1\, dz_1 \qquad (19.46)$$

$$\langle V_{Ne} \rangle = -\sum_{\alpha} \frac{Z_\alpha e^2}{4\pi\varepsilon_0} \int_{-\infty}^{\infty}\int_{-\infty}^{\infty}\int_{-\infty}^{\infty} \frac{\rho(x, y, z)}{r_\alpha}\, dx\, dy\, dz \qquad (19.47)$$

where $r_\alpha = [(x - x_\alpha)^2 + (y - y_\alpha)^2 + (z - z_\alpha)^2]^{1/2}$ is the distance from point (x, y, z) to nucleus α, located at $(x_\alpha, y_\alpha, z_\alpha)$. If we pretended that the electrons were smeared out into a static continuous distribution of charge whose electron density [defined as dn/dV, where dn is the infinitesimal fractional number of electrons in the infinitesimal volume dV located at point (x, y, z) in space] is ρ, then a term in the sum in (19.47) is the potential energy of attraction between nucleus α and the smeared-out electron charge cloud. J is given by (Prob. 19.50)

$$J \equiv \frac{1}{2} e^2 \int_{-\infty}^{\infty}\int_{-\infty}^{\infty}\int_{-\infty}^{\infty}\int_{-\infty}^{\infty}\int_{-\infty}^{\infty}\int_{-\infty}^{\infty} \frac{\rho(x_1, y_1, z_1)\rho(x_2, y_2, z_2)}{4\pi\varepsilon_0 r_{12}}\, dx_1\, dy_1\, dz_1\, dx_2\, dy_2\, dz_2 \qquad (19.48)$$

Finding the Kohn–Sham Orbitals

As noted in the discussion after (19.43), the quantities $\langle K_{e,s} \rangle$, $\langle V_{Ne} \rangle$, J, and E_{xc} can all be calculated if we know the orbitals θ_i^{KS}, and V_{NN} is easily calculated from the locations of the nuclei. Thus, we can find the molecular electronic ground-state energy E_e if the Kohn–Sham orbitals are known. How are the Kohn–Sham orbitals θ_i^{KS} found? Hohenberg and Kohn proved that the true ground-state electron probability density ρ minimizes the functional $E_e[\rho]$ for the energy. Since ρ is determined by the KS orbitals θ_i^{KS} [Eq. (19.42)], one can vary the orbitals θ_i^{KS} so as to minimize E_e in (19.43).

It turns out (see *Parr and Yang,* Sec. 7.2 for the proof) that the orthogonal and normalized KS orbitals that minimize E_e satisfy

$$\hat{h}_i^{\text{KS}}\theta_i^{\text{KS}} = \varepsilon_i^{\text{KS}}\theta_i^{\text{KS}}$$

[Eq. (19.41)] with the one-electron KS Hamiltonian $\hat{h}_i^{\text{KS}}$ being the sum of the following four one-electron terms: (*a*) the one-electron kinetic-energy operator $-(\hbar^2/2m_e)\nabla_1^2$; (*b*) the potential energy $-\Sigma_\alpha Z_\alpha e^2/4\pi\varepsilon_0 r_{1\alpha}$ of attractions between electron 1 and the nuclei; (*c*) the potential energy of repulsion between electron 1 and a hypothetical charge cloud of electron density ρ due to smeared out electrons; (*d*) an *exchange–correlation potential* $v_{xc}(x_1, y_1, z_1)$ whose form is discussed below. [The sum of terms *b*, *c*, and *d* equals what is called v_s in (19.40).]

The first three terms in $\hat{h}_i^{\text{KS}}$ are in fact identical to the first three terms *a*, *b*, and *c* of Secs. 18.9 and 19.5 in the Fock operator $\hat{F}$ of the Hartree–Fock equations (18.57). The only difference between the Hartree–Fock equations (18.57) for the Hartree–Fock orbitals and the Kohn–Sham equations (19.41) for the Kohn–Sham orbitals is that the exchange operator (term *d*) of the Hartree–Fock equations is replaced by the **exchange–correlation potential** v_{xc}. The expression for v_{xc} is found to be

$$v_{xc}(x, y, z) = \frac{\delta E_{xc}[\rho]}{\delta \rho} \qquad (19.49)$$

where the notation $\delta E_{xc}/\delta\rho$ indicates the *functional derivative* of the functional E_{xc} that occurs in the energy equation (19.43). The functional derivative $\delta E_{xc}/\delta\rho$ at (x, y, z) depends on how much the functional $E_{xc}[\rho]$ changes when the function ρ changes by a tiny amount in a tiny region centered at (x, y, z). We shall not worry about the precise definition of the functional derivative but simply note that if $E_{xc}[\rho]$ is known, its functional derivative v_{xc} can be easily found (see Prob. 19.51). The fourth term (term *d*) in the Fock operator $\hat{F}$ of the Hartree–Fock equations allows for the effects of electron exchange (the antisymmetry requirement) but does not allow for electron correlation. The fourth term v_{xc} in the KS operator $\hat{h}_i^{\text{KS}}$ allows for both exchange and correlation.

Carrying Out a Density-Functional Calculation

Assuming that we have a reasonable approximation for the functional $E_{xc}[\rho]$, how do we do a density-functional calculation? One starts with an initial guess for the molecule's electron density $\rho(x, y, z)$ found by superimposing calculated electron densities of the individual atoms at the nuclear geometry chosen for the calculation. From the initial ρ, one finds $E_{xc}[\rho]$ and then finds its functional derivative to get an initial estimate of v_{xc} [Eq. (19.49)]. This v_{xc} is used in the Kohn–Sham equations $\hat{h}_i^{\text{KS}}\theta_i^{\text{KS}} = \varepsilon_i^{\text{KS}}\theta_i^{\text{KS}}$ to solve for initial estimates of the orbitals θ_i^{KS}. (As is done in solving the Hartree–Fock equations, one usually expands the unknown orbitals using a basis set. Many of the same basis sets used for Hartree–Fock calculations are also used for KS calculations.) The initial orbitals θ_i^{KS} are used to calculate an improved probability density ρ from (19.42). This improved ρ is used to find an improved $E_{xc}[\rho]$, from which an improved v_{xc} is found. The improved v_{xc} is used in the KS equations (19.41) to find improved orbitals; etc. One continues the iterations until no further significant change is found from one cycle to the next. The molecular energy is then found from (19.43) using the final orbitals and ρ.

The Exchange–Correlation Energy Functional E_{xc}

Kohn and Sham suggested the use of a certain form for $E_{xc}[\rho]$ called the *local (spin) density approximation* (LDA or LSDA) that theory shows to be accurate when the electron density ρ varies very slowly with position. In a molecule, ρ does not vary very slowly with position. One finds that LSDA KS DFT calculations give good results for molecular geometries, dipole moments, and vibrational frequencies, but rather poor results for atomization energies. The LSDA E_{xc} is a definite integral of a certain function of ρ.

For convenience in devising approximations to E_{xc}, E_{xc} is usually split into an exchange part and a correlation part:

$$E_{xc}[\rho] = E_x[\rho] + E_c[\rho] \qquad (19.50)$$

and people devise separate approximations for E_x and E_c.

In the late 1980s, Becke showed that by taking E_{xc} as an integral of a certain function of ρ *and* the derivatives $\partial\rho/\partial x$, $\partial\rho/\partial y$, $\partial\rho/\partial z$ (these derivatives constitute the *gradient* of ρ), one gets greatly improved results for molecular atomization energies. Such a functional is called a *gradient-corrected functional* and use of a gradient-corrected functional gives the **generalized-gradient approximation** (GGA). In 1993, Becke proposed a further improvement in E_{xc}^{GGA} by adding to it a term aE_x^{HF}, where E_x^{HF} has the form of the expression used for the exchange energy in Hartree–Fock calculations but is evaluated using KS rather than Hartree–Fock orbitals, and a is an empirical parameter whose value was chosen to optimize the performance of E_{xc} in calculations on a test series of molecules. A GGA E_{xc} that includes a contribution from E_x^{HF} is called a **hybrid** GGA functional.

For a GGA functional, E_{xc} is taken as an integral of a function of ρ and its derivatives. An improvement on GGA functionals is gotten by taking E_{xc} as an integral of a function of ρ, the derivatives of ρ, and a quantity called the kinetic-energy density τ, where τ is a certain function of the derivatives of the Kohn–Sham orbitals θ_i^{KS}. Such a functional is called a **meta-GGA** functional. A contribution from E_{xc}^{HF} can also be added to a meta-GGA functional to give a hybrid meta-GGA functional. Thus, currently widely used functionals in quantum chemistry include GGA, meta-GGA, hybrid GGA, and hybrid meta-GGA functionals.

The most widely used functional in DFT calculations done in the period 1995–2005 has been the hybrid GGA functional called B3LYP, where B indicates that it includes a term for E_x^{GGA} devised by Becke, LYP indicates a term for E_c^{GGA} devised by Lee, Yang, and Parr, and the 3 indicates that it contains three empirical parameters whose values were chosen to optimize its performance. The hybrid functional B3PW91 is similar to B3LYP except that it uses the Perdew–Wang 1991 expression for E_c instead of the LYP formula.

Over 100 functionals have been proposed. Many of them contain substantial numbers of parameters whose values were chosen to make the functional yield good values for molecular properties. (For example, the M06-L meta-GGA functional contains 37 parameters; the M06 stands for University of Minnesota 2006 and the L stands for local, meaning this is not a hybrid functional.) Many of these functionals give superior performance to the widely used B3LYP functional. There is no one functional that can be designated as best overall, since functionals that give accurate performance for one property (such as dissociation energies) may not do so well for another property (such as activation energies for chemical reactions), and functionals that work well for organic compounds may not work as well for inorganic compounds. Some tests of large numbers of functionals are given in Y. Zhao and D. G. Truhlar, *J. Chem. Phys.*, **125,** 194101 (2006); Y. Zhao et al., *J. Chem. Theory Comput.*, **2,** 364 (2006); Y. Zhao and D. Truhlar, *J. Chem. Theory Comput.*, **3,** 289 (2007); J. Zheng et al., *J. Chem. Theory Comput.*, **3,** 569 (2007); Zhao and Truhlar, *Acc. Chem. Res.*, **41,** 157 (2008).

Performance of DFT

The time needed to do a DFT calculation is roughly the same as the time needed to do a Hartree–Fock calculation on the same molecule with the same basis set. A Hartree–Fock calculation (which restricts the molecular wave function to be a Slater determinant of spin-orbitals) yields only an approximate wave function and energy, no matter how large the basis set is. In contrast, a DFT calculation is, in principle, capable of yielding the exact values of the energy and other molecular properties. (No approximations were made in the equations of KS DFT.) In practice, because the true functional $E_{xc}[\rho]$ is unknown, DFT calculations yield approximate results. The quality of the results depends on how good the E_{xc} used in the calculation is. With present-day E_{xc} functionals, DFT calculations yield substantially more-accurate results than Hartree–Fock calculations. DFT calculations should not be done with basis sets smaller than 6-31G*.

One finds that Hartree–Fock calculations are often unreliable for transition-metal compounds. In contrast, DFT structures and relative energies of transition-metal compounds are usually reliable and "Computational transition metal chemistry today is almost synonymous with DFT for medium-sized molecules" [E. R. Davidson, *Chem. Rev.,* **100,** 351 (2000)].

For a sample of 108 molecules, average absolute errors in bond lengths, bond angles, dipole moments, and atomization energies were [A. C. Scheiner et al., *J. Comput. Chem.,* **18,** 775 (1997)]:

	HF/6-31G**	MP2/6-31G**	B3PW91/6-31G**
Bond lengths	0.021 Å	0.015 Å	0.011 Å
Bond angles	1.3°	1.1°	1.0°
Dipole moments	0.23 D	0.20 D	0.16 D
$\Delta_{at}U$/(kcal/mol)	119.2	22.0	6.8

The hybrid DFT method outperforms the Hartree–Fock (HF) and the MP2 methods. Considerably better results for atomization energies can be found with newer density functionals than with B3PW91. For example, for a set of 109 main-group atomization energies, calculations using a very large basis set found the following average absolute errors in atomization energies [Y. Zhao et al., *J. Chem. Theory Comput.,* **2,** 364 (2006)]: 1.9 kcal/mol for the hybrid meta-GGA functional PW6B95 and 2.2 kcal/mol for the hybrid meta-GGA functional BMK, which are superior to the results 3.1 kcal/mol for B3PW91 and 4.3 kcal/mol for the popular B3LYP functional.

Although the Kohn–Sham orbitals are calculated for the fictitious reference system (and not for the real molecule), one finds that Kohn–Sham orbitals closely resemble Hartree–Fock SCF MOs calculated for the real molecule [R. Stowasser and R. Hoffmann, *J. Am. Chem. Soc.,* **121,** 3414 (1999)]. Hartree–Fock MOs have been used to provide qualitative explanations and predictions of many chemical phenomena, and Kohn–Sham orbitals can be used for the same purpose.

Despite its many successes, DFT does have some drawbacks. DFT is basically a ground-state theory. People are currently working to extend it to excited states. Because an approximate E_{xc} is used, it is possible for a DFT calculation to give an energy that is less than the ground-state energy. In ab initio calculations (this term is defined in Sec. 19.11), one knows in principle how to improve the accuracy of the calculation, namely, one uses larger basis sets and goes to higher levels of theory (for example, HF, MP2, MP4, . . . or HF, CISD, CISDT, CISDTQ, . . . or HF, CCSD, CCSDT, . . .). (In practice, high-level ab initio calculations are limited to rather small molecules.) In DFT, the performance of a given functional cannot be predicted and one must try it out on a variety of molecules and properties to assess its performance. There is no systematic way in DFT to devise better functionals. Many of the currently available functionals do not give good accuracy when dealing with van der Waals intermolecular interactions; nor can they match the performance of very high-level ab initio methods, such as CCSD(T), when dealing with small molecules. DFT predictions for activation energies of chemical reactions are fairly often inaccurate.

For more on DFT, see *Parr and Yang; Levine,* sec. 16.4; W. Koch and M. C. Holthausen, *A Chemist's Guide to Density Functional Theory,* Wiley-VCH, 2000.

19.11 SEMIEMPIRICAL METHODS

Classification of Methods

Quantum-mechanical methods of treating molecules are classified as ab initio, density-functional, or semiempirical. An **ab initio** calculation uses the true molecular Hamiltonian and does not use empirical data in the calculation. (Ab initio means "from the beginning" in Latin.) The Hartree–Fock method calculates the antisymmetrized product Φ of spin-orbitals that minimizes the variational integral $\int \Phi^* \hat{H} \Phi \, d\tau$, where $\hat{H}$ is the true molecular Hamiltonian. Therefore, a Hartree–Fock calculation is an ab initio one (as is an SCF calculation that gives only an approximation to the Hartree–Fock wave function because of the limited size of the basis set). Of course, because of the restricted form of Φ, the Hartree–Fock method does not give the true wave function. CI, MP, and CC calculations (Sec. 19.9) are also ab initio methods and are capable in principle of converging to the exact wave function and energy provided the basis set is large enough and the level of the calculation is high enough. Note that the term "ab initio" does not guarantee high accuracy. For example, ab initio Hartree–Fock calculations give wildly inaccurate molecular dissociation energies.

A **semiempirical** method uses a simpler Hamiltonian than the true one, uses empirical data to assign values to some of the integrals that occur in the calculation, and neglects some of the integrals. The reason for resorting to semiempirical methods is that accurate ab initio calculations on large molecules cannot be done at present. Semiempirical methods were originally developed for conjugated organic molecules and later were extended to encompass all molecules.

Density-functional calculations are hard to classify as either ab initio or semiempirical and are usually considered as a third category of quantum-chemistry methods.

Semiempirical Methods for Conjugated Molecules

For a planar or near-planar conjugated organic compound, each MO can be classified as σ or π. Each σ MO is unchanged on reflection in the molecular plane (which is not a nodal plane for a σ MO), whereas each π MO changes sign on reflection in the molecular plane (which is a nodal plane for each π MO). (Recall the discussion of benzene in Sec. 19.6.) The σ MOs have electron probability density strongly concentrated in the region of the molecular plane. The π MOs have blobs of probability density above and below the molecular plane. The σ MOs can be taken as either delocalized or localized. However, the π MOs in a conjugated compound are best taken as delocalized. In conjugated molecules, the highest-energy occupied MOs are usually π MOs.

Because of the different symmetries of σ and π MOs, one can make the *approximation* of treating the π electrons separately from the σ electrons. One imagines the σ electrons to produce some sort of effective potential in which the π electrons move.

The simplest semiempirical treatment of conjugated molecules is the *free-electron molecular-orbital (FE MO) method.* The FE MO method deals only with the π electrons. It assumes that each π electron is free to move along the length of the molecule (potential energy $V = 0$) but cannot move beyond the ends of the molecule (potential energy $V = \infty$). This is the particle-in-a-box potential energy, and the FE MO method feeds the π electrons into particle-in-a-one-dimensional-box MOs, each such occupied

MO holding two electrons of opposite spin. The FE MO method is extremely crude and is of only historical interest. Problem 19.52 outlines an application of the FE MO method.

A somewhat more sophisticated method than the FE MO method is the **Hückel method** for conjugated hydrocarbons (developed in the 1930s). The Hückel MO method deals only with the π electrons. It takes each π MO as a linear combination of the $2p_z$ AOs of the conjugated carbon atoms (where the z axis is perpendicular to the molecular plane). These linear combinations are used in the variational integral, which is expressed as a sum of integrals involving the various $2p_z$ AOs. The Hückel method approximates many of these integrals as zero and leaves others as parameters whose values are picked to give the best fit to experimental data. Details may be found in most quantum-chemistry texts. The Hückel method was a mainstay of theoretically inclined organic chemists for many years but, because of the development of improved semiempirical methods (discussed below), is now only rarely used.

For certain purposes, all one is interested in is the relative signs of the AOs that contribute to the π MOs. (An example is the Woodward–Hoffmann rules for deducing the steric course of certain organic reactions; see any modern organic chemistry text.) To deduce these signs, we use the idea that the π MOs will show a pattern of nodes perpendicular to the molecular plane that will resemble the nodal pattern for the particle in a one-dimensional box and the harmonic oscillator.

Consider butadiene, $CH_2{=}CH{-}CH{=}CH_2$, for example. We take the π MOs as linear combinations of the four $2p_z$ carbon AOs, where the z axis is perpendicular to the molecular plane. (This is a minimal-basis-set treatment; Sec. 19.5.) Let p_1, p_2, p_3, p_4 denote these AOs. Figure 17.10 shows that the lowest π MO will have no nodes perpendicular to the molecular plane, the next lowest π MO will have one such node (located at the midpoint of the molecule), etc. To form the lowest π MO, we must therefore combine the four $2p_z$ AOs all with the same signs: $c_1p_1 + c_2p_2 + c_3p_3 + c_4p_4$, where the c's are all positive. For the purpose of determining the relative signs we won't worry about the fact that c_1 and c_2 differ in value (since the end and interior carbons are not equivalent); we shall simply write $p_1 + p_2 + p_3 + p_4$ for the lowest π MO. To have a single node in the center of the molecule, we take $p_1 + p_2 - p_3 - p_4$ as the second lowest π MO; this is the highest occupied π MO in the ground state, since there are four π electrons in butadiene. To get two symmetrically placed nodes, we take $p_1 - p_2 - p_3 + p_4$ as the third lowest π MO. To get three nodes, we take $p_1 - p_2 + p_3 - p_4$ as the fourth lowest π MO. See Fig. 19.37. Butadiene has four π electrons, two in each double bond. We put two electrons of opposite spin into each π MO. Hence the ground electronic state of butadiene will have the lowest two MOs of Fig. 19.37 occupied and the other two vacant.

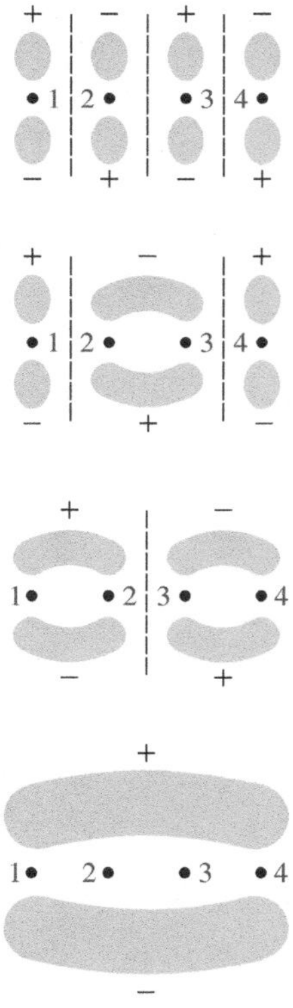

Figure 19.37

Rough sketches of the four lowest π MOs in butadiene.

General Semiempirical Methods

An improved version of the Hückel method, applicable to both conjugated and nonconjugated molecules, is the *extended Hückel (EH) method,* developed in the 1950s and 1960s by Wolfsberg and Helmholz and by Hoffmann. The EH method treats all the valence electrons of a molecule and neglects fewer integrals than the Hückel method. The calculations of the EH method are relatively easy to perform (thanks to the many simplifying approximations made). The quantitative predictions of the EH method are generally rather poor, and the main value of the method is the qualitative insights it provides into chemical bonding.

The Hückel and extended Hückel methods are quite crude, in that they use a very simplified Hamiltonian that contains no repulsion terms between electrons. Several improved semiempirical theories have been developed that include some of the electron repulsions in the Hamiltonian operator and that apply to both conjugated and nonconjugated molecules. Two widely used semiempirical methods are AM1 and PM3.

The AM1 (Austin Model 1) method (named after the University of Texas at Austin, where the method was developed) was devised by Dewar and co-workers [M. J. S. Dewar et al., *J. Am. Chem. Soc.*, **107,** 3902 (1985)]. AM1 treats only the valence electrons. It solves equations resembling the Hartree–Fock equations (18.57) to find self-consistent MOs, but since an approximate Hamiltonian is used and rather drastic approximations are made for many of the integrals that occur, the MOs found are only rough approximations to Hartree–Fock MOs. Dewar's aim was not to have a method giving approximations to Hartree–Fock results but one giving accurate molecular geometries and molecular dissociation energies. It might seem unreasonable to expect a theory that involves more approximations than the Hartree–Fock method to succeed in an area (calculation of dissociation energies) where the Hartree–Fock method fails. However, by choosing the values of the parameters in the AM1 method so as to reproduce known heats of atomization of many compounds, Dewar was able to build in compensation for the neglect of electron correlation that occurs in the Hartree–Fock theory.

AM1 has a different set of parameters for each element. The number of parameters varies somewhat from element to element but is on the order of 15 per element. The parameter values were fixed by varying the parameters so that the theory gives a good least-squares fit to known gas-phase $\Delta_f H^\circ_{298}$ values (found from AM1-calculated atomization energies and experimental heats of formation of gaseous atoms) and values of molecular geometries and dipole moments. PM3 (parametric method 3) is similar to AM1 except that it uses a different set of parameters [J. J. P. Stewart, *J. Comput. Chem.*, **10,** 209, 221 (1989); **11,** 543 (1990); **12,** 329 (1991)].

The AM1 and PM3 methods usually give satisfactory bond lengths and bond angles, but their predictions of dihedral angles are not so good. Calculated dipole moments are fairly reliable. AM1 and PM3 predictions of gas-phase $\Delta_f H^\circ_{298}$ values are of only fair accuracy. For samples of 460 bond lengths (R_{AB}), 196 bond angles ($\angle$ABC), 16 dihedral angles ($\angle$ABCD), 125 dipole moments (μ), and 886 gas-phase $\Delta_f H^\circ_{298}$ values, average absolute errors are [J. J. P. Stewart, *op. cit.*]:

	R_{AB}	$\angle$ABC	$\angle$ABCD	μ	$\Delta_f H^\circ_{298}$
AM1	0.051 Å	3.8°	12.5°	0.35 D	14.2 kcal/mol
PM3	0.037 Å	4.3°	14.9°	0.38 D	9.6 kcal/mol

Geometries and dipole moments are less accurate than for ab initio and DFT methods. AM1 and PM3 show large percentage errors in predicting barriers to internal rotation and do not give accurate predictions of conformational-energy differences. When AM1 or PM3 $\Delta_f H^\circ_{298}$ values are combined to predict ΔH° of a reaction, errors for each compound are multiplied by the stoichiometric coefficients in the reaction and errors for different compounds can add in a random manner, so the prediction of ΔH° is unreliable. Thus, the quantitative accuracy of AM1 and PM3 is mediocre.

A re-evaluation of the parameters of the AM1 method using additional data gave the RM1 method (Recife Model 1, developed at a university in Recife, Brazil); G. B. Rocha et al., *J. Comput. Chem.*, **27,** 1101 (2006).

A revision of the PM3 method gave the PM5 method. A further revision of PM3 gave the PM6 method, which has been parametrized for 70 elements [J. J. P. Stewart, *J. Mol. Model.*, **13,** 1173 (2007)]. PM6 has substantially more parameters than its predecessors. For a set of 1774 compounds containing no elements other than H, C, N, O, F, P, S, Cl, Br, and I, average absolute errors in gas-phase $\Delta_f H^\circ_{298}$ values in kcal/mol are

AM1	PM3	PM5	RM1	PM6
12.6	8.0	6.8	6.6	5.0

For a set of 712 bond lengths involving these elements, average absolute errors in Å are 0.03 for PM6 and about 0.04 for the other 4 methods. For bond angles involving these elements, the average absolute errors ranged from 3° to 4°, but when compounds of all main group elements were included the average absolute errors were around 8°. For compounds of H, C, N, O, F, P, S, Cl, Br, and I, the average absolute errors in dipole moments were about 0.4 D.

Because of the many approximations made, semiempirical methods are much faster than ab initio and DFT methods and can deal with larger molecules (up to a few thousand atoms) than can ab initio and DFT methods (up to a few hundred atoms). The molecular-mechanics method (Sec. 19.13) can deal with molecules with tens of thousands of atoms.

19.12 PERFORMING QUANTUM CHEMISTRY CALCULATIONS

The ab initio HF (Hartree–Fock) and MP methods, the DFT method, and semiempirical methods are widely used in chemistry, not just by theoretical chemists but by all kinds of chemists as an aid to predict and interpret experimental results.

In describing a molecular electronic-structure calculation, one specifies the method used followed by the basis set. For example, HF/6-31G* denotes a Hartree–Fock calculation done with the 6-31G* basis set. (Any calculation that solves the Hartree–Fock equations (18.57) is called a Hartree–Fock calculation even though the basis set might be substantially smaller than that needed to give a truly accurate approximation to the Hartree–Fock wave function.) B3LYP/6-31G** denotes a DFT calculation done with the B3LYP exchange–correlation functional and the 6-31G** basis set. The AM1 and PM3 methods use a fixed minimal-basis set of Slater-type orbitals. Since one cannot vary the basis set, no basis set is specified when describing an AM1 or PM3 calculation.

A **single-point calculation** is one done only at a single fixed molecular geometry specified by the user. In a **geometry-optimization,** the quantum-mechanical program will vary the locations of the nuclei so as to locate a minimum in the electronic energy E_e of (19.7). A geometry-optimization calculation consists of many single-point calculations, with each single-point energy calculation followed by an energy-gradient calculation (Sec. 19.8) to help the program decide on the next geometry to try. The geometry-optimization calculation continues until the magnitude of the gradient is very close to zero, indicating that an energy minimum has been found. In a **vibrational-frequency calculation,** the program calculates the molecular vibration frequencies (Section 20.8). A vibrational-frequency calculation must be preceded by a geometry-optimization, since vibrational frequencies calculated for a geometry that is not at an energy minimum are meaningless. A **transition-state optimization** attempts to find the geometry and electronic energy of the transition state in a chemical reaction (Sec. 22.4).

Geometry-optimization calculations for large molecules are too time consuming to be done with high-level methods. Since an accurate geometry can usually be found with a low-level method, a common procedure is to do a low-level calculation to find the geometry and then use this geometry in a single-point high-level calculation of the energy.

The geometry-optimization process locates the energy minimum that is closest to the starting geometry. For example, if one does a geometry-optimization calculation of butane (Figs. 19.33 and 19.34) and inputs an initial geometry that has a CCCC dihedral angle close to 60°, the program will converge to the geometry of the gauche conformer, whereas if one starts with the CCCC dihedral angle close to 180°, the program converges to the trans conformer. The trans conformer is the **global minimum** for butane, meaning that it has the lowest energy of all the conformers, whereas the gauche conformer is only a **local minimum,** meaning that its energy is the minimum energy for all geometries in its immediate vicinity. Every conformer (Sec. 19.8) lies at a local minimum.

Large molecules may have huge numbers of conformers, making it extraordinarily difficult to find the global minimum and those local minima whose energy is low enough to have them significantly populated at room temperature. For example, for the cycloalkane $C_{17}H_{34}$, a study found 262 conformers with energy in the range 0–3 kcal/mole above the global minimum, 1368 conformers in the range 3–5 kcal/mol, 8165 conformers in the range 5–10 kcal/mol, and 2718 conformers in the range 10–20 kcal/mol [I. Kolossváry and W. C. Guida, *J. Am. Chem. Soc.,* **118,** 5011 (1996)]. The backbone of each amino acid residue in a polypeptide chain has two dihedral angles, and each such dihedral angle has three likely potential-energy minima. Hence a polypeptide with 40 amino acid residues has at least $3^{2(40)} = 3^{80} \approx 10^{38}$ possible conformers. Many special methods of **conformational searching** exist, whose aim is to find low-energy conformers (see *Leach,* chap. 9; *Levine,* sec. 15.11). Because of the huge numbers of conformers involved, the energy calculations in conformational searching on large molecules are usually done with the nonquantum-mechanical molecular-mechanics method (Sec. 19.13).

Programs

Many programs for molecular electronic-structure calculations exist.

Gaussian (www.gaussian.com) is the most widely used program for ab initio and density functional calculations, and can also do semiempirical calculations. *Gaussian* exists in versions for UNIX workstations and Windows and Macintosh personal computers. The first version of *Gaussian* was released in 1970 and *Gaussian 03* was released in 2003. *Gaussian* was developed by John Pople and co-workers and has been a key force in the growing use of quantum-chemistry calculations by chemists, since it is an easy-to-use program that allows a very wide variety of calculations to be done by virtually every available quantum-mechanical method.

Some other quantum-chemistry programs are the free program GAMESS (www.msg.ameslab.gov/GAMESS), Q-Chem (www.q-chem.com), Jaguar (www.schrodinger.com), SPARTAN (www.wavefun.com), and HyperChem (www.hyper.com), and the semiempirical programs MOPAC2007 (www.openmopac.net/MOPAC2007.html) and AMPAC (www.semichem.com). The free Windows program ArgusLab (www.planaria-software.com) does Hartree–Fock, AM1, and PM3 calculations. ORCA (ewww.mpi-muelheim.mpg.de/bac/logins/neese/description.php) is a UNIX and Windows ab initio, density functional, and semiempirical program that is free to academic users.

Input

The input to a quantum-chemistry program specifies the kind of calculation to be done, the method and basis set to be used, and the initial set of locations for the nuclei. Figure 19.38 shows the input to the Windows version of *Gaussian* for a calculation on the methanol molecule, CH_3OH. In the Route Section, HF/3-21G tells the program to use the Hartree–Fock (SCF) method and to use the 3-21G basis set. The keyword Opt specifies a geometry optimization calculation by minimizing the energy. (If Opt is omitted, a single-point calculation is done.) The information in the Title Section has no effect on the program. The 0 and 1 tell the program that the molecule is uncharged and has spin multiplicity $2S + 1$ equal to 1; that is, $S = 0$.

The Molecule Specification section gives the initial geometry specified by the user. In Fig. 19.38, the geometry is specified by an array called a **Z-matrix.** To write down the Z-matrix, we first sketch the molecule (Fig. 19.39). The first column of the Z-matrix specifies the atoms present in the molecule. The numbers 1 to 6 after the element symbols are for the convenience of the user and can be omitted. The first row of the Z-matrix says that the first atom is a carbon atom. The second row tells the program to place an oxygen atom at a distance 1.43 Å from atom 1 (meaning the atom on row 1 of the Z-matrix). The value 1.43 Å was chosen to be the sum of the C and O

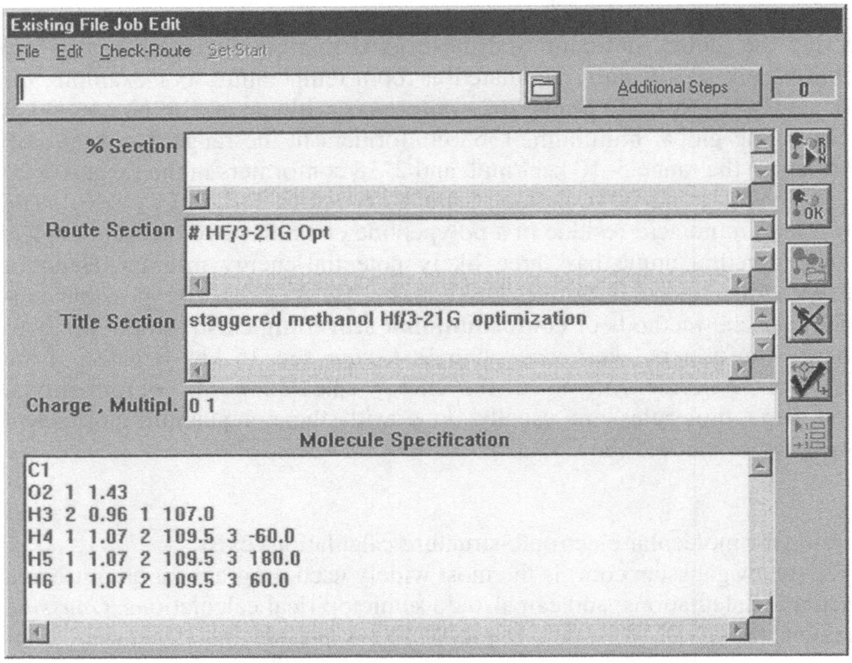

Figure 19.38

Input to Windows version of *Gaussian* for HF/3-21G geometry optimization of CH_3OH.

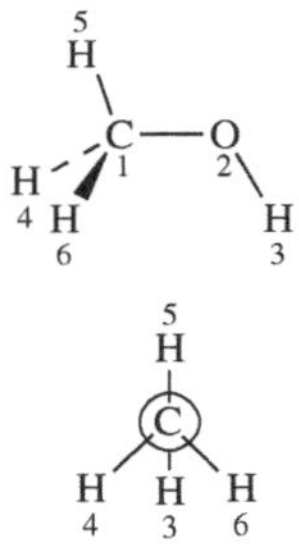

Figure 19.39

The methanol molecule. The lower figure is a Newman projection with the oxygen behind the carbon.

single-bond radii in Sec. 19.1; other bond distances in the Z-matrix were chosen in the same way. The third row of the Z-matrix places an H atom at a distance 0.96 Å from the row-2 atom (oxygen) with the angle H3–O2–C1 equal to 107.0° (note the 3, 2, and 1 in row 3).

The guess 107.0° for the COH bond angle was found using the VSEPR method (Sec. 19.1). The CH_3OH Lewis structure has four valence electron pairs on O. The VSEPR method arranges these pairs tetrahedrally. Since two of the four pairs on oxygen are lone pairs, the COH angle is predicted to be a bit less than the tetrahedral angle of 109.5°. Hence the 107° guess.

Row 4 of the Z-matrix places atom H4 1.07 Å from C1 with an angle of 109.5° (the tetrahedral angle predicted by VSEPR) for H4–C1–O2 (note the 4, 1, 2 on row 4). The -60.0 on row 4 means that the dihedral angle defined by atoms 4–1–2–3 is $-60.0°$. It takes one-third of a complete rotation of the CH_3 group to bring H4 to where H6 was and one-third of a rotation is $360°/3 = 120°$. To bring H4 into alignment with H3, we need one-half of the rotation just described, namely, 60°. The sign of the dihedral angle 4–1–2–3 is defined as negative if a counterclockwise rotation of the first-listed atom (atom 4) is needed to make it align with atom 3. Rows 5 and 6 of the Z-matrix are constructed similar to row 4. Of course, the Z-matrix for a molecule is not unique and many different valid Z-matrices can be written for methanol. In Fig. 19.39, the OH hydrogen was staggered with respect to the methyl hydrogens in accord with the rule that the conformation about a bond that joins two atoms each of which has tetrahedral bond angles is usually staggered.

Some rules in constructing a Z-matrix are: Bond angles must be greater than 0° and less than 180° (see Prob. 19.55). All angles and lengths must contain a decimal point. The atom numbers in columns 2, 4, and 6 must be atoms whose locations were specified in previous rows.

Note that although the bond distances were used as a convenience in specifying the Z-matrix, the Z-matrix does not actually specify which atoms are bonded to each other. All it does is tell *Gaussian* the atomic number and location of each nucleus. In fact, instead of using a Z-matrix, one can specify the initial geometry by giving the symbol for each nucleus followed by the three cartesian coordinates of that nucleus on the same line as the symbol.

In some programs (for example, SPARTAN, HyperChem, and ArgusLab) the user inputs the starting structure using a molecule builder that builds the model on the computer screen. The user selects the desired atoms and bonds and the program uses built-in typical bond lengths and angles and rules for conformations to produce the initial structure. The user can modify the conformation if desired. Although *Gaussian* itself does not have this feature, there are programs that can serve as graphical interfaces to *Gaussian*. Two such programs are GaussView (www.gaussian.com) and Chem3D (www.camsoft.com).

You can use the program CORINA to generate a molecular structure at no charge. Go to www.mol-net.de/online_demos/corina_demo.html and enter the structure as a SMILES string. A SMILES string omits hydrogens and uses an equals sign to denote a double bond. For example, CC=O denotes CH_3CHO and C1CC1 denotes cyclopropane, where the numeral 1's indicate atoms bonded together. (Instead of a SMILES string, you can also draw the structure using a molecular editor provided on the Website.) The program will generate a rotatable model of the molecule. If you right-click on the model, you get a host of options. By choosing Show and then Extract Mol Data, you will get a list of the nuclear cartesian coordinates; these can then be used as input to a quantum-chemistry program.

Output

Coming back to the methanol calculation, *Gaussian* begins the SCF calculation by using a semiempirical method to generate an initial guess for the MOs. It then solves the Hartree–Fock equations (18.57) using the specified basis set. The words SCF Done in the output are followed by E(RHF) = −114.3962071 A.U., which is the Hartree–Fock energy in atomic units (hartrees—Sec. 18.3). The orbital energies are given. The dipole moment is also calculated for this initial structure. The program calculates derivatives of the energy with respect to the nuclear coordinates. If these derivatives are all very small, the initial structure is close enough to a local minimum in the energy and the calculation is done. If not, the nuclei are moved to new locations (whose values are chosen based on the values of the energy derivatives and estimates of the energy second derivatives) and the SCF calculation is repeated at the new geometry. For this calculation, the energy found at the new geometry is −114.3978271 A.U., which is lower than the initial value, indicating we are closer to a minimum. The next two geometries yield energies of −114.3980171 and −114.3980192 A.U., respectively. For this last energy the energy derivatives are all small enough to declare that the optimization is completed and *Gaussian* reports the final optimized bond distances, bond angles, and dihedral angles and the final dipole moment. The optimized HF/3-21G geometry compared with the initially guessed geometry is:

	R(CO)	*R*(OH)	*R*(CH6)	*R*(CH5)	∠COH	∠OCH6	∠OCH5	∠H6COH3	∠H5COH3
Init.	1.43	0.96	1.07	1.07	107°	109.5°	109.5°	60°	180°
Opt.	1.441	0.966	1.085	1.079	110.3°	112.2°	106.3°	61.4°	180°

19.13 THE MOLECULAR-MECHANICS (MM) METHOD

The **molecular-mechanics** (or *empirical-force-field*) **method** can handle very large organic and organometallic ground-state molecules (10^4 atoms). Molecular mechanics (MM) is an empirical nonquantum-mechanical method and does not use a Hamiltonian or wave function. Instead, the molecule is viewed as atoms held together by bonds, and the molecular electronic energy is expressed as the sum of bond-stretching, bond-bending, and other kinds of energies.

The Steric Energy

Rather than dealing with the molecular electronic energy E_e of Eqs. (19.7) and (19.8), the MM method uses a quantity called the **steric energy** V_{steric} that (as will be explained below) differs from E_e by an unknown constant C, whose value may differ for different molecules: $V_{\text{steric}} = E_e + C$. Recall that in the Schrödinger equation (19.9) for nuclear motion, the electronic energy E_e serves as the potential energy. As noted after Eq. (2.17), the potential energy always has an arbitrary additive constant. Therefore, V_{steric}, which differs from E_e by only a constant, is also a valid potential energy for nuclear motion.

The MM expression for V_{steric} is

$$V_{\text{steric}} = V_{\text{str}} + V_{\text{bend}} + V_{\text{tors}} + V_{\text{cross}} + V_{\text{vdW}} + V_{\text{es}} \qquad (19.51)$$

The specific forms for each term in (19.51) and the parameter values used in these terms define a molecular-mechanics **force field,** since the forces on the nuclei can be found from the derivatives of the potential energy V_{steric} [Eq. (2.17)].

In (19.51), V_{str} is the change in electronic energy due to bond stretching. In the simplest force fields, V_{str} is taken as the sum of harmonic-oscillator terms [Eqs. (17.71) and (20.18)] for each pair of bonded atoms:

$$V_{\text{str}} = \sum_{1,2} V_{\text{str},ij} \qquad \text{where} \quad V_{\text{str},ij} = \tfrac{1}{2}k_{IJ}(l_{ij} - l_{IJ}^0)^2 \qquad (19.52)$$

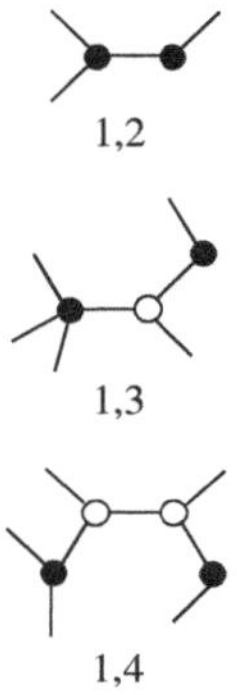

Figure 19.40

1,2, 1,3, and 1,4 atoms.

Atoms bonded to each other are called 1,2 atoms (Fig. 19.40). Atoms separated by two bonds are 1,3 atoms, and atoms separated by three bonds are 1,4 atoms. The sum in (19.52) is over all pairs of 1,2 atoms. The parameter k_{IJ} is the force constant for stretching the bond between atoms i and j, the parameter l_{IJ}^0 is a fixed *reference length,* and l_{ij} is the distance between the bonded atoms i and j for a particular molecular geometry. The harmonic-oscillator expression in (19.52) is not highly accurate for molecular vibration and some force fields use a more complicated and more accurate expression for $V_{\text{str},ij}$.

An MM force field classifies each atom in the molecule into an **atom type,** depending on its atomic number and on how it is bonded in the molecule. Some typical atom types are sp^3 carbon (carbon bonded to four atoms) nonaromatic sp^2 carbon, sp carbon, aromatic carbon, hydrogen bonded to carbon, hydrogen bonded to oxygen, hydrogen bonded to nitrogen, etc. A typical force field for organic compounds has 60 to 70 atom types. In the expression $V_{\text{str},ij} = \frac{1}{2}k_{IJ}(l_{ij} - l_{IJ}^0)^2$, the capital letters I and J stand for the atom types of atoms i and j in the molecule. Thus in $CH_3CH_2CH_2OCH{=}CH_2$, the $CH_3{-}CH_2$ bond and the $CH_2{-}CH_2$ bond have the same k_{IJ} and the same l_{IJ}^0, whereas the $CH_3{-}CH_2$ bond and the $CH{=}CH_2$ bond have different k_{IJ} values and different l_{IJ}^0 values. The reference length l_{IJ}^0 is close to the typical bond length between atoms of types I and J.

The parameters k_{IJ} and l_{IJ}^0 in V_{str} (and the parameters in the other terms) are found as follows. One picks initial values of the parameters guided by experimental data (such as bond-length data and force constants found from analysis of vibrational spectra) and by the results of ab initio calculations. One then varies the parameter values in the force field so as to minimize the errors in molecular geometries and other properties predicted by the force field for a set of test molecules (the training set). Finding the optimum parameters in a force field is similar to finding the best values for the parameters in a semiempirical theory (Sec. 19.11), to finding the nuclear coordinates that minimize the electronic energy in a molecular-geometry optimization calculation

(Sec. 19.12), and to finding parameters in a function so as to minimize the sum of squares of the deviations from experimental data points (Example 7.6 in Sec. 7.3). Similar mathematical procedures are used for all these *optimization* processes.

The term V_{bend} in the steric energy (19.51) is the energy change due to bending of bonds, and in the simplest force fields has the form

$$V_{\text{bend}} = \sum V_{\text{bend},ijk} \qquad \text{where} \qquad V_{\text{bend},ijk} = \tfrac{1}{2}k_{IJK}(\theta_{ijk} - \theta^0_{IJK})^2$$

The sum is over all bond angles; θ_{ijk} is the ijk bond angle, and k_{IJK} and θ^0_{IJK} are parameters.

The term V_{tors} accounts for the change in electronic energy with rotation (torsion) about a bond. (Figure 19.34 shows V_{tors} for rotation about the central CC bond in butane.) The form of V_{tors} is $V_{\text{tors}} = \Sigma_{1,4} V_{\text{tors},ijkl}$, where the sum is over all 1,4 atom pairs and $V_{\text{tors},ijkl}$ is a trigonometric function of the torsion dihedral angle defined by atoms $ijkl$. $V_{\text{tors},ijkl}$ contains one or more parameters. In CH_3CH_3, each H atom on the left carbon has a 1,4 relation with each of the three H atoms on the right carbon, so there are nine terms in V_{tors} for ethane.

V_{cross} contains cross terms that allow for interactions between stretching, bending, and torsion.

V_{es} allows for electrostatic attractions and repulsions between nonbonded atoms and usually has the form

$$V_{\text{es}} = \sum_{1,\geq 4} V_{\text{es},ij} \quad \text{where} \quad V_{\text{es},ij} = \frac{Q_iQ_j}{4\pi\varepsilon_0 R_{ij}}$$

where the sum goes over all 1,4, 1,5, 1,6, . . . atom pairs. R_{ij} is the distance between atoms i and j, and Q_i and Q_j are the (partial) electrical charges on atoms i and j in the molecule. The partial atomic charge Q_i on an atom in a molecule is an ill-defined quantity that can neither be measured experimentally nor calculated theoretically in a unique manner. Rather, many different theoretical methods have been proposed for arriving at Q_i values. A force field might use Q_i values based on the atom type and what atoms are bonded to atom i, or it might take Q_i values found on similar atoms in ab initio calculations on small molecules using some scheme for partitioning the electronic charge among the atoms.

The term V_{vdW} is the contribution of van der Waals nonbonded interactions (Sec. 21.10) between atoms in the molecule and is taken as $V_{\text{vdW}} = \Sigma_{1,\geq 4} V_{\text{vdW},ij}$, where $V_{\text{vdW},ij}$ usually has the form of the Lennard-Jones 6–12 potential [Eq. (21.136)].

A typical force field for organic molecules might specify values for 5000 parameters. The bond-bending parameters k_{IJK} and θ^0_{IJK} involve three atom types and the torsion parameters involve four atom types, so with perhaps 60 different atom types in the force field, a huge number of parameters (far greater than 10^4) are needed to cover all possible situations. If one applies molecular mechanics to somewhat unusual molecules, one often runs into the problem of needing parameters whose values are missing from the force field. When a parameter is missing, the MM program will estimate its value, based on a parameter whose value is likely to be similar to the missing value. Such estimates limit the accuracy of MM calculations.

Performing an MM Calculation

Unlike ab initio, semiempirical, and density-functional calculations, the input to a molecular-mechanics calculation must specify not only the initial locations of the nuclei but also which atoms are bonded to each other and how they are bonded (the atom types). This is most conveniently done by building a molecular model on a computer screen.

After the molecule and its initial geometry are specified, the molecular-mechanics program varies the molecular geometry to minimize the steric energy V_{steric} (just as quantum-mechanical methods minimize E_e in a geometry-optimization calculation). Thus one obtains the geometry and V_{steric} for the conformer closest to the starting geometry. Many MM programs can do automatic conformational searches to locate many different conformers and find their V_{steric} values.

The most time-consuming step in calculating V_{steric} of a large molecule is the evaluation of V_{es} and V_{vdW}. For a molecule with 10^4 atoms, V_{es} and V_{vdW} are each the sum of about $10^8/2$ $V_{\text{es},ij}$ and $V_{\text{vdW},ij}$ terms, and these terms must be recalculated for each geometry change in the geometry-optimization process. To reduce the calculation time, people often use cutoffs, meaning that the $V_{\text{es},ij}$ and $V_{\text{vdW},ij}$ terms are omitted for atoms separated by more than a specified cutoff distance. For $V_{\text{vdW},ij}$ a cutoff of 8 Å has often been used in the past. Although van der Waals interactions are weak and short range, the number of such interactions increases rapidly with increasing distance from a given atom, and it turns out that to obtain valid results, the cutoff for $V_{\text{vdW},ij}$ should be at least 18 Å. Although people have used cutoffs of 10 or 15 Å for $V_{\text{es},ij}$, electrostatic interactions are long-range and no cutoff should be used for them. Various special techniques have been devised to speed up evaluation of V_{es}.

In a quantum-mechanical calculation, the molecular electronic energy E_e has a well-defined meaning, namely, it is the energy for a given fixed configuration of the nuclei where the zero level of energy corresponds to a situation with all the electrons and nuclei at infinite separations from one another and at rest. In contrast, V_{steric} in a molecular-mechanics calculation does not have a well-defined meaning, since it refers to a hypothetical molecule in which all the bond distances and angles have their reference values and all torsional, electrostatic, and van der Waals interactions are absent. Thus V_{steric} differs from E_e by an unknown constant C. Assuming accurate modeling of V_{steric} and accurate quantum-mechanical calculation of E_e, minimization of V_{steric} will give the same geometry as minimization of E_e.

The constant C will be the same for different conformers of the same molecule since the zero level of V_{steric} is the same for such species. Therefore, the differences between V_{steric} for different conformers of the same molecule are valid estimates of the differences in E_e for these conformers, provided one uses the same force field to calculate all the V_{steric} values. (V_{steric} values of different force fields cannot be meaningfully compared.)

The difference in V_{steric} for the isomers CH_3CH_2OH and CH_3OCH_3 cannot be used to estimate ΔE_e for these molecules, since the nature of the bonds differs. However, we could use ΔV_{steric} to estimate ΔE_e for $CH_3CH_2CH_2OH$ and $CH_3CH(OH)CH_3$, since here the bonding is the same.

After finding the equilibrium geometry by an MM calculation, one can use the partial atomic charges Q_i to estimate the dipole moment $\boldsymbol{\mu} = \Sigma_i Q_i \mathbf{r}_i$. From the second derivatives of V_{steric}, one can find molecular vibration frequencies. By combining V_{steric} with empirical bond-energy parameters, one can estimate the gas-phase $\Delta_f H^\circ_{298}$, but most MM programs don't do this.

Force Fields and Programs

Some widely used MM force fields are MM2 for small and moderate-size organic compounds; MM3 for organic compounds, polypeptides, and proteins [N. L. Allinger and L. Yan, *J. Am. Chem. Soc.,* **115,** 11918 (1993) and references cited therein]; MMFF94 for small organic compounds, proteins, and nucleic acids [T. A. Halgren, *J. Comput. Chem.,* **20,** 730 (1999) and references cited therein]; and AMBER [W. D. Cornell et al., *J. Am. Chem. Soc.,* **117,** 5179 (1995); www.amber.ucsf.edu] and CHARMM [A. D. Mackerell et al., *J. Am. Chem. Soc.,* **117,** 11946 (1995); *J. Phys. Chem.,* **102,** 3586 (1998); yuri.harvard.edu] for polypeptides, proteins, and nucleic acids.

Many MM programs are available for personal computers. The program HyperChem has the force fields MM+ (a version of MM2), AMBER, BIO+ (a version of CHARMM), and OPLS. Spartan has the MMFF94 force field. ChemBio3D Ultra (www.camsoft.com/) is a Windows molecular-modeling program with MM2; a free trial version that lasts for two weeks is available (scistore.cambridgesoft.com/). BALLView (www.ballview.org) is a free molecular modeling Windows and Macintosh program with the AMBER and CHARMM force fields [A. Moll et al., *J. Comput.-Aided Mol. Des.,* **19,** 791 (2005); *Bioinformatics,* **22,** 365 (2006)].

Performance and Applications

A properly parametrized MM force field applied to compounds similar to those used in the parametrization will give very good results. For example, for a sample of 30 organic compounds, MMFF94 gave root-mean-square errors of 0.014 Å in bond lengths and 1.2° in bond angles [T. A. Halgren, *J. Comput. Chem.,* **10,** 982 (1989)]. MM3 gave an average absolute error of 0.6 kcal/mol in gas-phase $\Delta_f H^\circ_{298}$ values for a sample of 45 alcohols and ethers [N. L. Allinger et al., *J. Am. Chem. Soc.,* **112,** 8293 (1990)]. However, when, as often happens, missing parameters must be estimated, the accuracy of the results is impaired.

Because of its ability to handle large molecules, molecular mechanics is widely used to deal with biological molecules. For example, folding of a small protein in solution has been modeled using the AMBER force field (see Sec. 23.14).

Molecular mechanics is not suitable for dealing with chemical reactions in which bonds are broken. Several versions of combined quantum-mechanical and molecular mechanics methods (QM/MM) have been developed. To treat an enzyme-catalyzed reaction by a QM/MM method, one uses a quantum-mechanical method to treat the substrate and the active site of the enzyme and uses MM for the rest of the enzyme. Computational methods of studying enzyme reactions are discussed in J. Gao et al., *Chem. Rev.,* **106,** 3188 (2006).

19.14 FUTURE PROSPECTS

The use of electronic computers has brought remarkable advances in the ability of quantum chemists to deal with problems of real chemical interest. For example, quantum-chemistry calculations are now being used to study chemisorption on metal catalysts and hydration of ions in solution. Whereas quantum-mechanical calculations used to be confined to small molecules and were published in journals read mainly by physical chemists and chemical physicists, such calculations now deal with medium-sized and even fairly large molecules and appear regularly in the *Journal of the American Chemical Society,* read by all kinds of chemists.

The accuracy of quantum-mechanical predictions of molecular properties and the size of molecules treatable will increase as faster computers are developed and as new calculational methods are devised. Quantum-mechanical calculations will clearly play an expanding role in physical chemistry.

The 1998 Nobel Prize for chemistry was awarded to Walter Kohn, one of the developers of density-functional theory, and John A. Pople, one of the developers of the *Gaussian* series of programs. The Nobel committee stated that computational quantum chemistry "is revolutionising the whole of chemistry."

19.15 SUMMARY

Covalent bond distances can be estimated as the sum of atomic covalent radii. ΔH° of gas-phase reactions can be estimated from bond energies. Molecular dipole moments can be estimated as vector sums of bond dipole moments. The electronegativity of an element is a measure of the ability of an atom of that element to attract the electrons in a chemical bond.

Since nuclei are much heavier than electrons, one can deal separately with electronic and nuclear motions in a molecule (the Born–Oppenheimer approximation). One first solves an electronic Schrödinger equation with the nuclei in fixed positions. This gives an electronic energy and wave function that depend on the nuclear positions as parameters. One then uses this electronic energy as the potential energy in the Schrödinger equation for the nuclear motion (rotation and vibration).

Approximate electronic wave functions for H_2^+ can be written as linear combinations of atomic orbitals of each H atom. These one-electron wave functions can be used as molecular orbitals for homonuclear diatomic molecules. Each MO of a diatomic molecule is classified as σ, π, δ, ϕ, . . . according to whether m, the quantum number for the electronic orbital-angular-momentum component along the molecular axis has absolute value 0, 1, 2, 3, ... Homonuclear diatomic MOs are further classified as g or u according to whether the orbital has the same or the opposite sign on diagonally opposite sides of the molecule's center. In the LCAO MO approximation, the H_2 electronic ground-state wave function is given by $N[1s_A(1) + 1s_B(1)][1s_A(2) + 1s_B(2)] \cdot [\alpha(1)\beta(2) - \alpha(2)\beta(1)]/2^{1/2}$.

The requirement that electronic wave functions be antisymmetric leads to an apparent extra repulsion between electrons of like spin (Pauli repulsion).

An SCF wave function is one in which each electron is assigned to a single spin-orbital, the spatial part of each spin-orbital is expressed as a linear combination of basis functions, and the coefficients in these linear combinations are found by solving the Hartree–Fock equations; the SCF wave function is an antisymmetrized product (Slater determinant) of spin-orbitals. If the basis set is large enough so that the MOs found are the best possible orbitals, the SCF wave function is the Hartree–Fock wave function. To reach the true wave function, one must go beyond the Hartree–Fock wave function, for example, by using configuration interaction or Møller–Plesset perturbation theory.

MOs that satisfy the Hartree–Fock equations are called canonical MOs. Canonical MOs have the symmetry of the molecule and are spread out (delocalized) over the molecule. For example, the bonding MOs of BeH_2 in Fig. 19.25 extend over all three atoms. By taking linear combinations of the canonical MOs, one can form localized MOs. Each localized MO is classifiable as an inner-shell, lone-pair, or bonding MO. The MO wave function that uses localized MOs is equal to the wave function that uses canonical MOs. In an unconjugated molecule, each localized bonding MO is largely (but not completely) confined to two bonded atoms (molecules with three-center bonds are an exception). The AOs in a bonding MO are often hybridized extensively. For example, the hybridizations in BeH_2, C_2H_4, and CH_4 are approximately sp, sp^2, and sp^3, respectively.

Ab initio SCF wave functions give generally accurate values for molecular geometries, dipole moments, ionization energies, rotational barriers, relative energies of isomers, and electron probability densities (provided an adequate-size basis set is used), but give poor values of molecular dissociation energies.

Gaussian basis sets are usually used for molecular electronic-structure calculations.

The main methods for improving ab initio SCF wave functions are Møller–Plesset (MP) perturbation theory, which takes the unperturbed wave function as the SCF wave function; configuration interaction (CI), which is usually limited to inclusion of functions with one or two electrons in virtual orbitals; and the coupled-cluster (CC) method.

Density-functional theory (DFT) is based on the Hohenberg–Kohn theorem that the ground-state electron probability density ρ determines the ground-state energy and all other molecular properties. The Kohn–Sham (KS) version of DFT uses a fictitious reference system of noninteracting electrons whose probability density ρ is the same as that of the ground-state molecule. The KS theory is exact in principle, but approximate in practice, since the true form of the key quantity the exchange–correlation functional E_{xc} is unknown. Gradient-corrected approximations to E_{xc} (especially hybrid functionals) provide generally accurate results for small and medium-sized molecules, but are not as accurate as high-level ab initio methods such as CC. DFT is currently the most widely used calculation method in quantum chemistry.

The AM1, PM3, and PM6 semiempirical methods have a fair degree of accuracy for many (but not all) molecular properties and can be applied to molecules too large to treat by density-functional or ab initio methods.

The input to a quantum-mechanical calculation includes the identity and location of each nucleus. A geometry-optimization calculation finds the local energy minimum that is nearest to the starting geometry. One way to specify the starting geometry is with a Z-matrix.

The molecular-mechanics method is an empirical nonquantum-mechanical method that treats the molecule as atoms held together by bonds and writes the molecular steric energy as the sum of contributions from bond bending, bond stretching, bond torsion, and van der Waals and electrostatic interactions of nonbonded atoms. It can be applied to very large molecules and yields good results when applied to molecules similar to those for which the force field was parametrized.

FURTHER READING

Karplus and Porter, chaps. 5, 6; *Atkins and Friedman,* chaps. 8, 9; *Levine*, chaps. 13–17; *Lowe and Peterson,* chaps. 8, 10, 11, 14; *McQuarrie* (1983), chap. 9; *DeKock and Gray; Leach; Jensen; Ratner and Schatz,* chaps. 10–14.

PROBLEMS

Section 19.1

19.1 Estimate the bond lengths in (*a*) CH_3OH; (*b*) HCN.

19.2 Explain why the observed boron–fluorine bond length in BF_3 is substantially less than the sum of the B and F single-bond radii.

19.3 Predict the shape and bond angles of (*a*) $TeBr_2$: (*b*) $HgCl_2$; (*c*) $SnCl_2$; (*d*) XeF_2; (*e*) ClO_2^-.

19.4 Predict the shape of (*a*) BrF_3; (*b*) GaI_3; (*c*) H_3O^+; (*d*) PCl_3.

19.5 Predict the shape of (*a*) SnH_4; (*b*) SeF_4; (*c*) XeF_4; (*d*) BH_4^-; (*e*) BrF_4^-.

19.6 Predict the shape of (*a*) $AsCl_5$; (*b*) BrF_5; (*c*) $SnCl_6^{2-}$.

19.7 Predict the shape of (*a*) O_3; (*b*) NO_3^-; (*c*) SO_3; (*d*) SO_2; (*e*) SO_2Cl_2; (*f*) $SOCl_2$; (*g*) IO_3^-; (*h*) SOF_4; (*i*) XeO_3; (*j*) $XeOF_4$.

19.8 Estimate the bond angles in (*a*) CH_3CN; (*b*) $CH_2{=}CHCH_3$; (*c*) CH_3NH_2; (*d*) CH_3OH; (*e*) FOOF.

19.9 Use the rules given in Sec. 19.1 to draw a Newman projection of the expected lowest-energy conformation of each of the following and give the approximate value of the listed dihedral angle. Where there are several hydrogens, pick the ones that give the smallest positive dihedral angle. (*a*) HCOOH, *D*(HOCH); (*b*) NH_2NH_2, *D*(HNNH); (*c*) CH_2CHOH (vinyl alcohol); *D*(HOCC). Have the CO bond perpendicular to the plane of the paper in the Newman projection.

19.10 Does O_3 have a dipole moment? Explain your answer.

19.11 Use average bond energies to estimate ΔH°_{298} for the following gas-phase reactions: (*a*) $C_2H_2 + 2H_2 \rightarrow C_2H_6$; (*b*) $N_2 + 3H_2 \rightarrow 2NH_3$. Compare with the true values found from data in the Appendix.

19.12 The dipole moments of CH_3F and CH_3I are 1.85 and 1.62 D, respectively. Use the H—C bond moment listed in Sec. 19.1 to estimate the C—F and C—I bond moments.

19.13 Use bond moments to estimate the dipole moments of (*a*) CH_3Cl; (*b*) CH_3CCl_3; (*c*) $CHCl_3$; (*d*) $Cl_2C{=}CH_2$. Assume tetrahedral angles at singly bonded carbons and 120° angles at doubly bonded carbons. Compare with the experimental values, which are (*a*) 1.87 D; (*b*) 1.78 D; (*c*) 1.01 D; (*d*) 1.34 D.

19.14 What would the bond moment of C≡N be if one assumed the H—C moment was 0.4 but had the polarity H^-—C^+?

19.15 Use Table 19.1 to compute the Pauling electronegativity differences for the following pairs of elements: (*a*) C, H; (*b*) C, O; (*c*) C, Cl. Compare with the electronegativity differences found from the Pauling values in Table 19.2. The discrepancies are due to use of average bond energies different from those listed in Table 19.1.

19.16 Calculate the Allen-scale electronegativities of (*a*) H; (*b*) Li; (*c*) Be; (*d*) Na.

19.17 Some values of $\alpha/(4\pi\varepsilon_0 \text{ Å}^3)$ are 0.667 for H, $24._3$ for Li, 5.6_0 for Be, 3.0_3 for B, 1.7_6 for C, 1.1_0 for N, 0.80_2 for O, and 0.55_7 for F. Calculate the Nagle-scale electronegativities of these elements.

19.18 (*a*) If A, B, and C are elements whose electronegativities satisfy $x_A > x_B > x_C$, show that if the Pauling electronegativity scale is valid, then $\Delta_{AC}^{1/2} = \Delta_{AB}^{1/2} + \Delta_{BC}^{1/2}$. (*b*) Test this relation for C, N, and O.

19.19 (*a*) Write a Lewis dot formula for H_2SO_4 that has eight electrons around S. (*b*) What formal charge does this dot formula give the S atom? (The *formal charge* is found by dividing the electrons of each bond equally between the two bonded atoms.) How reasonable is this formal charge? (*c*) Write the Lewis dot formula for SF_6. (*d*) Write a dot formula for H_2SO_4 that gives S a zero formal charge. (*e*) Explain why the observed sulfur–oxygen bond lengths in SO_4^{2-} in metal sulfates are 1.5 to 1.6 Å, whereas the sum of the single-bond radii of S and O is 1.70 Å.

19.20 Draw the Lewis dot structure for CO. What is the formal charge (Prob. 19.19*b*) on carbon?

19.21 Predict the sign of $\Delta H°$ of each of the following reactions without using any thermodynamic data or bond-energy data: (*a*) $H_2(g) + Cl_2(g) \rightarrow 2HCl(g)$; (*b*) $CH_4(g) + Cl_2(g) \rightarrow CH_3Cl(g) + HCl(g)$.

Section 19.2

19.22 True or false? (*a*) In the electronic Schrödinger equation, V_{NN} is a constant. (*b*) $\hat{V}_{ee}$ is absent from the electronic Schrödinger equation. (*c*) $\hat{K}_e$ is absent from the electronic Schrödinger equation. (*d*) $\hat{K}_N$ is absent from the electronic Schrödinger equation. (*e*) $\hat{K}_N$ is present in the Schrödinger equation for nuclear motion.

19.23 For the H_2 molecule, give the explicit forms for each of the operators $\hat{K}_e$, $\hat{K}_N$, $\hat{V}_{NN}$, $\hat{V}_{Ne}$, and $\hat{V}_{ee}$. Use capital letters for the nuclei and numbers for the electrons. Use r_{1A} to denote the distance between electron 1 and nucleus A.

19.24 (*a*) The KF molecule has $R_e = 2.17$ Å. The ionization potential of K is 4.34 V, and the electron affinity of F is 3.40 eV. Use the model of nonoverlapping spherical ions to estimate D_e for KF. (The experimental value is 5.18 eV.) (*b*) Estimate the dipole moment of KF. (The experimental value is 8.60 D.) (*c*) Explain why KCl has a larger dipole moment than KF.

19.25 For an ionic molecule like NaCl, the electronic energy $E_e(R)$ equals the Coulomb's law potential energy $-e^2/4\pi\varepsilon_0 R$ plus a term that allows for the Pauli repulsion due to the overlap of the ions' probability densities. This repulsion term can be very crudely estimated by the function B/R^{12}, where B is a positive constant. (See the discussion of the Lennard-Jones potential in Sec. 21.10.) Thus, $E_e \approx B/R^{12} - e^2/4\pi\varepsilon_0 R$ for an ionic molecule. (*a*) Use the fact that E_e is a minimum at $R = R_e$ to show that $B = R_e^{11}e^2/48\pi\varepsilon_0$. (*b*) Use the above expression for E_e and the Na and Cl ionization potential and electron affinity to estimate D_e for NaCl ($R_e = 2.36$ Å). (*c*) D_e for NaCl is 4.25 eV. Does the function B/R^{12} overestimate or underestimate the Pauli repulsion? What value of m gives agreement with the observed D_e if the Pauli repulsion is taken as A/R^m, where A and m are constants?

Section 19.3

19.26 True or false? (*a*) There is no interelectronic repulsion in the H_2^+ molecule. (*b*) The wave function $N(1s_A + 1s_B)$ is the exact electronic wave function for the ground electronic state of H_2^+. (*c*) The ground electronic state of H_2^+ is a bound state and the first excited electronic state of H_2^+ is an unbound state. (*d*) The plane perpendicular to the molecular axis at the midpoint between the nuclei in H_2^+ is a node for the first excited state.

19.27 In one dimension, an even function satisfies $f(-x) = f(x)$ and an odd function satisfies $f(-x) = -f(x)$. State whether each of the following functions is even, odd, or neither. (*a*) $3x^2 + 4$; (*b*) $2x^2 + 2x$; (*c*) x; (*d*) the $v = 0$ harmonic-oscillator wave function; (*e*) the $v = 1$ harmonic-oscillator wave function.

19.28 For the ground electronic state of H_2^+ with the nuclei at their equilibrium separation, use the approximate wave function in (19.15) to calculate the probability of finding the electron in a tiny box of volume 10^{-6} Å^3 if the box is located (*a*) at one of the nuclei; (*b*) at the midpoint of the internuclear axis; (*c*) on the internuclear axis and one-third of the way from nucleus A to nucleus B. Use Table 19.3 and the equation $S = e^{-R/a_0}(1 + R/a_0 + R^2/3a_0^2)$ (*Levine,* sec. 13.5).

Section 19.4

19.29 Write down the MO wave function for the (repulsive) ground electronic state of He_2.

19.30 Give the ground-state MO electronic configuration for (*a*) He_2^+; (*b*) Li_2; (*c*) Be_2; (*d*) C_2; (*e*) N_2; (*f*) F_2. Which of these species are paramagnetic?

19.31 Give the bond order of each molecule in Prob. 19.30.

19.32 Use the MO electron configurations to predict which of each of the following sets has the highest D_e; (*a*) N_2 or N_2^+; (*b*) O_2, O_2^+, or O_2^-.

19.33 Let ψ be an eigenfunction of $\hat{H}$; that is, let $\hat{H}\psi = E\psi$. Show that $(\hat{H} + c)\psi = (E + c)\psi$, where c is any constant. Hence, ψ is an eigenfunction of $\hat{H} + c$ with eigenvalue $E + c$.

19.34 For each of the species NCl, NCl^+, and NCl^-, use the MO method to (*a*) write the valence-electron configuration; (*b*) find the bond order; (*c*) decide whether the species is paramagnetic.

Section 19.5

19.35 (*a*) List the minimal-basis-set AOs for a calculation on CO. (*b*) If the internuclear axis is the z axis, which of the AOs in (*a*) can contribute to σ MOs? To π MOs?

Section 19.6

19.36 Sketch the two antibonding MOs (19.31) of BeH_2.

19.37 For the linear BeH_2 molecule with the z axis taken as the molecular axis, classify each of the following AOs as g or u and as σ, π, or δ: (*a*) $Be3d_{z^2}$; (*b*) $Be3d_{x^2-y^2}$; (*c*) $Be3d_{xy}$; (*d*) $Be3d_{xz}$.

19.38 Let the line between atom A and atom B of a polyatomic molecule be the z axis. For each of the following atom A atomic orbitals, state whether it will contribute to a σ, π, or δ localized MO in the molecule: (*a*) s; (*b*) p_x; (*c*) p_y; (*d*) p_z; (*e*) d_{z^2}; (*f*) $d_{x^2-y^2}$; (*g*) d_{xy}; (*h*) d_{xz}; (*i*) d_{yz}.

19.39 (*a*) For H_2CO, list all the AOs that go into a minimal-basis-set MO calculation. (*b*) Use these AOs to form localized MOs for H_2CO. For each localized MO, state which AOs make the main contributions to it and state whether it is inner-shell, lone-pair, or bonding. State whether each localized bonding MO is σ or π. Take the z axis along the CO bond and the x axis perpendicular to the molecule. Use the σ-π description of the double bond.

19.40 (*a*) Use average-bond-energy data and data on C(*g*) and H(*g*) in the Appendix to estimate $\Delta_f H^\circ_{298}$ of $C_6H_6(g)$ on the assumption that benzene contains three carbon–carbon single bonds and three carbon–carbon double bonds. Compare the result with the experimental value. (*b*) Repeat (*a*) for cyclohexene(*g*) (one double bond).

19.41 Let $p_1, \ldots, p_6$ be the $2p_z$ AOs of the carbons in benzene. The unnormalized forms of the occupied π MOs in benzene are $p_1 + p_2 + p_3 + p_4 + p_5 + p_6$, $p_2 + p_3 - p_5 - p_6$, and $2p_1 + p_2 - p_3 - 2p_4 - p_5 + p_6$. (*a*) Sketch these MOs. (*b*) Which of the three is lowest in energy?

19.42 Plot the value along the z axis of the $2s + 2p_z$ hybrid AO in (19.35) versus z/a. Take the nuclear charge as 1. Note that the outer portion of the 2*s* AO in (19.35) is assumed positive (as in Fig. 19.28). This convention is opposite that used in Table 18.1, so multiply the 2*s* AO in Table 18.1 by -1 before adding it to $2p_z$.

Section 19.8

19.43 True or false? (*a*) D_e is always greater than D_0. (*b*) The molecular electron probability density is experimentally observable. (*c*) The molecular wave function is experimentally observable.

19.44 For each of the following, state whether specification of all bond distances and all bond angles fully specifies the molecular geometry. (*a*) H_2O; (*b*) H_2O_2; (*c*) NH_3; (*d*) $ClCH_2CH_2Cl$; (*e*) $CH_2{=}CH_2$.

19.45 Find the particle probability density function $\rho(x)$ for a system of two identical noninteracting particles each of which has spin $s = 0$ in a stationary state of a one-dimensional box of length a if the quantum numbers for the two particles are unequal ($n_1 \neq n_2$). Also find $\int_{-\infty}^{\infty} \rho(x)\,dx$. (*Hint*: Make sure you use a symmetric wave function.)

Section 19.10

19.46 Which of these are functionals? (*a*) $\int_3^8 [g(x)]^2\,dx$; (*b*) $df(x)/dx$; (*c*) $df(x)/dx|_{x=0}$; (*d*) $\int_0^2 [f(x) + d^2f/dx^2]\,dx$.

19.47 Which of the following are numbers and which are functions of x, y, and z? (*a*) E_e; (*b*) ρ; (*c*) E_{xc}; (*d*) v_{xc}.

19.48 Show that multiplication of the electronic Schrödinger equation $\hat{H}_e\psi_e = E_e\psi_e$ by ψ_e^* followed by integration over all space gives Eq. (19.45) for E_e.

19.49 True or false? (*a*) A DFT calculation does not find the wave function of the molecule. (*b*) The KS orbitals are for the reference system of noninteracting electrons. (*c*) The KS reference system has the same ground-state energy as the molecule. (*d*) The KS reference system has the same ground-state probability density ρ as the molecule. (*e*) The exact form of E_{xc} is unknown.

19.50 Show that J in (19.48) is the classical expression for the electrostatic self-repulsion energy of a continuous distribution of electrical charge whose charge density (charge per unit volume) is $-e\rho$. Start by writing the repulsion between two infinitesimal elements of charge dQ_1 and dQ_2 of the continuous distribution.

19.51 For the functional

$$F[\rho] \equiv \int_e^t \int_c^d \int_a^b g(x, y, z, \rho, \rho_x, \rho_y, \rho_z)\,dx\,dy\,dz$$

where $\rho_x \equiv \partial\rho/\partial x$, etc., one can show that the functional derivative is

$$\frac{\delta F}{\delta \rho} = \frac{\partial g}{\partial \rho} - \frac{\partial}{\partial x}\frac{\partial g}{\partial \rho_x} - \frac{\partial}{\partial y}\frac{\partial g}{\partial \rho_y} - \frac{\partial}{\partial z}\frac{\partial g}{\partial \rho_z}$$

In the LDA approximation, the exchange part of E_{xc} is given by

$$E_x^{\text{LDA}} = -\frac{3}{4}\left(\frac{3}{\pi}\right)^{1/3} \int_{-\infty}^{\infty}\int_{-\infty}^{\infty}\int_{-\infty}^{\infty} [\rho(x, y, z)]^{4/3}\,dx\,dy\,dz$$

Find $v_x^{\text{LDA}} = \delta E_x^{\text{LDA}}/\delta\rho$.

Section 19.11

19.52 To illustrate the FE MO method, consider the ions

$$(CH_3)_2\overset{+}{N}{=}CH({-}CH{=}CH)_k{-}\ddot{N}(CH_3)_2 \qquad (19.53)$$

where k, the number of —CH=CH groups in the ion, can be 0, 1, 2, ... Each ion has an equivalent Lewis structure with the charge on the right-hand nitrogen and all carbon–carbon single and double bonds interchanged. All the carbon–carbon bond lengths are equal, and the π electrons, which form the second bond of each double bond, are reasonably free to move along the molecule. (*a*) Use the FE MO method to calculate the wavelength of the longest-wavelength electronic absorption band of the ion (19.53) with $k = 1$. Assume the carbon–carbon and conjugated carbon–nitrogen bond distances are 1.40 Å (as in benzene) and add in one extra bond length at each end of the ion to get the "box" length. Begin by deciding how many π electrons there are. (Note that the lone pair on nitrogen takes part in the π bonding.) Assign two π electrons to each π MO and use the quantum numbers of the highest-occupied and lowest-vacant π MOs. Compare your result with the experimental value 312 nm.

(*b*) For the ion (19.53), show that the FE MO method predicts the longest-wavelength absorption to be at $\lambda = (2k + 4)^2 \cdot (64.6 \text{ nm})/(2k + 5)$.

19.53 Although ab initio SCF calculations do not give accurate molecular dissociation energies, whereas some semiempirical methods do give pretty good estimates of gas-phase $\Delta_f H°$ values, one can combine the results of an ab initio SCF calculation with empirical parameters to obtain a gas-phase $\Delta_f H°_{298}$ value that is usually more accurate than those obtained from semiempirical calculations. The equation used is

$$\Delta_f H°_{298} = N_A\left(E_e - \sum_a n_a \alpha_a\right)$$

where the Avogadro constant N_A converts from a per molecule to a per mole basis, E_e is the ab initio SCF electronic energy of the molecule, the sum goes over the different kinds of atoms in the molecule, n_a is the number of a atoms in the molecule, and the α_a's are empirical parameters for the various kinds of atoms. The α_a values are found by fitting known $\Delta_f H°$ values and depend somewhat on the basis set used. For HF/6-31G* calculations, some α_a values in hartrees (1 hartree = 27.2114 eV = 4.35974×10^{-18} J) are

$$\alpha_H = -0.57077, \quad \alpha_C = -37.88449,$$

$$\alpha_N = -54.46414, \quad \alpha_O = -74.78852$$

Some HF/6-31G* energies in hartrees are -40.19517 for CH_4 and -150.76479 for H_2O_2. Use these results to estimate $\Delta_f H°_{298}$ for $CH_4(g)$ and $H_2O_2(g)$. The experimental values are -74.8 kJ/mol for $CH_4(g)$ and -136.3 kJ/mol for $H_2O_2(g)$.

Section 19.12

19.54 Draw the conformation of the molecule described by the following Z-matrix.

C1						
C2	1	1.54				
H3	2	1.09	1	120.0		
H4	1	1.09	2	109.5	3	180.0
H5	1	1.09	2	109.5	3	−60.0
H6	1	1.09	2	109.5	3	60.0
O7	2	1.23	1	120.0	4	0.0

19.55 (*a*) Devise a Z-matrix for each of the two planar conformers of formic acid, HC(=O)OH. (*b*) To avoid 180° angles (which are forbidden in a Z-matrix), one includes a dummy atom, symbolized by X, in the Z-matrix. *Gaussian* uses the dummy atom to define the locations of the nondummy atoms but ignores X in the quantum-mechanical calculations. Devise a Z-matrix for CO_2 that has a dummy atom placed 1.0 Å from C with the line XC making a 90° angle with the molecular axis. (A dihedral angle 4–2–1–3 with atoms 4 and 3 both bonded to atom 2 is allowed.)

19.56 (*a*) Using suitable software, do HF/6-31G* and HF/6-31G** geometry optimizations for methanol and compare with the HF/3-21G geometry. Also compare the dipole moments found with these three basis sets. (*b*) Do a B3LYP/6-31G* geometry optimization for CH_3OH.

19.57 (*a*) Using suitable software, do HF/6-31G* geometry optimizations for each of the two planar conformers of HCOOH and find the HF/6-31G* energy difference in kcal/mol (omitting zero-point energy). (*b*) Repeat (*a*) using B3LYP/6-31G* calculations.

Section 19.13

19.58 Using suitable software and as many different force fields as you have access to, do MM geometry optimizations for CH_3OH.

19.59 (*a*) Do MM geometry optimizations to find the predicted energy difference between the gauche and trans conformers of butane.

19.60 (*a*) Do MM geometry optimizations to find the predicted energy difference between the cis and trans isomers of 1,2-difluoroethylene.

19.61 For C_2Cl_6, how many terms are there in each of the following: (*a*) V_{str}; (*b*) V_{bend}; (*c*) V_{tors}; (*d*) V_{vdW}; (*e*) V_{es}?

General

19.62 Arrange the following molecular energies in order of increasing magnitude: (*a*) the typical energy of a covalent single bond; (*b*) the average molecular kinetic energy in a fluid at room temperature; (*c*) the rotational barrier in C_2H_6; (*d*) the typical energy of a double bond; (*e*) the ionization energy of H.

19.63 True or false? (*a*) The maximum electron probability density in the ground electronic state of H_2^+ occurs at each nucleus. (*b*) If sufficient basis functions are used, the Hartree–Fock wave function of a many-electron molecule will reach the true wave function. (*c*) In a homonuclear diatomic molecule, combining two 2*p* AOs always yields a π MO. (*d*) An ab initio calculation will give an accurate prediction of every molecular property. (*e*) In homonuclear diatomic molecules, all *u* MOs are antibonding. (*f*) The H_2^+ ground electronic state has spin quantum number $S = 0$. (*g*) The H_2 ground electronic state has spin quantum number $S = 0$. (*h*) Two electrons with the same spin have zero probability of being at the same point in space. (*i*) All triatomic molecules are planar.

19.64 Which scientist mentioned in Chapter 19 is described by each of the following? (*a*) He headed the U.S. atom-bomb project in World War II. He is one of the main characters in John Adams' opera *Doctor Atomic.* (*b*) He won the Nobel Prize for chemistry and the Nobel Prize for peace. In 1952, he was unable to attend a scientific meeting in Britain when the U.S. State Department refused to issue him a passport. In 1953, he published an erroneous model for DNA. Some people have speculated that had he gone to the 1952 meeting, he might have seen Rosalind Franklin's x-ray diffraction photos of DNA crystals, and, being an expert on crystallography, might have been able to beat out Watson and Crick in arriving at the correct DNA structure.

REVIEW PROBLEMS

R19.1 Give the values of the n, l, and m quantum numbers of a 5d electron.

R19.2 True or false? (*a*) If the state function of a system is an eigenfunction of the operator $\hat{B}$ with eigenvalue b, then a measurement of property B must give b as the result. (*b*) If we know what the state function of a system is, then we can predict what value will be obtained when we measure the property B. (*c*) The probability density for a stationary state is independent of time.

R19.3 Which of these MOs are eigenfunctions of the Fock operator $\hat{F}$? (*a*) The canonical MOs. (*b*) The energy-localized MOs.

R19.4 (*a*) Give an example of a linear operator. (*b*) Give an example of a nonlinear operator.

R19.5 (*a*) For a 5d electron, give the magnitude of the electronic orbital angular momentum. A numerical answer is required. (*b*) Give the magnitude of the spin angular momentum of an electron.

R19.6 Write down the forms of the two-electron spin functions and state whether each function is symmetric or antisymmetric. For each of these functions, state the values of each of the spin quantum numbers S and M_s.

R19.7 (*a*) Write down the Hamiltonian operator for the helium atom. (*b*) The perturbation-theory of He in Chapter 18 omitted one term from the Hamiltonian operator to obtain the zeroth-order wave functions and energies. Which term was omitted? (*c*) Write down the zeroth-order wave function for the ground state of He, including the spin part.

R19.8 Predict the approximate bond angles in each of the following: (*a*) CH_3OH; (*b*) XeF_4; (*c*) SF_4; (*d*) HCOOH.

R19.9 Name three molecular properties that are reasonably accurately predicted by Hartree–Fock calculations, and name one property that is very poorly predicted by such calculations.

R19.10 If $\hat{A} = \partial^2/\partial x^2$ and $\hat{B} = x^3 \times$, find $(\hat{A}\hat{B} - \hat{B}\hat{A})(x^2 y^2 z^2)$.

R19.11 Which of the following equations is always true? (*a*) $\hat{B}(f+g) = \hat{B}f + \hat{B}g$; (*b*) $(\hat{B}+\hat{C})f = \hat{B}f + \hat{C}f$; (*c*) $\hat{B}\hat{C}f = \hat{C}\hat{B}f$.

R19.12 The lowest term of the ground-state electron configuration of the carbon atom is ^3P. What is the magnitude of the total orbital electronic angular momentum for the states of this term? What is the magnitude of the total spin angular momentum for the states of this term?

R19.13 The classical-mechanical expression for the z component of the orbital angular momentum of a particle is $L_z = xp_y - yp_x$. What is the expression for the $\hat{L}_z$ operator?

R19.14 State the variation theorem. Define all quantities in any expression you write.

R19.15 For a particle in a one-dimensional box of length l and with the origin at the left end of the box, apply the variation function $\phi = x^2(l - x)^2$ inside the box and $\phi = 0$ outside the box and find the percent error in the ground-state energy.

R19.16 Use the CCBDB at srdata.nist.gov/cccbdb to find the following quantities as calculated by each of the methods HF/6-31G*, HF/6-31G**, MP2/6-31G*, MP2/cc-pVTZ, CCSD(T)/cc-pVTZ, PM3, B3LYP/6-31G*. (The method precedes the slash and the basis set follows the slash.) In some cases, results for one or more methods might not be available in the CCCBDB. Give the MP2FC results; the FC stands for the frozen-core approximation, and means that terms involving excitation of inner-shell electrons are omitted from the calculation. Compare the results with the experimental values given in the CCCBDB. (*a*) The dipole moment of CH_3OH; (*b*) the bond angle in NH_3; (*c*) the bond lengths in CH_3OH; (*d*) the barrier to internal rotation in C_2H_6.

R19.17 Sketch (*a*) the potential energy function for the particle in a one-dimensional box; (*b*) the potential energy function for the harmonic oscillator; (*c*) the electronic energy including nuclear repulsion as a function of internuclear distance for the ground electronic state of a stable diatomic molecule, showing R_e and D_e.

R19.18 Define (*a*) degeneracy; (*b*) linear operator.

R19.19 Classify each of these methods as ab initio, density functional, semiempirical, or molecular mechanics: (*a*) MMFF94; (*b*) CCSD(T); (*c*) MP2; (*d*) PM6; (*e*) B3LYP.

Chapter 19

19.1 The table of bond radii in Sec. 19.1 gives the following estimates.

(a) 0.30 Å + 0.77 Å = 1.07 Å for CH, 0.77 Å + 0.66 Å = 1.43 Å for CO, and 0.66 Å + 0.30 Å = 0.96 Å for OH.

(b) 0.30 Å + 0.77 Å = 1.07 Å for HC and 0.60 Å + 0.55 Å = 1.15 Å for CN.

19.2 Each BF bond has some double-bond character, as shown by the Lewis structure 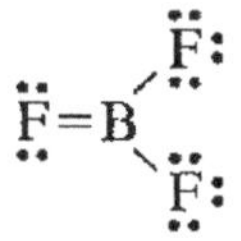 and two others.

19.3 **(a)** The $TeBr_2$ Lewis dot formula has four electron pairs around Te and two lone pairs on Te. The geometry is bent with bond angle somewhat less than 109½°.

(b) Hg has electron configuration $\cdots 5d^{10}6s^2$ and has 2 valence electrons. So Hg has two valence pairs in the Lewis structure, and $HgCl_2$ is linear.

(c) With 3 pairs on Sn, $SnCl_2$ is bent with angle a bit less than 120°.

(d) With 5 pairs on Xe, XeF_2 is linear.

(e) The dot formula has four pairs on Cl, so the ion is bent with angle somewhat less than 109½°.

19.4 **(a)** With 5 pairs on Br, BrF_3 is T-shaped (Fig. 19.2b).

(b) Three pairs on Ga. Trigonal planar.

(c) 4 pairs on O. Trigonal pyramidal with angles a bit less than 109½°.

(d) Four pairs on P. Trigonal pyramidal with angles a bit less than 109½°.

19.5 **(a)** 4 pairs on Sn. Tetrahedral.

(b) 5 pairs on Se. Seesaw shape.

(c) 6 pairs on Xe. Square planar.

(d) 4 pairs on B. Tetrahedral.

(e) 6 pairs on Br. Square planar.

19.6 **(a)** 5 pairs on As. Trigonal bipyramidal.

(b) 6 pairs on Br. Square-based pyramid.

(c) 6 pairs on Sn. Octahedral.

19.7 Each multiple bond is counted as one pair.

(a) "3" pairs on O. Bent, with angle close to 120°.

(b) "3" pairs on N. Trigonal planar.

(c) "3" pairs on S. Trigonal planar.

(d) "3" pairs on S. Bent. Angle close to 120°.

(e) "4" pairs on S. Approximately tetrahedral.

(f) "4" pairs on S. Pyramidal with angles close to 109½°.

(g) "4" pairs on I. Pyramidal with angles close to 109½°.

(h) "5" pairs on S. Trigonal bipyramidal.

(i) "4" pairs on Xe. Trigonal pyramidal with angles close to 109½°.

(j) "6" pairs on Xe. Square-based pyramid.

19.8 **(a)** There are four electron pairs around the methyl carbon, so the HCH and HCC bond angles will be close to 109½°. (Because the four groups attached to the methyl carbon are not all the same, we cannot expect the angles at this C to be exactly 109½°.) There are "2" pairs around the CN carbon, so the CCN angle will be 180°.

(b) There are "3" pairs around the CH_2 carbon and the bond angles at this carbon will be close to 120°. Because of the greater repulsions exerted by the double-bond's electrons, the HCH angle will be a bit less than 120° and the HCC angles a bit more than 120°. The bond angles at the CH carbon will be close to 120°. The angles at the CH_3 carbon will be close to 109½°.

(c) The HCH and HCN angles at the methyl carbon are close to 109½°; There are 4 pairs around the N and the HNC angles are a bit less than 109½°.

(d) Near 109½° for the HCH and HCO angles. A bit less than 109½° for the HOC angle.

(e) A bit less than 109½° for each FOO angle.

19.9 **(a)** By rule 2, the OH hydrogen will eclipse the CO double bond, which makes *D*(HOCH) equal to 180°. The molecule is planar. A Newman projection with the O behind the C is

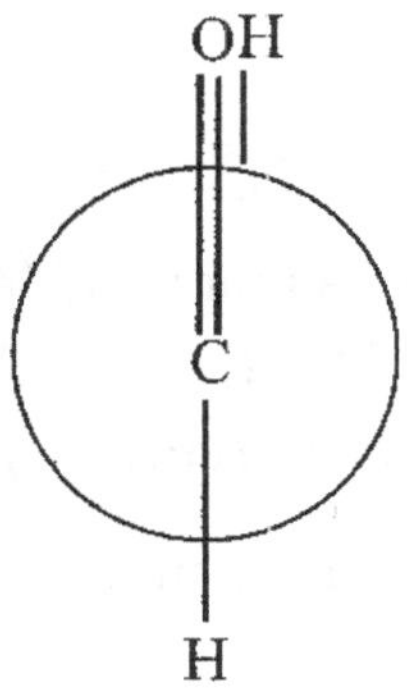

(b) By rule 1, the bonds are staggered. The two possible conformations are

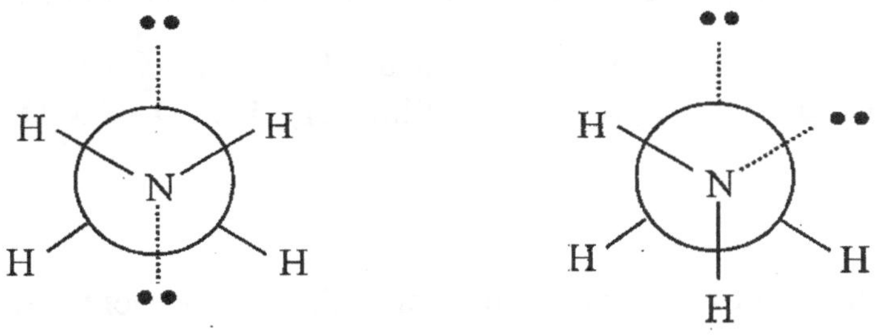

The rules don't allow us to decide whether I or II is more stable. The smallest *D*(HNNH) angle is 60° in both I and II. Experiment shows that in fact II is more stable than I.

(c) Rule 2 says the OH hydrogen eclipses the double bond in this planar molecule. *D*(HOCC) = 0. A Newman projection with O behind C is

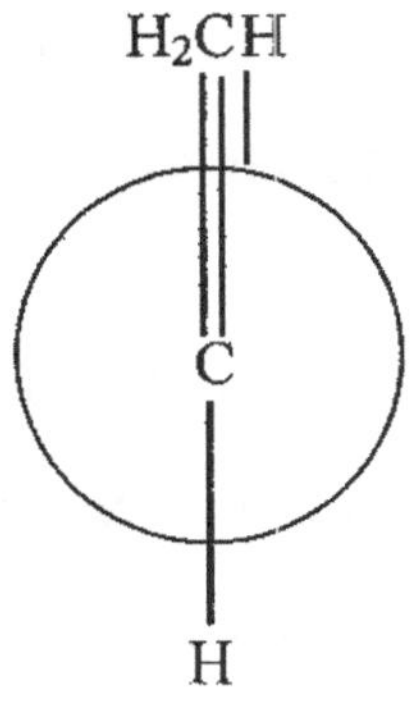

19.10 The Lewis dot structure is $:\ddot{\underset{\cdot\cdot}{O}}-\ddot{O}=\ddot{O}: \longleftrightarrow :\ddot{O}=\ddot{O}-\ddot{\underset{\cdot\cdot}{O}}:$. The bond angle is close to 120°. The lone pair on the central atom makes the dipole moment nonzero.

19.11 **(a)** $\Delta_{at}H^{\circ}_{298} \approx \Delta_{at}H^{\circ}_{298,re} - \Delta_{at}H^{\circ}_{298,pr} =$
[2(415) + 812 + 2(436)] kJ/mol – [344 + 6(415)] kJ/mol = –320 kJ/mol.
The true value is [–84.68 – 226.73 – 2(0)] kJ/mol = –311.4 kJ/mol.

(b) $\Delta H^{\circ}_{298} \approx$ [946 + 3(436) – 2(3)391] kJ/mol = –92 kJ/mol.
In truth, ΔH°_{298} = 2(–46.1) kJ/mol = –92.2 kJ/mol.

19.12 We assume tetrahedral angles. As noted in Sec. 19.1, the vector sum of three CH moments in a CH_3 group equals the moment of one CH bond. The H_3CF dipole moment is thus the sum of the moments $\overset{+}{H}-\overset{-}{C}$ and $\overset{+}{C}-\overset{-}{F}$. We have 1.85 D = 0.4 D + μ_{CF} and μ_{CF} = 1.4$_5$ D. Similarly, 1.62 D = 0.4 D + μ_{CI} and μ_{CI} = 1.2 D.

19.13 **(a)** The net moment of the CH_3 group equals the CH moment and the dipole moment is the sum of the moments $\overset{+}{H}-\overset{-}{C}$ and $\overset{+}{C}-\overset{-}{Cl}$.
So $\mu \approx$ 0.4 D + 1.5 D = 1.9 D.

(b) $\mu \approx$ 0.4 D + 1.5 D = 1.9 D.

(c) $\mu \approx$ 0.4 D + 1.5 D = 1.9 D. (Here, agreement with experiment is poor.)

(d) 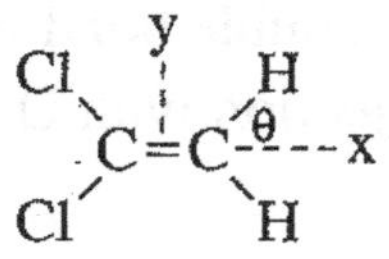

$\mu_x \approx 2\mu_{ClC} \cos\theta + 2\mu_{CH} \cos\theta =$

2(1.5 D) cos 60° + 2(0.4 D) cos 60° = 1.9 D. μ_y = 0. Hence $\mu \approx$ 1.9 D.

19.14 The moments listed in Sec. 19.1 give the H_3CCN moment as the sum of the $\overset{+}{H}-\overset{-}{C}$ and $\overset{+}{C}\equiv\overset{-}{N}$ bond moments, namely, as μ = 0.4 D + 3.5 D = 3.9 D. If we now assume the polarity $\overset{-}{H}-\overset{+}{C}$, then the H_3C moment is oppositely directed from the CN moment, and we would have 3.9 D = –0.4 D + $\mu_{C\equiv N}$ and $\mu_{C\equiv N}$ = 4.3 D (instead of 3.5 D).

19.15 **(a)** $\Delta_{CH}/(\text{kJ/mol}) = 415 - \frac{1}{2}(344 + 436) = 25$. $|x_C - x_H| = 0.102(25)^{1/2} = 0.5$ (compared with 0.3 in Table 19.2).

(b) $\Delta_{CO}/(\text{kJ/mol}) = 350 - \frac{1}{2}(344 + 143) = 106\frac{1}{2}$. $|x_C - x_O| = 0.102(106)^{1/2} = 1.0_5$ (compared with 0.9 in Table 19.2).

(c) $\Delta_{CCl}/(\text{kJ/mol}) = 328 - \frac{1}{2}(344 + 243) = 34\frac{1}{2}$. $|x_C - x_{Cl}| = 0.102(34\frac{1}{2})^{1/2} = 0.6$ (compared with 0.7 in Table 19.2).

19.16 **(a)** For H, there is only one valence electron and $\langle E_{i,\text{val}} \rangle = 13.6$ eV (Sec. 18.3), so $x_H = 0.169(13.6) = 2.30$.

(b) Li has one valence electron whose ionization energy is given by the table in Sec. 19.8 as 5.4 eV, so $x_{Li} = 0.169(5.4) = 0.91$.

(c) Be has electron configuration $1s^2 2s^2$ and each $2s$ electron has ionization energy 9.3 eV (Sec. 18.8), so $x_{Be} = 0.169(9.3) = 1.5_7$.

(d) $x_{Na} = 0.169(5.1) = 0.86$.

19.17 $x_H = 1.66(1/0.667)^{1/3} + 0.37 = 2.27$. $x_{Li} = 1.66(1/24.3)^{1/3} + 0.37 = 0.94$. $x_{Be} = 1.66(2/5.60)^{1/3} + 0.37 = 1.55$. $x_B = 1.66(3/3.0_3)^{1/3} + 0.37 = 2.0_2$. $x_C = 1.66(4/1.7_6)^{1/3} + 0.37 = 2.5_5$. N: 3.1_2. O: 3.6_2. F: 4.2_3.

19.18 **(a)** $|x_A - x_B| + |x_B - x_C| = x_A - x_B + x_B - x_C = x_A - x_C = |x_A - x_C|$. Substitution of Eq. (19.3) into this equation gives $0.102[\Delta_{AB}/(\text{kJ/mol})]^{1/2} + 0.102[\Delta_{BC}/(\text{kJ/mol})]^{1/2} = 0.102[\Delta_{AC}/(\text{kJ/mol})]^{1/2}$, so $\Delta_{AB}^{1/2} + \Delta_{BC}^{1/2} = \Delta_{AC}^{1/2}$.

(b) Table 19.1 gives $\Delta_{ON}/(\text{kJ/mol}) = 175 - \frac{1}{2}(143 + 159) = 24$; $\Delta_{NC}/(\text{kJ/mol}) = 292 - \frac{1}{2}(344 + 159) = 40.5$; $\Delta_{OC}/(\text{kJ/mol}) = 350 - \frac{1}{2}(143 + 344) = 106.5$. $\Delta_{ON}^{1/2} + \Delta_{NC}^{1/2} = 11._3\ (\text{kJ/mol})^{1/2}$; $\Delta_{OC}^{1/2} = 10._3\ (\text{kJ/mol})^{1/2}$. The relation in (a) is obeyed fairly well.

19.19 **(a)** 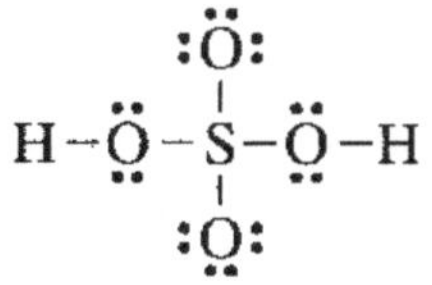

(b) In the dot formula of (a), the S has $\frac{1}{2}(8) = 4$ valence electrons, as compared with 6 valence electrons in a free S atom. The formal charge on S is +2 for this dot formula. This is an unlikely value for a nonmetallic element.

(c)

$$:\ddot{F}: \ :\ddot{F}:$$
$$:\ddot{F}-S-\ddot{F}:$$
$$:\ddot{F}: \ :\ddot{F}:$$

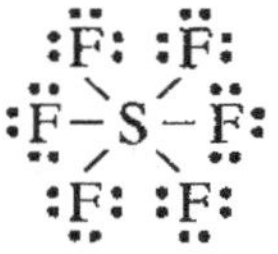

(d) The SF_6 dot formula shows that S can share as many as 12 valence electrons. (This is due to the presence of $3d$ orbitals on S.) A dot

formula for H_2SO_4 that gives S a zero formal charge is $H-\ddot{O}-\underset{\underset{:O:}{\parallel}}{\overset{\overset{:O:}{\parallel}}{S}}-\ddot{O}-H$.

Here S has $\frac{1}{2}(12) = 6$ valence electrons, as in a free S atom.

(e) A dot formula for SO_4^{2-} that gives S a zero formal charge is

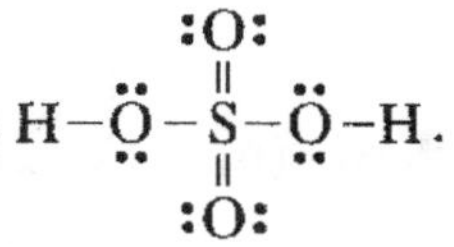

$$\left[:\ddot{O}-\underset{\underset{:O:}{\parallel}}{\overset{\overset{:O:}{\parallel}}{S}}-\ddot{O}: \right]^{2-}$$

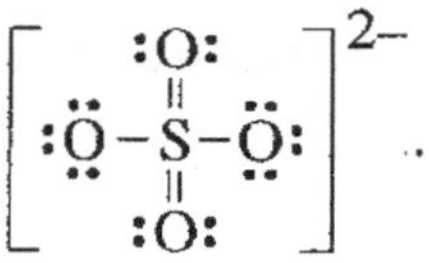

. In addition, there are other resonance structures in which

the double bonds and single bonds are permuted. Each sulfur–oxygen bond is intermediate between a single bond and a double bond.

19.20 :C≡O: The carbon has $2 + \frac{1}{2}(6) = 5$ valence electrons, as compared with 4 in the free C atom. The formal charge on C is −1. (This formal charge opposes the greater electronegativity of O and produces a dipole moment with the polarity $\overset{-}{C}\overset{+}{O}$.)

19.21 (a) Because of the electronegativity difference between H and Cl, we expect the H – Cl bond energy to be larger than the average of the H – H and Cl – Cl bond energies, so $\Delta H°$ is negative.

(b) $\Delta H° < 0$, for reasons similar to those in (a).

19.22 (a) T **(b)** F; **(c)** F; **(d)** T; **(e)** T.

19.23 $\hat{K}_e = -(\hbar^2/2m_e)\nabla_1^2 - (\hbar^2/2m_e)\nabla_2^2$; $\hat{K}_N = -(\hbar^2/2m_A)\nabla_A^2 - (\hbar^2/2m_B)\nabla_B^2$; $\hat{V}_{NN} = Z_A Z_B e^2/4\pi\varepsilon_0 R_{AB}$; $\hat{V}_{Ne} = -Z_A e^2/4\pi\varepsilon_0 r_{1A} - Z_B e^2/4\pi\varepsilon_0 r_{1B} - Z_A e^2/4\pi\varepsilon_0 r_{2A} - Z_B e^2/4\pi\varepsilon_0 r_{2B}$; $\hat{V}_{ee} = e^2/4\pi\varepsilon_0 r_{12}$.

19.24 (a) $KF \xrightarrow{1} K^+ + F^- \xrightarrow{2} K + F$. According to the model, the energy needed to dissociate KF to $K^+ + F^-$ is $\Delta E_1 \approx e^2/4\pi\varepsilon_0 R_e = (1.602 \times 10^{-19}\ C)^2/4\pi(8.854 \times 10^{-12}\ C^2/N\text{-}m^2)(2.17 \times 10^{-10}\ m) = 1.06_3 \times 10^{-18}\ J = 6.63\ eV$. The energy change for step 2 is $\Delta E_2 = -4.34\ eV + 3.40\ eV = -0.94\ eV$. The net ΔE is 6.63 eV − 0.94 eV = 5.69 eV.

(b) According to the model, $\mu \approx eR_e = (1.602 \times 10^{-19}\ C)(2.17 \times 10^{-10}\ m) = 3.48 \times 10^{-29}\ C\ m = 10.4\ D$, where (19.2) was used.

(c) Both compounds are essentially completely ionic with charges of +1 and −1 on the cation and anion. The larger size of Cl^- as compared with F^- makes R_e greater in KCl and gives KCl the greater dipole moment (which is approximately eR_e).

19.25 (a) At R_e, $\partial E_e/\partial R = 0 = -12B/R^{13} + e^2/4\pi\varepsilon_0 R^2$ and $B = e^2 R_e^{11}/12(4\pi\varepsilon_0)$.

(b) At equilibrium, the electronic energy is $E_{e,eq} = B/R_e^{12} - e^2/4\pi\varepsilon_0 R_e = e^2 R_e^{11}/12(4\pi\varepsilon_0)R_e^{12} - e^2/4\pi\varepsilon_0 R_e = -11e^2/12(4\pi\varepsilon_0)R_e = -11(1.602 \times 10^{-19}\ C)^2/12(4\pi)(8.854 \times 10^{-12}\ C^2/N\text{-}m^2)(2.36 \times 10^{-10}\ m) = -8.96 \times 10^{-19}\ J = -5.59\ eV$. According to the model, it requires 5.59 eV to dissociate the NaCl molecule to $Na^+ + Cl^-$. ΔE for $Na^+ + Cl^- \to Na + Cl$ is −5.14 eV + 3.61 eV = −1.53 eV (where data was taken from Example 19.1 in the text). The model gives D_e of NaCl as 5.59 eV − 1.53 eV = 4.06 eV.

(c) The Pauli repulsion decreases D_e. Since 4.06 eV is less than the true D_e, the function B/R^{12} overestimates the Pauli repulsion. For $E_e = A/R^m - e^2/4\pi\varepsilon_0 R$, we have at R_e, $\partial E_e/\partial R = 0 = -mA/R^{m+1} + e^2/4\pi\varepsilon_0 R^2$ and $A = e^2 R_e^{m-1}/4\pi\varepsilon_0 m$. So $E_{e,eq} = A/R_e^m - e^2/4\pi\varepsilon_0 R_e = e^2 R_e^{m-1}/4\pi\varepsilon_0 m R_e^m - e^2/4\pi\varepsilon_0 R_e = -(1 - 1/m)e^2/4\pi\varepsilon_0 R_e$. Then $4.25\ eV = (1 - 1/m)e^2/4\pi\varepsilon_0 R_e - 5.14\ eV + 3.61\ eV$. We have $e^2/4\pi\varepsilon_0 R_e = (1.602 \times 10^{-19}\ C)^2/4\pi(8.854 \times 10^{-12}\ C^2/N\text{-}m^2)(2.36 \times 10^{-10}\ m) = 9.77 \times 10^{-19}\ J = 6.10\ eV$, so $1 - 1/m = 0.948$ and $m = 19$.

19.26 **(a)** T; **(b)** F; **(c)** T; **(d)** T.

19.27 **(a)** Even; **(b)** neither; **(c)** odd; **(d)** even; **(e)** odd.

19.28 The box size is small enough to be considered "infinitesimal." The probability is $|\psi|^2\,dV \approx (2+2S)^{-1}(1s_A + 1s_B)^2(10^{-6}\text{ Å}^3)$, where $S = e^{-R/a_0}(1 + R/a_0 + R^2/3a_0^2)$. At the equilibrium separation of $R = 1.06$ Å $= 2.00a_0$, we find $S = e^{-2}(1 + 2 + 4/3) = 0.586$ and $(2 + 2S)^{-1} = 0.315$.

(a) At nucleus A, $r_A = 0$ and $r_B = R = 2.00a_0$, so $1s_A = \pi^{-1/2}(1/a_0)^{3/2}e^0 = 1.466\text{ Å}^{-3/2}$ and $1s_B = \pi^{-1/2}(1/a_0)^{3/2}e^{-2.00} = 0.198\text{ Å}^{-3/2}$. Then $|\psi|^2\,dV \approx 0.315(1.466 + 0.198)^2\text{ Å}^{-3}(10^{-6}\text{ Å}^3) = 8.7 \times 10^{-7}$.

(b) At the midpoint of the internuclear axis, $r_A = r_B = R/2 = 1.00a_0$, $1s_A = 1s_B = \pi^{-1/2}(1/a_0)^{3/2}e^{-1.00} = 0.539\text{ Å}^{-3/2}$ and $|\psi|^2\,dV \approx 0.315(0.539 + 0.539)^2(10^{-6}) = 3.7 \times 10^{-7}$.

(c) $r_A = R/3 = 2a_0/3$, $r_B = 2R/3 = 4a_0/3$, $1s_A = 0.753\text{ Å}^{-3/2}$, $1s_B = 0.386\text{ Å}^{-3/2}$, $|\psi|^2\,dV \approx 4.1 \times 10^{-7}$.

19.29 The MO electron configuration is $(\sigma_g 1s)^2(\sigma_u^* 1s)^2$. To achieve antisymmetry, we use a Slater determinant. Analogous to Eqs. (18.54) and (19.20) and the Be wave function in Prob. 18.50, we write

$$N\begin{vmatrix} \sigma_g 1s(1)\alpha(1) & \sigma_g 1s(1)\beta(1) & \sigma_u^* 1s(1)\alpha(1) & \sigma_u^* 1s(1)\beta(1) \\ \sigma_g 1s(2)\alpha(2) & \sigma_g 1s(2)\beta(2) & \sigma_u^* 1s(2)\alpha(2) & \sigma_u^* 1s(2)\beta(2) \\ \sigma_g 1s(3)\alpha(3) & \sigma_g 1s(3)\beta(3) & \sigma_u^* 1s(3)\alpha(3) & \sigma_u^* 1s(3)\beta(3) \\ \sigma_g 1s(4)\alpha(4) & \sigma_g 1s(4)\beta(4) & \sigma_u^* 1s(4)\alpha(4) & \sigma_u^* 1s(4)\beta(4) \end{vmatrix}$$

where N is a normalization constant.

19.30 We use the homonuclear diatomic MOs in Fig. 19.15.

(a) $(\sigma_g 1s)^2(\sigma_u^* 1s)$.

(b) $(\sigma_g 1s)^2(\sigma_u^* 1s)^2(\sigma_g 2s)^2$.

(c) $(\sigma_g 1s)^2(\sigma_u^* 1s)^2(\sigma_g 2s)^2(\sigma_u^* 2s)^2$.

(d) $(\sigma_g 1s)^2(\sigma_u^* 1s)^2(\sigma_g 2s)^2(\sigma_u^* 2s)^2(\pi_u 2p)^4$.

(e) $(\sigma_g 1s)^2(\sigma_u^* 1s)^2(\sigma_g 2s)^2(\sigma_u^* 2s)^2(\pi_u 2p)^4(\sigma_g 2p)^2$. He_2^+ has an unpaired electron and so is paramagnetic. All the others have filled shells and are not paramagnetic.

(f) (e) with $(\pi_g^* 2p)^4$ added.

19.31 **(a)** $(2-1)/2 = ½$. **(b)** $(4-2)/2 = 1$. **(c)** $(4-4)/2 = 0$. **(d)** $(8-4)/2 = 2$.
(e) $(10-4)/2 = 3$ (in agreement with the dot formula :N≡N:). **(f)** 1.

19.32 **(a)** The N_2 MO configuration is given in Prob. 19.30e. The highest occupied N_2 MO is a bonding MO, so N_2^+ has one less bonding electron than N_2. Therefore N_2 has the higher D_e.

(b) The O_2 MO configuration is shown in Fig. 19.18. The highest-occupied shell, $\pi_g^* 2p$, is half-filled and is antibonding. Therefore O_2^+ has one fewer antibonding electron than O_2 and has two fewer antibonding electrons than O_2^-, so O_2^+ has the highest D_e.

19.33 $(\hat{H} + c)\psi = \hat{H}\psi + c\psi = E\psi + c\psi = (E + c)\psi$, where the definition of the sum of operators was used.

19.34 **(a)** We feed the valence electrons into the MOs of Fig. 19.22, where $n = 2$ and $n' = 2$. NCl has $5 + 7 = 12$ valence electrons and has the valence-electron configuration $(\sigma s)^2(\sigma^* s)^2(\pi p)^4(\sigma p)^2(\pi^* p)^2$. NCl^+ and NCl^- have 11 and 13 electrons, respectively, and have the configurations $(\sigma s)^2(\sigma^* s)^2(\pi p)^4(\sigma p)^2(\pi^* p)$ and $(\sigma s)^2(\sigma^* s)^2(\pi p)^4(\sigma p)^2(\pi^* p)^3$, respectively.

(b) $(8-4)/2 = 2$ for NCl (which is similar to O_2); $(8-3)/2 = 2.5$ for NCl^+; $(8-5)/2 = 1.5$ for NCl^-.

(c) Each of these species has a partly filled π^* shell, so each has $S \neq 0$ and each is paramagnetic.

19.35 **(a)** C1s, C2s, C2p_x, C2p_y, C2p_z, O1s, O2s, O2p_x, O2p_y, O2p_z.

(b) C1s, C2s, C2p_z, O1s, O2s, O2p_z contribute to σ MOs. C2p_x, C2p_y, O2p_x, O2p_y contribute to π MOs.

19.36

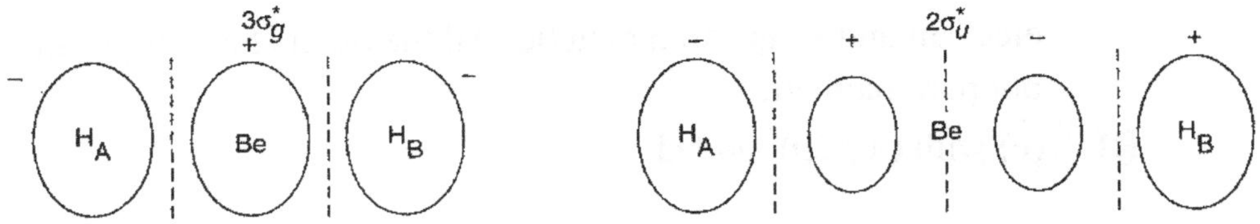

19.37 **(a)** g, σ. **(b)** g, δ. **(c)** g, δ. **(d)** g, π.

19.38 **(a)** σ, since it has no nodal planes that contain the internuclear (z) axis. **(b)** π, since it has one nodal plane containing the internuclear axis. **(c)** π. **(d)** σ. **(e)** σ (see Fig. 18.6.) **(f)** δ, since it has two nodal planes containing the internuclear axis. **(g)** δ. **(h)** π. **(i)** π.

19.39 **(a)** H_A1s, H_B1s, C1s, C2s, $C2p_x$, $C2p_y$, $C2p_z$, O1s, O2s, $O2p_x$, $O2p_y$, $O2p_z$.

(b) The dot formula 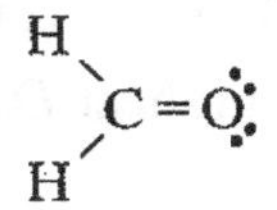 suggests the following description of the occupied localized MOs. We use sp^2 hybrid AOs on C to form the CH bonds and the σ bond of the double bond. These sp^2 hybrids are formed from C2s and the in-plane p orbitals $C2p_y$ and $C2p_z$. The bonding σ MO between H_A and C is a linear combination of H_A1s, C2s, $C2p_y$, and $C2p_z$. The bonding σ MO between H_B and C is a linear combination of H_B1s, C2s, $C2p_y$, and $C2p_z$. As we did with carbon, we form in-plane sp^2 hybrids on oxygen, using O2s, $O2p_y$, and $O2p_z$; these hybrids go to form the σ bond of the double bond and the lone-pair AOs on oxygen. Overlap between $C2p_x$ and $O2p_x$ forms the π bond of the CO double bond. The σ bond of the CO double bond is formed by overlap of those sp^2 hybrids on C and O that point along the z (CO) axis; the $C2p_y$ and $O2p_y$ AOs do not contribute to these sp^2 hybrids—the $C2p_y$ and $O2p_y$ AOs each have one nodal plane containing the z axis and cannot contribute to the CO σ MO. Therefore the CO localized σ bonding MO is a linear combination of C2s, $C2p_z$, O2s, and $O2p_z$. The lone-pair localized MOs on O are formed from the O2s, $O2p_y$, and $O2p_z$ AOs. There is an inner-shell localized MO on C that is essentially identical to the C1s AO and an inner-shell localized MO on O that is identical to O1s.

19.40 **(a)** $3H_2(g) + 6C(\text{graphite}) \xrightarrow{a} 6H(g) + 6C(g) \xrightarrow{b} C_6H_6(g)$. Appendix data give $\Delta H^\circ_{a,298} = [6(217.96) + 6(716.68) - 3(0) - 6(0)]$ kJ/mol = 5607.8 kJ/mol. Viewing C_6H_6 as containing three CC single bonds and three CC double bonds, we use the bond energies in Table 19.1 to get $\Delta H^\circ_{b,298} \approx -[6(415) + 3(344) + 3(615)]$ kJ/mol = –5367 kJ/mol. Then $\Delta_f H^\circ_{298} = \Delta H^\circ_{a,298} + \Delta H^\circ_{b,298} \approx 241$ kJ/mol. The Appendix gives the experimental value as 83 kJ/mol, so benzene is far more stable than it would be if it were composed of isolated single and double bonds.

(b) $5H_2(g) + 6C(\text{graphite}) \xrightarrow{a} 10H(g) + 6C(g) \xrightarrow{b} C_6H_{10}(g)$. Appendix data give $\Delta H^\circ_{a,298} = [10(217.96) + 6(716.68) - 5(0) - 6(0)]$ kJ/mol = 6479.7 kJ/mol. The bond-energy table gives $\Delta H^\circ_{b,298} \approx -[10(415) + 5(344) + 615]$ kJ/mol = –6485 kJ/mol. Then $\Delta_f H^\circ_{298} = \Delta H^\circ_{a,298} + \Delta H^\circ_{b,298} \approx -5._3$ kJ/mol. The Appendix gives the experimental value as –5.4 kJ/mol.

19.41 **(a)** The following sketches show a view from above the molecular plane. The dashed lines denote nodal planes perpendicular to the molecular plane. The molecular plane is a nodal plane for each MO. The lobes below the molecular plane have signs opposite those of corresponding lobes above the plane.

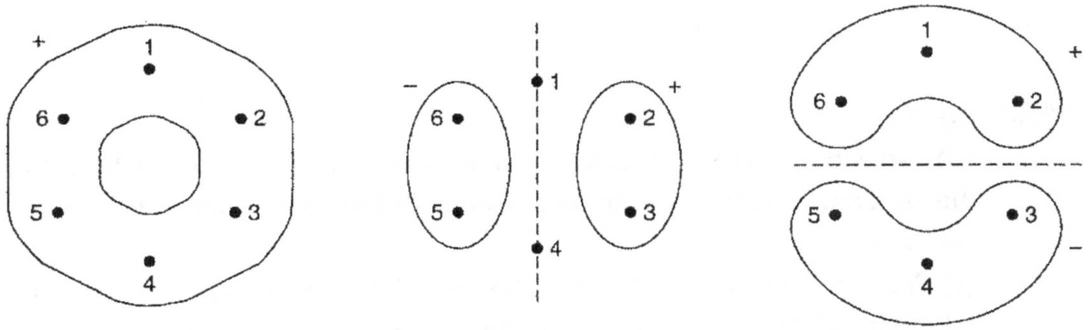

(b) The MO $p_1 + p_2 + p_3 + p_4 + p_5 + p_6$ has the fewest nodes and builds up the most electron probability density between the nuclei; this MO is lowest in energy.

19.42 Let $f \equiv (2s + 2p_z)/\sqrt{2}$. Along the z axis we have $x = 0$, $y = 0$, $r = (x^2 + y^2 + z^2)^{1/2} = (z^2)^{1/2} = |z|$ and $f = 2^{-1/2}(1/4)(2\pi)^{-1/2} \times [a^{-3/2}(|z|/a - 2)e^{-|z|/2a} + a^{-5/2}ze^{-|z|/2a}] = (1/8\pi^{1/2}a^{3/2})e^{-|w|/2}(w + |w| - 2)$, where $w \equiv z/a$. We have $1/8\pi^{1/2}a^{3/2} = 0.183$ Å$^{-3/2}$ and we find

f	−0.050	−0.082	−0.135	−0.222	−0.366	−0.143	0
z/a	−4	−3	−2	−1	0	0.5	1

f	0.086	0.135	0.163	0.149	0.120	0.091	0.066
z/a	1.5	2	3	4	5	6	7

The graph shows a sharp negative peak at $z/a = 0$ and a broad positive region for $z/a > 1$.

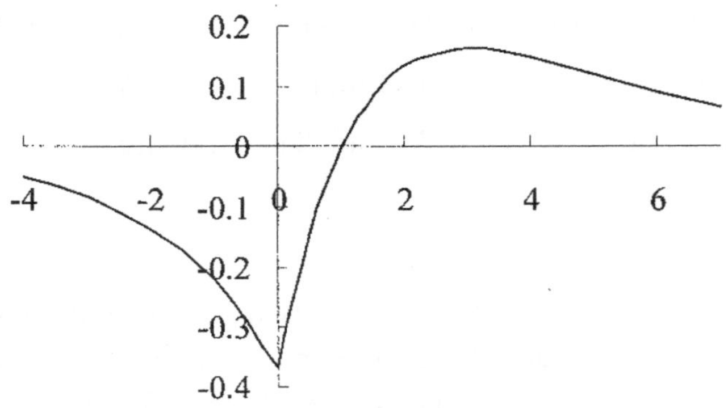

19.43 **(a)** T; **(b)** T; **(c)** F.

19.44 **(a)** Yes.

(b) No, since the HOOH dihedral angle must also be specified. Changing this dihedral angle without changing any bond distances or angles changes the structure.

(c) Yes. Imagine adding the H atoms one a time. When H_1 is added to N, the NH_1 bond distance specifies the structure. When H_2 is now added, the structure is fully specified by the two bond distances and the HNH angle (as is true for H_2O). When the third H is added, the values of the N-H_3 bond distance, the H_1NH_3 bond angle and the H_2NH_3 bond angle give three conditions that determine the three spatial coordinates of atom H_3.

(d) No. The ClCCCl dihedral angle must be specified.

(e) No. Twisting one CH_2 group with respect to the other generates different structures but does not change any bond distances or angles.

19.45 Let $f_{n_1}(x_1) \equiv (2/a)^{1/2}\sin(n_1\pi x_1/a)$ (for $0 \le x_1 \le a$) denote a particle-in-a-box wave function, where n_1 is the quantum number and x_1 is the coordinate. For noninteracting particles, the wave function is the product of wave functions for each particle, and one's first impulse might be to take ψ as $f_{n_1}(x_1)f_{n_2}(x_2)$. However, this function is not symmetric when $n_1 \ne n_2$, and spin-0 particles are bosons and require symmetric wave functions. Therefore, the normalized wave function [analogous to (18.45)] is $\psi = 2^{-1/2}[f_{n_1}(x_1)f_{n_2}(x_2) + f_{n_1}(x_2)f_{n_2}(x_1)]$. To get ρ, we integrate $|\psi|^2$ over the coordinates of all particles but one and multiply by the number of particles: $\rho =$

$$2(2^{-1})[\int_0^a [f_{n_1}^2(x_1)f_{n_2}^2(x_2) + 2f_{n_1}(x_1)f_{n_2}(x_2)f_{n_1}(x_2)f_{n_2}(x_1) + f_{n_1}^2(x_2)f_{n_2}^2(x_1)]\,dx_2.$$

$$= f_{n_1}^2(x_1)\int_0^a f_{n_2}^2(x_2)\,dx_2 + 2f_{n_1}(x_1)f_{n_2}(x_1)\int_0^a f_{n_1}(x_2)f_{n_2}(x_2)\,dx_2 + f_{n_2}^2(x_1)\int_0^a f_{n_1}^2(x_2)\,dx_2$$

Use of the normalization condition for f and the orthogonality property (17.36) gives $\rho(x_1) = f_{n_1}^2(x_1) + 0 + f_{n_2}^2(x_1)$. Dropping the unnecessary subscript 1, we have $\rho(x) = f_{n_1}^2(x) + f_{n_2}^2(x) = (2/a)\sin^2(n_1\pi x/a) + (2/a)\sin^2(n_2\pi x/a)$ for $0 \le x \le a$, and $\rho = 0$ elsewhere. Also, $\int_{-\infty}^{\infty}\rho(x)\,dx = \int_0^a \rho(x)\,dx = 1 + 1 = 2$, because of normalization.

19.46 **(a)** Yes; **(b)** no; **(c)** yes; **(d)** yes.

19.47 **(a)** E_e in (19.8) depends in the coordinates of the nuclei; for fixed nuclear positions, E_e is a constant.

(b) The probability density ρ is a function of the spatial coordinates x, y, z. (In a molecule, it also depends parametrically on the nuclear locations.)

(c) E_{xc} in (19.44) is the sum of definite integrals (which are numbers) and E_{xc} is a constant for a fixed nuclear configuration. (Of course, it changes when the nuclear positions change.)

(d) υ_{xc} in (19.49) is a function of x, y, z (and also depends parametrically on the nuclear coordinates).

19.48 Multiplication of (19.7) by ψ_e^* and integration over all space gives $\int \psi_e^* \hat{H}_e \psi_e \, d\tau = E_e \int \psi_e^* \psi_e \, d\tau$. Because of normalization, the integral on the right side equals 1 and $\int \psi_e^* \hat{H}_e \psi_e \, d\tau = E_e$. Use of (19.6) gives $E_e =$ $\int \psi_e^* (\hat{K}_e + \hat{V}_{Ne} + \hat{V}_{ee} + \hat{V}_{NN}) \psi_e \, d\tau = \int \psi_e^* \hat{K}_e \psi_e \, d\tau + \int \psi_e^* \hat{V}_{Ne} \psi_e \, d\tau +$ $\int \psi_e^* \hat{V}_{ee} \psi_e \, d\tau + \int \psi_e^* \hat{V}_{NN} \psi_e \, d\tau = \langle K_e \rangle + \langle V_{Ne} \rangle + \langle V_{ee} \rangle + \langle V_{NN} \rangle$.

19.49 **(a)** T; **(b)** T; **(c)** F; **(d)** T; **(e)** T.

19.50 The potential energy of interaction between the infinitesimal charge elements dQ_1 and dQ_2 of the continuous distribution is given by Eq. (18.1) as $dQ_1\, dQ_2/4\pi\varepsilon_0 r_{12}$, where r_{12} is the distance between dQ_1 and dQ_2. If ρ is the electron probability density, then $-e\rho$ is the charge density (the charge per unit volume) and multiplication by the infinitesimal volume gives $dQ_1 =$ $-e\rho(x_1, y_1, z_1)\, dx_1\, dy_1\, dz_1$ and $dQ_2 = -e\rho(x_2, y_2, z_2)\, dx_2\, dy_2\, dz_2$. Integration of $dQ_1\, dQ_2/4\pi\varepsilon_0 r_{12}$ over the coordinates x_2, y_2, z_2 of dQ_2 gives the energy of interaction of dQ_1 with the entire continuous charge distribution. If we then integrate over the coordinates x_1, y_1, z_1 of dQ_1, we get the total energy of interaction of all the infinitesimal elements of charge with one another, except that we must divide by 2 to avoid counting each interaction twice. Thus the interaction energy is given by Eq. (19.48).

19.51 We have $g = -(3/4)(3/\pi)^{1/3}\rho^{4/3}$ so $\partial g/\partial \rho = -(3/4)(3/\pi)^{1/3}(4/3)\rho^{1/3}$ $-(3/\pi)^{1/3}\rho^{1/3}$ and $\partial g/\partial \rho_x = 0$. Then, since $F = E_x^{\text{LDA}}$, we have $\delta E_x^{\text{LDA}}/\delta\rho = -(3/\pi)^{1/3}\rho^{1/3}$.

19.52 There are 2 pi electrons in each double bond, plus the lone pair on N, for a total of $2k + 4$ pi electrons. These fill the lowest $k + 2$ pi MO's and the transition is from $n = k + 2$ to $n = k + 3$. There are $2k + 2$ conjugated single and double bonds and addition of an extra bond length at each end gives a box length of

$(2k + 4)(1.40\ \text{Å}) = a$. Then $|\Delta E| = (h^2/8ma^2)[(k+3)^2 - (k+2)^2] =$ $(2k+5)(h^2/8ma^2)$. $|\Delta E| = h\nu = hc/\lambda$ and $\lambda = hc/|\Delta E| = 8ma^2c/(2k+5)h =$ $8m(2k+4)^2(1.40\ \text{Å})^2c/(2k+5)h = 8(9.11 \times 10^{-31}\ \text{kg})(2k+4)^2(1.40)^2 \times$ $(10^{-10}\ \text{m})^2(3.00 \times 10^8\ \text{m/s})/(2k+5)(6.63 \times 10^{-34}\ \text{J s}) =$ $(64.6\ \text{nm})(2k+4)^2/(2k+5)$. For $k = 1$, $\lambda = (64.6\ \text{nm})(2+4)^2/(2+5) = 332$ nm.

19.53 For $CH_4(g)$, $\Delta_f H^\circ_{298} \cong N_A(E_e - \sum_a n_a\alpha_a) =$ $(6.022 \times 10^{23}/\text{mol})[-40.19517 - 1(-37.88449) - 4(-0.57077)]$ hartrees $\times$ $(4.35974 \times 10^{-18}\ \text{J/hartree}) = -72.5$ kJ/mol. For $H_2O_2(g)$, $E_e - \sum_a n_a\alpha_a =$ $[-150.76479 - 2(-0.57077) - 2(-74.78852)]$ hartrees $= -0.04621$ hartrees and $\Delta_f H^\circ_{298} \cong -121.3$ kJ/mol.

19.54 The molecule is CH_3CHO, with H4, H5, and H6 bonded to C1 and H3 and O bonded to C2. The atoms C1, C2, H3, and O lie in the same plane. Atom H4 eclipses the O atom and H3 is staggered between H5 and H6.

19.55 **(a)** The OH hydrogen lies on the same side of the C—O bond as the carbonyl oxygen in one conformer and on the opposite side in the other conformer. For the first-mentioned conformer, a possible Z-matrix is

```
C1
O2  1  1.43
O3  1  1.23  2  120.0
H4  2  0.96  1  108.0  3    0.0
H5  1  1.07  2  120.0  4  180.0
```

where bond lengths were estimated from the bond radii in Sec. 19.1. The Z-matrix of the second conformer is the same except that the dihedral angles in row four and row five are changed from 0.0 and 180.0 to 180.0 and 0.0, respectively.

(b) One possible answer is

```
C1
X2  1  1.0
O3  1  1.23  2  90.0
O4  1  1.23  2  90.0  3  180.0
```

19.56 For the Windows version of *Gaussian*, the input in Fig. 19.38, appropriately modified, can be used. For part (a), only the basis set needs to be changed. For part (b), HF/3-21G in the Route Section should be changed to B3LYP/6-31G*. In using *Gaussian*, you are asked to specify a filename for the output. Note that filenames with an asterisk are forbidden. Results are (atoms as in Fig. 19.39):

	HF/3-21G	HF/6-31G*	HF/6-31G**	B3LYP/6-31G*
μ/D	2.12	1.87	1.83	1.69
R(CO)/Å	1.441	1.399	1.398	1.419
R(OH)/Å	0.966	0.946	0.942	0.969
R(CH4)/Å	1.085	1.088	1.088	1.101
R(CH5)/Å	1.079	1.081	1.082	1.093
∠COH/°	110.3	109.4	109.6	107.7
∠OCH4/°	112.2	112.0	112.1	112.7
∠OCH5/°	106.3	107.2	107.3	106.7
D(H6COH3)	61.4°	61.2°	61.2°	61.5°
D(H5COH3)	180°	180°	180°	180°

19.57 **(a)** The Z-matrices of Prob. 19.55a are used. One finds −188.7623096 hartrees for the conformer with the OH hydrogen near the carbonyl O and −188.7525454 hartrees for the other conformer. This is an energy difference of 0.0097642 hartrees. From Eq. (19.1) and the equation after (18.18), one hartree corresponds to 27.211(23.061 kcal/mol) = 627.51 kcal/mol, so the conformer with the OH hydrogen close to the CO oxygen is predicted to be more stable by 6.1 kcal/mol.

(b) The calculated energies are −189.7554562 hartrees for the conformer with the OH hydrogen near the carbonyl O and −189.7471663 hartrees for the other conformer. This is an energy difference of 0.0082899 hartrees or 5.2 kcal/mol.

19.58 MM2 in CS Chem 3D Ultra 9.0 and MMFF94s in Spartan (values in parentheses) give *R*(CO) = 1.419 Å (1.416 Å), *R*(OH) = 0.960 Å (0.972 Å), all CH distances = 1.112 Å (1.098 Å), ∠COH = 109.2° (107.1°), ∠HCO = 108.6° and 108.4° (109.5°, 109.0°), *D*(HCOH) = 180° and ±59.8° (180°, ±60.4°).

19.59 MM2 in CS Chem 3D Ultra 9.0 gives a steric energy of 2.173 kcal/mol for the geometry-optimized trans (anti) conformer and 3.034 kcal/mol for the optimized gauche conformer. The predicted energy difference is 0.86 kcal/mol, with the trans being of lower energy. MMFF94s in Spartan gives a steric energy of −21.238 kJ/mol for the anti form and −17.965 kJ/mol for gauche; the energy difference is 0.78 kcal/mol.

19.60 MM2 in CS Chem 3D Ultra 9.0 gives a steric energy of 1.371 kcal/mol for the cis isomer and 2.524 kcal/mol for the trans isomer. The cis isomer is predicted to be more stable than the trans by 1.2 kcal/mol, which is not what one would expect from chemical intuition. [For references on this "cis effect," see N. C. Craig et al., *J. Phys. Chem. A*, **102**, 6745 (1998).] MMFF94s in Spartan erroneously predicts the trans isomer to be more stable.

19.61 **(a)** There are seven bonds and hence 7 terms in V_{str}.

(b) There are 6 bond angles at each carbon atom and therefore a total of 12 bond angles and 12 terms in V_{bend}. [Each carbon is bonded to four atoms and the bond angle at a carbon is described by specifying the two atoms at the ends of the angle; the number of ways of selecting two objects from four objects is $\frac{1}{2}(4)(3) = 6$.]

(c) Each Cl at one carbon has a 1,4 relation with three Cl atoms on the second carbon. There are thus a total of 3(3) = 9 terms in V_{tors}.

(d) There are no 1,5 or higher interactions and as in (c), there are 9 pairs of atoms that have a 1,4 relation, so there are 9 terms in V_{vdW}.

(e) 9, as in (d).

19.62 **(a)** $\approx$ 400 kJ/mol; **(b)** $=(3/2)RT \approx 4$ kJ/mol; **(c)** $\approx$ 12 kJ/mol; **(d)** $\approx$ 600 kJ/mol; **(e)** = 1300 kJ/mol, which corresponds to 13.6 eV/molecule. So (b) < (c) < (a) < (d) < (e).

19.63 **(a)** T. **(b)** F. **(c)** F. **(d)** F. **(e)** F. **(f)** F. **(g)** T. **(h)** T. **(i)** T.

R19.1 $n = 5$, $l = 2$, m can be −2, −1, 0, 1, or 2.

R19.2 **(a)** T. **(b)** F (As shown by the discussion of measuring the position, only probabilities can be predicted, unless the state function happens to be an eigenfunction of $\hat{B}$.) **(c)** T.

R19.3 **(a)** Yes. **(b)** No.

R19.4 **(a)** d/dx; **(b)** $(\ \ )^2$.

R19.5 **(a)** $[2(3)]^{1/2}\hbar = 6^{1/2}\hbar = 6^{1/2}(6.626 \times 10^{-34}\text{ J s})/2\pi = 2.58 \times 10^{-34}$ J s.
(b) $[\frac{1}{2}(\frac{3}{2})]^{1/2}\hbar = 3^{1/2}\hbar/2 = 3^{1/2}(6.626 \times 10^{-34}\text{ J s})/4\pi = 9.13 \times 10^{-35}$ J s.

R19.6 See Eqs. (18.40), (18.41), the paragraph that follows them, and Fig. 18.13.

R19.7 **(a)** See Eq. (18.35). **(b)** $e^2/4\pi\varepsilon_0 r_{12}$. **(c)** See Eq. 18.42.

R19.8 **(a)** The angles at C are close to 109½°; the angle at O is a bit less than 109½°.
(b) 90°.
(c) A bit less than 90° and a bit less than 120°.
(d) A bit less than 109½° at O; a bit more than 120° for ∠OCO and for ∠HCO(carbonyl O), a bit less than 120° for ∠HCO(hydroxyl O).

R19.9 Well predicted: geometries, dipole moments, barriers to internal rotation, electron probability densities; poorly predicted: dissociation energies.

R19.10 $\hat{A}\hat{B}(x^2y^2z^2) = (\partial^2/\partial x^2)(x^5y^2z^2) = 20x^3y^2z^2$
$\hat{B}\hat{A}(x^2y^2z^2) = x^3(\partial^2/\partial x^2)(x^2y^2z^2) = x^3 2y^2z^2 = 2x^3y^2z^2$.
$\hat{A}\hat{B}(x^2y^2z^2) - \hat{B}\hat{A}(x^2y^2z^2) = 18x^3y^2z^2$

R19.11 Only (b).

R19.12 $L = 1$ and $|\mathbf{L}| = (1 \cdot 2)^{1/2}\hbar = 2^{1/2}\hbar$. $S = 1$ and $|\mathbf{S}| = (1 \cdot 2)^{1/2}\hbar = 2^{1/2}\hbar$.

R19.13 $(\hbar/i)(x\,\partial/\partial y - y\,\partial/\partial x)$.

R19.14 If ϕ is any normalized, well-behaved function of the coordinates of the system, then $\int \phi^* \hat{H} \phi \, d\tau \geq E_{gs}$, where E_{gs} is the ground-state energy of the system with Hamiltonian operator $\hat{H}$.

R19.15 (This is the same problem as 17.59a.)

$\int \phi^* \phi \, d\tau = \int_0^l x^4 (l - x)^4 \, dx = \int_0^l (l^4 x^4 - 4l^3 x^5 + 6l^2 x^6 - 4lx^7 + x^8) \, dx =$
$l^9/5 - 4l^9/6 + 6l^9/7 - 4l^9/8 + l^9/9 = l^9/630$

$\int \phi^* \hat{H} \phi \, d\tau = -(\hbar^2/2m) \int_0^l x^2 (l - x)^2 (d^2/dx^2)(l^2 x^2 - 2lx^3 + x^4) \, dx =$

$-(\hbar^2/2m) \int_0^l x^2 (l - x)^2 (2l^2 - 12lx + 12x^2) \, dx =$

$-(\hbar^2/2m) \int_0^l (2l^4 x^2 - 16l^3 x^3 + 38l^2 x^4 - 36lx^5 + 12x^6) \, dx =$

$-(\hbar^2/2m)(2l^7/3 - 16l^7/4 + 38l^7/5 - 36l^7/6 + 12l^7/7) = (\hbar^2 l^7/m)\frac{1}{105}$

$\int \phi^* \hat{H} \phi \, d\tau \Big/ \int \phi^* \phi \, d\tau = (\hbar^2/ml^2)(630/105) = 3h^2/2\pi^2 ml^2 = 0.152h^2/ml^2$. The exact ground-state energy is $(1/8)\,h^2/ml^2 = 0.125\,h^2/ml^2$, so the percent error is 22%. The tedious algebra is best done using a computer algebra program such as Mathematica, MathCad, or Maple or using a graphing calculator with such capabilities.

R19.17 **(a)** See Fig. 17.7. **(b)** See Fig. 17.16. **(c)** See the lowest curve in Fig. 19.5.

R19.18 **(a)** When more than one state has the same energy, that energy level is said to be degenerate, and the degree of degeneracy of the level is the number of states that have that energy.

(b) A linear operator $\hat{B}$ is one that obeys the relations $\hat{B}(cf) = c\hat{B}f$ and $\hat{B}(f + g) = \hat{B}f + \hat{B}g$ for all functions f and g and all constants c.

R19.19 **(a)** molecular mechanics; **(b)** ab initio; **(c)** ab initio; **(d)** semiempirical; **(e)** density functional.

CHAPTER 20 Spectroscopy and Photochemistry

CHAPTER OUTLINE

Most of our experimental information on the energy levels of atoms and molecules comes from spectroscopy, the study of the absorption and the emission of electromagnetic radiation (light) by matter. Section 20.1 examines the nature of light. Section 20.2 is a general discussion of spectroscopy. This is followed by Secs. 20.3 to 20.10 on the rotational and vibrational spectra of diatomic and polyatomic molecules. Electronic spectra are considered in Sec. 20.11, and magnetic resonance spectra in Secs. 20.12 and 20.13. Closely related to spectroscopy is photochemistry (Sec. 20.15), the study of chemical reactions caused or catalyzed by light. The full application of molecular symmetry to spectroscopy and other areas of quantum chemistry is based on the branch of mathematics called group theory, which is discussed in Sec. 20.16.

Spectroscopy is the major experimental tool for investigations at the molecular level. Spectroscopy is used to find molecular structures (conformations, bond lengths, and angles) and molecular vibration frequencies. Organic chemists use nuclear-magnetic-resonance spectroscopy in structural investigations. Analytical chemists use spectroscopy to find the composition of a sample. In kinetics, spectroscopy is used to follow the concentrations of reacting species as functions of time, and to detect reaction intermediates. Biochemists use spectroscopy extensively to study the structure and dynamics of biological molecules. Astronomers use spectroscopy to investigate the composition of stars, planets, and interstellar dust. Spectroscopy is used to determine the levels of pollutants in air.

20.1 ELECTROMAGNETIC RADIATION

In 1801, Thomas Young observed interference of light when a light beam was diffracted at two pinholes, thereby showing the wave nature of light. A wave involves a vibration in space and in time, and so the question arises: What physical quantity is vibrating in a light wave? The answer was provided by Maxwell.

In the 1860s, Maxwell showed that the known laws of electricity and magnetism can all be derived from a set of four differential equations. These equations (called *Maxwell's equations*) interrelate the electric and magnetic field vectors **E** and **B**, the electric charge, and the electric current. Maxwell's equations are the fundamental equations of electricity and magnetism, just as Newton's laws are the fundamental equations of classical mechanics.

In addition to containing all the laws of electricity and magnetism known in the 1860s, Maxwell's equations predicted something that was unknown at that time, namely, that an accelerated electric charge will emit energy in the form of electromagnetic waves traveling at a speed v_{em} in vacuum, where

$$v_{em} = (4\pi\varepsilon_0 \times 10^{-7}\ \mathrm{N\ s^2\ C^{-2}})^{-1/2} \qquad (20.1)$$

Recall that ε_0 (the electric constant) occurs in Coulomb's law (13.1). Substitution of the experimental value (13.2) of ε_0 gives $v_{em} = 2.998 \times 10^8$ m/s, which equals the experimentally observed speed of light in vacuum. Maxwell therefore proposed that light consists of electromagnetic waves.

Maxwell's prediction of the existence of electromagnetic waves was confirmed by Hertz in 1887. Hertz produced electromagnetic waves by the oscillations of electrons in the metal wires of a tuned ac circuit; he detected these waves using a loop of wire (just as the antenna in your TV set detects electromagnetic waves emitted by the transmitters of TV stations). The oscillating electric field of the electromagnetic wave exerts a time-varying force on the electrons in the wires of the detector circuit, thereby producing an alternating current in these wires.

Maxwell's equations show that **electromagnetic waves** consist of oscillating electric and magnetic fields. The electric and magnetic field vectors **E** and **B** are perpendicular to each other and are perpendicular to the direction of travel of the wave. Figure 20.1 shows an electromagnetic wave traveling in the y direction. The vectors shown give the values of **E** and **B** at points on the y axis at one instant of time. As time passes, the crests (peaks) and troughs (valleys) move to the right. This figure is not a complete description of the wave, since it gives the values of **E** and **B** only at points lying on the y axis. To describe an electromagnetic wave fully we must give the values of six numbers (the three components of **E** and the three components of **B**) at every point in the region of space through which the wave is moving.

The wave shown in Fig. 20.1 is **plane-polarized,** meaning that the **E** vectors all lie in the same plane. Such a plane-polarized wave would be produced by the back-and-forth oscillation of electrons in a straight wire. The light emitted by a collection of heated atoms or molecules (for example, sunlight) is *unpolarized,* with the electric-field vectors pointing in different directions at different points in space. This is because the molecules act independently of one another and the radiation produced has random orientations of the **E** vector at various points in space. For an unpolarized wave, **E** is still perpendicular to the direction of travel.

The **wavelength** λ of a wave is the distance between successive crests. One **cycle** is the portion of the wave that lies between two successive crests (or between any two successive points having the same phase). The **frequency** ν (nu) of the wave is the number of cycles passing a given point per unit time. If ν cycles pass a given point in unit time, the time it takes one cycle to pass a given point is $1/\nu$, and a crest has therefore traveled a distance λ in time $1/\nu$. If c is the speed of the wave, then since distance = rate $\times$ time, we have $\lambda = c(1/\nu)$ or

$$\lambda\nu = c \qquad \textbf{(20.2)*}$$

The frequency is commonly given in the units s^{-1}. The SI system of units defines the frequency unit of one reciprocal second as one **hertz** (Hz): $1 \text{ Hz} \equiv 1 \text{ s}^{-1}$. Various multiples of the hertz are also used, for example, the kilohertz (kHz), megahertz (MHz), and gigahertz (GHz):

$$1 \text{ Hz} \equiv 1 \text{ s}^{-1}, \quad 1 \text{ kHz} = 10^3 \text{ Hz}, \quad 1 \text{ MHz} = 10^6 \text{ Hz}, \quad 1 \text{ GHz} = 10^9 \text{ Hz}$$

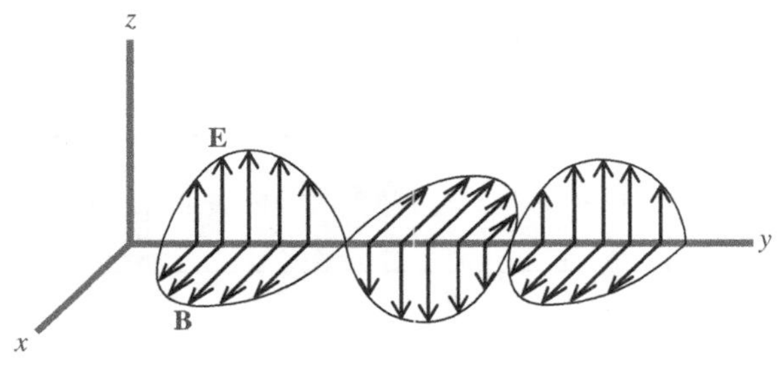

Figure 20.1

A portion of a plane-polarized electromagnetic wave. $1\frac{1}{2}$ cycles are shown.

Common units of λ include the **angstrom** (Å), the **micrometer** (μm), and the **nanometer** (nm):

$$1 \text{ Å} \equiv 10^{-8} \text{ cm} = 10^{-10} \text{ m}, \quad 1 \ \mu\text{m} = 10^{-6} \text{ m}, \quad 1 \text{ nm} = 10^{-9} \text{ m} = 10 \text{ Å}$$

The **wavenumber** $\tilde{\nu}$ of a wave is the reciprocal of the wavelength:

$$\tilde{\nu} \equiv 1/\lambda \qquad \textbf{(20.3)*}$$

Most commonly, $\tilde{\nu}$ is expressed in cm^{-1}. The wavelength λ is the length of one cycle of the wave. The wavenumber $\tilde{\nu}$ expressed in cm^{-1} is the number of cycles of the wave in a 1-cm length.

The human eye is sensitive to electromagnetic radiation with λ in the range 400 nm (violet light) to 750 nm (red light), but there is no upper or lower limit to the values of λ and ν of an electromagnetic wave. Figure 20.2 shows the **electromagnetic spectrum,** the range of frequencies and wavelengths of electromagnetic waves. For convenience, the electromagnetic spectrum is divided into various regions, but there is no sharp boundary between adjacent regions.

All frequencies of electromagnetic radiation (light) travel at the same speed $c = 3 \times 10^8$ m/s in vacuum. The speed of light c_B in substance B depends on the nature of B and on the frequency of the light. The ratio c/c_B for a given frequency of light is the **refractive index** n_B of B for that frequency:

$$n_B \equiv c/c_B \qquad (20.4)$$

Some values of n_B at 25°C and 1 atm for yellow light of vacuum wavelength 589.3 nm ("sodium D light") follow:

Air	H_2O	C_6H_6	C_2H_5OH	CS_2	CH_2I_2	NaCl	Glass
1.0003	1.33	1.50	1.36	1.63	1.75	1.53	1.5–1.9

For quartz at 18°C and 1 atm, n decreases from 1.57 to 1.45 as the vacuum wavelength of the light increases from 185 to 800 nm. Organic chemists use n as a conveniently measured property to help characterize a liquid.

When a light beam passes obliquely from one substance to another, it is bent or *refracted,* because of the difference in speeds in the two substances. (Since $c = \lambda\nu$, if c changes, either λ or ν or both must change. It turns out that λ changes but ν stays the same as the wave goes from one medium to another.) The amount of refraction depends on the ratio of the speeds of the light in the two substances and so depends on the refractive indices of the substances. Because n of a substance depends on ν, light of different frequencies is refracted by different amounts. This allows us to separate or **disperse** an electromagnetic wave containing many frequencies into its component frequencies. An example is the dispersion of white light into the colors red, orange, yellow, green, blue, and violet by a glass prism.

So far in this section, we have presented the classical picture of light. However, in 1905, Einstein proposed that the interaction between light and matter could best be

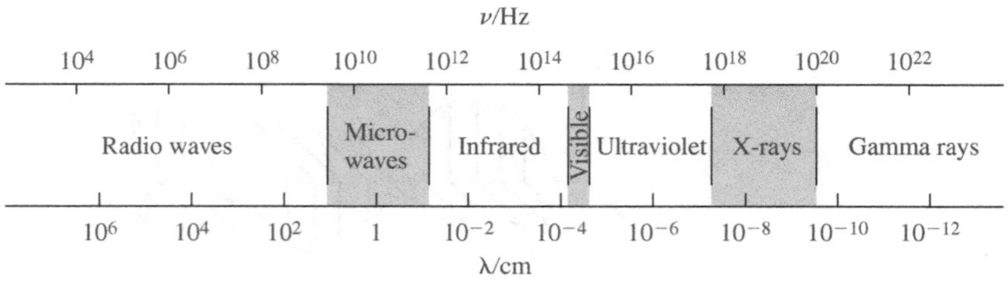

Figure 20.2

The electromagnetic spectrum. The scale is logarithmic.

understood by postulating a particlelike aspect of light, each quantum (photon) of light having an energy $h\nu$ (Sec. 17.2). The direction of increasing frequency in Fig. 20.2 is thus the direction of increasing energy of the photons.

Although the classical picture of electromagnetic radiation as being produced by an accelerated charge is appropriate for the production of radio waves by electrons moving more or less freely in a metal wire (such electrons have a continuous range of allowed energies), the emission and absorption of radiation by atoms and molecules can generally be understood only by using quantum mechanics. The quantum theory of radiation pictures a photon as being produced or absorbed when an atom or molecule makes a transition between two allowed energy levels.

20.2 SPECTROSCOPY

Spectroscopy

In **spectroscopy,** one studies the absorption and emission of electromagnetic radiation (light) by matter. In a broader sense, spectroscopy deals with all interactions of electromagnetic radiation and matter and so also includes scattering of light (Sec. 20.10) and rotation of the plane of polarization of polarized light by optically active substances (Sec. 20.14).

The set of frequencies absorbed by a sample is its **absorption spectrum;** the frequencies emitted constitute the **emission spectrum.** A **line spectrum** contains only discrete frequencies. A **continuous spectrum** contains a continuous range of frequencies. A heated solid commonly gives a continuous emission spectrum. An example is the blackbody radiation spectrum (Fig. 17.1*b*). A heated gas that is not at high pressure gives a line spectrum, corresponding to transitions between the quantum-mechanically allowed energy levels of the individual molecules of the gas.

When a sample of molecules is exposed to electromagnetic radiation, the electric field of the radiation exerts a time-varying force on the electrical charges (electrons and nuclei) of each molecule. To treat the interaction of radiation and matter, one uses quantum mechanics, in particular, the time-dependent Schrödinger equation (17.10). Since the mathematics is complicated, we shall just quote the results and omit the derivations.

The quantum-mechanical treatment shows that when a molecule in the stationary state m is exposed to electromagnetic radiation, it may **absorb** a photon of frequency ν and make a transition to a higher-energy state n if the radiation's frequency satisfies $E_n - E_m = h\nu$ (Fig. 20.3*a*). This agrees with Eq. (17.7), given by Bohr. A molecule in stationary state n in the absence of radiation can spontaneously go to a lower stationary state m, emitting a photon whose frequency satisfies $E_n - E_m = h\nu$. This is **spontaneous emission** of radiation (Fig. 20.3*b*).

Exposing a molecule in state n to electromagnetic radiation whose frequency satisfies $E_n - E_m = h\nu$ will increase the probability that the molecule will undergo a transition to the lower state m with emission of a photon of frequency ν. Emission due to exposure to electromagnetic radiation is called **stimulated emission** (Fig. 20.3*c*).

We shall see in Sec. 20.3 that electronic states of a molecule are more widely spaced than vibrational states, which in turn are more widely spaced than rotational states. Transitions between molecular electronic states correspond to absorption in the ultraviolet (UV) and visible regions. Vibrational transitions correspond to absorption in the infrared (IR) region. Rotational transitions correspond to absorption in the microwave region. (See Fig. 20.2.)

The experimental techniques for absorption spectroscopy in the UV, visible, and IR regions are similar. Here, one passes a beam of light containing a continuous range of frequencies through the sample, disperses the radiation using a prism or diffraction

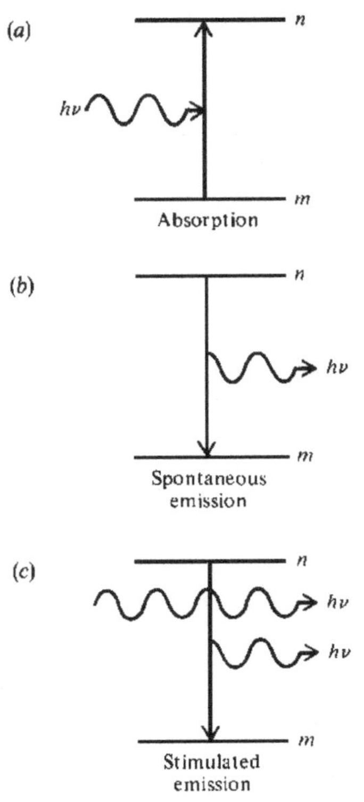

Figure 20.3

Absorption, spontaneous emission, and stimulated emission of radiation between states m and n.

grating, and at each frequency compares the intensity of the transmitted light with the intensity of a reference beam that did not pass through the sample.

The techniques of microwave spectroscopy are described in Sec. 20.7.

The radiation falling on the sample in a spectroscopy experiment not only causes absorption from the lower level of a transition but also produces stimulated emission from the upper level. This stimulated emission travels in the same direction as the incident radiation beam and so decreases the observed absorption signal. (Spontaneous emission is sent out in all directions and need not be considered.) Therefore the intensity of an absorption is proportional to the population difference between the lower and upper levels.

The energy of absorbed radiation is usually dissipated by intermolecular collisions to translational, rotational, and vibrational energies of the molecules, thereby increasing the temperature of the sample. Some of the absorbed energy may be radiated by the excited molecules (fluorescence and phosphorescence; Sec. 20.15). This occurs especially in low-pressure gases, where the average time between collisions is much greater than in liquids. The relative amount of emission depends on the average time between collisions compared with the average lifetimes of the various excited states. Sometimes the absorbed radiation leads to a chemical reaction (Sec. 20.15).

Selection Rules

The quantum-mechanical treatment of the interaction between radiation and matter shows that the probability of absorption or emission between the stationary states m and n is proportional to the square of the magnitude of the integral

$$\boldsymbol{\mu}_{mn} \equiv \int \psi_m^* \hat{\boldsymbol{\mu}} \psi_n \, d\tau \qquad \text{where } \hat{\boldsymbol{\mu}} = \sum_i Q_i \mathbf{r}_i \tag{20.5}$$

where the integration is over the full range of electronic and nuclear coordinates and the sum goes over all the charged particles in the molecule. In the Born–Oppenheimer approximation (Sec. 19.2), the stationary-state wave functions ψ_m and ψ_n are each the product of electronic and nuclear wave functions. $\hat{\boldsymbol{\mu}}$ is the electric dipole-moment operator and occurs in Eq. (19.38). $\mathbf{r}_i$ is the displacement vector of charge Q_i from the origin. The integral $\boldsymbol{\mu}_{mn}$ is called the **transition (dipole) moment.** For pairs of states for which $\boldsymbol{\mu}_{mn} = 0$, the probability of a radiative transition is zero, and the transition is said to be **forbidden.** When $\boldsymbol{\mu}_{mn} \neq 0$, the transition is **allowed.** The allowed changes in quantum number(s) of a system constitute the **selection rule(s)** for the system. [The sum $\Sigma_i Q_i \mathbf{r}_i$ in (20.5) arises from the interaction of the electric field of the electromagnetic radiation with the charges of the molecule.]

EXAMPLE 20.1 Particle-in-a-box selection rule

Find the selection rule for a particle of charge Q in a one-dimensional box of length a.

To find the selection rule, we shall evaluate $\boldsymbol{\mu}_{mn}$ and see when it is nonzero. There is only one particle, so the sum in (20.5) has only one term. For this one-dimensional problem, the displacement from the origin equals the x coordinate. The wave functions are given by (17.35) as $(2/a)^{1/2} \sin (n\pi x/a)$. Therefore

$$\mu_{mn} = \frac{2Q}{a} \int_0^a x \sin \frac{m\pi x}{a} \sin \frac{n\pi x}{a} \, dx$$

Using a table of integrals or integrals.wolfram.com, one finds (Prob. 20.6)

$$\mu_{mn} = \frac{Qa}{\pi^2}\left\{\frac{\cos[(m-n)\pi]-1}{(m-n)^2} - \frac{\cos[(m+n)\pi]-1}{(m+n)^2}\right\} \quad (20.6)$$

We have $\cos\theta = 1$ for $\theta = 0, \pm 2\pi, \pm 4\pi, \pm 6\pi, \ldots$ and $\cos\theta = -1$ for $\theta = \pm\pi, \pm 3\pi, \pm 5\pi, \ldots$. If m and n are both even numbers or both odd numbers, then $m - n$ and $m + n$ are even numbers and μ_{mn} equals zero. If m is even and n is odd, or vice versa, then $m - n$ and $m + n$ are odd and μ_{mn} is nonzero. Hence, radiative transitions between particle-in-a-box states m and n are allowed only if the change $m - n$ in quantum numbers is an odd number. A particle in a box in the $n = 1$ ground state can absorb radiation of appropriate frequency and go to $n = 2$ or $n = 4$, etc., but cannot make a radiative transition to $n = 3$ or $n = 5$, etc. The particle-in-a-box selection rule is $\Delta n = \pm 1, \pm 3, \pm 5, \ldots$.

Exercise

For a hydrogen atom, evaluate the x, y, and z components of $\boldsymbol{\mu}_{mn}$ for m and n being the 1s and 2s states; take the origin at the nucleus. (*Answer:* 0, 0, 0.)

For the one-dimensional harmonic oscillator (Sec. 17.12), one finds that the transition moment is zero unless $\Delta v = \pm 1$, and this is the harmonic-oscillator selection rule. Thus, a harmonic oscillator in the $v = 2$ state can go only to $v = 3$ or $v = 1$ by absorption or emission of a photon. The selection rule for the two-particle rigid rotor (Sec. 17.14) is found to be $\Delta J = \pm 1$.

EXAMPLE 20.2 Rotational absorption frequencies

For a collection of identical two-particle rigid rotors, find the expression for the allowed rotational absorption frequencies, assuming that many rotational levels are populated.

Spectroscopy ordinarily deals with a collection of molecules distributed among states according to the Boltzmann distribution law (Sec. 14.9). The selection rule for absorption is $\Delta J = +1$. Some of the rigid rotors in the $J = 0$ level will absorb radiation and make a transition to the $J = 1$ level when exposed to radiation of the appropriate frequency. Some of the rotors in the $J = 1$ level will absorb radiation of appropriate frequency and go to $J = 2$; etc. See Fig. 20.13. To find the allowed transition frequencies, we set the photon energy $h\nu$ equal to the energy *difference* between the upper and lower rotational levels involved in the transition:

$$E_{\text{upper}} - E_{\text{lower}} = h\nu \quad \textbf{(20.7)*}$$

Sometimes students fail to distinguish between the *energy* of a state and the *energy difference* between states and erroneously set the photon energy $h\nu$ equal to the energy of a stationary state. This error is similar to the failure to distinguish between the enthalpy H of a system in a thermodynamic state and the enthalpy change ΔH for a process.

Let J_1 and J_2 be the lower and upper rotational quantum numbers for the transition. The rotational levels are $E_{\text{rot}} = J(J+1)\hbar^2/2I$ [Eq. (17.81)]. Equation (20.7) gives

$$h\nu = J_2(J_2+1)\hbar^2/2I - J_1(J_1+1)\hbar^2/2I$$

The rotational selection rule $\Delta J = 1$ (stated just before Example 20.2) gives $J_2 = J_1 + 1$, so

$$h\nu = (J_1 + 1)(J_1 + 2)\hbar^2/2I - J_1(J_1 + 1)\hbar^2/2I = (2J_1 + 2)\hbar^2/2I$$
$$\nu = (J_1 + 1)h/4\pi^2 I \quad \text{where } J_1 = 0, 1, 2, \ldots$$

The absorption frequencies are thus $h/4\pi^2 I$, $2h/4\pi^2 I$, $3h/4\pi^2 I$, ...

Exercise

A hypothetical one-dimensional quantum-mechanical system has the energy levels $E = (k + 2)b$, where b is a positive constant and the quantum number k has the values $k = 1, 2, 3, 4, \ldots$. For a collection of such systems with many energy levels populated and each system having the same value of b, find the allowed absorption frequencies of radiation if the selection rule is $\Delta k = \pm 1$. (*Answer: b/h.*)

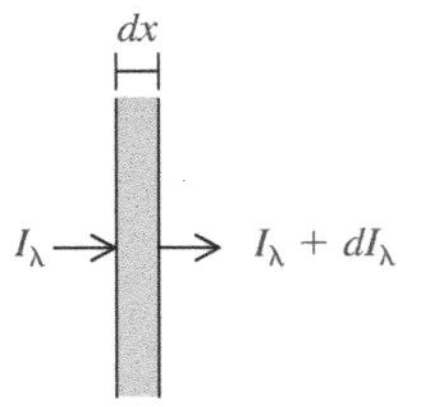

Figure 20.4

Radiation incident on and emerging from a thin slice of sample.

The Beer–Lambert Law

The absorption of UV, visible, and IR light by a sample is often described by the Beer–Lambert law, which we now derive.

Consider a beam of light passing through a sample of pure substance B or of B dissolved in a solvent that neither absorbs radiation nor interacts strongly with B. The beam may contain a continuous range of wavelengths, but we shall focus attention on the radiation whose vacuum wavelength lies in the very narrow range from λ to $\lambda + d\lambda$.

Let $I_{\lambda,0}$ be the intensity of the radiation with wavelength in the range λ to $\lambda + d\lambda$ that is incident on the sample, and let I_λ be the intensity of this radiation after it has gone through a length x of the sample. The **intensity** is defined as the energy per unit time that falls on unit area perpendicular to the beam. The intensity is proportional to the number of photons incident on unit area in unit time. Let N_λ photons of wavelength between λ and $\lambda + d\lambda$ fall on the sample in unit time, and let dN_λ be the number of such photons absorbed by a thickness dx of the sample (Fig. 20.4). The probability that a given photon will be absorbed in the thickness dx is dN_λ/N_λ. This probability is proportional to the number of B molecules that a photon encounters as it passes through the layer of thickness dx. The number of B molecules encountered is proportional to the molar concentration c_B of B and to the layer thickness dx. Therefore, $dN_\lambda/N_\lambda \propto c_B\, dx$.

Let dI_λ be the change in light intensity at wavelength λ due to passage through the layer of thickness dx. Then $dI_\lambda \propto -dN_\lambda$. (The minus sign arises because dN_λ was defined as positive and dI_λ is negative.) Also, $I_\lambda \propto N_\lambda$. Hence, $dI_\lambda/I_\lambda \propto -dN_\lambda/N_\lambda$, and $dI_\lambda/I_\lambda \propto -c_B\, dx$. Letting α_λ be the proportionality constant and integrating along the length of the sample, we have

$$\frac{dI_\lambda}{I_\lambda} = -\alpha_\lambda c_B\, dx \quad \text{and} \quad \int_{I_{\lambda,0}}^{I_{\lambda,l}} \frac{dI_\lambda}{I_\lambda} = -\alpha_\lambda c_B \int_0^l dx \qquad (20.8)$$

$$\ln \frac{I_{\lambda,l}}{I_{\lambda,0}} = 2.303 \log_{10} \frac{I_{\lambda,l}}{I_{\lambda,0}} = -\alpha_\lambda c_B l \qquad (20.9)$$

where l is the sample's length and $I_{\lambda,l}$ is the intensity of the radiation transmitted by the sample. Letting $\varepsilon_\lambda \equiv \alpha_\lambda/2.303$ and defining the **absorbance** A_λ at wavelength λ as $\log_{10}(I_{\lambda,0}/I_{\lambda,l})$, we have

$$A_\lambda \equiv \log_{10}(I_{\lambda,0}/I_{\lambda,l}) = \varepsilon_\lambda c_B l \qquad (20.10)$$

which is the **Beer–Lambert law.** Equations (20.9) and (20.10) can be written as

$$I_{\lambda,l} = I_{\lambda,0}\, e^{-\alpha_\lambda c_B l} = I_{\lambda,0} 10^{-\varepsilon_\lambda c_B l} \qquad (20.11)$$

The intensity I_λ of the radiation decreases exponentially along the sample cell.

The fraction of incident radiation transmitted is the **transmittance** T_λ of the sample at wavelength λ. Thus $T_\lambda \equiv I_{\lambda,l}/I_{\lambda,0}$. Since $I_{\lambda,0}/I_{\lambda,l} = 10^{A_\lambda}$, we have

$$T_\lambda \equiv I_{\lambda,l}/I_{\lambda,0} = 10^{-A_\lambda}$$

For $A_\lambda = 1$, T_λ is $10^{-1} = 0.1$, and 90% of the radiation at λ is absorbed. For $A_\lambda = 2$, T_λ is 10^{-2}, and 99% of the radiation is absorbed. Figure 20.5 plots A_λ and T_λ versus l for $\varepsilon_\lambda = 10$ dm^3 mol^{-1} cm^{-1} and $c_B = 0.1$ mol/dm^3.

The quantity ε_λ is the **molar absorption coefficient** or *molar absorptivity* (formerly called the molar extinction coefficient) of substance B at wavelength λ. Most commonly, the concentration c_B is expressed in mol/dm^3 and the path length l in centimeters, so ε is commonly given in dm^3 mol^{-1} cm^{-1}. Figure 20.37 shows ε as a function of λ for gas-phase benzene over a range of UV frequencies. Since the vertical scale in this figure is logarithmic, ε varies over an enormous range.

If several absorbing species B, C, . . . are present and there are no strong interactions between the species, then $dI_\lambda/I_\lambda = -(\alpha_B c_B + \alpha_C c_C + \cdots)\,dx$ and (20.10) becomes

$$A_\lambda \equiv \log_{10}(I_{\lambda,0}/I_{\lambda,l}) = (\varepsilon_{\lambda,B}c_B + \varepsilon_{\lambda,C}c_C + \cdots)l \tag{20.12}$$

where $\varepsilon_{\lambda,B}$ is the molar absorption coefficient of B at wavelength λ. If the molar absorption coefficients are known for B, C, . . . at several wavelengths, measurement of $I_{\lambda,0}/I_{\lambda,l}$ at several wavelengths allows a mixture of unknown composition to be analyzed. Recall the use of spectroscopy to determine reaction rates—Sec. 16.2.

An application of (20.12) is the pulse oximeter, a device that continuously monitors the oxygenation of the blood of critically ill or anesthetized patients without drawing blood. Light of wavelengths 650 and 940 nm is sent through a fingertip of the patient, and the time-varying (pulsatile) components of the absorbances at these wavelengths are measured to give the absorption due to arterial blood. Reduced and oxygenated hemoglobin have different ε's at each of these wavelengths, and this allows the percentage of oxygenated hemoglobin to be found.

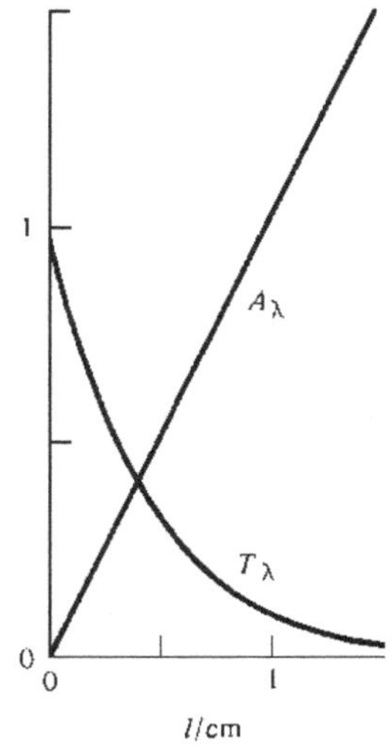

Figure 20.5

Absorbance and transmittance versus cell length for a sample with $\varepsilon_\lambda = 10$ dm^3 mol^{-1} cm^{-1} and $c_B = 0.1$ mol/dm^3.

Lasers

The use of lasers has made possible many new types of spectroscopic experiments and has greatly improved the resolution and precision of spectroscopic investigations. The word *laser* is an acronym for "light amplification by stimulated emission of radiation."

To achieve laser action, one must first produce a **population inversion** in the system. This is a nonequilibrium situation with more molecules in an excited state than in a lower-lying state. Let the populations and energies of the two states involved be N_2 and N_1 and E_2 and E_1, with $E_2 > E_1$. Suppose we have a population inversion with $N_2 > N_1$. Photons of frequency $\nu_{12} = (E_2 - E_1)/h$ spontaneously emitted as molecules drop from state 2 to 1 will stimulate other molecules in state 2 to emit photons of frequency ν_{12} and fall to state 1 (Fig. 20.3*c*). Photons of frequency ν_{12} will also induce absorption from state 1 to 2, but because the system has $N_2 > N_1$, stimulated emission will predominate over absorption and we will get a net amplification of the radiation of frequency ν_{12}. A stimulated-emission photon is emitted in phase with the photon that produces its emission and travels in the same direction as this photon.

To see how a population inversion and laser action are produced, consider Fig. 20.6, in which the states 1, 2, and 3 are those involved in the laser action. Using either an electric discharge or light emitted from a flashlamp, one excites some molecules from the ground state, state 1, to state 3, a process called *pumping*. Suppose that the most probable fate of the molecules in state 3 is to rapidly give up energy to surrounding molecules and fall to state 2 without emitting radiation. Further suppose that the probability of spontaneous emission from state 2 to the ground state 1 is extremely

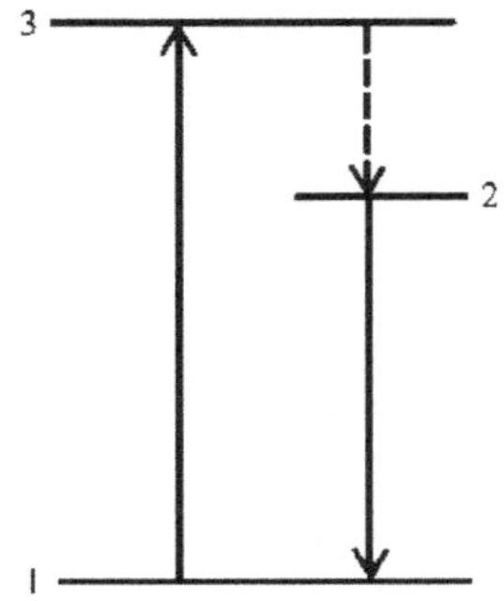

Figure 20.6

States involved in laser action.

small, so the population of state 2 builds up rapidly and eventually exceeds that of state 1, thereby giving a population inversion between states 2 and 1.

The laser system is contained in a cylindrical cavity whose ends have parallel mirrors. A few photons are spontaneously emitted as molecules go from state 2 to state 1. Those emitted at an angle to the cylinder's axis pass out of the system and play no part in the laser action. Those emitted along the laser axis travel back and forth between the end mirrors and stimulate emission of further photons of frequency $(E_2 - E_1)/h$. The presence of the end mirrors makes the laser a resonant optical cavity in which a standing-wave pattern is produced. If l is the distance between the mirrors, only light of wavelength λ such that $n\lambda/2$ is equal to l, where n is an integer, will resonate in the cavity. This makes the laser radiation nearly monochromatic (single-frequency). (Ordinarily, a given transition in a collection of molecules is spread over a range of frequencies, as a result of various effects; see *Hollas,* sec. 2.3.) One of the end mirrors is made partially transmitting to allow some of the laser radiation to leave the cavity.

The laser output is highly monochromatic, highly directional, intense, and coherent. *Coherent* means the phase of the radiation varies smoothly and nonrandomly along the beam. These properties make possible many applications in spectroscopy and kinetics. Thousands of different lasers exist. The material in which the laser action occurs may be a solid, liquid, or gas. The frequency emitted may lie in the infrared, visible, or ultraviolet region. The laser light may be emitted as a brief pulse (recall the use of lasers in flash photolysis—Sec. 16.14), or it may be continuously emitted, giving a CW (continuous wave) laser. Most lasers emit light of fixed frequency, but by using a tunable dye laser or tunable semiconductor laser, one can vary the frequency that will resonate in the cavity.

Kinds of lasers. A *solid-state metal-ion laser* contains a transparent crystal or glass to which a small amount of an ionic transition-metal or rare-earth compound has been added. For example, the *ruby laser* contains an Al_2O_3 crystal with a small amount of Cr_2O_3; the Cr^{3+} ions substitute for some of the Al^{3+} ions in the crystal structure. The Nd:YAG laser contains an yttrium aluminum garnet (YAG) crystal ($Y_3Al_5O_{12}$) with impurity Nd^{3+} ions substituting for some Y^{3+} ions. The laser action involves electronic energy levels of the Cr^{3+} or Nd^{3+} ions in the crystal (electric) field of the host crystal. These lasers are usually pumped by a surrounding flashlamp to achieve population inversion. Small Nd:YAG CW lasers are pumped by semiconductor-laser light.

In a *semiconductor laser* (also called a *diode laser*), the population inversion is achieved in a semiconducting solid. The electronic energy levels of solids occur in continuous bands rather than discrete energy levels (see Fig. 23.27 and the accompanying discussion). The highest occupied and lowest vacant energy bands in a semiconductor are called the *valence band* and the *conduction band,* respectively. A semiconductor laser contains a junction between an n-type semiconductor and a p-type semiconductor. An n-type (n is for negative) semiconductor contains impurity atoms that have more valence electrons than the atoms they replace, whereas a p-type (p is for positive) semiconductor contains impurity atoms with fewer valence electrons than the atoms they replace. For example, the semiconductor GaAs can be made an n-type semiconductor by adding excess As atoms, which replace some of the Ga atoms in the crystal structure; addition of excess Ga gives a p-type semiconductor. When a voltage is applied across the junction of a semiconductor laser, electrons drop from the conduction band of the n-type semiconductor to the valence band of the p-type semiconductor, emitting laser radiation, most commonly in the infrared. Semiconductor lasers can be tuned over a small frequency range by changing the temperature. Semiconductor lasers are tiny (a few millimeters in length) and are used in compact disk players and for high-resolution infrared spectroscopy.

A *gas laser* contains a gas at low pressure. An electric discharge through the gas leads to population inversion. In the *helium–neon laser,* an electric discharge is passed through a mixture of helium and neon with helium mole fraction 0.9. Collisions with electrons excite He atoms to the ^1S and ^3S terms of the $1s2s$ configuration. Collisions of excited He atoms with Ne atoms transfer energy and excite Ne atoms to a high electronic energy level, producing a population inversion in the Ne atoms. The He–Ne laser is a CW laser and is used in optical scanners such as supermarket bar-code readers. The CO_2 laser contains a mixture of CO_2, N_2, and He gases. An electric discharge raises N_2 molecules to the $v = 1$ first excited vibrational level. Collisions between N_2 and CO_2 molecules raise CO_2 molecules to excited vibration–rotation levels, producing a population inversion that gives laser emission.

In chemical lasers, a gas-phase exothermic chemical reaction produces products in excited vibrational levels; this population inversion leads to laser emission. The reactants flow continually through the cell of a chemical laser. See Sec. 22.3 for an example.

Another kind of gas-phase laser is an *exciplex laser*. An *exciplex* (excited complex) is a species that is formed from two atoms or molecules and that is stable (with respect to dissociation) in an excited electronic state but is unstable in its ground electronic state. When the two atoms or molecules that form such a species are identical, the species is called an *excimer* (excited dimer). (Very commonly, the word "excimer" is used whether or not the atoms or molecules forming the species are identical.) A simple excimer is He_2, which is unstable in its ground electronic state (no minimum in its potential-energy curve of electronic energy versus internuclear distance, except for a *very* slight minimum) but has bound excited electronic states that correspond to excitation of an electron from the antibonding σ_u^*1s MO to a higher MO that is bonding.

The excimers and exciplexes used in lasers are diatomic molecules. In a commonly used excimer laser, a continuous electric discharge is passed through a mixture of He, Kr, and F_2. The discharge produces Kr^+ and F^- ions according to $Kr + e^- \rightarrow Kr^+ + 2e^-$ and $F_2 + e^- \rightarrow F^- + F$. These ions combine to form the exciplex KrF* ($F^- + Kr^+ + He \rightarrow$ KrF* + He), where the star indicates the KrF species is formed in an excited electronic state (one with a minimum in its potential-energy curve), and the He is present as a third body to carry away energy and allow the formation of KrF* (Sec. 16.12). The KrF* (which has a 2-ns lifetime) rapidly drops to its ground electronic state emitting ultraviolet laser radiation. The unstable KrF ground state immediately dissociates to Kr and F. Recombination of F atoms forms the starting material F_2. This is a two-level laser in which population inversion is maintained because the lower level is unstable. Exciplex lasers are used in surgery.

In a *dye laser,* a solution of an organic dye flows through the laser cell. The dye is pumped to an excited electronic level S_1 by light from a flashlamp or from another laser. The lifetime for spontaneous emission from S_1 to vibrationally excited levels of the ground electronic state S_0 is of the order of 10^{-9} s. Intermolecular collisions in the liquid then cause extremely rapid (10^{-11} s) decay from the excited vibrational levels of S_0 to the ground vibrational level of S_0, thereby producing a population inversion between S_1 and the vibrationally excited levels of S_0, and we get laser emission. The vibration–rotation levels of S_0 are broadened by intermolecular collisions in the solution to give a continuous band of energy, so emission from S_1 to S_0 occurs over a continuous range of frequencies. By using a diffraction grating and changing the angle of the grating with respect to the laser beam, one can select the frequency of the laser radiation. The tunability of dye lasers makes them valuable in spectroscopy.

20.3 ROTATION AND VIBRATION OF DIATOMIC MOLECULES

We now consider nuclear motion in an isolated diatomic molecule. By "isolated" we mean that interactions with other molecules are slight enough to neglect. This condition is well met in a gas at low pressure.

Translation, Rotation, and Vibration

Recall from Sec. 19.2 that in the Born–Oppenheimer approximation, the potential energy for motion of the nuclei in a molecule is E_e, the electronic energy (including nuclear repulsion) as a function of the spatial configuration of the nuclei. Each electronic state of a diatomic molecule has its own E_e curve (for example, see Fig. 19.5). The Schrödinger equation (19.9) for nuclear motion in a particular electronic state is

$$(\hat{K}_N + E_e)\psi_N \equiv \hat{H}_N\psi_N = E\psi_N$$

where E is the total energy of the molecule, $\hat{K}_N$ is the operator for the kinetic energies of the nuclei, and ψ_N is the wave function for nuclear motion.

For a diatomic molecule composed of atoms A and B with masses m_A and m_B, we have $\hat{K}_N = -(\hbar^2/2m_A)\nabla_A^2 - (\hbar^2/2m_B)\nabla_B^2$ and $E_e = E_e(R)$, where R is the internuclear distance. The potential energy $E_e(R)$ in the Schrödinger equation for nuclear motion is a function of only the relative coordinates of the two nuclei. Therefore the results of Sec. 17.13 apply, and the center-of-mass motion and the internal motion can be dealt with separately. The total molecular energy is the sum of E_{tr}, the translational energy of the molecule as a whole, and E_{int}, the energy of the relative or internal motion of the two atoms:

$$E = E_{tr} + E_{int} \tag{20.13}$$

The allowed translational energies E_{tr} can be taken as those of a particle in a three-dimensional box, Eq. (17.47). The box is the container in which the gas molecules are confined.

From Sec. 17.13, the internal energy levels E_{int} are found from the Schrödinger equation for internal motion, which is

$$[-(\hbar^2/2\mu)\nabla^2 + E_e(R)]\psi_{int} = E_{int}\psi_{int} \tag{20.14}$$

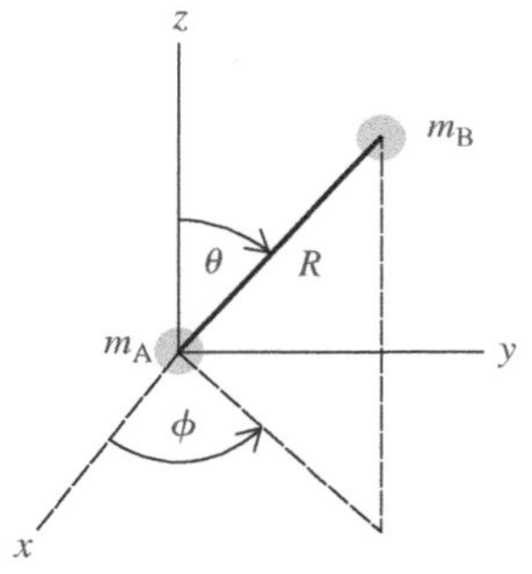

Figure 20.7

Coordinates R, θ, and ϕ for the internal nuclear motion in a diatomic molecule.

The operator ∇^2 equals $\partial^2/\partial x^2 + \partial^2/\partial y^2 + \partial^2/\partial z^2$, where x, y, and z are the coordinates of one nucleus relative to the other; ψ_{int} is a function of x, y, and z. The reduced mass μ is given by Eq. (17.79) as $\mu = m_A m_B/(m_A + m_B)$. Instead of the relative cartesian coordinates x, y, and z, it is more convenient to use the spherical coordinates R, θ, and ϕ of one nucleus relative to the other (Fig. 20.7).

The kinetic energy of internal motion can be divided into kinetic energy of rotation and kinetic energy of vibration. Rotation changes the orientation of the molecular axis in space (that is, changes θ and ϕ) while the internuclear distance R remains fixed. Vibration changes the internuclear distance. Since R is fixed for rotation, the potential energy $E_e(R)$ is associated with the vibration of the molecule. The rotational motion involves the coordinates θ and ϕ, whereas the vibrational motion involves R. It is therefore reasonable to treat the rotational and vibrational motions separately. This is actually an approximation, since there is interaction between rotation and vibration, as discussed later in this section.

The rotational energies of a rigid (fixed interparticle distance) two-particle rotor were given in Sec. 17.14. A diatomic molecule does not actually remain rigid while it rotates, because there is always some vibrational motion. Even the ground vibrational state has a zero-point vibrational energy. However, the nuclei vibrate about the equilibrium internuclear distance R_e, so to a good approximation, we can treat the molecule as a two-particle rigid rotor with separation R_e between the nuclei. From (17.81) and (17.82) the rotational energy of a *diatomic* molecule is (approximately)

$$E_{rot} \approx \frac{J(J+1)\hbar^2}{2I_e}, \qquad I_e = \mu R_e^2, \qquad \mu \equiv \frac{m_A m_B}{m_A + m_B}, \qquad J = 0, 1, 2, \ldots \qquad \textbf{(20.15)*}$$

$$E_{rot} \approx B_e h J(J+1) \qquad \text{where } B_e \equiv h/8\pi^2 I_e \tag{20.16}$$

I_e is the **equilibrium moment of inertia.** B_e is the **equilibrium rotational constant.** Note that R_e differs for different electronic states of the same molecule (see, for example, Fig. 19.5). The rotational wave function ψ_{rot} depends on the angles θ and ϕ defining the orientation of the molecule in space (Fig. 20.7). The rotational levels (20.15) are $(2J + 1)$-fold degenerate, since the value of the quantum number M_J [Eq. (17.83)] does not affect E_{rot}.

Having dealt with rotation by using the rigid-rotor approximation, we now consider vibration of a diatomic molecule. The vibrational motion is along the internuclear separation R, so the vibrational part of the nuclear kinetic-energy operator $(-\hbar^2/2\mu)\nabla^2$ in (20.14) is $(-\hbar^2/2\mu)\, d^2/dR^2$. (This is an oversimplification. See *Levine,* sec. 13.2, for a fuller discussion.) The potential energy for vibration is $E_e(R)$, as noted after (20.14). The vibrational wave function ψ_{vib} is a function of R. We have separated the rotational energy E_{rot} from the internal energy E_{int} in (20.14), so the energy that occurs in the vibrational Schrödinger equation is $E_{\text{int}} - E_{\text{rot}}$. The vibrational Schrödinger equation is therefore

$$\left[-\frac{\hbar^2}{2\mu}\frac{d^2}{dR^2} + E_e(R)\right]\psi_{\text{vib}}(R) = (E_{\text{int}} - E_{\text{rot}})\psi_{\text{vib}}(R) \tag{20.17}$$

The potential-energy function $E_e(R)$ in (20.17) differs for each different electronic state. It is useful to expand $E_e(R)$ in a Taylor series about the equilibrium distance R_e. Equation (8.32) gives

$$E_e(R) = E_e(R_e) + E_e'(R_e)(R - R_e) + \tfrac{1}{2}E_e''(R_e)(R - R_e)^2 + \tfrac{1}{6}E_e'''(R_e)(R - R_e)^3 + \cdots$$

where $E_e'(R_e)$ equals $dE_e(R)/dR$ evaluated at $R = R_e$, with similar definitions for $E_e''(R_e)$, etc. Since we are considering a bound state with a minimum in E_e at $R = R_e$ (Fig. 19.5), the first derivative $E_e'(R_e)$ equals 0. This eliminates the second term in the expansion. The nuclei vibrate about their equilibrium separation R_e. For the lower-energy vibrational levels, the vibrations will be confined to distances R reasonably close to R_e, so the terms involving $(R - R_e)^3$ and higher powers will be smaller than the term involving $(R - R_e)^2$; we shall neglect these terms. Therefore

$$E_e(R) \approx E_e(R_e) + \tfrac{1}{2}E_e''(R_e)(R - R_e)^2 \tag{20.18}$$

We have found that in the region near R_e, the electronic energy curve $E_e(R)$ is approximately a parabolic (quadratic) function of $R - R_e$, the deviation of R from R_e. This is evident in Figs. 19.5 and 20.9. Figure 20.8 plots the approximation (20.18) and the exact E_e for the ground electronic state of H_2.

The quantity $E_e(R_e)$ in (20.18) is a constant for a given electronic state and will be called the **equilibrium electronic energy** E_{el} of the electronic state:

$$E_{\text{el}} \equiv E_e(R_e) \tag{20.19}$$

For the H_2 ground electronic state, the equilibrium dissociation energy is $D_e = 4.75$ eV, so $E_e(R_e) = E_{\text{el}}$ is 4.75 eV below the energy of the dissociated molecule. The energy of two ground-state H atoms is given by (18.18) as $2(-13.60 \text{ eV}) = -27.20$ eV, so $E_{\text{el}} = -31.95$ eV for the H_2 ground electronic state (Figs. 20.8 and 19.5).

Let $x \equiv R - R_e$. Then $d^2/dx^2 = d^2/dR^2$. Substitution of (20.18) and (20.19) into (20.17) and rearrangement gives as the approximate vibrational Schrödinger equation for a diatomic molecule

$$\left[-\frac{\hbar^2}{2\mu}\frac{d^2}{dx^2} + \frac{1}{2}E_e''(R_e)x^2\right]\psi_{\text{vib}} = (E_{\text{int}} - E_{\text{rot}} - E_{\text{el}})\psi_{\text{vib}} \tag{20.20}$$

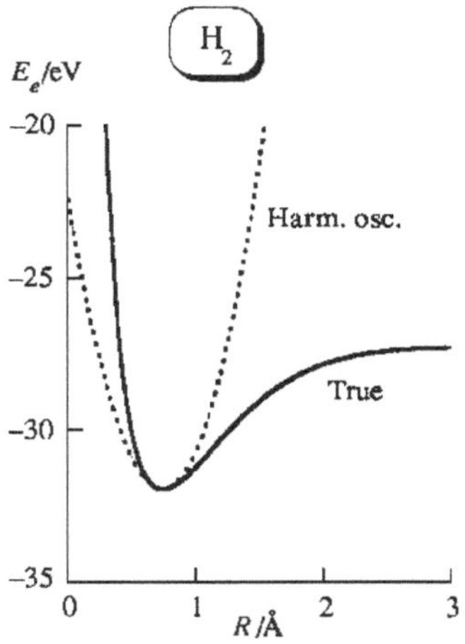

Figure 20.8

Harmonic-oscillator approximation to the potential-energy curve for nuclear motion in the H_2 ground electronic state compared with the true curve. The curve $E_e(R)$ is found by solving the electronic Schrödinger equation (19.7) at many internuclear distances R.

Since (20.20) has the same form as the Schrödinger equation (17.75) for a one-dimensional harmonic oscillator, the solutions to (20.20) are the harmonic-oscillator wave functions. The quantity $E_{int} - E_{rot} - E_{el}$ in (20.20) corresponds to the harmonic-oscillator energy E in (17.75). Therefore $E_{int} - E_{rot} - E_{el} = (v + \frac{1}{2})h\nu_e$. Combining this equation with $E = E_{tr} + E_{int}$ [Eq. (20.13)], we have as the molecular energy E:

$$E = E_{tr} + E_{rot} + E_{vib} + E_{el} \qquad \textbf{(20.21)*}$$

$$E_{vib} \approx (v + \tfrac{1}{2})h\nu_e, \quad v = 0, 1, 2, \ldots \quad \text{diatomic molecule} \qquad \textbf{(20.22)*}$$

The force constant k in (17.75) corresponds to $E_e''(R_e)$ in (20.20). Also, m in (17.75) corresponds to μ in (20.20). Therefore the **equilibrium** (or **harmonic**) **vibrational frequency** ν_e of the diatomic molecule is [Eq. (17.73)]

$$\nu_e = \frac{1}{2\pi}\left(\frac{k_e}{\mu}\right)^{1/2} = \frac{1}{2\pi}\left[\frac{E_e''(R_e)}{\mu}\right]^{1/2} \qquad (20.23)$$

From (17.71), the **equilibrium force constant** k_e has units of force over length.

In summary, we have seen that to a good approximation, the energy E of an isolated diatomic molecule is $E = E_{tr} + E_{rot} + E_{vib} + E_{el}$, where the translational energy levels E_{tr} are those of a particle in a box, the rotational energy levels E_{rot} can be approximated by the rigid-rotor energies $J(J + 1)\hbar^2/2I_e$, the vibrational levels E_{vib} can be approximated by the harmonic-oscillator levels $(v + \frac{1}{2})h\nu_e$, and the equilibrium electronic energy E_{el} is the energy at the minimum in the electronic energy curve $E_e(R)$. The translational, rotational, and vibrational energies represent energy over and above the energy at the minimum in the E_e curve. The molecule's internal energy [Eq. (20.13)] is $E_{rot} + E_{vib} + E_{el}$:

$$E_{int} \approx B_e hJ(J + 1) + (v + \tfrac{1}{2})h\nu_e + E_{el} \qquad (20.24)$$

Anharmonicity

The approximation (20.18) for the potential energy of nuclear motion leads to harmonic-oscillator vibrational energy levels. Figure 20.8 shows that for $R \gg R_e$, the parabolic approximation (20.18) is poor. The potential energy $E_e(R)$ is not really a harmonic-oscillator potential energy, and this **anharmonicity** adds correction terms to the approximate vibrational-energy expression (20.22). One finds that the main correction term to (20.22) is

$$-h\nu_e x_e(v + \tfrac{1}{2})^2$$

where the *anharmonicity constant* $\nu_e x_e$ is nearly always positive and depends on $E_e'''(R_e)$ and $E_e^{(iv)}(R_e)$. The quantity x_e (which is a constant and not a coordinate) usually lies in the range 0.002 to 0.02 for diatomic molecules (see Table 20.1 in Sec. 20.4). Therefore, $\nu_e x_e \ll \nu_e$. As v increases, the magnitude of the term $-h\nu_e x_e(v + \frac{1}{2})^2$ becomes larger relative to $(v + \frac{1}{2})h\nu_e$.

With inclusion of the anharmonicity correction term, the spacing between the adjacent levels v and $v + 1$ becomes

$$(v + \tfrac{3}{2})h\nu_e - h\nu_e x_e(v + \tfrac{3}{2})^2 - [(v + \tfrac{1}{2})h\nu_e - h\nu_e x_e(v + \tfrac{1}{2})^2]$$
$$= h\nu_e - 2h\nu_e x_e(v + 1)$$

Thus the spacing between adjacent diatomic-molecule vibrational levels decreases as v increases. This is in contrast to the equally spaced levels of a harmonic oscillator. A harmonic oscillator has an infinite number of vibrational levels. However, a diatomic-molecule bound electronic state has only a finite number of vibrational levels, since once the vibrational energy exceeds the equilibrium dissociation energy D_e, the molecule dissociates. Figure 20.9 shows the vibrational levels of the H_2 ground electronic

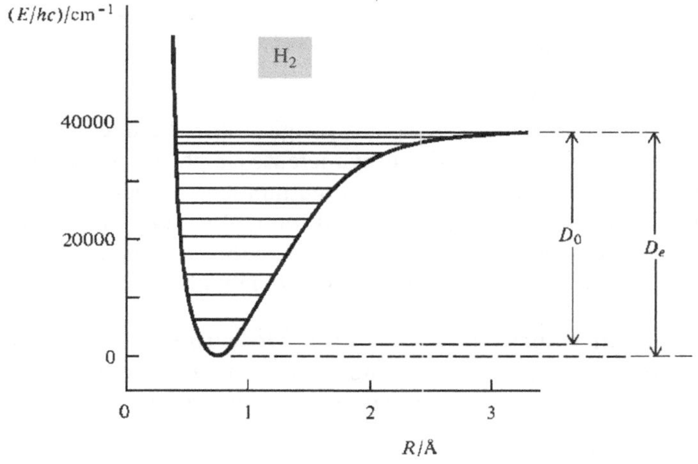

Figure 20.9

Ground-electronic-state vibrational levels of H_2. [Data from S. Weissman et al., *J. Chem. Phys.*, **39,** 2226 (1963).]

state, which has 15 bound vibrational levels ($v = 0$ through 14). Note the decreasing spacing as v increases. The zero level of energy has been taken at the minimum of the potential-energy curve. The $v = 14$ level lies a mere 150 cm^{-1} below $E_e(\infty)/hc$.

Ground-Vibrational-State Dissociation Energy

The lowest molecular rotational energy is zero, since $E_{rot} = 0$ for $J = 0$ in (20.15). However, the lowest vibrational energy is nonzero. In the harmonic-oscillator approximation, the $v = 0$ ground vibrational state has the energy $\frac{1}{2}h\nu_e$. With inclusion of the anharmonicity term $-h\nu_e x_e(v + \frac{1}{2})^2$, the vibrational ground-state energy is $\frac{1}{2}h\nu_e - \frac{1}{4}h\nu_e x_e$. The dissociation energy measured from the lowest vibrational state is called the **ground-vibrational-state dissociation energy** D_0 (Fig. 20.9) and is somewhat less than the equilibrium dissociation energy D_e, because of the zero-point vibrational energy:

$$D_e = D_0 + \tfrac{1}{2}h\nu_e - \tfrac{1}{4}h\nu_e x_e \qquad (20.25)$$

For the H_2 ground electronic state, $D_0 = 4.48$ eV and $D_e = 4.75$ eV.

What is the relation between the molecular quantity D_0 and thermodynamic quantities? Consider the dissociation reaction $H_2(g) \rightarrow 2H(g)$. The standard-state thermodynamic properties refer to ideal gases, which have no intermolecular interactions. At the temperature absolute zero, the molecules will all be in the lowest available energy level. For H_2, this is the $v = 0$, $J = 0$ level of the ground electronic state. Therefore the thermodynamic quantity ΔU_0° for $H_2(g) \rightarrow 2H(g)$ will equal $N_A D_0$, where N_A is the Avogadro constant. Spectroscopic determination of D_0 for the H_2 ground electronic state gives $D_0 = 4.4781$ eV. Using (19.1), we get $\Delta U_0^\circ = 432.07$ kJ/mol. ΔU_{298}° differs from ΔU_0° because of the increase in translational energies of the 2 moles of H atoms and 1 mole of H_2 molecules and the increase in H_2 rotational energy on going from 0 to 298 K; see Prob. 21.70. (Excited vibrational and electronic levels of H_2 are not significantly occupied at 298 K.)

Vibration–Rotation Interaction

Equation (20.21) neglects the interaction between vibration and rotation. Because of the anharmonicity of the potential-energy curve $E_e(R)$, the average distance R_{av} between the nuclei increases as the vibrational quantum number increases (see Fig. 20.9). This increase in R_{av} increases the effective moment of inertia $I = \mu R_{av}^2$ and decreases the rotational energies, which are proportional to $1/I$. To allow for this effect, one adds the term $-h\alpha_e(v + \frac{1}{2})J(J + 1)$ to the energy, where the *vibration–rotation coupling constant* α_e is a positive constant that is much smaller than the rotational constant B_e in Eq. (20.16) (see Table 20.1).

Centrifugal Distortion

Since the molecule is not really a rigid rotor, the internuclear distance increases very slightly as the rotational energy increases, a phenomenon called *centrifugal distortion*. Centrifugal distortion increases the moment of inertia and hence decreases the rotational energy below that of a rigid rotor. One finds that the term $-hDJ^2(J+1)^2$ must be added to the rotational energy, where the *centrifugal-distortion constant D* is a very small positive constant. Don't confuse D with the dissociation energy.

Internal Energy of a Diatomic Molecule

With inclusion of anharmonicity and vibration–rotation interaction, the expression (20.24) for the internal energy of a diatomic molecule becomes

$$E_{\text{int}} = E_{\text{el}} + h\nu_e(v + \tfrac{1}{2}) - h\nu_e x_e(v + \tfrac{1}{2})^2 + hB_eJ(J+1) - h\alpha_e(v + \tfrac{1}{2})J(J+1) \qquad (20.26)$$

Since the centrifugal-distortion term is tiny, it is omitted from E_{int}. The total energy equals $E_{\text{int}} + E_{\text{tr}}$. If vibration–rotation interaction is neglected, E_{int} is approximated as the sum of electronic, vibrational, and rotational energies:

$$E_{\text{int}} \approx E_{\text{el}} + h\nu_e(v + \tfrac{1}{2}) - h\nu_e x_e(v + \tfrac{1}{2})^2 + hB_eJ(J+1) \qquad (20.27)$$

Since ν_e [which depends on $E_e''(R_e)$] and B_e (which depends on R_e) each differ for different electronic states of the same molecule, each electronic state of a molecule has its own set of vibrational and rotational levels. Figure 20.10 shows some of the vibration–rotation levels of one electronic state of a diatomic molecule. The dots indicate higher rotational levels of each vibrational level. (Figure 20.9 shows only the energy levels with $J = 0$.)

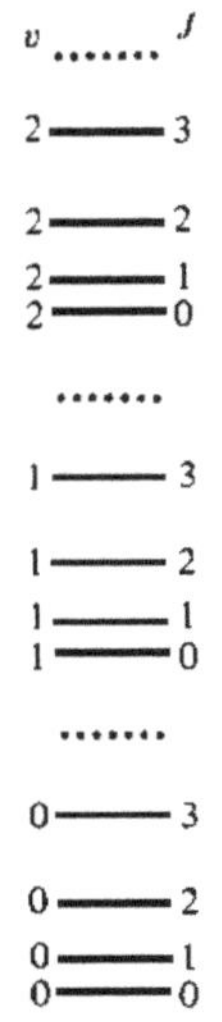

Figure 20.10

Vibration–rotation energy levels for a diatomic molecule. For each vibrational state (defined by the quantum number v), there is a set of rotational levels (defined by the quantum number J). For the ground electronic state of H_2, v goes from 0 to 14 (Fig. 20.9).

Level Spacings

The translational energy levels for a particle in a cubic box of volume V are [Eq. (17.50)] $(n_x^2 + n_y^2 + n_z^2)h^2/8mV^{2/3}$. The spacing between adjacent translational levels with quantum numbers n_x, n_y, n_z and $n_x + 1, n_y, n_z$ is $(2n_x + 1)h^2/8mV^{2/3}$, since $(n_x + 1)^2 - n_x^2 = 2n_x + 1$. A typical molecule has translational energy of roughly $\frac{1}{2}kT$ associated with motion in each direction (Chapter 14). The equation $\frac{1}{2}kT = n_x^2h^2/8mV^{2/3}$ gives for a molecule of molecular weight 100 in a 1-cm³ box at room temperature: $n_x = 8 \times 10^8$; this gives a spacing of 5×10^{-30} J $= 3 \times 10^{-11}$ eV between adjacent translational levels. This energy gap is so tiny that, for all practical purposes, we can consider the translational energy levels of the gas molecules to be continuous rather than discrete. The very close spacing results from the macroscopic size of the container's volume V.

The spacing between the rotational levels with quantum numbers J and $J + 1$ is given by Eq. (20.15) as $(J+1)\hbar^2/\mu R_e^2$, since $(J+1)(J+2) - J(J+1) = 2(J+1)$. For the CO ground electronic state, R_e is 1.1 Å. The reduced mass μ equals $m_A m_B/(m_A + m_B)$; the atomic masses m_A and m_B are found by dividing the molar masses by the Avogadro constant. Thus

$$\mu = \frac{[(12 \text{ g mol}^{-1})/N_A][(16 \text{ g mol}^{-1})/N_A]}{(28 \text{ g mol}^{-1})/N_A} = 1.1 \times 10^{-23} \text{ g}$$

This gives $\hbar^2/\mu R_e^2 = 8 \times 10^{-23}$ J $= 0.0005$ eV. The spacing between adjacent CO rotational levels is $J + 1$ times this number, where $J = 0, 1, 2, \ldots$.

The spacing between the harmonic-oscillator vibrational levels is $h\nu_e$. Observations on vibrational spectra show that ν_e is typically 10^{13} to 10^{14} s^{-1}, so the vibrational spacing is typically 7×10^{-21} to 7×10^{-20} J (0.04 to 0.4 eV).

Electronic absorption spectra show that the spacing $E_{el,2} - E_{el,1}$ between the ground and first excited electronic levels is typically 2 to 6 eV.

Thus, *electronic energy differences are substantially greater than vibrational energy differences, which in turn are much greater than rotational energy differences.* Figure 20.11 shows the typical ranges of energy differences between the lowest two rotational states, vibrational states, and electronic states of diatomic molecules.

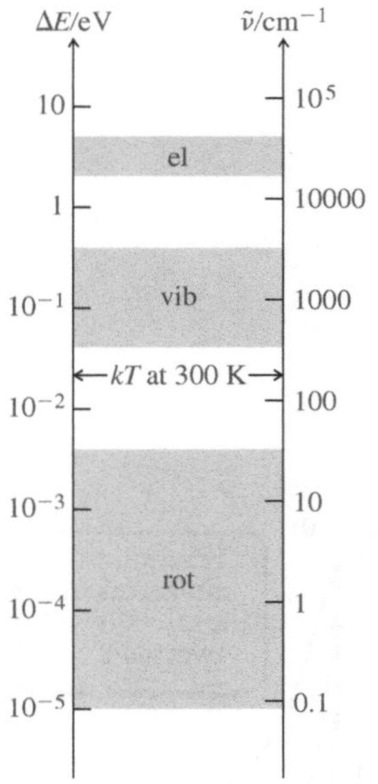

Figure 20.11

Typical ranges of spacings between electronic, vibrational, and rotational energy levels for diatomic molecules. $\tilde{\nu} = \Delta E/hc$. The scale is logarithmic.

The Boltzmann Distribution

In spectroscopy, one observes absorption or emission of radiation by a collection of many molecules in a gas, liquid, or solid. The molecules populate the various possible quantum states in accord with the Boltzmann distribution law. Absorption from a given quantum state will be observed only if there are a significant number of molecules in that state, so we want to examine the population of rotational, vibrational, and electronic states under typical conditions. The Boltzmann distribution law (14.74) gives the population ratio for states i and j as $N_i/N_j = e^{-(E_i-E_j)/kT}$. At room temperature, $kT = 0.026$ eV.

Note that if states i and j belong to a degenerate energy level, then to find the population of an energy level, we must multiply the population of a state of that level by the degeneracy of the level (which is the number of states having that energy). When applied to energy levels, rather than states, the Boltzmann distribution law becomes

$$\frac{N(E_r)}{N(E_s)} = \frac{g_r}{g_s} e^{-(E_r-E_s)/kT}$$

where g_r and g_s are the degeneracies of energy levels E_r and E_s with populations $N(E_r)$ and $N(E_s)$. For a diatomic-molecule rotational energy level, the degeneracy is $2J + 1$ (Sec. 17.14).

The typical molecular rotational-level spacing ΔE_{rot} found above is substantially less than kT at room temperature, so $e^{-\Delta E_{rot}/kT}$ is close to 1, and many excited rotational levels are populated at room temperature.

The vibrational frequency ν_e equals $(1/2\pi)(k_e/\mu)^{1/2}$. For heavy diatomic molecules (for example, Br_2 and I_2), the large value of μ gives a ν_e of roughly 10^{13} s^{-1} and a spacing $h\nu_e$ of roughly 0.04 eV, which is comparable to kT at room temperature. Hence, for heavy diatomic molecules, there is significant occupation of one or more excited vibrational levels at room temperature. However, relatively light diatomic molecules (for example, H_2, HCl, CO, and O_2) have ν_e values of roughly 10^{14} s^{-1} and vibrational spacings of roughly 0.4 eV. This is substantially greater than kT at room temperature, so $e^{-\Delta E_{vib}/kT}$ is very small and nearly all the molecules are in the ground vibrational level at room temperature. Figure 20.12 plots the fraction of molecules in the ground rotational state and in the ground vibrational state versus T for some diatomic molecules.

Since ΔE_{el} is substantially greater than kT at room temperature, excited electronic states are generally not occupied at room temperature.

The Effect of Nuclear Spin

We shall see in Sec. 20.12 that an atomic nucleus has a spin angular momentum **I**, whose magnitude is $|\mathbf{I}| = [I(I + 1)]^{1/2}\hbar$, where the nuclear spin quantum number I can be 0, 1/2, 1, 3/2, 2 . . ., and has a fixed value for a given kind of nucleus. The z component of nuclear spin has the possible values $M_I\hbar$, where $M_I = -I, -I + 1, \ldots, I - 1, I$. The number of M_I values is $2I + 1$. Values of I for various isotopes are listed in a table inside the back cover. The molecular wave function should include a nuclear-spin factor (just as electron spin must be included in the electronic wave function of an atom or molecule). For a homonuclear diatomic molecule, the spin–statistics theorem requires that the complete molecular wave function be symmetric with respect to interchange of the identical nuclei if the nuclei have integral spin and must be antisymmetric if the nuclei have half-integral spin.

The molecular wave function includes electronic, vibrational, rotational, and nuclear-spin factors. We saw that for $s = \frac{1}{2}$, there are 3 symmetric and 1 antisymmetric two-electron spin functions [Eqs. (18.40) and (18.41)]. For nuclear spin quantum I, the number of symmetric two-nuclei spin wave functions ψ_{ns} is $(I + 1)(2I + 1)$ and the number of antisymmetric nuclear spin functions is $I(2I + 1)$ (Prob. 20.127):

$$(I + 1)(2I + 1) \text{ symmetric } \psi_{ns} \quad \text{and} \quad I(2I + 1) \text{ antisymmetric } \psi_{ns}$$

The vibrational factor ψ_{vib} in the molecular wave function is a function of the internuclear distance R, and R does not change when the nuclei are interchanged, so ψ_{vib} is a symmetric function. It turns out that interchange of the nuclei multiplies the rotational factor ψ_{rot} by $(-1)^J$, where J is the rotational quantum number, so ψ_{rot} is symmetric for $J = 0, 2, 4, \ldots$ and is antisymmetric for $J = 1, 3, 5, \ldots$. Surprisingly, interchange of the nuclei affects the electronic factor ψ_{el}. This is because the electronic spatial coordinates are defined relative to coordinate axes that are attached to the nuclei and move with the nuclei. Whether ψ_{el} is symmetric or antisymmetric with respect to interchange of the nuclei depends on the nature of the electronic state. (More precisely, it depends on what the electronic *term* is. We discussed electronic terms for atoms in Sec. 18.7. For electronic terms of diatomic molecules, see *Levine,* sec. 13.8.) For most diatomic molecules, ψ_{el} of the ground electronic state is symmetric with respect to nuclear exchange. An important exception is O_2, where the ground-state ψ_{el} is antisymmetric for exchange of the nuclei.

Consider $^{14}N_2$, which has a symmetric ψ_{el}. Since the ^{14}N nucleus has $I = 1$ and is a boson, the $^{14}N_2$ wave function must be symmetric. Each of the symmetric $J = 0, 2, 4, \ldots$ ψ_{rot} functions of $^{14}N_2$ must be multiplied by one of the $(I + 1)(2I + 1)$ symmetric ψ_{ns} functions to give the complete wave function. Each of the antisymmetric $J = 1, 3, 5, \ldots$ ψ_{rot} functions must be multiplied by one of the $I(2I + 1)$ antisymmetric ψ_{ns} functions. The nuclear-spin function does not affect the energy (unless an external magnetic field is present), so, in addition to the $(2J + 1)$-fold rotational degeneracy, we have a factor due to the nuclear-spin degeneracy. Thus for the ground electronic state of N_2, the degeneracy of the $J = 0, 2, 4, \ldots$ rotational levels is $(2J + 1)(I + 1)\cdot(2I + 1) = 6(2J + 1)$ and the degeneracy of the $J = 1, 3, 5, \ldots$ levels is $(2J + 1)I\cdot(2I + 1) = 3(2J + 1)$. These different degeneracies of the odd-J and even-J rotational levels affect the line intensities in the Raman spectrum (Sec. 20.10) of $^{14}N_2$.

The ^{12}C nucleus has $I = 0$ and $^{12}C_2$ has a symmetric ground-state ψ_{el}. Here $(I + 1)\cdot(2I + 1) = 1$ and $I(2I + 1) = 0$, so there are no antisymmetric nuclear-spin functions. Since ^{12}C is a boson, the one symmetric nuclear-spin function must be combined with the symmetric $J = 0, 2, 4, \ldots$ rotational levels to give an overall ψ that is symmetric. Since there is no antisymmetric nuclear-spin function to combine with the antisymmetric $J = 1, 3, 5, \ldots$ rotational levels, the $J = 1, 3, 5, \ldots$ rotational levels *do not exist* for the ground electronic state of $^{12}C_2$. For O_2, each ^{16}O nucleus has $I = 0$ and the ground-state ψ_{el} is antisymmetric. The situation is complicated by the fact that the nonzero electronic spin angular momentum of the O_2 ground electronic state combines with the rotational angular momentum, so a detailed discussion is omitted. However, just as with the zero-nuclear-spin molecule $^{12}C_2$, half the rotational levels of ground-electronic-state $^{16}O_2$ do not exist.

For heteronuclear diatomic molecules, there is no symmetry or antisymmetry requirement for interchange of the nonidentical nuclei. The nuclear-spin degeneracy is $(2I_A + 1)(2I_B + 1)$, where I_A and I_B are the spins of the two nuclei, and is the same for all rotational levels. Since it is the same for all levels, it usually can be ignored.

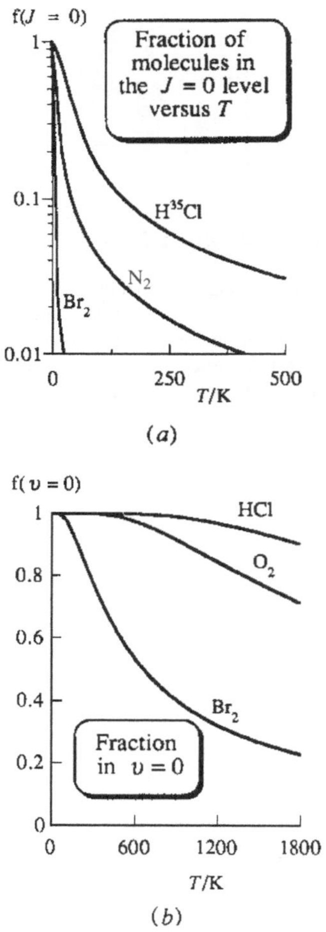

Figure 20.12

The fraction f of molecules in (*a*) the ground rotational state and (*b*) the ground vibrational state of some diatomic molecules plotted versus T. The vertical scale in (*a*) is logarithmic. The temperature scales in (*a*) and (*b*) differ.

20.4 ROTATIONAL AND VIBRATIONAL SPECTRA OF DIATOMIC MOLECULES

This section deals with radiative transitions between different vibration–rotation levels of the same electronic state of a diatomic molecule. Transitions between different electronic states are considered in Sec. 20.11.

Selection Rules

To find the spectrum line frequencies, we need the selection rules.

The Born–Oppenheimer approximation gives the molecular wave function as the product of electronic and nuclear wave functions: $\psi = \psi_e\psi_N$. For a transition between two states ψ and ψ' with no change in electronic state ($\psi_e = \psi_e'$), the transition-moment integral (20.5) determining the selection rules is

$$\int\int \psi_e^*\psi_N^*\hat{\boldsymbol{\mu}}\psi_e\psi_N'\,d\tau_e\,d\tau_N = \int \psi_N^*\psi_N'\left(\int \psi_e^*\hat{\boldsymbol{\mu}}\psi_e\,d\tau_e\right)d\tau_N = \int \psi_N^*\psi_N'\boldsymbol{\mu}\,d\tau_N$$

where (19.38) was used to introduce the electric dipole moment $\boldsymbol{\mu}$ of the electronic state. The magnitude μ of $\boldsymbol{\mu}$ depends on the internuclear distance. (The experimentally observed value is an average over the molecular zero-point vibration.) One expands μ in a Taylor series about its value at R_e:

$$\mu(R) = \mu(R_e) + \mu'(R_e)(R - R_e) + \tfrac{1}{2}\mu''(R_e)(R - R_e)^2 + \cdots$$

where $\mu(R_e)$ is virtually the same as the experimentally observed dipole moment. One then substitutes this expansion and the expressions for ψ_N and ψ_N' into the above integral and evaluates this integral. The details are omitted.

The selection rules for diatomic-molecule transitions with no change in electronic state turn out to be

$$\Delta J = \pm 1 \qquad \textbf{(20.28)*}$$

$$\Delta v = 0, \pm 1\ (\pm 2, \pm 3, \ldots) \qquad \textbf{(20.29)*}$$

$$\Delta v = 0 \text{ not allowed if } \mu(R_e) = 0 \qquad \Delta v = \pm 1 \text{ not allowed if } \mu'(R_e) = 0 \qquad \textbf{(20.30)*}$$

where $\mu(R_e)$ is the molecule's electric dipole moment evaluated at the equilibrium bond distance and $\mu'(R_e)$ is $d\mu/dR$ evaluated at R_e. The parentheses in (20.29) indicate that the $\Delta v = \pm 2, \pm 3, \ldots$ transitions are far less probable than the $\Delta v = 0$ and ± 1 transitions. If the molecule were a harmonic oscillator, and if terms after the $\mu'(R_e)(R - R_e)$ term in the $\mu(R)$ expansion were negligible, only $\Delta v = 0, \pm 1$ transitions would occur.

Rotational Spectra

Transitions with no change in electronic state and with $\Delta v = 0$ give the **pure-rotation spectrum** of the molecule. These transitions correspond to photons with energies in the microwave and far-IR regions. (The far-IR region is the portion of the IR region bordering the microwave region in Fig. 20.2.) Equation (20.30) shows that *a diatomic molecule has a pure-rotation spectrum only if it has a nonzero electric dipole moment.* A homonuclear diatomic molecule (for example, H_2, N_2, Cl_2) has no pure-rotation spectrum.

The pure-rotation spectrum is usually observed as an absorption spectrum. The absorption transitions all have $\Delta J = +1$ [Eq. (20.28)]. Because the rotational levels are not equally spaced, and because many excited rotational levels are populated at room temperature, there will be several lines in the pure-rotation spectrum. From $E_{\text{upper}} - E_{\text{lower}} = h\nu$, the absorption frequencies are $\nu = (E_{J+1} - E_J)/h$, where the energy levels are given by (20.26). For a pure-rotational transition, E_{el} is unchanged and v is unchanged, so the only terms in (20.26) that change are the last two terms, which involve the rotational quantum number J. The sum of these two terms is $hJ(J + 1)\cdot[B_e - \alpha_e(v + \frac{1}{2})]$, and the frequency of a diatomic-molecule pure-rotational transition between levels J and $J + 1$ is (recall Example 20.2 in Sec. 20.2)

$$\nu = (J + 1)(J + 2)[B_e - \alpha_e(v + \tfrac{1}{2})] - J(J + 1)[B_e - \alpha_e(v + \tfrac{1}{2})]$$

$$\nu = 2(J + 1)[B_e - \alpha_e(v + \tfrac{1}{2})] \equiv 2(J + 1)B_v \qquad \text{where } J = 0, 1, 2, \ldots \qquad (20.31)$$

and where the *mean rotational constant* B_v for states with vibrational quantum number v is defined as

$$B_v \equiv B_e - \alpha_e(v + \tfrac{1}{2}) \tag{20.32}$$

Tables sometimes list $B_0 = B_e - \frac{1}{2}\alpha_e$, instead of B_e. The rotational constant B_v allows for the increase in average internuclear distance and consequent decrease in rotational energies due to anharmonic vibration.

The wavenumbers of the pure-rotational transitions are $\tilde{\nu} = 1/\lambda = \nu/c = 2(J + 1)B_v/c$. *We shall use a tilde to indicate division of a molecular constant by c.* Thus, $\tilde{\nu} = 2(J + 1)\tilde{B}_v$, where $\tilde{B}_v \equiv B_v/c$. In particular [Eq. (20.16)],

$$\tilde{B}_0 \equiv B_0/c = h/8\pi^2 I_0 c \tag{20.33}$$

where I_0 is the moment of inertia averaged over the zero-point vibration. In the research literature, the tilde is often omitted.

For the majority of diatomic molecules, only the $v = 0$ vibrational level is significantly populated at room temperature, and the pure-rotation frequencies (20.31) become $2(J + 1)B_0$. The pure-rotation spectrum is a series of equally spaced lines at $2B_0$, $4B_0$, $6B_0$, . . . (if centrifugal distortion is neglected). The line at $2B_0$ is due to absorption by molecules in the $J = 0$ level; that at $4B_0$ is due to absorption by $J = 1$ molecules, etc. See Fig. 20.13.

If excited vibrational levels are appreciably populated, each rotational transition shows one or more nearby satellite lines due to transitions between rotational levels of the $v = 1$ vibrational level, between rotational levels of the $v = 2$ vibrational level, etc. These satellites are much weaker than the main line because the population of vibrational levels falls off rapidly as v increases.

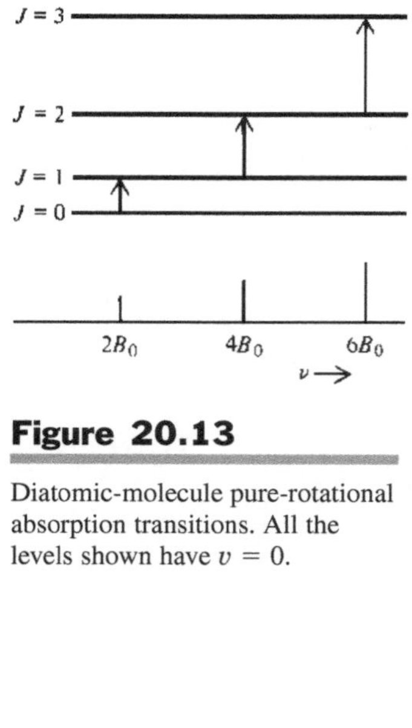

Figure 20.13

Diatomic-molecule pure-rotational absorption transitions. All the levels shown have $v = 0$.

Since the moments of inertia of different isotopic species of the same molecule differ, each isotopic species has its own pure-rotation spectrum. For $^{12}C^{16}O$, some observed pure-rotational transitions (all for $v = 0$) are (1 GHz $= 10^9$ Hz):

$J \to J + 1$	$0 \to 1$	$1 \to 2$	$2 \to 3$	$3 \to 4$	$4 \to 5$
ν/GHz	115.271	230.538	345.796	461.041	576.268

The slight decreases in the successive spacings are due to centrifugal distortion. For $^{13}C^{16}O$ and $^{12}C^{18}O$, the $J = 0 \to 1$ pure-rotation transitions occur at 110.201 and 109.782 GHz, respectively.

From the observed frequency of a pure-rotational transition of a heteronuclear diatomic molecule, one can use (20.31) to calculate B_0. From B_0 one gets I_0, and from $I_0 = \mu R_0^2$ one gets R_0, the internuclear distance averaged over the zero-point vibration. If vibrational satellites are observed, they can be used to find α_e and then (20.32) allows calculation of B_e and hence of the equilibrium bond distance R_e.

EXAMPLE 20.3 Calculating a bond distance from the rotational spectrum

Calculate R_0 for $^{12}C^{16}O$ using the $J = 0 \to 1$ frequency of 115.271 GHz for $v = 0$.

From (20.31) with $J = 0$, we have

$$\nu = 2B_0 = \frac{2h}{8\pi^2 I_0} = \frac{h}{4\pi^2 \mu R_0^2} \quad \text{and} \quad R_0 = \left(\frac{h}{4\pi^2 \mu \nu}\right)^{1/2}$$

This result also follows from $h\nu = E_{\text{upper}} - E_{\text{lower}} = 1(2)\hbar^2/2\mu R_0^2 - 0$, where the rotational levels (20.15) were used. An approximate value of μ for $^{12}C^{16}O$

was found in Sec. 20.3, and use of the accurate atomic masses in the table inside the back cover gives the accurate value $\mu = 1.13850 \times 10^{-23}$ g. Substitution in $R_0 = (h/4\pi^2\mu\nu)^{1/2}$ gives

$$R_0 = \left[\frac{6.62607 \times 10^{-34}\text{ J s}}{4\pi^2(1.13850 \times 10^{-26}\text{ kg})(115.271 \times 10^9\text{ s}^{-1})}\right]^{1/2} = 1.13089 \times 10^{-10}\text{ m}$$

Exercise

The $J = 2 \rightarrow 3$ pure-rotational transition for the $v = 0$ state of $^{39}K^{37}Cl$ occurs at 22410 MHz. Neglecting centrifugal distortion, find R_0 for $^{39}K^{37}Cl$. (*Answer:* 2.6708 Å.)

Vibration–Rotation Spectra of Diatomics

Transitions with a change in vibrational state ($\Delta v \neq 0$) and with no change in electronic state give the **vibration–rotation spectrum** of the molecule. These transitions involve infrared photons. Equation (20.30) shows that a diatomic molecule has a vibration–rotation spectrum only if the change in dipole moment $d\mu/dR$ is nonzero at R_e. When a homonuclear diatomic molecule vibrates, its μ remains zero. For a heteronuclear diatomic molecule, vibration changes μ. Therefore, a *diatomic molecule shows an IR vibration–rotation spectrum only if it is heteronuclear.*

The vibration–rotation spectrum is usually observed as an absorption spectrum. From (20.29) the absorption transitions have $\Delta v = +1$ $(+2, +3, \ldots)$, where the transitions in parentheses are much less probable. Since $\Delta J = 0$ is not allowed by the selection rule (20.28), there is no pure-vibration spectrum for a diatomic molecule; when v changes, J also changes.

Vibrational levels are much more widely spaced than rotational levels, so the IR vibration–rotation spectrum consists of a series of bands. Each **band** corresponds to a transition between two particular vibrational levels v'' and v' and consists of a series of lines, each line corresponding to a different change in rotational state. (See Fig. 20.14.) We shall first ignore the rotational structure of the bands and shall calculate the frequency of a *hypothetical* transition where v changes but J is 0 in both the initial and final states; this gives the position of the **band origin.**

IR spectroscopists commonly work with wavenumbers rather than frequencies. The wavenumber $\tilde{\nu}_{\text{origin}}$ of the band origin for an absorption transition from vibrational level v'' to v' is $\tilde{\nu}_{\text{origin}} = 1/\lambda = \nu/c = (E_{v'} - E_{v''})/hc$. The use of the energy expression (20.26) with $J = 0$ gives

$$\tilde{\nu}_{\text{origin}} = \tilde{\nu}_e(v' - v'') - \tilde{\nu}_e x_e[v'(v' + 1) - v''(v'' + 1)] \qquad (20.34)$$

$$\tilde{\nu}_e \equiv \nu_e/c, \qquad \tilde{\nu}_e x_e \equiv \nu_e x_e/c \qquad (20.35)$$

where ν_e is the molecule's equilibrium vibrational frequency.

Throughout this chapter, *a double prime will be used on quantum numbers of the lower state and a single prime on quantum numbers of the upper state.*

Usually, most of the molecules are in the $v = 0$ vibrational level at room temperature, and the strongest band is the $v = 0 \rightarrow 1$ band, called the **fundamental band.** The $v = 0 \rightarrow 2$ band (the **first overtone**) is much weaker than the fundamental band. From (20.34), the fundamental, first overtone, second overtone, . . . bands occur at $\tilde{\nu}_e - 2\tilde{\nu}_e x_e$, $2\tilde{\nu}_e - 6\tilde{\nu}_e x_e$, $3\tilde{\nu}_e - 12\tilde{\nu}_e x_e$, . . . :

$v'' \rightarrow v'$	$0 \rightarrow 1$	$0 \rightarrow 2$	$0 \rightarrow 3$
$\tilde{\nu}_{\text{origin}}$	$\tilde{\nu}_e - 2\tilde{\nu}_e x_e$	$2\tilde{\nu}_e - 6\tilde{\nu}_e x_e$	$3\tilde{\nu}_e - 12\tilde{\nu}_e x_e$

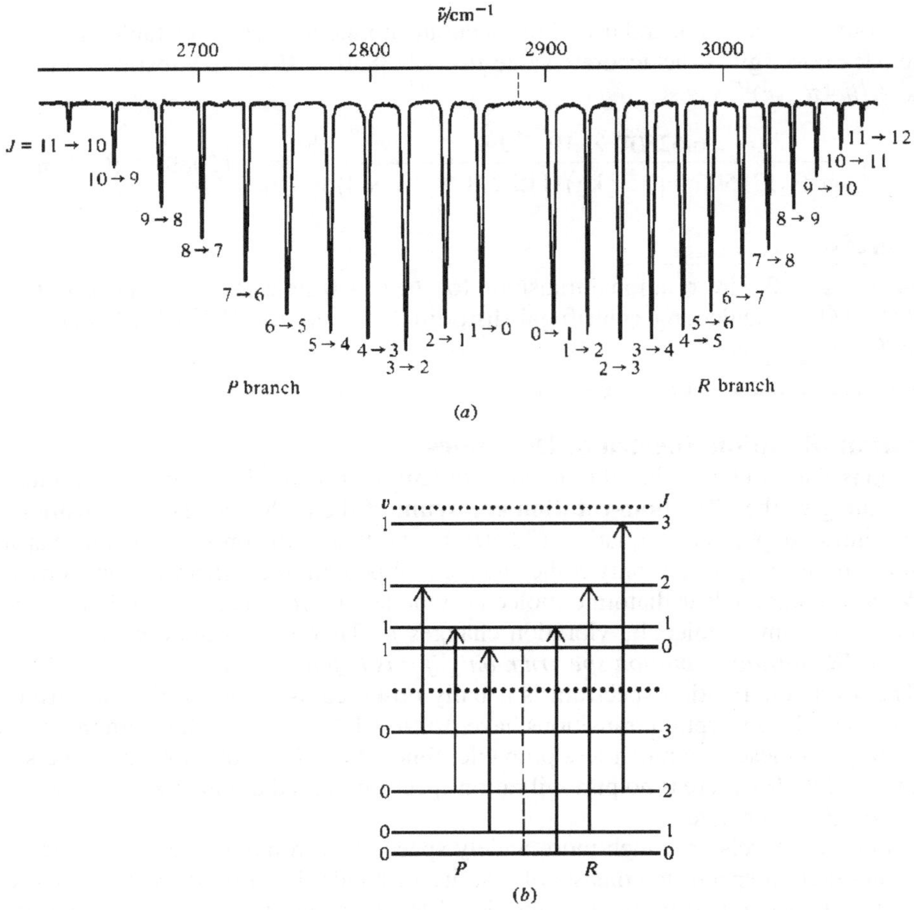

Figure 20.14

(*a*) The gas-phase low-pressure $v = 0 \rightarrow 1$ vibration band of $^1H^{35}Cl$ at room temperature. The dashed line indicates the position of the band origin ($\tilde{\nu}_{origin} = 2886\ cm^{-1}$). (In a sample of naturally occurring HCl, each line is a doublet because of the presence of $^1H^{37}Cl$.) It is conventional to display infrared absorption spectra with absorption intensity increasing downward. (*b*) The first few *P*-branch and *R*-branch transitions on an energy-level diagram. Each arrow in (*b*) is positioned directly below the spectrum line it corresponds to in (*a*).

Because the anharmonicity term $\nu_e x_e$ is much less than ν_e, the frequencies of the fundamental and overtone bands are *approximately* ν_e, $2\nu_e$, $3\nu_e$, etc. The wavenumber of the band origin of the fundamental band is called $\tilde{\nu}_0$:

$$\tilde{\nu}_0 = \tilde{\nu}_e - 2\tilde{\nu}_e x_e \quad \text{and} \quad \nu_0 = \nu_e - 2\nu_e x_e \tag{20.36}$$

The symbols ω_e, ω_0, and $\omega_e x_e$ are commonly used for $\tilde{\nu}_e$, $\tilde{\nu}_0$, and $\tilde{\nu}_e x_e$, respectively.

Some band origins of $^1H^{35}Cl$ IR bands are:

$v'' \rightarrow v'$	$0 \rightarrow 1$	$0 \rightarrow 2$	$0 \rightarrow 3$	$0 \rightarrow 4$	$0 \rightarrow 5$
$\tilde{\nu}_{origin}/cm^{-1}$	2886.0	5668.0	8346.8	10922.8	13396.2

From these data, $\tilde{\nu}_e$ and $\tilde{\nu}_e x_e$ can be found (Prob. 20.31).

If the vibrational frequency of the molecule is relatively low, or if the gas is heated, one observes absorption transitions originating from states with $v'' > 0$. These are called *hot bands*.

Having calculated the locations of IR band origins of diatomic molecules, we now deal with the rotational structure of each band. From the selection rule $\Delta J = \pm 1$ [Eq. (20.28)], each line in an IR band involves either $\Delta J = +1$ or $\Delta J = -1$. Vibration–rotation transitions with $\Delta J = +1$ give the ***R* branch** lines of the band; transitions with $\Delta J = -1$ give the ***P* branch.** Let J'' be the rotational quantum number of the lower vibration–rotation level and J' the rotational quantum number of the upper level (Fig. 20.15). The wavenumbers of the vibration–rotation lines are $\tilde{\nu} = (E_{J'v'} - E_{J''v''})/hc$.

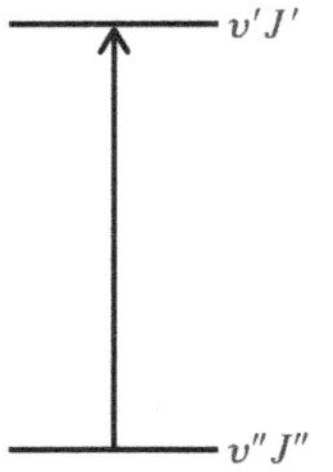

Figure 20.15

Energy levels of a vibration–rotation transition.

If we use the approximate energy expression (20.27) that neglects vibration–rotation interaction, then

$$\tilde{\nu} \approx \tilde{\nu}_e(v' - v'') - \tilde{\nu}_e x_e[v'(v' + 1) - v''(v'' + 1)] + \tilde{B}_e J'(J' + 1) - \tilde{B}_e J''(J'' + 1)$$

$$\tilde{\nu} \approx \tilde{\nu}_{\text{origin}} + \tilde{B}_e J'(J' + 1) - \tilde{B}_e J''(J'' + 1)$$

where (20.34) for $\tilde{\nu}_{\text{origin}}$ was used. For R branch lines, we have $J' = J'' + 1$. The use of this relation in the $\tilde{\nu}$ expression gives $\tilde{\nu}_R \approx \tilde{\nu}_{\text{origin}} + 2\tilde{B}_e(J'' + 1)$. Similarly, one finds (Prob. 20.28) $\tilde{\nu}_P \approx \tilde{\nu}_{\text{origin}} - 2\tilde{B}_e J''$. Thus

$$\tilde{\nu}_R \approx \tilde{\nu}_{\text{origin}} + 2\tilde{B}_e(J'' + 1) \qquad \text{where } J'' = 0, 1, 2, \ldots \tag{20.37}$$

$$\tilde{\nu}_P \approx \tilde{\nu}_{\text{origin}} - 2\tilde{B}_e J'' \qquad \text{where } J'' = 1, 2, 3, \ldots \tag{20.38}$$

$J'' = 0$ is excluded for the P branch lines because J' cannot be -1. The approximate equations (20.37) and (20.38) predict equally spaced lines on either side of $\tilde{\nu}_{\text{origin}}$ with no line at $\tilde{\nu}_{\text{origin}}$, since $J = 0 \to 0$ is forbidden for diatomic molecules.

The accurate expressions for $\tilde{\nu}_R$ and $\tilde{\nu}_P$ found using (20.26) (which includes vibration–rotation interaction) for E_{int} are (Prob. 20.29)

$$\tilde{\nu}_R = \tilde{\nu}_{\text{origin}} + [2\tilde{B}_e - \tilde{\alpha}_e(v' + v'' + 1)](J'' + 1) - \tilde{\alpha}_e(v' - v'')(J'' + 1)^2 \tag{20.39}$$

$$\tilde{\nu}_P = \tilde{\nu}_{\text{origin}} - [2\tilde{B}_e - \tilde{\alpha}_e(v' + v'' + 1)]J'' - \tilde{\alpha}_e(v' - v'')J''^2 \tag{20.40}$$

where $\tilde{\nu}_{\text{origin}}$ is given by (20.34) and the J'' values are given by (20.37) and (20.38). The vibration–rotation interaction constant α_e makes the R branch line spacings decrease as J'' increases and the P branch spacings increase as J'' increases.

Figure 20.14 shows the $v = 0 \to 1$ vibration–rotation band of $^1H^{35}Cl$.

Note that the line intensities in each branch in Fig. 20.14 first increase as J increases and then decrease at high J. The explanation for this is as follows. The Boltzmann distribution law $N_i/N_j = e^{-(E_i - E_j)/kT}$ gives the populations of the quantum-mechanical *states* i and j and not the populations of the quantum-mechanical energy levels. For each rotational energy level, there are $2J + 1$ rotational states, corresponding to the $2J + 1$ values of the M_J quantum number (Sec. 17.14), which does not affect the energy. Thus, the population of each rotational energy level J is $2J + 1$ times the population of the quantum-mechanical rotational state with quantum numbers J and M_J. For low J, this $2J + 1$ degeneracy factor outweighs the decreasing exponential in the Boltzmann distribution law, so the populations of rotational levels at first increase as J increases. For high J, the exponential dominates the $2J + 1$ factor, and the rotational populations decrease as J increases (Fig. 20.16).

Measurement of the wavenumbers of the lines of a vibration–rotation band allows $\tilde{\nu}_{\text{origin}}$, $\tilde{B}_e$, and $\tilde{\alpha}_e$ to be found from Eqs. (20.39) and (20.40). Knowing $\tilde{\nu}_{\text{origin}}$ for at least two bands, we can use (20.34) to find $\tilde{\nu}_e$ and $\tilde{\nu}_e x_e$. The rotational constant $\tilde{B}_e$ allows the moment of inertia and hence the internuclear distance R_e to be found. The vibrational frequency ν_e gives the force constant of the bond [Eq. (20.23)].

Vibrational and rotational constants for some diatomic molecules are listed in Table 20.1. Homonuclear diatomics have no pure-rotational (microwave) or vibration–rotation (IR) spectra, and their constants are found from electronic spectra (Sec. 20.11) or Raman spectra (Sec. 20.10). The values listed are for the ground electronic states, except that the CO* values are for one of the lower excited electronic states of CO. Note that $D_e/hc > \tilde{\nu}_e \gg \tilde{B}_e$, in accord with the earlier conclusion that electronic energies are greater than vibrational energies, which are greater than rotational energies. Note also the large force constants for N_2 and CO, which have triple bonds. Other things being equal, *the stronger the bond, the larger the force constant and the shorter the bond length.*

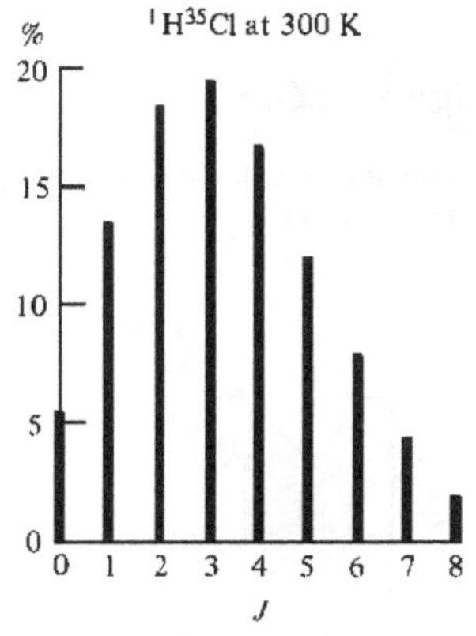

Figure 20.16

Percentage populations of $^1H^{35}Cl$ $v = 0$ rotational energy levels at 300 K.

TABLE 20.1

Constants of Some Diatomic Molecules[a]

Molecule	D_e/hc / cm^{-1}	R_e / Å	k_e / N/m	$\tilde{\nu}_e$ / cm^{-1}	$\tilde{B}_e$ / cm^{-1}	$\tilde{\alpha}_e$ / cm^{-1}	$\tilde{\nu}_e x_e$ / cm^{-1}
$^{1}H_2$	38297	0.741	576	4403.2	60.85	3.06	121.3
$^{1}H^{35}Cl$	37240	1.275	516	2990.9	10.593	0.31	52.8
$^{14}N_2$	79890	1.098	2295	2358.6	1.998	0.017	14.3
$^{12}C^{16}O$	90544	1.128	1902	2169.8	1.931	0.018	13.3
$^{12}C^{16}O$*	29424	1.370	555	1171.9	1.311	0.018	10.6
$^{127}I_2$	12550	2.666	172	214.5	0.0374	0.0001	0.6
$^{23}Na^{35}Cl$	34300	2.361	109	366	0.2181	0.0016	2.0
$^{12}C^{1}H$	29400	1.120	448	2858.5	14.457	0.53	63.0

[a]Data mainly from K. P. Huber and G. Herzberg, *Molecular Spectra and Molecular Structure,* vol. IV, *Constants of Diatomic Molecules,* Van Nostrand Reinhold, 1979.

The centrifugal-distortion constant D is of significant magnitude only for light molecules. Some values of $\tilde{D} \equiv D/c$ for ground electronic states are 0.05 cm^{-1} for $^{1}H_2$, 0.002 cm^{-1} for $^{1}H^{19}F$, 6×10^{-6} cm^{-1} for $^{14}N_2$, and 5×10^{-9} cm^{-1} for $^{127}I_2$.

20.5 MOLECULAR SYMMETRY

In Chapter 19, we used the symmetry of AOs to decide which AOs contribute to a given MO. We now discuss molecular symmetry elements as preparation for studying the rotational motion of polyatomic molecules. (The full application of symmetry to quantum chemistry requires group theory; see Sec. 20.16.)

Symmetry Elements

A molecule has an ***n*-fold axis of symmetry,** symbolized by C_n, if rotation by 360°/n (1/nth of a complete rotation) about this axis results in a nuclear configuration indistinguishable from the original one. For example, the bisector of the bond angle in HOH is a C_2 axis, since rotation by 360°/2 = 180° about this axis merely interchanges the two equivalent H atoms (Fig. 20.17). The hydrogens in the figure have been labeled H_a and H_b, but in reality they are physically indistinguishable from each other.

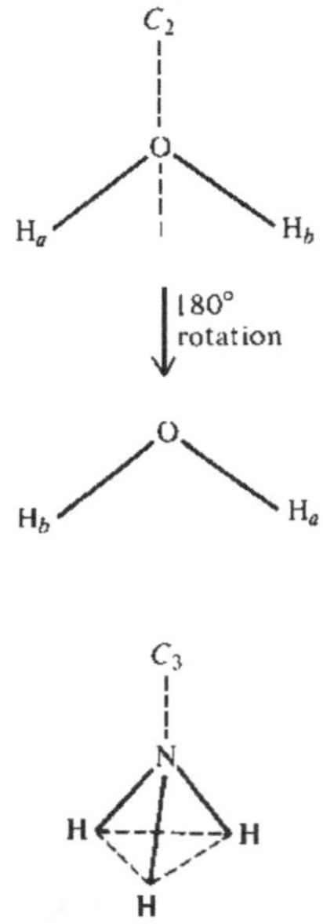

Figure 20.17

Symmetry axes (the dashed lines) in H_2O and NH_3.

The NH_3 molecule has a C_3 axis passing through the N nucleus and the midpoint of the triangle formed by the three H nuclei (Fig. 20.17). Rotation by 360°/3 = 120° about this axis sends equivalent hydrogens into one another. The hexagonal molecule benzene (C_6H_6) has a C_6 axis perpendicular to the molecular plane and passing through the center of the molecule. Rotation by 360°/6 about this axis sends equivalent nuclei into one another. This C_6 axis is also a C_3 axis and a C_2 axis, since 120° and 180° rotations about it send equivalent atoms into one another. Benzene has six other C_2 axes, each lying in the molecular plane; three of these go through two diagonally opposite carbons, and three bisect opposite pairs of carbon–carbon bonds.

A molecule has a **plane of symmetry,** symbolized by σ, if reflection of all nuclei through this plane sends the molecule into a configuration physically indistinguishable from the original one. Any planar molecule (for example, H_2O) has a plane of symmetry coinciding with the molecular plane, since reflection in the molecular plane leaves all nuclei unchanged in position. H_2O also has a second plane of symmetry; this lies perpendicular to the molecular plane (Fig. 20.18). Reflection through this second plane interchanges the equivalent hydrogens.

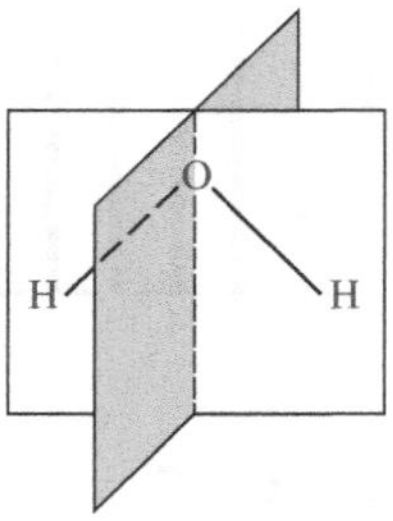

Figure 20.18

Symmetry planes in H_2O.

NH_3 has three planes of symmetry. Each symmetry plane passes through the nitrogen and one hydrogen and bisects the angle formed by the nitrogen and the other two hydrogens. Reflection in one of these planes leaves N and one H unchanged in position and interchanges the other two H's.

A molecule has a **center of symmetry,** symbol i, if inversion of each nucleus through this center results in a configuration indistinguishable from the original one. By **inversion** through point P we mean moving a nucleus at x, y, z to the location $-x$, $-y$, $-z$ on the opposite side of P, where the origin is at P. Neither H_2O nor NH_3 has a center of symmetry; p-dichlorobenzene has a center of symmetry (at the center of the benzene ring), but m-dichlorobenzene does not (Fig. 20.19). A molecule with a center of symmetry cannot have a dipole moment.

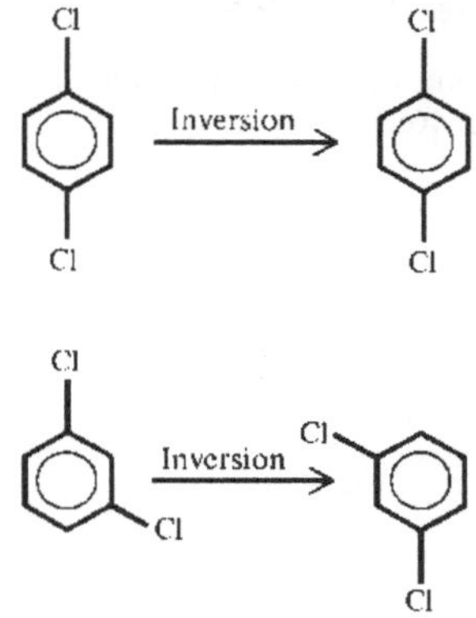

Figure 20.19

Inversion in p-dichlorobenzene and in m-dichlorobenzene. The coordinate origin is taken at the center of the ring.

A molecule has an ***n*-fold improper axis** (or **rotation--reflection axis**), symbol S_n, if rotation by 360°/n about the axis followed by reflection in a plane perpendicular to this axis sends the molecule into a configuration indistinguishable from the original one. Figure 20.20 shows a 90° rotation about an S_4 axis in CH_4 followed by a reflection in a plane perpendicular to this axis. The final configuration has hydrogens at the same locations in space as the original configuration. Note that the 90° rotation alone produces a configuration of the hydrogens that is distinguishable from the original one, so the S_4 axis in methane is not a C_4 axis. Methane has two other S_4 axes. Each S_4 axis goes through the centers of opposite faces of the cube in which the molecule has been inscribed. The C_6 axis in benzene is also an S_6 axis. (An S_2 axis is equivalent to a center of symmetry; see Prob. 20.40.)

Symmetry Operations

Associated with each symmetry element (C_n, S_n, σ, or i) of a molecule is a set of **symmetry operations.** For example, consider the C_4 axis of the square-planar molecule XeF_4. We can rotate the molecule by 90°, 180°, 270°, and 360° about this axis to give configurations indistinguishable from the original one (Fig. 20.21). The operation of rotating the molecule by 90° about the C_4 axis is symbolized by $\hat{C}_4$. Note the circumflex. Since a 180° rotation can be viewed as two successive 90° rotations, we symbolize a 180° rotation by $\hat{C}_4^2$, which equals $\hat{C}_4\hat{C}_4$. (Recall that multiplication of two operators means applying the operators in succession.) Similarly, 270° and 360° rotations are denoted by $\hat{C}_4^3$ and $\hat{C}_4^4$. Since a 360° rotation brings every nucleus back to its original location, we write $\hat{C}_4^4 = \hat{E}$, where $\hat{E}$ is the **identity operation,** defined as "do nothing." The operation $\hat{C}_4^5$, which is a 450° rotation about the C_4 axis, is the same as the 90° $\hat{C}_4$ rotation and is not counted as a new symmetry operation.

Since two successive reflections in the same plane bring the nuclei back to their original locations, we have $\hat{\sigma}^2 = \hat{E}$. Likewise, $\hat{\imath}^2 = \hat{E}$.

Since each symmetry operation must leave the location of the molecular center of mass unchanged, the symmetry elements of a molecule all intersect at the center of mass. Since a symmetry rotation must leave the orientation of the molecular dipole moment unchanged, the dipole moment of a molecule with a C_n axis must lie on this axis. A molecule with two or more noncoincident C_n axes has no dipole moment.

The set of symmetry operations of a molecule forms what mathematicians call a *group*.

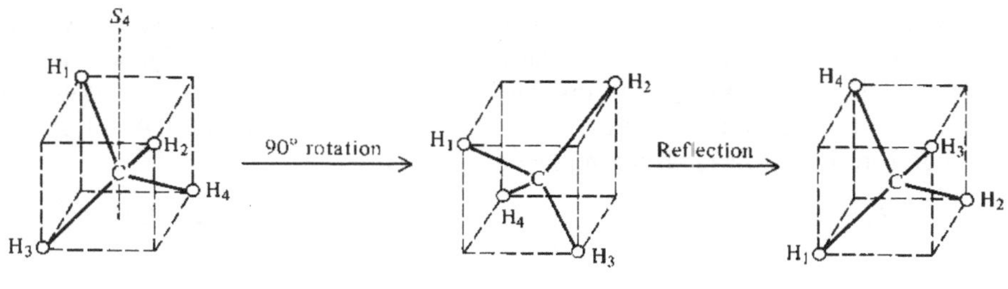

Figure 20.20

An S_4 axis in CH_4. The carbon atom is at the center of the cube.

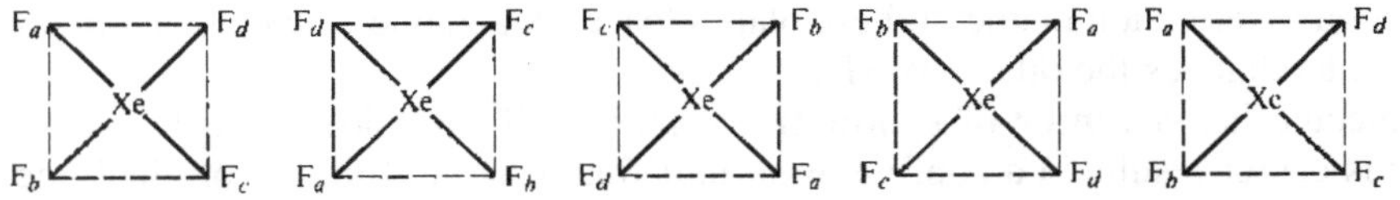

Figure 20.21

Some of the symmetry rotations for XeF_4.

20.6 ROTATION OF POLYATOMIC MOLECULES

As with diatomic molecules, it is usually a good approximation to take the energy of a polyatomic molecule as the sum of translational, rotational, vibrational, and electronic energies. We now consider the rotational energy levels.

Classical Mechanics of Rotation

The classical-mechanical treatment of the rotation of a three-dimensional body is complicated, and this section gives the results of such a treatment without giving proofs.

The **moment of inertia** I_x of a system of mass points $m_1, m_2, \ldots$ about an arbitrary axis x is defined by

$$I_x \equiv \sum_i m_i r_{x,i}^2 \qquad (20.41)$$

where $r_{x,i}$ is the perpendicular distance from mass m_i to the x axis. Consider a set of three mutually perpendicular axes x, y, and z. The *products of inertia* I_{xy}, I_{xz}, and I_{yz} for the x, y, z system are defined by the sums $I_{xy} \equiv -\Sigma_i m_i x_i y_i$, etc., where x_i and y_i are the x and y coordinates of mass m_i. Any three-dimensional body possesses three mutually perpendicular axes a, b, and c passing through the center of mass and having the property that the products of inertia I_{ab}, I_{ac}, and I_{bc} are each zero for these axes; these three axes are called the **principal axes of inertia** of the body. The moments of inertia I_a, I_b, and I_c calculated with respect to the principal axes are the **principal moments of inertia** of the body.

Symmetry aids in locating the principal axes. One can show that: *A molecular symmetry axis coincides with one of the principal axes. A molecular symmetry plane contains two of the principal axes and is perpendicular to the third.*

EXAMPLE 20.4 Principal axes and principal moments

The Xe—F bond length in the square-planar molecule XeF_4 (Fig. 20.22) is 1.94 Å. Locate the principal axes of inertia, and calculate the principal moments of inertia of XeF_4.

The center of mass is at the Xe nucleus, and the three principal axes of inertia must intersect at this point. One of the principal axes coincides with the C_4 symmetry axis perpendicular to the molecular plane. The other two principal axes are perpendicular to the C_4 axis and lie in the molecular plane. They can be taken to coincide with the two C_2 axes that pass through the four F's, or they can be taken to coincide with the two C_2 axes that bisect the FXeF angles. (For highly symmetric molecules, the orientation of the principal axes may not be unique.) Let us take the two in-plane principal axes (which we label the a and b axes) to go through the F atoms. The perpendicular distances of the atoms from the a principal axis are then 0 for Xe, 0 for two F's, and 1.94 Å for two F's. Therefore Eq. (20.41) with x replaced by a gives

$$I_a = 2(19.0 \text{ amu})(1.94 \text{ Å})^2 = 143 \text{ amu Å}^2 = 2.37 \times 10^{-45} \text{ kg m}^2$$

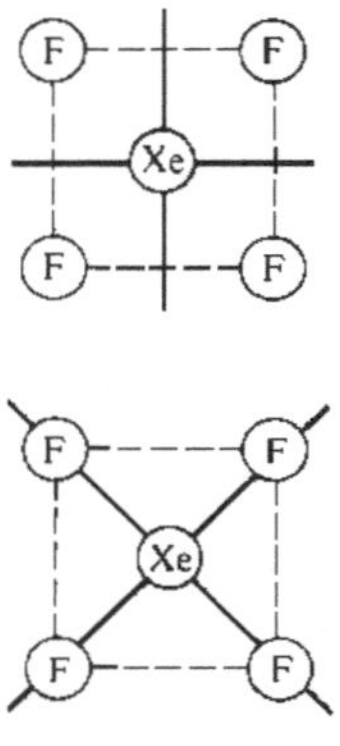

Figure 20.22

Two possible choices of in-plane principal axes in XeF_4.

since 1 amu = (1 g/mol)/N_A (Sec. 1.4). Clearly, $I_b = I_a$. The c axis is perpendicular to the molecular plane and goes through the Xe atom. Therefore, r_c in (20.41) is 1.94 Å for each F, and I_c = 4(19.0 amu)(1.94 Å)^2 = 286 amu Å^2.

Exercise

The octahedral molecule SF_6 has an S—F bond length of 1.56 Å. Find the principal moments of inertia of SF_6. (*Answers:* 185 amu Å^2, 185 amu Å^2, 185 amu Å^2.)

For molecules without symmetry, finding the principal axes is more complicated and is omitted.

The principal axes are important because the classical-mechanical rotational energy can be simply expressed in terms of the principal moments of inertia. The classical-mechanical rotational energy of a body turns out to be

$$E_{\text{rot}} = P_a^2/2I_a + P_b^2/2I_b + P_c^2/2I_c \qquad (20.42)$$

where P_a, P_b, and P_c are the components of the rotational angular momentum along the a, b, and c principal axes. Note the presence of three terms, each quadratic in a momentum, a fact mentioned in Sec. 14.10 on equipartition of energy.

If all the mass points lie on the same line (as in a linear molecule), this line is one of the principal axes, since it is a symmetry axis. The rotational angular momentum component of the body along this axis is zero, since the distance between all the masses and the axis is zero. Therefore, one of the three terms in (20.42) is zero, and a linear molecule has only two quadratic terms in its classical rotational energy.

The principal axes are labeled so that $I_a \le I_b \le I_c$.

A body is classified as a **spherical, symmetric,** or **asymmetric top,** according to whether three, two, or none of the principal moments I_a, I_b, I_c are equal:

Spherical top:	$I_a = I_b = I_c$
Symmetric top:	$I_a = I_b \neq I_c$ or $I_a \neq I_b = I_c$
Asymmetric top:	$I_a \neq I_b \neq I_c$

It can be shown that *a molecule with one C_n or S_n axis with $n \ge 3$ is a symmetric top. A molecule with two or more noncoincident C_n or S_n axes with $n \ge 3$ is a spherical top*. NF_3, with one C_3 axis, is a symmetric top. XeF_4, with one C_4 axis, is a symmetric top. CCl_4, with four noncoincident C_3 axes (one through each C—Cl bond), is a spherical top. H_2O, with no C_3 or higher axis, is an asymmetric top.

Quantum-Mechanical Rotational Energy Levels

(For proofs of the following results, see *Harmony,* chap. 7.)

The **rotational constants** A, B, C of a polyatomic molecule are defined by [similar to (20.16)]

$$A \equiv h/8\pi^2 I_a, \quad B \equiv h/8\pi^2 I_b, \quad C \equiv h/8\pi^2 I_c \qquad (20.43)$$

The rotational energy levels of a spherical top are

$$E_{\text{rot}} = J(J+1)\hbar^2/2I = BhJ(J+1), \quad J = 0, 1, 2, \ldots \qquad (20.44)$$

where $I \equiv I_a = I_b = I_c$. The spherical-top rotational wave functions involve the quantum numbers J, K, and M, where K and M each range from $-J$ to J in integral steps. Each spherical-top rotational level is $(2J+1)^2$-fold degenerate, corresponding to the $(2J+1)^2$ different choices for K and M for a fixed J.

The rotational levels of a symmetric top with $I_b = I_c$ are

$$E_{\text{rot}} = \frac{J(J+1)\hbar^2}{2I_b} + K^2\hbar^2\left(\frac{1}{2I_a} - \frac{1}{2I_b}\right) = BhJ(J+1) + (A-B)hK^2$$

$$J = 0, 1, 2, \ldots; \qquad K = -J, -J+1, \ldots, J-1, J \tag{20.45}$$

There is also an M quantum number that does not affect E_{rot}. If $I_a = I_b$, then I_a and A in (20.45) are replaced by I_c and C. The quantity $K\hbar$ is the component of the rotational angular momentum along the molecular symmetry axis.

The axis of a linear molecule is a C_∞ axis, since rotations by 360°/n, where n = 2, 3, . . . , ∞, about this axis are symmetry operations. Hence, a linear molecule is a symmetric top. (This is also obvious from the fact that the moments of inertia about all axes that pass through the center of mass and are perpendicular to the molecular axis are equal.) Because all the nuclei lie on the C_∞ symmetry axis, there can be no rotational angular momentum along this axis and K must be zero for a linear molecule. Hence, Eq. (20.45) gives

$$E_{\text{rot}} = J(J+1)\hbar^2/2I_b = BhJ(J+1) \qquad \text{linear molecule}$$

where I_b is the moment of inertia about an axis through the center of mass and perpendicular to the molecular axis. A special case is a diatomic molecule, Eq. (20.15).

The rotational levels for an asymmetric top are extremely complicated and follow no simple pattern.

As with diatomic molecules (Sec. 20.3), when a polyatomic molecule has identical nuclei that are interchanged by a rotation (for example NH_3), the effect of nuclear spin on the degeneracy of the rotational levels must be considered. For example, for the ground electronic state of $C^{16}O_2$, it turns out that for certain vibrational levels, the J = 1, 3, 5, . . . rotational levels do not exist, and for other vibrational levels, the even-J rotational levels do not exist. (For details, see *Herzberg,* vol. II, pp. 15–18, 26–29, 38–40, 52–55; I. N. Levine, *Molecular Spectroscopy,* Wiley, 1975, secs. 4.8 and 6.10.)

Selection Rules

The selection rules for the pure-rotational (microwave) spectra of polyatomic molecules are as follows.

Just as for diatomic molecules, *a polyatomic molecule must have a nonzero dipole moment to undergo a pure-rotational transition with absorption or emission of radiation.* Because of their high symmetry, all spherical tops (for example, CCl_4 and SF_6) and some symmetric tops (for example, C_6H_6 and XeF_4) have no dipole moment and exhibit no pure-rotation spectrum.

For a symmetric top with a dipole moment (for example, CH_3F), the pure-rotational selection rules are

$$\Delta J = \pm 1, \qquad \Delta K = 0 \tag{20.46}$$

The use of (20.46), (20.45), and $E_{\text{upper}} - E_{\text{lower}} = h\nu$ gives the frequency of the pure-rotational $J \rightarrow J+1$ absorption transition as

$$\nu = 2B(J+1), \qquad J = 0, 1, 2, \ldots \qquad \text{symmetric top} \tag{20.47}$$

The microwave spectrum of a symmetric top consists of a series of equally spaced lines at $2B$, $4B$, $6B$, . . . (provided centrifugal distortion is negligible).

The most populated vibrational state has all vibrational quantum numbers equal to zero, and the rotational constant B determined from the microwave spectrum is an average over the zero-point vibrations and is designated B_0. If excited vibrational levels are significantly populated at room temperature, vibrational satellites are observed, as discussed for diatomic molecules.

The selection rules for asymmetric tops are complicated, and are omitted.

20.7 MICROWAVE SPECTROSCOPY

Except for light diatomic molecules, pure-rotational spectra occur in the microwave portion of the electromagnetic spectrum. Figure 20.23 sketches a highly simplified version of a microwave spectrometer. A special electronic tube (either a klystron or a backward-wave oscillator) generates virtually monochromatic microwave radiation, whose frequency can be readily varied over a wide range. The radiation is transmitted through a hollow metal pipe called a *waveguide*. A portion of waveguide sealed at both ends with mica windows is the absorption cell. The cell is filled with low-pressure (0.01 to 0.1 torr) vapor of the molecule to be investigated. (At medium and high pressures, intermolecular interactions broaden the rotational absorption lines, giving an essentially continuous rotational absorption spectrum.) The microwave radiation is detected with a metal-rod antenna mounted in the waveguide and connected to a semiconductor diode. When the klystron frequency coincides with one of the absorption frequencies of the gas, there is a dip in the microwave power reaching the detector.

Any substance with a dipole moment and with sufficient vapor pressure can be studied. The use of a waveguide heated to 1000 K enables the rotational spectra of the alkali halides to be observed. Very large molecules are hard to study. Such molecules have many low-energy vibrational levels that are significantly populated at room temperature; this produces a microwave spectrum with so many lines that it is extremely hard to figure out which rotational transitions the various lines correspond to. Some large molecules whose microwave spectra have been successfully investigated are azulene, β-fluoronaphthalene, and $C_6H_6Cr(CO)_3$.

The microwave spectra of molecular ions produced continuously by a glow discharge in the absorption cell and of free radicals produced by an electric discharge (or by reaction of products of an electric discharge with stable species) and continuously pumped through the absorption cell have been observed. Species studied include CO^+, HCO^+, CF_2, PH_2, and CH_3O. See E. Hirota, *Chem. Rev.,* **92,** 141 (1992). The microwave spectra of van der Waals molecules (Sec. 21.10) and hydrogen-bonded dimers such as $(H_2O)_2$ have been studied using molecular beams (Sec. 22.3).

The microwave spectrum of a symmetric top is very simple [Eq. (20.47)] and yields the rotational constant B_0. The microwave spectrum of an asymmetric top is quite complicated, but once one has assigned several lines to transitions between specific rotational levels, one can calculate the three rotational constants A_0, B_0, C_0 and the corresponding principal moments of inertia.

The principal moments of inertia depend on the bond distances and angles and the conformation. Knowledge of one moment of inertia for a symmetric top or three moments of inertia for an asymmetric top is generally not enough information to determine the structure fully. One therefore prepares isotopically substituted species of the molecule and observes their microwave spectra to find their moments of inertia. The molecular geometry is determined by the nuclear configuration that minimizes the energy E_e in the electronic Schrödinger equation $\hat{H}_e\psi_e = E_e\psi_e$. The terms in the electronic Hamiltonian $\hat{H}_e$ are independent of the nuclear masses. Therefore, isotopic substitution does not affect the equilibrium geometry. (This isn't quite true, because

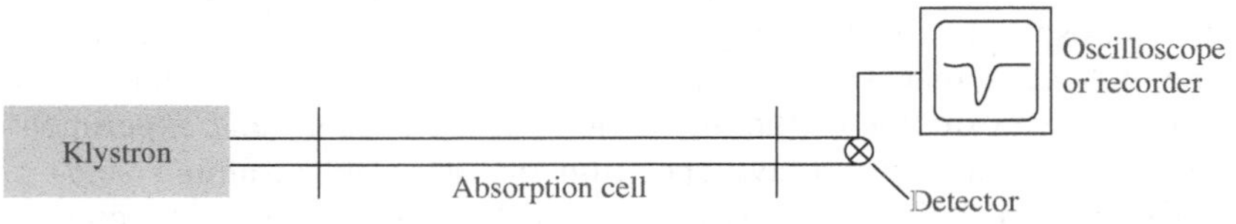

Figure 20.23

A microwave spectrometer.

of very slight deviations from the Born–Oppenheimer approximation.) When enough isotopically substituted species have been studied, the complete molecular structure can be determined. Some molecular structures found by microwave spectroscopy are:

CH_2F_2: $R(\mathrm{CH}) = 1.09$ Å, $R(\mathrm{CF}) = 1.36$ Å, $\angle\mathrm{HCH} = 113.7°$, $\angle\mathrm{FCF} = 108.3°$

CH_3OH: $R(\mathrm{CH}) = 1.09$ Å, $R(\mathrm{CO}) = 1.43$ Å, $R(\mathrm{OH}) = 0.94$ Å,

$\angle\mathrm{HCH} = 108.6°$, $\angle\mathrm{COH} = 108.5°$

C_6H_5F: average $R(\mathrm{CC}) = 1.39$ Å, $R(\mathrm{CH}) = 1.08$ Å, $R(\mathrm{CF}) = 1.35$ Å

H_2S: $R(\mathrm{SH}) = 1.34$ Å, $\angle\mathrm{HSH} = 92.1°$

SO_2: $R(\mathrm{SO}) = 1.43$ Å, $\angle\mathrm{OSO} = 119.3°$

O_3: $R(\mathrm{OO}) = 1.27$ Å, $\angle\mathrm{OOO} = 116.8°$

Molecular dipole moments can be found from microwave spectra. An insulated metal plate is inserted lengthwise in the waveguide. Application of a voltage to this plate subjects the gas molecules to an electric field, which shifts their rotational energies. (A shift in molecular energy levels due to an applied external electric field is called a *Stark effect*.) The magnitudes of these shifts depend on the components of the molecular electric dipole moment. Microwave spectroscopy gives quite accurate dipole moments. Moreover, it gives the components of the dipole-moment vector along the principal axes of inertia, so the orientation of the dipole-moment vector is known. Some dipole moments in debyes determined by microwave spectroscopy are

C_3H_8	$HC(CH_3)_3$	H_2O_2	H_2O	H_2S	azulene	NaCl	KCl	HCl	ClF	CH_3D
0.08	0.13	2.2	1.85	0.97	0.80	9.0	10.3	1.12	0.88	0.006

The nonzero dipole moments of the saturated hydrocarbons $H_2C(CH_3)_2$ and $HC(CH_3)_3$ are due to deviations from the $109\frac{1}{2}°$ tetrahedral bond angle and to polarization of the electron probability densities of the CH_3, CH_2, and CH groups by the unsymmetrical electric fields in the molecules [S. W. Benson and M. Luria, *J. Am. Chem. Soc.*, **97,** 704 (1975)]. The dipole moment of CH_3D is due to deviations from the Born–Oppenheimer approximation.

Besides molecular structures and dipole moments, microwave spectroscopy also yields values of barriers to internal rotation (Sec. 19.8) in polar molecules. Although internal rotation is a vibrational motion, it affects the pure-rotational spectrum, which allows barriers to be found. (See *Sugden and Kenney,* chap. 8.)

Many molecular species in interstellar space (in interstellar clouds of gas and dust or in circumstellar shells) have been detected mainly by observation of microwave emission lines using a radio telescope. Over 120 species have been found including OH, H_3O^+, H_2O, NH_3, SO_2, CH_3OH, CH_3CHO, C_2H_5OH, HCOOH, NH_2CHO, HC_9N, and NaCl. For a list of species found, see www.cv.nrao.edu/~awootten/allmols.html. Some evidence suggests the presence of the amino acid glycine, but this has not been confirmed [L. E. Snyder et al., *Astrophys. J.*, **619,** 914 (2005)].

Observed intensities of the microwave emission lines from ClO and O_3 in the Antarctic stratosphere show a strong correlation between increasing ClO and decreasing ozone, indicating that Antarctic ozone depletion is due to chlorofluorocarbons.

An impressive example of the interaction between theory and experiment is the study of the gas-phase microwave spectrum of the simplest amino acid, glycine, NH_2CH_2COOH. In 1978, the microwave spectrum of the glycine conformation labeled II in Fig. 20.24 was observed and rotational transitions were assigned, allowing its rotational constants to be found. However, ab initio SCF calculations predicted that conformation I in Fig. 20.24 would be 1 or 2 kcal/mol lower in energy than the

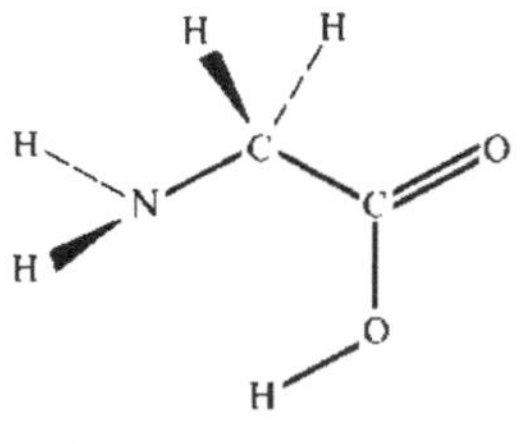

II

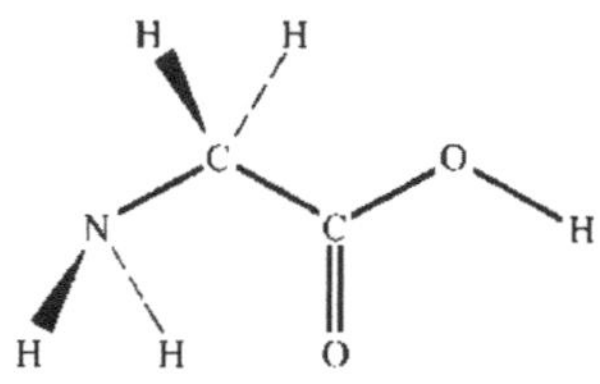

I

Figure 20.24

Two conformations of glycine. Conformer I has three intramolecular hydrogen bonds; II has one such bond.

observed conformation II. This prompted a further search of the microwave spectrum. Guided by the theoretically predicted rotational constants for conformation I, the experimentalists found transitions due to I and obtained its rotational constants, which are in excellent agreement with the theoretically predicted values. (The lowest-energy conformation I has a weaker microwave spectrum than II does because the dipole moment of I is much smaller.) The observed line intensities indicate that I is $1._4$ kcal/mol more stable than II. [See L. Schäfer et al., *J. Am. Chem. Soc.*, **102,** 6566 (1980); R. D. Suenram and F. J. Lovas, *ibid.*, **102,** 7180 (1980); T. F. Miller and D. C. Clary, *Phys. Chem. Chem. Phys.*, **6,** 2563 (2004).]

More than 1000 molecules have been studied by microwave spectroscopy.

20.8 VIBRATION OF POLYATOMIC MOLECULES

The nuclear wave function ψ_N of a molecule containing $\mathcal{N}$ atoms is a function of the $3\mathcal{N}$ spatial coordinates needed to specify the locations of the nuclei. The nuclear motions are translations, rotations, and vibrations. The translational wave function is a function of three coordinates, the x, y, and z coordinates of the center of mass. The rotational wave function of a linear molecule is a function of the two angles θ and ϕ (Fig. 20.7) needed to specify the spatial orientation of the molecular axis. To specify the orientation of a nonlinear molecule, one chooses some axis in the molecule, gives θ and ϕ for this axis, and gives the angle of rotation of the molecule itself about this axis. Thus, ψ_{rot} depends on three angles for a nonlinear molecule. The nuclear vibrational wave function therefore depends on $3\mathcal{N} - 3 - 2 = 3\mathcal{N} - 5$ coordinates for a linear molecule and $3\mathcal{N} - 3 - 3 = 3\mathcal{N} - 6$ coordinates for a nonlinear molecule. The number of independent coordinates needed to specify each kind of motion (translation, rotation, vibration) is called the number of **degrees of freedom** for that kind of motion (Fig. 20.25). Each such coordinate specifies a mode of motion, so the number of degrees of freedom is the number of modes of motion.

DEGREES OF FREEDOM

	Lin.	Nonlin.
Trans.	3	3
Rot.	2	3
Vib.	$3\mathcal{N} - 5$	$3\mathcal{N} - 6$
Total	$3\mathcal{N}$	$3\mathcal{N}$

Figure 20.25

Degrees of freedom of a molecule with $\mathcal{N}$ atoms.

We first discuss the classical-mechanical treatment of molecular vibration and then the quantum-mechanical treatment.

Classical Mechanics of Vibration

Consider a molecule with $\mathcal{N}$ nuclei vibrating about their equilibrium positions subject to the potential energy E_e, where E_e is a function of the nuclear coordinates and is found by solving the electronic Schrödinger equation (19.7). For small vibrations, we can expand E_e in a Taylor series and neglect terms higher than quadratic, as we did for a diatomic molecule [Eq. (20.18)]. Substituting this expansion into Newton's second law, one finds that any classical molecular vibration can be expressed as a linear combination of what are called normal modes of vibration.

In a given **normal mode,** all the nuclei vibrate about their equilibrium positions at the same frequency; the nuclei all vibrate in phase, meaning that each nucleus passes through its equilibrium position at the same time. However, the vibrational amplitudes of different nuclei may differ. A molecule has $3\mathcal{N} - 6$ or $3\mathcal{N} - 5$ normal modes, depending on whether it is nonlinear or linear, respectively.

For a diatomic molecule, $3\mathcal{N} - 5 = 1$. In the single normal mode, the two atoms vibrate along the internuclear axis. For a heteronuclear diatomic molecule, the vibrational amplitude of the heavier atom is less than that of the lighter atom.

Each normal mode has its own vibrational frequency. The forms and frequencies of the normal modes depend on the molecular geometry, the nuclear masses, and the force constants (the second derivatives of E_e with respect to the nuclear coordinates). If these quantities are known, one can find the normal modes and their frequencies.

(For each normal mode, the vibrational Hamiltonian has a kinetic-energy term that is quadratic in a momentum and a potential-energy term that is quadratic in a coordinate. This fact was used in the Sec. 14.10 discussion of equipartition.)

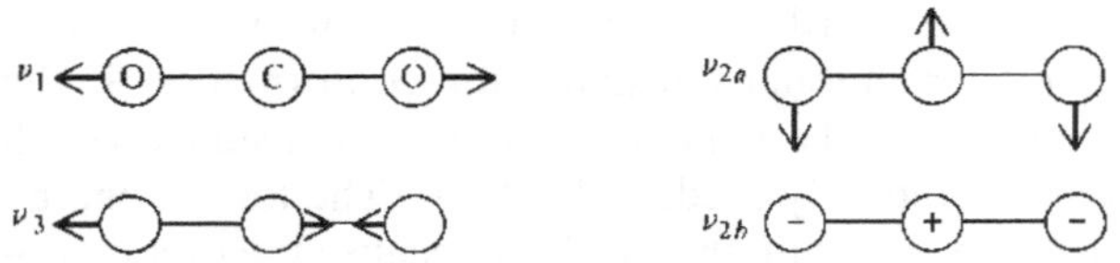

Figure 20.26

Normal vibrational modes of CO_2. The plus and minus signs indicate motion out of and into the plane of the paper.

Figure 20.26 shows the normal modes of the linear molecule CO_2, which has $3(3) - 5 = 4$ normal modes. In the *symmetric stretching* vibration labeled ν_1, the carbon nucleus is motionless. In the *asymmetric stretch* ν_3, all the nuclei vibrate. The *bending* mode ν_{2b} is the same as ν_{2a} rotated by 90° about the molecular axis. Clearly, these two modes have the same frequency. Bond-bending frequencies are generally lower than bond-stretching frequencies. For CO_2, observation of IR and Raman spectra gives $\tilde{\nu}_1 = 1340\ \text{cm}^{-1}$, $\tilde{\nu}_2 = 667\ \text{cm}^{-1}$, $\tilde{\nu}_3 = 2349\ \text{cm}^{-1}$, where $\tilde{\nu} = \nu/c$.

Figure 20.27 shows the three normal modes of the nonlinear molecule H_2O. Here, ν_1 and ν_3 are stretching vibrations, and ν_2 is a bending vibration.

In each of the diagrams of Figs. 20.26 and 20.27, the arrows show the directions of motion of the atoms at one moment during a normal mode of vibration. Since the atoms oscillate about their equilibrium positions, the directions of atomic motions change twice during each cycle of vibration. The relative magnitudes of the arrows indicate the relative amplitudes of the atomic vibrations. The amplitude of the heavy O atom is much less than that of an H atom.

Quantum Mechanics of Vibration

Neglecting terms higher than quadratic in the E_e expansion and expressing the vibrations in terms of normal modes, one finds that the Hamiltonian for vibration becomes the sum of harmonic-oscillator Hamiltonians, one for each normal mode. Hence, *the quantum-mechanical vibrational energy of a polyatomic molecule is approximately the sum of* $3\mathcal{N} - 6$ *or* $3\mathcal{N} - 5$ *harmonic-oscillator energies, one for each normal mode:*

$$E_{\text{vib}} \approx \sum_{i=1}^{3\mathcal{N}-6} (v_i + \tfrac{1}{2})h\nu_i \qquad \textbf{(20.48)*}$$

$$v_1 = 0, 1, 2, \ldots, \quad v_2 = 0, 1, 2, \ldots, \quad \cdots \quad v_{3\mathcal{N}-6} = 0, 1, 2, \ldots$$

where ν_i is the frequency of the ith normal mode and v_i is its quantum number. For linear molecules, the upper limit is $3\mathcal{N} - 5$. The $3\mathcal{N} - 6$ or $3\mathcal{N} - 5$ vibrational quantum numbers vary independently of one another. The ground vibrational level has all v_i's equal to zero and has the zero-point energy $\frac{1}{2}\sum_i h\nu_i$ (anharmonicity neglected).

One finds that the most probable vibration–rotation absorption transitions are those for which one vibrational quantum number v_j changes by $+1$ with all others (v_k, $k \neq j$) unchanged. However, for the transition $v_j \to v_j + 1$ to occur, *the jth normal-mode vibration must change the molecular dipole moment*. Since the harmonic-oscillator levels are spaced by $h\nu_j$, the transition $v_j \to v_j + 1$ produces a band in the IR spectrum at the frequency ν_j [$\nu_{\text{light}} = (E_{\text{upper}} - E_{\text{lower}})/h = h\nu_j/h = \nu_j$].

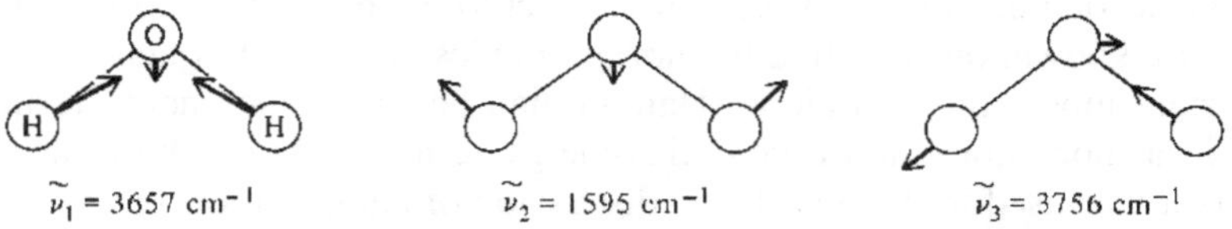

Figure 20.27

The normal modes of H_2O. The heavy oxygen atom has a much smaller vibrational amplitude than the light hydrogen atoms.

For CO_2, the symmetric stretch ν_1 leaves the dipole moment unchanged, and no IR band is observed at ν_1. One says this vibration is **IR-inactive.** The CO_2 vibrations ν_2 and ν_3 each change the dipole moment and are **IR-active.**

In a molecule with no symmetry elements, all the normal modes change the dipole moment and all are IR-active. The only molecules with no IR-active modes are homonuclear diatomics.

The most populated vibrational level has all the v_i's equal to 0. A transition from this level to a level with $v_j = 1$ and all other vibrational quantum numbers equal to zero gives an **IR fundamental band.** Transitions where one v_j changes by 2 or more and all others are unchanged give **overtone bands.** Transitions where two vibrational quantum numbers change give **combination bands.**

The selection rules determining the allowed overtone and combination bands are not so readily found as for the fundamental bands. For example, for the IR-active vibration ν_3 of CO_2, the first, third, fifth, . . . ($v_3' = 2, 4, 6$ . . .) overtones are IR-inactive, and the second, fourth, sixth, . . . ($v_3' = 3, 5, 7$ . . .) overtones are IR-active. An overtone of a forbidden fundamental may be allowed in certain cases. A combination band might involve a change in the quantum number of an IR-inactive vibration. The best way to determine which overtone and combination bands are allowed is to use group theory. (See *Herzberg,* vol. II, chap. III for the procedure.)

Overtone and combination bands are substantially weaker than fundamental bands.

The frequencies of the IR fundamental bands give (some of) the **fundamental vibration frequencies** of the molecule. Because of anharmonicity, the fundamental frequencies differ somewhat from the equilibrium vibrational frequencies [see Eq. (20.36)]. Usually, not enough data are available to determine the anharmonicity corrections, and so one works with the fundamental frequencies.

For H_2O, all three normal modes are IR-active. Wavenumbers of some IR band origins for H_2O vapor follow. Also listed are the band intensities (s = strong, m = medium, w = weak) and the quantum numbers $v_1' \, v_2' \, v_3'$ of the upper vibrational level. In all cases, the lower level is the 000 ground state.

$\tilde{\nu}_{origin}/cm^{-1}$	1595(s)	3152(m)	3657(s)	3756(s)	5331(m)	6872(w)
$v_1' \, v_2' \, v_3'$	010	020	100	001	011	021

The three strongest bands are the fundamentals and give the fundamental vibration wavenumbers as $\tilde{\nu}_1 = 3657\ cm^{-1}$, $\tilde{\nu}_2 = 1595\ cm^{-1}$, and $\tilde{\nu}_3 = 3756\ cm^{-1}$. Because of anharmonicity, the overtone transition 000 $\rightarrow$ 020 occurs at a bit less than twice the frequency of the 000 $\rightarrow$ 010 fundamental band. The combination band at 6872 cm^{-1} has $\tilde{\nu}_{origin} \approx 2\tilde{\nu}_2 + \tilde{\nu}_3 = 6946\ cm^{-1}$.

Molecules with more than, say, five atoms generally have one or more vibrational modes with frequencies low enough to have excited vibrational levels significantly populated at room temperature. This gives hot bands in the IR absorption spectrum. Figure 20.28 plots the fraction of molecules whose vibrational quantum number v_i is 0 versus the vibrational wavenumber $\tilde{\nu}_i$ for two temperatures. At room temperature, only vibrational modes with $\tilde{\nu}_i$ less than 700 cm^{-1} have excited vibrational levels significantly populated.

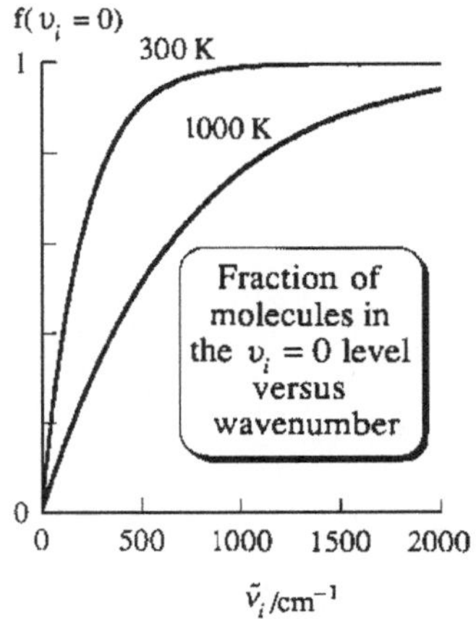

Figure 20.28

The fraction of molecules with vibrational quantum number $v_i = 0$ plotted versus vibrational wavenumber $\tilde{\nu}_i$ at two temperatures.

As with diatomic molecules, gas-phase IR bands of polyatomic molecules show a rotational structure. For certain vibrational transitions, the $\Delta J = 0$ transition is allowed, giving a line (called the ***Q* branch**) at the band origin.

The many modes of vibration of large molecules make it hard to correctly assign observed IR bands to various transitions. Rather accurate quantum-mechanical calculation of molecular vibrational frequencies is possible for molecules not too large; "it is virtually impossible to interpret and correctly assign the vibrational spectra of large polyatomic molecules without quantum-mechanical calculations" (P. Pulay in D. R. Yarkony, ed., *Modern Electronic Structure Theory,* World Scientific, 1995, Part II, chap. 19).

The absorption of IR radiation by atmospheric water vapor and CO_2 has a major effect on the earth's temperature. The sun's surface is at 6000 K, and most of the energy of its radiation (which is approximately that of a blackbody) lies in the visible region and the near-UV and near-IR regions bordering the visible. The earth's atmospheric gases do not have major absorptions in these regions, so most of the sun's radiation reaches the earth's surface. (Recall, however, the UV absorption by stratospheric O_3; Sec. 16.16.) The earth's surface is at 300 K, and most of the energy it radiates is in the mid-infrared. H_2O vapor and CO_2 in the atmosphere absorb a significant fraction of the IR radiation emitted by the earth and reradiate part of it back to the earth, making the earth warmer than if these gases were absent (the "greenhouse effect"). A *greenhouse gas* is one that absorbs in the IR region from 5 to 50 μm, where most of the earth's radiation is emitted. As noted earlier in this section, the most abundant atmospheric gases N_2, O_2, and Ar do not absorb IR radiation. H_2O and many minor atmospheric gases such as CO_2, CH_4, O_3, SF_6, and the chlorofluorocarbons (Sec. 16.16) are greenhouse gases.

The atmosphere's CO_2 content is steadily increasing because of the burning of fossil fuels. The Fourth Assessment Report "Climate Change 2007" of the Intergovernmental Panel on Climate Change (co-winner of the 2007 Nobel Peace Prize) examined six different scenarios for future economic and population growth, and for each of these scenarios, a best estimate and a likely range for the 1990 to 2100 increase in the Earth's average surface temperature (in the absence of additional climate policies) was given (www.ipcc.ch). The best estimates ranged from 1.8 to 4.0 °C (3.2 to 7.2 °F) and the minimum and maximum changes for the likely ranges were 1.1 and 6.4 °C (2.0 and 11.5 °F).

20.9 INFRARED SPECTROSCOPY

Figure 20.29 outlines a double-beam dispersion IR spectrometer. The radiation source is an electrically heated rod that emits continuous-frequency radiation. If the sample is in solution, the reference cell is filled with pure solvent; otherwise, it is empty. The chopper causes the sample beam and the reference beam to fall alternately on the prism. The prism disperses the radiation into its component frequencies. (The use of a diffraction grating instead of a prism gives higher resolution.) The mirror rotates, thereby changing the frequency of the radiation falling on the detector. The motor that drives this mirror also drives the chart paper on which the spectrum is recorded as a function of frequency. If the sample does not absorb at a given frequency, the sample and reference beams falling on the detector are equally intense and the ac amplifier receives no ac signal. At a frequency for which the sample absorbs,

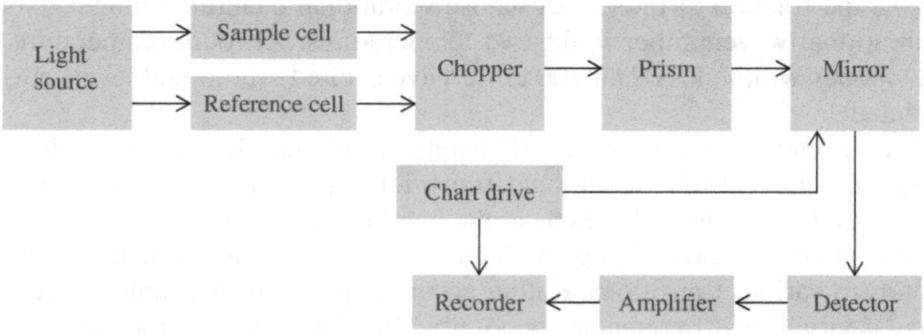

Figure 20.29

A double-beam infrared spectrometer.

the light intensity reaching the detector varies at a frequency equal to the chopper frequency; the detector then puts out an ac signal whose strength depends on the intensity of absorption. IR detectors used include thermocouples, temperature-dependent resistors, and photoconductive materials. Dispersion IR spectrometers have been largely replaced by Fourier-transform IR spectrometers, which are discussed at the end of this section.

Applications of IR Spectroscopy

The sample may be gaseous, liquid, or solid. Solids are ground up with a hydrocarbon or fluorocarbon oil to form a paste, or they are ground up with KBr and squeezed in a die to form a transparent disk. Samples of biopolymers and synthetic polymers are often studied as very thin films formed by evaporating to dryness a solution of the polymer on the absorption-cell window. IR spectra of solids at high pressures can be obtained by using a diamond-anvil cell (Sec. 7.4).

One can study the IR spectrum of a species chemisorbed on a metal catalyst, by compressing tiny metal crystals supported on an inert oxide such as SiO_2 to form a thin disk. The disk is mounted inside an IR absorption cell, the cell is evacuated, and the adsorbate gas is admitted to the cell (Fig. 20.30*a*). The spectrum of the clean metal and support is subtracted from that of the metal, support, and chemisorbed gas. Instead of transmitting the IR radiation through the sample, one can get the IR spectrum of a chemisorbed species by reflecting the radiation off a single metal crystal containing the adsorbate (Fig. 20.30*b*); this is *reflection–absorption infrared spectroscopy* (RAIS or RAIRS; also called IRAS). Another way to obtain vibrational frequencies of chemisorbed species is by *electron-energy-loss spectroscopy* (EELS). Here, a monoenergetic beam of electrons is reflected off the metal-crystal surface (Fig. 20.30*c*), and the energies of the reflected electrons are measured. The losses in electron energy give the vibrational energy changes in the adsorbate, from which the adsorbate vibrational frequencies can be found.

The IR spectra of many short-lived free radicals and molecular ions have been studied. One technique (*matrix isolation*) is to cool a mixture of Ar and a relatively small amount of a stable species to a few kelvins. Photolysis (decomposition by light) of the frozen solid then produces long-lived radicals trapped in an inert solid matrix. Alternatively, the gas-phase mixture can be photolyzed at room temperature and then condensed on a very cold surface. Gas-phase spectra of transient species can be studied by producing the species in the IR absorption cell using an electric discharge or laser photolysis or by flowing the species (produced by a discharge or chemical reaction) through the absorption cell. (Recall the IR determination of the structure of the vinyl cation; Sec. 19.9). By monitoring the IR absorption as a function of time, the reaction rates of radicals can be studied. For tabulations of species studied, see E. Hirota, *Chem. Rev.*, **92,** 141 (1992); P. F. Bernath, *Ann. Rev. Phys. Chem.*, **41,** 91 (1990).

Gas-phase IR bands under high resolution consist of closely spaced lines due to changes in the rotational quantum numbers. In liquids and solids, rotational structure is not observed, and each band appears as a broad absorption. In solids, the molecules do not rotate. In liquids (and in high-pressure gases), the high rate of intermolecular collisions shortens the lifetimes of vibrational–rotational states by inducing transitions. This lifetime shortening broadens the rotational fine-structure lines, since one form of the uncertainty principle is $\tau\ \Delta E \gtrsim h$, where ΔE is the uncertainty in the energy of a state whose lifetime is τ. The lines are also broadened due to the shifting of the rotational energy levels by the electric fields of intermolecular interactions. The net result is that the rotational lines are merged into a single broad absorption in liquids.

Use of a tunable laser as the IR source can dramatically improve the **resolution** (this is the minimum difference in wavenumber for which two nearby lines can be distinguished as separate lines) and the accuracy of line-frequency measurements. Use of

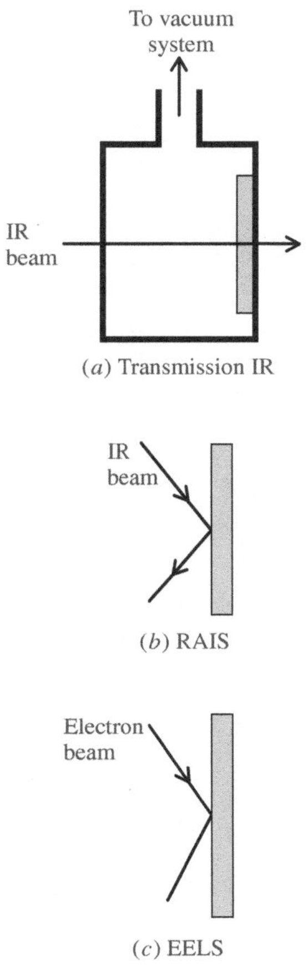

Figure 20.30

Methods for studying the vibrations of a chemisorbed species.

a laser-source IR spectrometer with a resolution of 0.0005 cm^{-1} allowed the rotational structure of vibrational bands of gas-phase cubane, C_8H_8, to be studied [A. S. Pine et al., *J. Am. Chem. Soc.*, **106,** 891 (1984)].

The spacings between the rotational lines of an IR band depend on the moments of inertia, so analysis of the rotational structure of vibration–rotation bands enables molecular structures to be determined. Usually, isotopic species must also be studied to give enough information for a structural determination. Structures determined from IR spectra are not as accurate as those found from microwave spectra, but IR spectroscopy has the advantage of allowing structures of nonpolar molecules to be determined; nonpolar molecules (other than diatomics) have some IR-active vibrations.

Some nonpolar-molecule structures determined from IR spectra are

$$C_2H_6\text{:}\quad R(\text{CH}) = 1.10\ \text{Å},\quad R(\text{CC}) = 1.54\ \text{Å},\quad \angle\text{HCC} = 110°$$

$$C_2H_4\text{:}\quad R(\text{CH}) = 1.09\ \text{Å},\quad R(\text{CC}) = 1.34\ \text{Å},\quad \angle\text{HCC} = 121°$$

$$C_2H_2\text{:}\quad R(\text{CH}) = 1.06\ \text{Å},\quad R(\text{CC}) = 1.20\ \text{Å},\quad \angle\text{HCC} = 180°$$

Analysis of the IR spectrum of gas-phase H_3O^+ (generated by an electric discharge through a mixture of H_2 and O_2) gave the structure of this pyramidal species as $R(\text{OH}) = 0.97$ Å and $\angle\text{HOH} = 114°$ [J. Tang and T. Oka, *J. Mol. Spec.*, **196,** 120 (1999)].

Even if rotational fine structure is not resolved, information about the symmetry of the molecular structure can be obtained from observation of the IR spectrum and the Raman spectrum (Sec. 20.10). For example, the planar molecule oxalyl fluoride, FC(O)C(O)F, has $3(6) - 6 = 12$ normal modes of vibration. Group theory shows that the cis conformation will have 10 IR-active fundamentals and 12 Raman-active fundamentals, whereas the trans conformation will have 6 IR-active fundamentals and 6 Raman-active fundamentals. (The trans conformation has a center of symmetry, and group theory shows that in a molecule with a center of symmetry, no normal mode can be both IR-active and Raman-active.) Observation of IR and Raman spectra of oxalyl fluoride showed that in the solid only the trans conformer is present, but the liquid and gas contain a mixture of the two conformers [J. R. Durig et al., *J. Chem. Phys.*, **54,** 4428 (1971)].

For most molecules larger than benzene, the large number of vibrational modes, the presence in the spectrum of many hot bands, overtones, and combination bands, and the line broadening due to intermolecular collisions give an IR spectrum in which rotational structure is not observable and broad vibrational bands overlap one another, making it difficult to obtain structural information. However, the use of characteristic group frequencies (see below), supplementation of IR spectral information with Raman spectral information (Sec. 20.10), and comparison with spectra of similar compounds of known structure often allow useful structural information to be obtained from the vibrational spectra of large molecules.

Organic chemists use IR spectroscopy to help identify an unknown compound. Although most or all of the atoms are vibrating in each normal mode, certain normal modes may involve mainly motion of only a small group of atoms bonded together, with the other atoms vibrating only slightly. For example, aldehydes and ketones have a normal mode which is mainly a C=O stretching vibration, and its frequency is approximately the same in most aldehydes and ketones. Normal-mode analysis shows that two bonded atoms A and B in a compound will exhibit a characteristic vibrational frequency provided that either the force constant of the A—B bond differs greatly from the other force constants in the molecule or there is a large difference in mass between A and B. Double and triple bonds have much larger force constants than single bonds, so one observes characteristic vibrational frequencies for C=O, C=C, C≡C, and C≡N bonds. Since hydrogens are much lighter than carbons, one observes characteristic vibrational frequencies for OH, NH, and CH groups. In contrast, vibrations involving C—C, C—N, and C—O single bonds occur over a very wide range of frequencies.

Some characteristic IR wavenumbers in cm^{-1} for stretching vibrations are:

OH	NH	CH	C≡C	C=C	C=O
3200–3600	3100–3500	2700–3300	2100–2250	1620–1680	1650–1850

For a molecule with internal rotation about a single bond, the changes in electronic energy E_e produced by changes in the dihedral angle of internal rotation give rise to the potential-energy function for torsion (twisting) about that bond. Figure 19.34 shows this function for butane. Usually, one of the molecule's normal modes is essentially a torsional motion about the single bond. Because rotational barriers for single-bond torsion are low, the force constant and vibrational frequency for this normal mode are low. Typically $\tilde{\nu}$ is in the range 50 to 400 cm^{-1} for torsion about a single bond.

If one can observe and reliably assign the fundamental, overtones, and hot bands for the torsional mode of each conformer present, one can then attempt to derive a potential function for the torsion that will fit the torsional vibrational-level data. Additional information, such as the enthalpy difference(s) between conformers (which can be found from the temperature dependence of line-intensity ratios) can help determine the potential function. The torsional-potential function of butane in Fig. 19.34 was found from observed torsional-vibration transition frequencies of the gauche and trans conformers. For 1,3-butadiene, two different potential functions (the dashed curves in Fig. 19.36) were found to give good fits to observed torsional transition frequencies.

The most commonly studied portion of the IR spectrum is from 4000 to 400 cm^{-1} (2.5 to 25 μm).

The complete IR spectrum is a highly characteristic property of a compound and has been compared to a fingerprint.

Quantitative analysis of mixtures can be done by applying the Beer–Lambert law (Sec. 20.2) to IR absorption.

Molecular vibrational frequencies are affected by changes in bonding, conformation, and hydrogen bonding, so IR spectroscopy can be used to study such changes. Hydrogen bonding reduces the frequencies of OH and NH stretching vibrations. For example, water vapor shows symmetric and asymmetric OH stretches at 3657 and 3756 cm^{-1} (Fig. 20.27); in liquid water these are replaced by a broad absorption band centered at 3400 cm^{-1}. Conformations and hydrogen bonding of biological molecules and synthetic polymers have been studied using IR spectroscopy. (See B. H. Stuart, *Infrared Spectroscopy,* chaps. 6 and 7, Wiley 2004.)

The amide group in the backbone of proteins gives rise to 9 characteristic IR bands, labeled A, B, and I through VII. The most useful is amide band I in the region 1600 to 1700 cm^{-1}, which is mainly a C=O stretching vibration, with small amounts of NH bending and CN stretching. To avoid interference from the liquid H_2O bending vibration at 1645 cm^{-1}, D_2O, with the bending vibration at 1215 cm^{-1}, is used as the solvent. (The 1645 cm^{-1} bending wavenumber differs from the 1595 cm^{-1} vapor-phase value in Fig. 20.27 because of intermolecular interactions in liquid water.) The frequency of the amide I band is sensitive to hydrogen bonds formed by the atoms of the amide group and so changes depending on the secondary structure (α-helix, β-sheet, turn, random coil) of the portion of the protein producing the band. Therefore the amide I band of a folded protein is a composite of overlapping bands of various frequencies. Analysis of this composite band allows the percentage of each kind of secondary structure in the protein to be estimated. Moreover, by applying a T-jump (Sec. 16.14) to a protein solution at whose initial temperature there is an equilibrium between folded and unfolded protein, one perturbs this equilibrium. By following the intensity of one or more frequencies of the amide I band as a function of time after the

T-jump, one can follow the kinetics of the relaxation to equilibrium, thereby finding the rate constant for folding of the protein.

IR spectroscopy and EELS are widely used to study the structure of species chemisorbed on solids. For $CH_2{=}CH_2$ chemisorbed on Pt, certain vibrational bands occur at about the same frequencies as IR bands of compounds containing *n*-alkyl groups bonded to metal atoms, thereby indicating the presence of σ-bonded ethylene ($PtCH_2CH_2Pt$), shown in Fig. 19.32*a*; other IR bands for C_2H_4 on Pt occur at about the same frequencies as IR bands of species known to contain C_2H_4 π bonded to a metal atom (for example, the complex ion $[Pt(C_2H_4)Cl_3]^-$), and they are evidence for the π complex shown in Fig. 19.32*b*. Still other bands occur at frequencies close to those of the compound $CH_3CCo_3(CO)_9$ and show the presence of the ethylidyne complex CH_3CPt_3 (Fig. 20.31). When adsorbed ethylidyne is formed, one H atom dissociates from C_2H_4 and is chemisorbed on the Pt surface and a second H atom shifts from one carbon to the other.

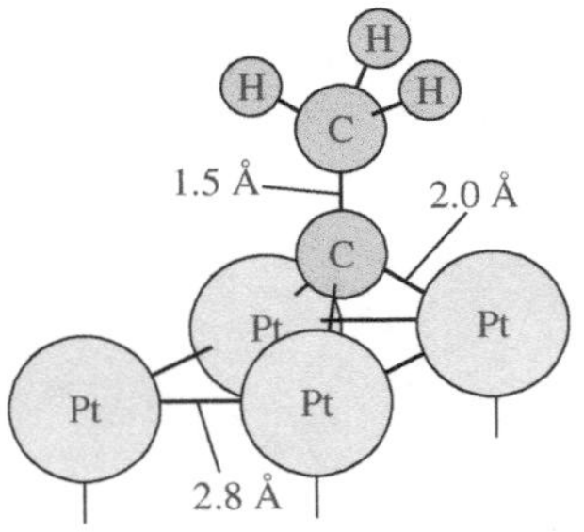

Figure 20.31

Structure of the ethylidyne complex formed when C_2H_4 (ethylene) is chemisorbed on the (111) surface of Pt at 300 K. (The notation is explained in Sec. 23.7.)

When $H_2(g)$ is chemisorbed on ZnO, one finds O—H and Zn—H IR bands at 3502 cm^{-1} and 1691 cm^{-1} due to dissociatively adsorbed hydrogen. In addition, there is a band at 4019 cm^{-1}, which is 142 cm^{-1} below the vibrational wavenumber $\tilde{\nu}_0$ = 4161 cm^{-1} of $H_2(g)$ [Table 20.1 and Eq. (20.36)]. For physically adsorbed diatomic molecules, one typically observes a downward frequency shift of about 20 cm^{-1} compared with the gas phase. The large shift for H_2 on ZnO is interpreted as evidence for nondissociatively chemisorbed H_2 (Fig. 19.32*d*); see C. C. Chang *et al., J. Phys. Chem.,* **77,** 2634 (1973). (Although H_2 gas does not absorb IR radiation, chemisorbed H_2 molecules do absorb; the bond to the surface induces a dipole moment whose value changes as the H_2 internuclear distance changes.)

Fourier-Transform IR Spectroscopy

A **Fourier-transform IR (FT-IR) spectrometer** gives improved *sensitivity* (the ability to detect weak signals), speed, and accuracy of wavelength measurement as compared with the dispersion IR spectrometer of Fig. 20.29. An FT-IR spectrometer (Fig. 20.32) does not disperse the radiation (and so has no prism or grating) but uses a Michelson interferometer to form an interferogram, which is mathematically manipulated by a microcomputer to give the IR absorption spectrum. Continuous-frequency radiation from the source strikes the beam splitter, a flat, partially transparent plate that reflects half the incident light to the fixed mirror (beam *b*) and transmits half to the moving mirror (beam *c*). The moving mirror is driven by a motor and moves parallel to itself at constant speed. The beams *d* and *e* reflected from the two mirrors meet at the beam splitter, which transmits part of *d* and reflects part of *e* to give the combined beam *f*. Beam *f* passes through the sample, where absorption occurs to give beam *g*.

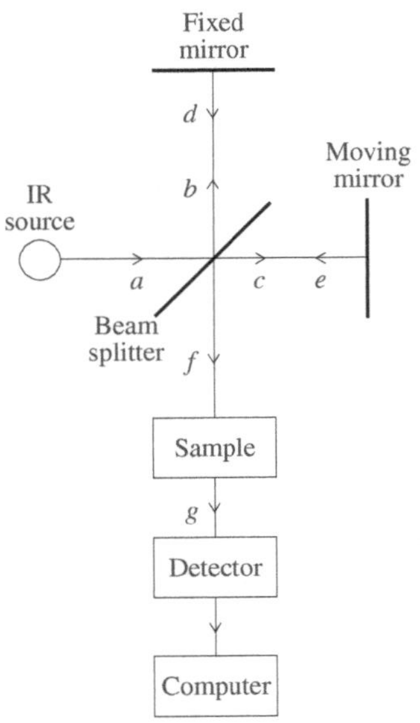

Figure 20.32

A Fourier-transform IR spectrometer.

Temporarily let us suppose that the light from the source contains only a single wavelength λ, whose wavenumber is $\tilde{\nu} = 1/\lambda$. Beams *d* and *e* interfere with each other when they meet to form beam *f*, and by adding the electric fields of beams *d* and *e*, one finds that the intensity I_f of beam *f* is (for a derivation, see F. A. Jenkins and H. E. White, *Fundamentals of Optics,* 4th ed., McGraw-Hill, 1976, sec. 12.1)

$$I_f = \tfrac{1}{2}B_f[1 + \cos(2\pi\delta/\lambda)] = \tfrac{1}{2}B_f[1 + \cos(2\pi\delta\tilde{\nu})]$$

where δ is the difference in path lengths traveled by beams *c* and *e* as compared with beams *b* and *d*, and B_f is the intensity beam *f* would have if there were no path difference ($\delta = 0$). If *x* is the difference in distances of the moving mirror and the fixed mirror from the beam splitter, then $\delta = 2x$. Both *x* and δ change with time. When δ is a whole-number multiple of the wavelength ($\delta = n\lambda$), then $\cos(2\pi\delta/\lambda) = 1$ and $I_f = B_f$, because the beams meeting to form the combined beam are in phase. When $\delta = (n + \tfrac{1}{2})\lambda$, then $\cos(2\pi\delta/\lambda) = -1$ and $I_f = 0$, since the beams meet out of phase.

When beam f passes through the sample, the sample's absorption changes I_f and B_f in the above equation to I_g and B_g.

Since the radiation actually contains a continuous range of wavenumbers, we must integrate over all wavenumbers $\tilde{\nu}$ to get I_f and I_g. Also, B_g is a function of $\tilde{\nu}$ [$B_g = B_g(\tilde{\nu})$], since the sample's absorbance and the light emitted by the source are functions of $\tilde{\nu}$. Therefore

$$I_g(\delta) = \frac{1}{2}\int_0^\infty B_g(\tilde{\nu})[1 + \cos(2\pi\delta\tilde{\nu})]\,d\tilde{\nu}$$

$$= \frac{1}{2}\int_0^\infty B_g(\tilde{\nu})\,d\tilde{\nu} + \frac{1}{2}\int_0^\infty B_g(\tilde{\nu})\cos(2\pi\delta\tilde{\nu})\,d\tilde{\nu}$$

As the mirror moves, δ changes and I_g changes, so I_g is a function of the path-length difference δ. The detector records I_g as a function of time, and since δ is known at each time, the detector measures the function $I_g(\delta)$. At $\delta = 0$, light of all frequencies in beams d and e meets in phase and the intensity I_g is a maximum. The maximum intensity $I_g(0)$ is given by the above equation with $\delta = 0$ as $I_g(0) = \int_0^\infty B_g(\tilde{\nu})\,d\tilde{\nu}$. Hence the $I_g(\delta)$ equation becomes $I_g(\delta) = \frac{1}{2}I_g(0) + \frac{1}{2}\int_0^\infty B_g(\tilde{\nu})\cos(2\pi\delta\tilde{\nu})\,d\tilde{\nu}$. Defining $F(\delta) \equiv I_g(\delta) - \frac{1}{2}I_g(0)$, we have

$$F(\delta) \equiv I_g(\delta) - \tfrac{1}{2}I_g(0) = \frac{1}{2}\int_0^\infty B_g(\tilde{\nu})\cos(2\pi\delta\tilde{\nu})\,d\tilde{\nu}$$

The *interferogram function* $F(\delta)$ is known, since $I_g(\delta)$ and $I_g(0)$ are known. The function $B_g(\tilde{\nu})$ is the intensity of radiation reaching the detector as a function of wavenumber (with no path difference between the beams) and is exactly the spectrum we want to find. A mathematical theorem of Fourier analysis shows that if $F(\delta)$ and $B_g(\tilde{\nu})$ are related by the above equation and if $F(\delta)$ has zero slope at $\delta = 0$ [which it does because $I_g(\delta)$ is a maximum at $\delta = 0$ and $F(\delta)$ differs from $I_g(\delta)$ only by a constant], then $B_g(\tilde{\nu})$ is given by (see M. J. Lighthill, *Introduction to Fourier Analysis and Generalized Functions,* Cambridge, 1958, sec. 1.4)

$$B_g(\tilde{\nu}) = 8\int_0^\infty F(\delta)\cos(2\pi\tilde{\nu}\delta)\,d\delta$$

For full accuracy in calculating $B_g(\tilde{\nu})$, one must make observations with δ extending to ∞. In practice, the path-length difference will have some maximum value δ_{max} (which is typically on the order of 1 to 20 cm), and the infinite limit in the integral (which is called the *Fourier cosine transform* of F) is approximated by δ_{max}. A computer, which is part of the spectrometer, calculates the spectrum $B_g(\tilde{\nu})$ from $F(\delta)$. Since only a single beam is used, one does a separate run with the sample cell empty and the computer combines the two spectra to give the sample's absorption spectrum. Because radiation of all wavenumbers reaches the detector at each instant and all this radiation is used to calculate the spectrum $B_g(\tilde{\nu})$, an FT-IR spectrometer gives much better signal-to-noise ratio than a dispersion instrument, where only a tiny portion of the radiation reaches the detector at each instant. Moreover, one can further improve the signal-to-noise ratio by making repeated runs and having the computer average the spectra. The noise then tends to cancel, since it is randomly positive and negative.

Currently, most commercially available IR spectrometers are Fourier-transform instruments.

20.10 RAMAN SPECTROSCOPY

Raman spectroscopy is quite different from absorption spectroscopy in that it studies light *scattered* by a sample rather than light absorbed or emitted. Suppose a photon collides with a molecule in state a. If the energy of the photon corresponds to the energy difference between state a and a higher level, the photon may be absorbed, the molecule making a transition to the higher level. No matter what the energy of the photon is, the photon–molecule collision may **scatter** the photon, meaning that the photon's direction of motion is changed. Although most of the scattered photons undergo no change in frequency and energy (*Rayleigh scattering*), a small fraction of the scattered photons exchange energy with the molecule during the collision. The resulting increase or decrease in energy of the scattered photons is the **Raman effect,** first observed by C. V. Raman in 1928.

Let ν_0 and ν_{scat} be the frequencies of the incident photon and the Raman-scattered photon, respectively, and let E_a and E_b be the energies of the molecule before and after it scatters the photon. Conservation of energy gives $h\nu_0 + E_a = h\nu_{scat} + E_b$, or

$$\Delta E \equiv E_b - E_a = h(\nu_0 - \nu_{scat}) \equiv h\,\Delta\nu \qquad (20.49)$$

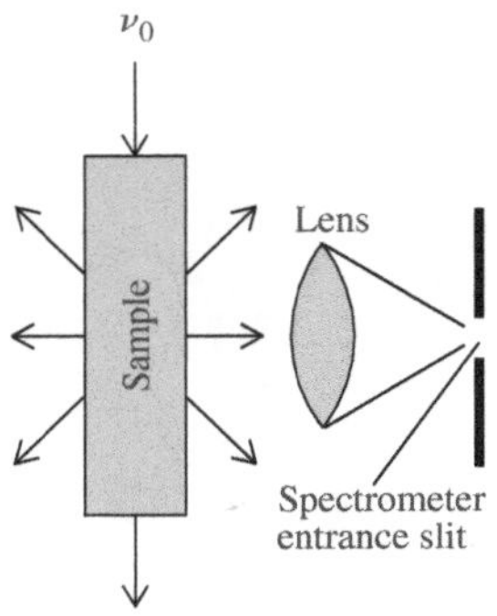

Figure 20.33

In Raman spectroscopy, one observes light scattered at right angles to the incident radiation.

The energy difference ΔE is the difference between two stationary-state energies of the molecule, so *measurement of the* **Raman shifts** $\Delta\nu \equiv \nu_0 - \nu_{scat}$ *gives molecular energy-level differences.*

In Raman spectroscopy, one exposes the sample (gas, liquid, or solid) to monochromatic radiation of any convenient frequency ν_0. Unlike absorption spectroscopy, ν_0 need have no relation to the difference between energy levels of the sample molecules. Usually ν_0 lies in the visible or near-UV region. The Raman-effect lines are extremely weak (only about 0.001% of the incident radiation is scattered, and only about 1% of the scattered radiation is Raman-scattered). Therefore, the very intense light of a laser beam is used as the exciting radiation. Scattered light at right angles to the laser beam is focused on the entrance slit of a spectrometer (Fig. 20.33), which disperses the radiation using a diffraction grating and records light intensity versus frequency, giving the Raman spectrum.

Figure 20.34 shows the pattern of Raman lines for a gas of diatomic molecules. The strong central line at ν_0 is due to light scattered with no frequency change. On either side of ν_0 and close to it are lines corresponding to pure-rotational transitions in the molecule; the lines with $\nu_{scat} > \nu_0$ result from transitions where J decreases, and those with $\nu_{scat} < \nu_0$ result from increases in J. On the low-frequency side of ν_0 is a band of lines corresponding to $v = 0 \rightarrow 1$ vibration–rotation transitions in the molecule. If the $v = 1$ vibrational level is significantly populated, there is a weak band of lines on the high-frequency side, corresponding to $v = 1 \rightarrow 0$ vibration–rotation transitions. For historical reasons, the Raman lines with $\nu_{scat} < \nu_0$ are called *Stokes lines* and those with $\nu_{scat} > \nu_0$ are *anti-Stokes lines.*

Investigation of the Raman selection rules shows that spherical tops exhibit no pure-rotational Raman spectra, but all symmetric and asymmetric tops show pure-rotational Raman spectra. Raman scattering is a different process than absorption, and the rotational Raman selection rules differ from those for absorption. One finds that for linear molecules the pure-rotational Raman frequencies on either side of ν_0 correspond to $\Delta J = \pm 2$. The rotational Raman selection rules for nonlinear molecules are more complicated and are omitted. The pure-rotational Raman spectrum can yield the structure of a nonpolar molecule (for example, F_2, C_6H_6) that is not a spherical top.

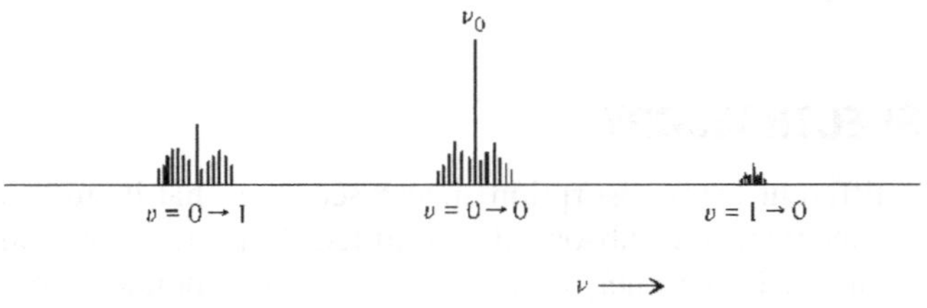

Figure 20.34

Raman spectrum of a gas of diatomic molecules.

For benzene, the rotational Raman spectrum of *sym*-$C_6H_3D_3$, together with spectroscopic data for C_6H_6 and C_6D_6, gave $R_0(CC) = 1.397$ Å and $R_0(CH) = 1.08_6$ Å [H. G. M. Edwards, *J. Mol. Struct.*, **161,** 23 (1987)].

For a polyatomic molecule, the Raman spectrum shows several $\Delta v_j = 1$ fundamental bands, each corresponding to a Raman-active vibration. The *k*th normal mode is Raman-active if $(\partial\alpha/\partial Q_k)_e \neq 0$, where α is the molecular polarizability (Sec. 13.14), Q_k is the normal coordinate for the *k*th vibrational mode, and the derivative is evaluated at the equilibrium nuclear configuration. (The normal coordinate Q_k is a certain linear combination of the cartesian coordinates of the nuclei that vibrate in the *k*th normal mode; Q_k measures the extent of departure from the equilibrium configuration of the nuclei.) As noted in Sec. 13.14, α is different in different directions in the molecule, and one must examine the derivative of each component of α when deciding on Raman activity. It is hard to determine whether $(\partial\alpha/\partial Q_k)_e$ is zero by examining the atomic motions in the *k*th normal mode. The best way to determine which modes are Raman-active is to use group theory (see *Herzberg,* vol. II, chap. III).

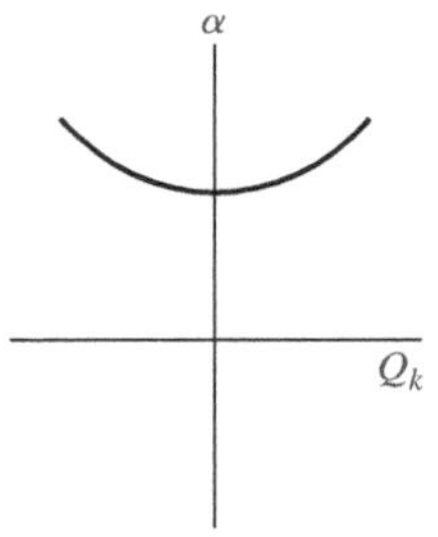

Figure 20.35

Behavior of each component of polarizability as a function of the coordinate describing the normal-mode vibrations ν_2 or ν_3 of CO_2.

Since the Raman selection rule differs from the IR selection rule, a given normal mode may be IR-active and Raman-inactive, IR-inactive and Raman-active, active in both the IR and Raman, or inactive in both the IR and Raman. For a molecule with no symmetry, all vibrational modes are Raman-active. Every molecule (including homonuclear diatomics) has at least one Raman-active normal mode.

The CO_2 symmetric stretching vibration ν_1 (Fig. 20.26), which is IR-inactive, is Raman-active. For the CO_2 vibrations ν_2 and ν_3, each component of α behaves as shown in Fig. 20.35, where $Q_k = 0$ corresponds to the equilibrium nuclear configuration. Although the ν_2 and ν_3 vibrations each change α, each of these vibrations is Raman-inactive because each has $(\partial\alpha/\partial Q_k)_e = 0$, as is evident from the zero slope at $Q_k = 0$.

Overtone and combination bands are usually too weak to be observed in Raman spectroscopy, so vibrational Raman spectra are simpler than IR spectra.

As in IR spectra, vibrational Raman spectra of gases show rotational structure, but rotational structure is absent in Raman spectra of liquids and solids.

It is hard (but not impossible) to obtain IR spectra below 100 cm^{-1}, but easy to obtain Raman spectra for the Raman shift $\Delta\tilde{\nu}$ in the range 10 to 100 cm^{-1}, so Raman spectroscopy is the preferred technique in this wavenumber range. For $\Delta\tilde{\nu}$ between 0 and 10 cm^{-1}, the Raman-shifted lines are hidden by the strong Rayleigh-scattered peak at $\tilde{\nu}_0$.

An important advantage of vibrational Raman spectroscopy over IR spectroscopy arises from the fact that liquid water shows only weak vibrational Raman scattering in the Raman-shift range 300–3000 cm^{-1} but has strong, broad IR absorptions in this range. Thus, vibrational Raman spectra of substances in aqueous solution can readily be studied without solvent interference. An aqueous solution of mercury(I) nitrate shows a strong peak at a Raman shift of 170 cm^{-1}, and this peak is absent from spectra of other nitrate salt solutions. The peak is due to vibration of the Hg—Hg bond and indicates that the mercury(I) ion is diatomic (Hg_2^{2+}). The vibrational Raman spectrum of CO_2 dissolved in water is almost the same as that of liquid CO_2 and shows no H_2CO_3 vibrational absorption bands, thereby confirming that CO_2 in water exists mainly as solvated CO_2 molecules, not as H_2CO_3. Vibrational Raman spectroscopy has been used to study ion pairing (Sec. 10.8) and solvation of ions. Vibrational Raman spectra of biological molecules in aqueous solution provide information on conformations and hydrogen bonding. (See P. R. Carey, *Biochemical Aspects of Raman and Resonance Raman Spectroscopy,* Academic Press, 1982; A. T. Wu, *Raman Spectroscopy and Biology,* Wiley, 1982; H.-U. Gremlich and B. Yan, eds., *Infrared and Raman Spectroscopy of Biological Materials,* Dekker, 2001.)

In **resonance Raman spectroscopy,** the exciting frequency ν_0 is chosen to coincide with an electronic absorption frequency (Sec. 20.11) of the species being studied. This dramatically increases the intensities of the Raman-scattered radiation for those vibrational modes that are localized in the portion of the molecule that is responsible for the electronic absorption at ν_0. Two important advantages of resonance Raman spectroscopy in the study of biological molecules are: (*a*) the increased scattering intensity allows study of solutions at the high dilutions (10^{-3} to 10^{-6} mol/dm^3) characteristic of biopolymers in organisms; (*b*) the selectivity of the intensity enhancement "samples" only the vibrations in one region of the molecule, simplifying the spectrum and allowing study of the bonding in that region. The resonance Raman spectra of the oxygen-carrying proteins hemoglobin and myoglobin give information on the bonding in the heme group; see J. M. Friedman et al., *Ann. Rev. Phys. Chem.,* **33,** 471 (1982).

When certain molecules or ions (mainly those containing O, N, or S atoms with lone pairs) are adsorbed on or are atomically close to a roughened surface or colloidal dispersion or vacuum-deposited thin film of certain metals (mainly Ag, Cu, or Au), the intensity of the molecules' Raman spectra is enormously increased, giving rise to *surface-enhanced Raman spectroscopy* (SERS). Interactions between the incident radiation and the electrons in the metal surface produce a great increase in the strength of the electric field of the electromagnetic radiation, thereby amplifying the Raman scattering. In some cases, chemical bonding between the molecule and the surface contributes to the effect.

20.11 ELECTRONIC SPECTROSCOPY

Electronic spectra involving transitions of valence electrons occur in the visible and UV regions and are studied in both absorption and emission. The detector is usually either a photomultiplier tube or a photodiode. These devices convert photons into an electrical signal. One can study emission spectra of solids, liquids, and gases by raising the molecules to an excited electronic state using photons from a high-intensity lamp or a laser and then observing the emission at right angles to the incident beam. Spontaneous emission of light by excited-state atoms or molecules that rapidly follows absorption of light is called **fluorescence** (see also Sec. 20.15).

Atomic Spectra

For the hydrogen atom, the selection rule for n is found to be Δn = any value. The use of $E_a - E_b = h\nu$ and the energy-level formula (18.14) gives the spectral wavenumbers as

$$\widetilde{\nu} = \frac{1}{\lambda} = \frac{\nu}{c} = \frac{\mu e^4}{8\varepsilon_0^2 ch^3}\left(\frac{1}{n_b^2} - \frac{1}{n_a^2}\right) \equiv R_{\text{H}}\left(\frac{1}{n_b^2} - \frac{1}{n_a^2}\right) \tag{20.50}$$

where the reduced mass depends on the electron and proton masses, $\mu = m_e m_p/(m_e + m_p)$ [Eq. (17.79)], and where the Rydberg constant for hydrogen is defined as $R_{\text{H}} \equiv \mu e^4/8\varepsilon_0^2 ch^3 = 109678\ \text{cm}^{-1}$.

The H-atom spectrum consists of several series of lines, each ending in a continuous band. Transitions involving the n-value changes $2 \to 1, 3 \to 1, 4 \to 1, \ldots$ give the Lyman series in the UV (Fig. 20.36). As n_a in Eq. (20.50) goes to infinity, the Lyman-series lines converge to the limiting value $\widetilde{\nu} = 1/\lambda = R_{\text{H}} = 109678\ \text{cm}^{-1}$, corresponding to $\lambda = 91.2$ nm. Beyond this limit is a continuous absorption or emission due to transitions between ionized H atoms and ground-state H atoms; the energy of

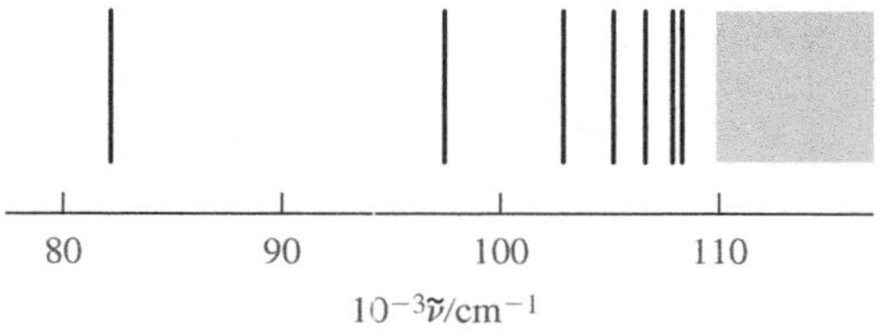

Figure 20.36

The Lyman series of H. Only the first seven lines are shown.

an ionized atom takes on a continuous range of positive values. The position of the Lyman-series limit allows the ionization energy of H to be found.

The H-atom transitions $3 \to 2$, $4 \to 2$, $5 \to 2$, . . . give the Balmer series, which lies in the visible region. The transitions $4 \to 3$, $5 \to 3$, $6 \to 3$, . . . give the Paschen series in the IR.

Spectra of many-electron atoms are quite complicated, because of the many terms and levels arising from a given electron configuration. Once the spectrum has been unraveled, the atomic energy levels can be found.

Since each element has lines at frequencies characteristic of that element, atomic absorption and emission spectra are used to analyze for most chemical elements. For example, Ca, Mg, Na, K, and Pb in blood samples can be determined by atomic absorption spectroscopy.

For inner-shell electrons, the effective nuclear charge Z_{eff} is nearly equal to the atomic number; Eq. (18.55) shows that the differences between these inner-shell energies increase rapidly with increasing atomic number. For atoms beyond the second period, these energy differences correspond to x-ray photons. X-rays are produced when a beam of high-energy electrons penetrates a metal target. The deceleration of the electrons as they penetrate the target produces a continuous x-ray emission spectrum. In addition, an electron in the beam that collides with an inner-shell electron of a target atom can knock this electron out of the atom. The spontaneous transition of a higher-level electron of the ionized atom into the vacancy thereby created will produce an x-ray photon of frequency corresponding to the energy difference, giving an x-ray emission line spectrum superimposed on the continuous emission.

Molecular Electronic Spectra

If ψ'' and ψ' are the lower and upper states in a molecular electronic transition, the transition frequencies are given by

$$h\nu = (E'_{\text{el}} - E''_{\text{el}}) + (E'_{\text{vib}} - E''_{\text{vib}}) + (E'_{\text{rot}} - E''_{\text{rot}}) \qquad (20.51)$$

Each term in parentheses is substantially larger than the following term.

The electronic selection rules are rather complicated. Perhaps the most important one is that the transition-moment integral (20.5) vanishes unless

$$\Delta S = 0$$

where S is the total electronic spin quantum number. Actually, this selection rule does not hold rigorously, and weak transitions with $\Delta S \neq 0$ sometimes occur. The ground electronic state of most molecules has all electron spins paired and so has $S = 0$ (a **singlet state**). Some exceptions are O_2 (a **triplet** ground level with $S = 1$) and NO_2 (the odd number of electrons gives $S = \frac{1}{2}$ and a doublet ground level).

An electronic transition consists of a series of **bands,** each band corresponding to a transition between a given pair of vibrational levels. Gas-phase spectra under high resolution may show each band to consist of closely spaced lines arising from transitions between different rotational levels. For relatively heavy molecules, the close spacings between the rotational levels usually makes it impossible to resolve the rotational lines. In liquids, no rotational structure is observed and the vibrational bands are

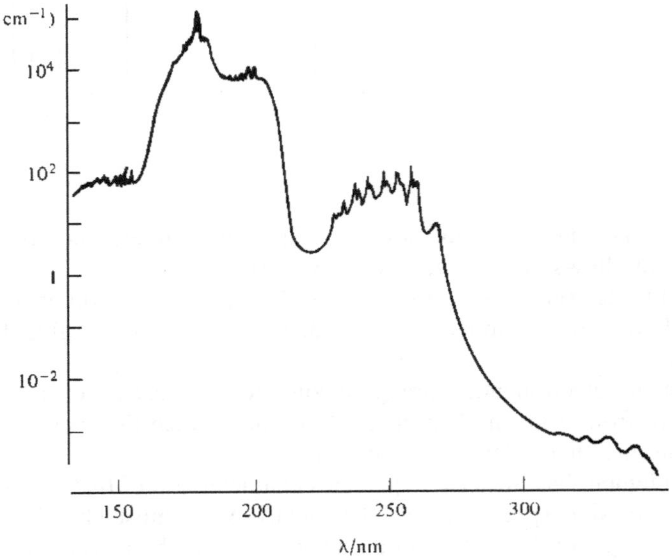

Figure 20.37

Sketch of the electronic absorption spectrum of gas-phase benzene. The vertical scale is logarithmic.

often broadened enough to merge them into a single broad absorption band for each electronic transition. Figure 20.37 shows the electronic absorption spectrum of gas-phase benzene.

Analysis of the rotational lines of an electronic absorption band allows the molecular vibrational and rotational constants to be obtained. Since all molecules show electronic absorption spectra, this allows one to obtain R_e and ν_e for homonuclear diatomics, which show no pure-rotational or vibration–rotation spectra. Excited electronic states often have geometries quite different from those of the ground electronic state. For example, HCN is nonlinear in some excited states.

Molecular dissociation energies and spacings between electronic energy levels are also obtained from electronic spectra. Figure 20.38 shows potential-energy curves and vibrational levels for the ground state and an excited electronic state of a diatomic molecule. Suppose we are able to observe transitions from the $v'' = 0$ vibrational level of the ground electronic state to each of the vibrational levels of the excited electronic state, as shown by the arrows in the figure. The bands corresponding to these transitions will come closer together as the excited-state vibrational quantum number v' increases, and these bands will be followed by a continuous absorption corresponding to transitions to states with energy above E'_{at}. Figure 20.38 gives

$$h\nu_{cont} = h\nu_{00} + D'_0 = D''_0 + E'_{at} - E''_{at} \tag{20.52}$$

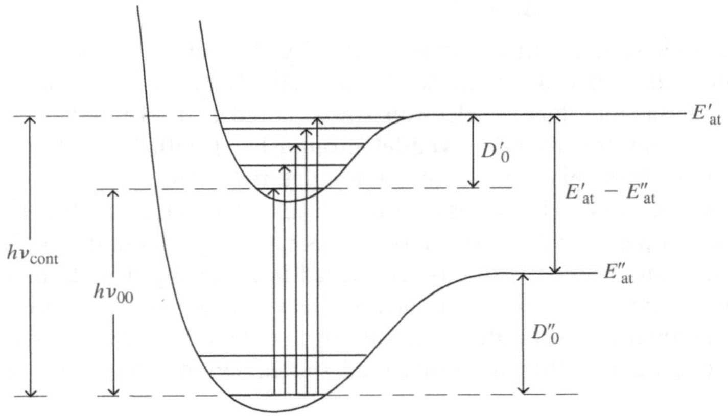

Figure 20.38

Determination of dissociation energies of electronic states of a diatomic molecule. E''_{at} and E'_{at} are the energies of the separated atoms produced by dissociation of the ground state and the first excited electronic state, respectively.

where ν_{cont} is the frequency at which continuous absorption begins, ν_{00} is the frequency of the $v'' = 0$ to $v' = 0$ transition, D_0'' and D_0' are the dissociation energies of the ground and excited electronic states, and E_{at}'' and E_{at}' are the energies of the separated atoms into which the ground and excited electronic states dissociate. Analysis of the molecule's electronic spectrum will usually tell us which atomic states the two molecular electronic states dissociate into, so $E_{at}' - E_{at}''$ can be found from known atomic energy levels (which are found from atomic spectra). Hence, Eq. (20.52) allows us to find the dissociation energies D_0'' and D_0'. Also, the quantity $h\nu_{00}$ gives the separation between the ground vibrational levels of the two electronic states.

Substances that absorb visible light are colored. Conjugated organic compounds often show electronic absorption in the visible region, due to excitation of a π electron to an antibonding π orbital. In the particle-in-a-box model of Prob. 19.52, the energy spacings are proportional to $1/a^2$, where a is the box length, so as the length of the conjugated chain increases, the lowest absorption frequency moves from the UV into the visible region. Transition-metal ions [for example, $Cu^{2+}(aq)$, $MnO_4^-(aq)$] frequently show visible absorption due to electronic transitions.

Certain groups called **chromophores** give characteristic electronic absorption bands. For example, the C=O group in the majority of aldehydes and ketones produces a weak electronic absorption in the region 270 to 295 nm with molar absorption coefficient $\varepsilon = 10$ to 30 dm^3 mol^{-1} cm^{-1} and a strong band at about 180 to 195 nm with $\varepsilon = 10^3$ to 10^4 dm^3 mol^{-1} cm^{-1}. The weak band is due to an $n \to \pi^*$ transition, and the strong band is due to a $\pi \to \pi^*$ transition, where n signifies an electron in a nonbonding (lone-pair) orbital on oxygen, π is an electron in the π MO of the double bond, and π^* is an electron in the corresponding antibonding π MO. The longer-wavelength electronic absorptions of benzene in Fig. 20.37 are $\pi \to \pi^*$ transitions. Proteins absorb in the near-UV region (200 to 400 nm) as a result of $\pi \to \pi^*$ transitions of electrons in the aromatic amino acid residues and $\pi \to \pi^*$ transitions in the amide groups. The proteins hemoglobin and myoglobin have strong visible absorptions because of $\pi \to \pi^*$ transitions in the heme group, which is a large conjugated chromophore.

Fluorescence Fluorescence spectroscopy is widely used in biochemistry, partly because it is capable of giving spectra from extremely small amounts of material. In absorption spectroscopy of a substance in solution, one compares the intensity of two beams, one that passed through the sample and one that went through pure solvent. If the absorber is present in a very small concentration, two nearly equal intensities are being compared, which is difficult. In contrast, in fluorescence spectroscopy, one compares the radiation emitted (at right angles to the exciting radiation beam) with darkness, which allows much greater sensitivity. Fluorescence spectroscopy allows quantitative analysis for substances present in trace amounts, for example, drugs, pesticides, and atmospheric pollutants. A molecule that shows fluorescence when exposed to the wavelength(s) used in the exciting radiation is called a **fluorophore.**

One can locate specific biomolecules in biological tissues and cells using *fluorescence microscopy*. Here, the specimen is illuminated with UV or visible light and is observed through a microscope using a filter that transmits only fluorescent light in a certain wavelength range. The biomolecules to be studied are either fluorescent or are bonded to a fluorescent labeling species.

In the technique of **laser-induced fluorescence** (LIF), one uses a laser to excite molecules to an excited electronic state and then observes the resulting fluorescence as the molecules drop to lower states. The near monochromaticity and tunability of the laser radiation allows one to control which vibrational state of the excited electronic state is populated, making it easier to study the fluorescence. LIF has been used to detect free radicals in kinetic studies of combustion.

Let S_0 denote the ground electronic state of a fluorophore, where the S stands for singlet. Let S_1 denote the lowest-lying electronically excited singlet state. Before absorption of the exciting radiation, most of the fluorophore molecules are in the ground vibrational state of S_0. Excited electronic states nearly always have equilibrium geometries that differ significantly from that of the ground electronic state. Figure 17.18 shows that the harmonic-oscillator ground-state wave function is concentrated near $x = 0$, which corresponds in Eqs. (20.18) and (20.20) to the equilibrium geometry. Because the ground vibrational wave functions of S_0 and S_1 have their probability densities concentrated in different regions of space, the magnitude of the transition dipole moment (20.5) is very small for this pair of vibrational states. Thus there is little probability for absorption of radiation (or emission of radiation) to occur between the ground vibrational states of S_0 and S_1. *The greatest probabilities are for excitation from the ground vibrational state of S_0 to highly excited vibrational states of S_1.*

After excitation, most of the molecules in these excited vibrational levels of S_1 will lose vibrational energy in collisions with neighboring molecules and drop to the ground vibrational level of S_1 before they fluoresce back to the ground state S_0. (This *vibrational relaxation* process is much faster than fluorescence.) Because the transition moment (20.5) is small for transitions between ground vibrational levels of S_0 and S_1, most of the fluorescence comes from molecules going from the ground vibrational level of S_1 to excited vibrational levels of S_0. The fluorescence emission transitions thus have smaller $|\Delta E|$ values than the excitation transitions. In fluorescence, the center of the fluorescence emission band almost always occurs at a longer wavelength (lower frequency and lower photon energy) than the exciting radiation absorbed. The difference in wavelength between the most strongly absorbed exciting wavelength and the most strongly emitted fluorescence wavelength is called the **Stokes shift.** The Stokes shift enables one to filter out scattered excitation radiation, making it easier to observe fluorescence.

DNA Sequencing Fluorescence plays a key role in automated DNA sequencing by capillary electrophoresis (Sec. 15.6). One mixes the single-stranded DNA to be sequenced with the following substances: (*a*) a *primer,* which is a short segment of nucleotide polymer whose base sequence is complementary to that of a known portion of the DNA to be sequenced (or is complementary to a known sequence that has been attached to the DNA); (*b*) DNA polymerase enzyme; (*c*) the four kinds of DNA nucleotide monomers, each consisting of deoxyribose, phosphate, and one of the four DNA bases bonded together; (*d*) small amounts of four kinds of modified DNA nucleotide monomers that each have dideoxyribose instead of deoxyribose and that have one of four dyes that fluoresce at different wavelengths. Which dye has been bonded to the modified nucleotide depends on which base the nucleotide contains. The primer molecules hydrogen bond to the single-stranded DNA molecules and then nucleotide monomers chemically bond to the primer molecules in an order that is complementary to the base sequence of the DNA template. When a modified nucleotide happens to bond to the lengthened primer, the structure of the modified nucleotide prevents further nucleotides bonding on. One thus ends up with a mixture of lengthened primers of all possible lengths, with the last nucleotide in each sequence having a base that corresponds to a particular fluorescent dye attached to the last nucleotide.

The lengthened primer molecules are separated from the DNA templates and the mixture undergoes electrophoresis, which separates the lengthened primers according to length. Near the end of the capillary, a laser excites fluorescence of the dye molecules and the wavelength of the emitted fluorescent radiation reports on which base is at the end of the lengthened primer that is passing that point in the capillary. Each lengthened primer is one nucleotide longer than the one that preceded it past the observation point, so one has determined the base sequence that is complementary to that in the DNA.

Jet Cooling A technique used to simplify the electronic spectra of gas-phase radicals, ions, and large molecules is **jet cooling.** Here a mixture of a small amount of the species to be studied (gas B) and a large amount of He(g) or Ar(g) at high pressure (1 to 100 atm) expands through a small hole of diameter d_{hole} into a vacuum chamber. Conditions are such that $d_{\text{hole}} \gg \lambda$, where λ is the mean free path of gas molecules in the immediate vicinity of the hole. As explained after Eq. (14.58), when $\lambda \ll d_{\text{hole}}$, collisions in the region of the hole produce a bulk collective flow through the hole, and the expanding helium forms a flowing jet, which has substantial macroscopic kinetic energy [the term K in Eq. (2.35)]. The source of this macroscopic kinetic energy is the random molecular kinetic energy [the term $\frac{3}{2}RT$ in Eq. (2.89)] of the He atoms, so the He is cooled to a very low temperature, on the order of 1 K or less. Collisions of gas B molecules with the cold He gas in the region immediately beyond the hole cool the B molecules.

Translational energy of B molecules is transferred most easily, and the random translational motion of B molecules corresponds to a translational temperature T_{tr} of about 1 K. Rotational energy is transferred somewhat less easily than translational energy, and the rotation of the B molecules corresponds to a rotational temperature T_{rot} of perhaps 2 K. (T_{rot} is the temperature in the Boltzmann distribution law that corresponds to the observed distribution of molecules among rotational energy levels; see Prob. 20.62.) Vibrational cooling also occurs, but usually to a lesser extent. Typically, T_{vib} is on the order of 100 K in the jet. The low value of T_{tr} in the jet makes the speed of sound in the jet [given in Sec. 14.5 as $(C_P RT_{\text{tr}}/C_V M)^{1/2}$] far less than the flow rate of the jet, so the jet is described as *supersonic*.

Because there is a reduction in the number of rotational and vibrational states that are populated, and because line broadening due to collisions is reduced in the low-density jet, the appearance of the spectrum is greatly simplified, making it easier to interpret. Vibrational bands that overlap to form a broad, featureless spectrum at room temperature are narrowed and become well separated. Rotational structure becomes more easily resolved. Molecules as large as zinc tetrabenzoporphine ($ZnN_4C_{36}H_{20}$) have been studied [U. Even, J. Jortner, and J. Friedman, *J. Phys. Chem.*, **86,** 2273 (1982)]. Because the B molecules are at a low density, a very sensitive technique such as LIF is used to study the electronic spectrum. By using an electric discharge or photolysis, radicals and ions (such as CH_2, CH_3, C_2H_5, C_5H_5, H_2O^+) can be produced and their spectra observed. For more on spectroscopy of jet-cooled species, see M. Ito et al., *Ann. Rev. Phys. Chem.*, **39,** 123 (1988); P. C. Engelking, *Chem. Rev.*, **91,** 399 (1991).

20.12 NUCLEAR-MAGNETIC-RESONANCE SPECTROSCOPY

Nuclear-magnetic-resonance (NMR) spectroscopy is "the most important spectroscopic technique in chemistry" [J. Jonas and H. S. Gutowsky, *Ann. Rev. Phys. Chem.*, **31,** 1 (1980)]. Before discussing NMR, we review the physics of magnetic fields.

The Magnetic Field

A magnetic field is produced by the motion of electric charge. Examples include the motion of electrons in a wire, the motion of electrons in free space, and the "spinning" of an electron about its own axis. The fundamental magnetic field vector **B** is called the **magnetic induction** or the **magnetic flux density.** There is a second magnetic field vector **H**, named the *magnetic field strength.* It used to be thought that **H** was the fundamental magnetic vector, but it is now known that **B** is. Really, **B** ought to be called the magnetic field strength, but it's too late to correct this injustice by giving **B** its proper name.

The definition of **B** is as follows. Imagine a positive test charge Q_t moving through point P in space with velocity **v**. If for arbitrary directions of **v** we find that a force $\mathbf{F}_\perp$

perpendicular to **v** acts on Q_t at P, we say that a magnetic field **B** is present at point P. One finds that there is one direction of **v** that makes $\mathbf{F}_\perp$ equal zero, and the direction of **B** is defined to coincide with this particular direction of **v**. The magnitude of **B** at point P is then defined by $B \equiv F_\perp/(Q_t v \sin\theta)$, where θ is the angle between **v** and **B**. (For $\theta = 0$, $F_\perp$ becomes 0.) Thus

$$F_\perp = Q_t vB \sin\theta \tag{20.53}$$

(In terms of vectors, $\mathbf{F}_\perp = Q_t \mathbf{v} \times \mathbf{B}$, where $\mathbf{v} \times \mathbf{B}$ is the vector cross-product. The magnetic force is perpendicular to both **v** and **B**.) If an electric field **E** is also present, it exerts a force $Q_t\mathbf{E}$ in addition to the magnetic force $\mathbf{F}_\perp$.

Equation (20.53) is written in SI units. The SI unit of B is the **tesla** (T), also called the Wb/m^2, where Wb stands for weber. From (20.53) and (2.7)

$$1 \text{ T} \equiv 1 \text{ N C}^{-1}\text{ m}^{-1}\text{ s} = 1 \text{ kg s}^{-1}\text{ C}^{-1} \tag{20.54}$$

(The strength of the earth's magnetic field at its surface varies from 26 to 60 μT, depending on location, and has declined about 10% in the last 100 years.)

Magnetic fields are produced by electric currents. Experiment shows that the magnetic field at a distance r from a very long straight wire in vacuum carrying a current I is proportional to I and inversely proportional to r; that is, $B = kI/r$, where k is a constant. The unit of electric current, the ampere (A), is defined (Sec. 15.5) so as to give k the value 2×10^{-7} T m A^{-1}. The constant k is also written as $\mu_0/2\pi$, where μ_0 is called the **magnetic constant** or the **permeability of vacuum.** Thus

$$\mu_0 \equiv 4\pi \times 10^{-7} \text{ T m A}^{-1} = 4\pi \times 10^{-7} \text{ N C}^{-2}\text{ s}^2 \tag{20.55}$$

where (20.54) and 1 A = 1 C/s were used.

Equation (20.1) for the speed of light in vacuum can be written as

$$c = (\varepsilon_0\mu_0)^{-1/2} \tag{20.56}$$

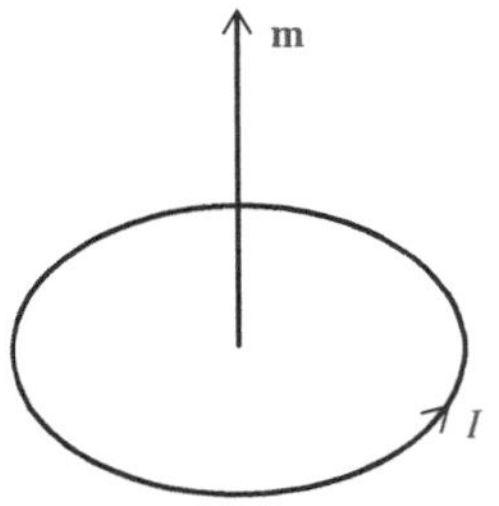

Figure 20.39

The magnetic dipole moment **m** is perpendicular to the plane of the current loop.

Consider a tiny loop of current I flowing in a circle of area A. At distances large compared with the radius of the loop, one finds (*Halliday and Resnick,* sec. 34-6) that the magnetic field produced by this current loop has the same mathematical form as the electric field of an electric dipole (Sec. 13.14) except that the electric dipole moment $\boldsymbol{\mu}$ is replaced by the **magnetic dipole moment m,** where **m** is a vector with magnitude $|\mathbf{m}| = IA$ and direction perpendicular to the plane of the current loop (Fig. 20.39); thus

$$|\mathbf{m}| \equiv IA \tag{20.57}$$

The tiny current loop is called a **magnetic dipole.** (The symbol $\boldsymbol{\mu}$ is often used for the magnetic dipole moment, but this can be confused with the symbol for electric dipole moment.)

A magnetic dipole acts like a tiny magnet with a north pole on one side of the current loop and a south pole on the other side. A bar magnet suspended in an external magnetic field has a preferred minimum-energy orientation in the field. Thus, a compass needle orients itself in the earth's magnetic field with one particular end of the needle pointing toward the earth's north magnetic pole. To turn the needle away from this orientation requires an input of energy. In general, a magnetic dipole **m** has a minimum-energy orientation in an externally applied magnetic field **B**. The potential energy V of interaction between **m** and the external field **B** can be shown to be (*Halliday and Resnick,* sec. 33-4)

$$V = -|\mathbf{m}|B\cos\theta \equiv -\mathbf{m}\cdot\mathbf{B} \tag{20.58}$$

where θ is the angle between **m** and **B**. In (20.58), $\mathbf{m}\cdot\mathbf{B} \equiv |\mathbf{m}|B\cos\theta$ is the **dot product** of **m** and **B**. The minimum-energy orientation has **m** and **B** in the same direction, so that $\theta = 0$ and $V = -|\mathbf{m}|B$. The maximum-energy orientation has **m** and **B** in opposite directions ($\theta = 180°$) and $V = |\mathbf{m}|B$. The zero of potential energy has been arbitrarily chosen to make $V = 0$ at $\theta = 90°$.

Nuclear Spins and Magnetic Moments

Nuclei, like electrons, have a **spin angular momentum I**. A nucleus has two spin quantum numbers, I and M_I. These are analogous to s and m_s for an electron. The magnitude of the nuclear spin angular momentum is

$$|\mathbf{I}| = [I(I+1)]^{1/2}\hbar \qquad \textbf{(20.59)*}$$

and the possible values of the z component of **I** are [recall Eq. (18.32)]

$$I_z = M_I\hbar \qquad \text{where } M_I = -I, -I+1, \ldots, I-1, I \qquad \textbf{(20.60)*}$$

The nuclear spin is the resultant of the spin and orbital angular momenta of the neutrons and protons that compose the nucleus. (See Prob. 18.46 for a discussion of how angular momenta combine in quantum mechanics.) The neutron and the proton each have a spin quantum number $\frac{1}{2}$. A nucleus with an odd mass number A has a half-integral I value ($\frac{1}{2}$ or $\frac{3}{2}$ or . . .). A nucleus with A even has an integral value of I. A nucleus with A even and atomic number Z even has $I = 0$. Some values of I are listed in a table inside the back cover. For a nucleus with $I = \frac{1}{2}$ (for example, ^{1}H), the possible orientations of the spin vector **I** are the same as shown in Fig. 18.11 for an electron. For a nucleus with $I = 1$, the possible orientations of **I** are those shown in Fig. 18.10.

A moving charge produces a magnetic field. We can crudely picture spin as due to a particle rotating about one of its own axes. Hence, we expect a charged particle with spin to act as a tiny magnet. The magnetic properties of a particle with spin can be described in terms of the particle's magnetic dipole moment **m**.

Consider a particle of charge Q and mass m moving in a circle of radius r with speed v. The time for one complete revolution is $t = 2\pi r/v$, and the current flow is $I = Q/t = Qv/2\pi r$. From (20.57), the magnetic moment is $|\mathbf{m}| = \pi r^2 I = Qvr/2$. The particle's orbital angular momentum (Sec. 18.4) is $L = mvr$, and so the magnetic moment can be written as $|\mathbf{m}| = QL/2m$. The angular-momentum vector **L** is perpendicular to the circle, as is the magnetic-moment vector **m**. Therefore

$$\mathbf{m} = Q\mathbf{L}/2m \qquad (20.61)$$

A nucleus has a spin angular momentum **I**, and its magnetic dipole moment is given by an equation resembling (20.61). However, instead of using the charge and mass of the nucleus, it is more convenient to use the proton charge and mass e and m_p. Moreover, because of the composite structure of the nucleus, an extra numerical factor g_N must be included. Thus the **magnetic (dipole) moment m** of a nucleus is

$$\mathbf{m} = g_N \frac{e}{2m_p}\mathbf{I} \equiv \gamma\mathbf{I} \qquad (20.62)$$

where g_N is the **nuclear *g* factor** and the **magnetogyric** (or **gyromagnetic) ratio** γ of the nucleus is defined as

$$\gamma \equiv \frac{e}{2m_p}g_N = (4.78942 \times 10^7\ \text{Hz/T})g_N \qquad (20.63)$$

where the table of physical constants and 1 C/kg = 1 s^{-1} T^{-1} = 1 Hz/T [see Eq. (20.54)] were used. Present theories of nuclear structure cannot predict g_N values. They must be determined experimentally. Values of g_N, I, atomic mass, and percent abundance are given for some isotopes in a table inside the back cover. The fact that g_N for ^{1}H is not a simple number indicates that the proton has an internal structure.

From (20.62) and (20.59), the magnitude of the magnetic moment of a nucleus is

$$|\mathbf{m}| = |g_N|(e/2m_p)[I(I + 1)]^{1/2}\hbar = |\gamma|[I(I + 1)]^{1/2}\hbar$$

This equation is often written as $|\mathbf{m}| = |g_N| \mu_N[I(I + 1)]^{1/2}$, where the **nuclear magneton** μ_N is a physical constant defined as $\mu_N \equiv e\hbar/2m_p = 5.0508 \times 10^{-27}$ J/T. NMR spectroscopists prefer to write equations in terms of γ rather than μ_N and g_N, so we won't use μ_N.

Nuclear-Magnetic-Resonance (NMR) Spectroscopy

In NMR spectroscopy, one applies an external magnetic field **B** to a sample containing nuclei with nonzero spin. For simplicity, we initially consider a single isolated nucleus with magnetic dipole moment **m**. The energy of the nuclear magnetic dipole in the applied field **B** depends on the orientation of **m** with respect to **B** and is given by (20.58) and (20.62) as

$$E = -\mathbf{m}\cdot\mathbf{B} = -\gamma\mathbf{I}\cdot\mathbf{B} = -\gamma|\mathbf{I}|B\cos\theta \qquad (20.64)$$

where $|\mathbf{I}|$ is the magnitude (length) of the spin-angular-momentum vector **I** and θ is the angle between **B** and **I**. Let the direction of the applied field be called the z direction. Figure 20.40*a* shows that $|\mathbf{I}|\cos\theta$ equals I_z, the z component of **I**. Hence, $E = -\gamma I_z B$. However, only certain orientations of **I** (and the associated magnetic moment **m**) in the field are allowed by quantum mechanics (Fig. 20.40*b*); I_z is quantized with the possible values $I_z = M_I\hbar$ [Eq. (20.60)] so $E = -\gamma M_I\hbar B$. The nuclear magnetic moment in the applied magnetic field therefore has the following set of quantized energy levels:

$$E = -\gamma\hbar B M_I, \qquad M_I = -I, \ldots, +I \qquad (20.65)$$

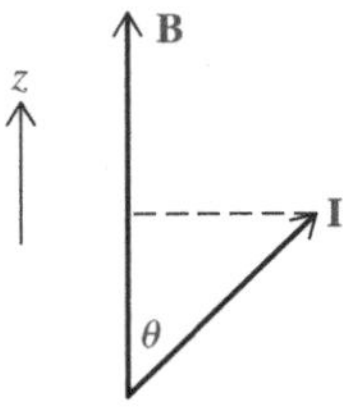

(*a*)

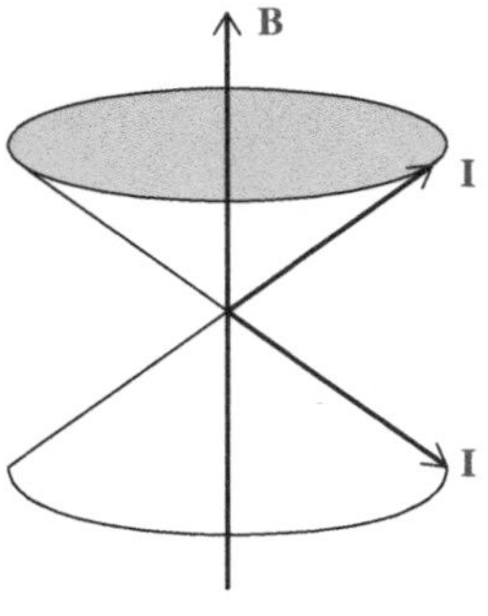

(*b*)

Figure 20.40

(*a*) Projection of **I** on the z axis. (*b*) The possible orientations of **I** for $I = \frac{1}{2}$ lie on the surfaces of two cones.

Figure 20.41 shows the allowed nuclear-spin energy levels of the nuclei ^1H (with $I = \frac{1}{2}$) and ^2H (with $I = 1$) as a function of the applied magnetic field. As B increases, the spacing between levels increases. In the absence of an external magnetic field, all orientations of the spin have the same energy.

By exposing the sample to electromagnetic radiation of appropriate frequency, one can observe transitions between these nuclear-spin energy levels. The selection rule is found to be

$$\Delta M_I = \pm 1 \qquad (20.66)$$

The NMR absorption frequency satisfies $h\nu = |\Delta E| = |\gamma|\hbar B|\Delta M_I| = |\gamma|\hbar B$, where (20.65) was used. Hence

$$\nu = \frac{|\gamma|}{2\pi}B \qquad \textbf{(20.67)*}$$

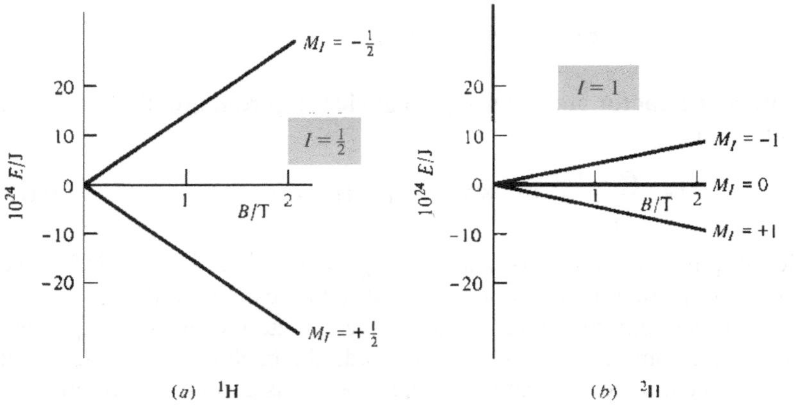

Figure 20.41

Nuclear-spin energy levels versus applied magnetic field for (*a*) ^1H; (*b*) ^2H.

[γ is negative for some nuclei since g_N in (20.63) is negative for some nuclei.] Although there are $2I + 1$ different energy levels for the nuclear magnetic dipole in the field, the selection rule (20.66) allows only transitions between adjacent levels, which are equally spaced. A collection of identical noninteracting nuclei therefore gives a single NMR absorption frequency.

Using g_N values from inside the back cover in (20.63), one finds the following $\gamma/2\pi$ values:

nucleus	^1H	^{13}C	^{15}N	^{19}F	^{31}P
$(\gamma/2\pi)/(MHz/T)$	42.577	10.708	−4.317	40.078	17.251

A typical magnetic field easily attained in the laboratory is 1 T. For the ^1H nucleus (a proton) in this field, Eq. (20.67) and the $\gamma/2\pi$ table give the NMR absorption frequency as 42.577 MHz. This is in the radio-frequency (rf) portion of the electromagnetic spectrum. (FM radio stations broadcast from 88 to 108 MHz.)

NMR in bulk matter was first observed by Bloch and by Purcell in 1945. In his Nobel Prize acceptance speech, Purcell said: "I remember, in the winter of our first experiments, just seven years ago, looking on snow with new eyes. There the snow lay around my doorstep—great heaps of protons quietly precessing in the earth's magnetic field. To see the world for a moment as something rich and strange is the private reward of many a discovery." [*Science,* **118,** 431 (1953).]

Nuclei with $I = 0$ (for example, $^{12}_{6}C$, $^{16}_{8}O$, $^{32}_{16}S$) have no magnetic moment and no NMR spectrum. Nuclei with $I \geq 1$ have something called an electric quadrupole moment, which broadens the NMR absorption lines, tending to obscure the chemically interesting details. Thus nuclei with $I = \frac{1}{2}$ (such as ^1H, ^{13}C, ^{15}N, and ^{31}P) are the most suitable for study. Some $I \neq \frac{1}{2}$ nuclei such as ^2H, ^{11}B, and ^{14}N have been studied. The most studied nuclei are ^1H and ^{13}C.

In spectroscopy, one usually varies the frequency ν of the incident electromagnetic radiation until absorption is observed. In NMR spectroscopy, one has the alternative of keeping ν fixed and varying the spacing between the levels by varying the magnitude B of the applied field until (20.67) is satisfied and absorption occurs. Both alternatives have been used in NMR.

Figure 20.42 shows a simplified version of an NMR spectrometer. The sample is usually a liquid. An electromagnet or permanent magnet applies a uniform magnetic field B_0. The value of B_0 is varied over a narrow range by varying the current through

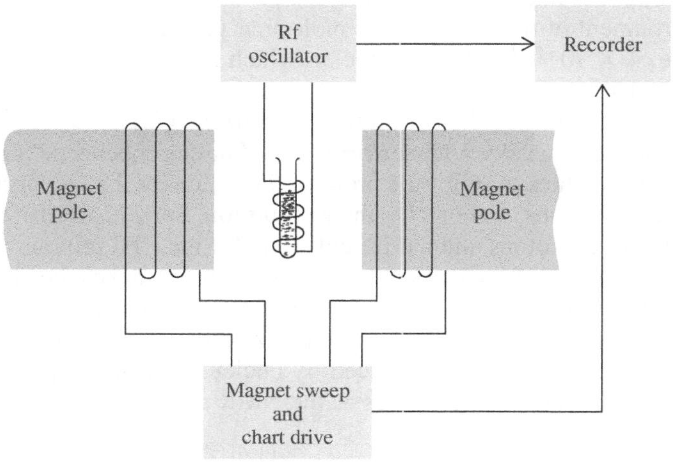

Figure 20.42

An NMR spectrometer.

coils around the magnet poles. This current also drives the chart paper on which the spectrum is recorded. The sample tube is spun rapidly to average out any inhomogeneities in $\mathbf{B}_0$. A coil connected to an rf transmitter (oscillator) and wound around the sample exposes the sample to electromagnetic radiation of fixed frequency. The coil's inductance depends in part on what is inside the coil. The coil is part of a carefully tuned rf circuit in the transmitter. When the sample absorbs energy, its characteristics change, thereby detuning the transmitter circuit and decreasing the transmitter output. This decrease is recorded as the NMR signal. An alternative setup uses a separate detector coil at right angles to the transmitter coil. The flipping of the spins caused by absorption induces a current in the detector coil.

A spectrometer like that of Fig. 20.42 in which the sample is continuously exposed to rf radiation while one varies B_0 or ν in order to scan through the spectrum is called a *continuous-wave* (CW) spectrometer. CW NMR spectrometers have been made obsolete by Fourier-transform NMR spectrometers (these are discussed later in this section).

Recall that the intensity of an absorption line is proportional to the population difference between the levels involved (Sec. 20.2). In NMR spectroscopy, the separation ΔE between the energy levels (Fig. 20.41) is much less than kT at room temperature, and the Boltzmann distribution law $N_2/N_1 = e^{-(E_2-E_1)/kT}$ shows that there is only a slight difference between the populations of the two levels of the transition (Prob. 20.71). Hence the NMR absorption signal is quite weak, and NMR spectroscopy is hard to apply to very small samples.

Chemical Shifts

If Eq. (20.67) gave the NMR absorption frequencies for nuclei in molecules, NMR would be of no interest to chemists. However, the actual magnetic field experienced by a nucleus in a molecule differs very slightly from the applied field B_0, due to the magnetic field produced by the molecular electrons. Most ground-state molecules have all electrons paired, which makes the total electronic spin and orbital angular momenta equal to zero. With zero electronic angular momentum, the electrons produce no magnetic field. However, when an external magnetic field is applied to a molecule, this changes the electronic wave function slightly, thereby producing a slight contribution of the electronic motions to the magnetic field at each nucleus. (This effect is similar to the polarization of a molecule produced by an applied electric field.)

The magnetic field of the electrons usually opposes the applied magnetic field B_0 and is proportional to B_0. The electronic contribution to the magnetic field at a given nucleus i is $-\sigma_i B_0$, where the proportionality constant σ_i is called the **shielding constant** for nucleus i. The value of σ_i at a given nucleus depends on the electronic environment of the nucleus. For molecular protons, σ_i usually lies in the range 1×10^{-5} to 4×10^{-5}. For heavier nuclei (which have more electrons than H), σ_i may be 10^{-4} or 10^{-3}.

For each of the six protons in benzene (C_6H_6), σ_i is the same, since each proton has the same electronic environment. For chlorobenzene (C_6H_5Cl), there are three different values of σ_i for the protons, one value for the two ortho protons, one for the two meta protons, and one for the para proton. For CH_3CH_2Br, there is one value of σ_i for the CH_3 protons and a different value for the CH_2 protons. The low barrier to internal rotation about the carbon–carbon single bond makes the electronic environment of all three methyl protons the same (except at extremely low temperatures).

Addition of the electronic contribution $-\sigma_i B_0$ to the applied field B_0 gives the magnetic field B_i experienced by nucleus i as $B_i = B_0(1 - \sigma_i)$. Substitution in $\nu = |\gamma|B/2\pi$ [Eq. (20.67)] gives as the NMR absorption frequencies of a molecule

$$\nu_i = (|\gamma_i|/2\pi)(1 - \sigma_i)B_0 \tag{20.68}$$

where γ_i is the magnetogyric ratio of nucleus i. If one holds the frequency fixed and varies B_0, the values of the applied field B_0 at which absorption occurs are

$$B_{0,i} = \frac{2\pi\nu_{\text{spec}}}{|\gamma_i|(1 - \sigma_i)} \tag{20.69}$$

where ν_{spec} is the fixed spectrometer frequency.

Different kinds of nuclei (^{1}H, ^{13}C, ^{19}F, etc.) have very different γ values, so their NMR absorption lines occur at very different frequencies. In a given NMR experiment, one examines the NMR spectrum of only one kind of nucleus, and γ_i in (20.69) has a single value. We now consider proton (^{1}H) NMR spectra.

Each chemically different kind of proton in a molecule has a different value of σ_i and a different NMR absorption frequency. Thus, C_6H_6 shows one NMR peak, C_6H_5Cl shows three NMR peaks, and CH_3CH_2Cl shows two NMR peaks (spin–spin coupling neglected; see the next subsection). The relative intensities of the peaks are proportional to the number of protons producing the absorption. For CH_3CH_2OH, the three peaks have a 3:2:1 ratio (Fig. 20.43). Nuclei in a molecule that have the same shielding constant as a result of either molecular symmetry (the protons in C_6H_6) or internal rotation about a single bond (the methyl protons in CH_3OH) are called **chemically equivalent.**

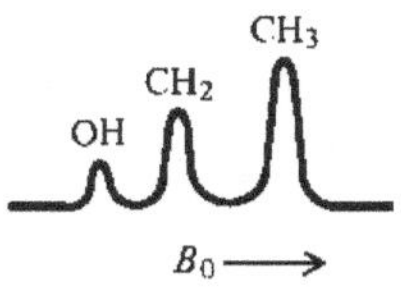

Figure 20.43

The low-resolution proton NMR spectrum of pure liquid CH_3CH_2OH.

The variation in ν_i in (20.68) or $B_{0,i}$ in (20.69) due to variation in the chemical (that is, electronic) environment of the nucleus is called the chemical shift. The **chemical shift** δ_i of proton i is defined by

$$\delta_i \equiv (\sigma_{\text{ref}} - \sigma_i) \times 10^6 \tag{20.70}$$

where σ_{ref} is the shielding constant for the protons of the reference compound tetramethylsilane (TMS), $(CH_3)_4Si$. All the TMS protons are equivalent, and TMS shows a single proton NMR peak. The factor 10^6 is included to give δ a convenient magnitude. Note that σ and δ are dimensionless. Also, the proportionality constants σ_i and σ_{ref} in (20.70) are molecular properties that are independent of the applied field B_0 and the spectrometer frequency ν_{spec}. Hence, *δ_i is independent of B_0 and ν_{spec}.* For a spectrometer in which B_0 is held fixed, the chemical shift δ can be expressed to a high degree of accuracy as (Prob. 20.81)

$$\delta_i = \frac{\nu_i - \nu_{\text{ref}}}{\nu_{\text{ref}}} \times 10^6 \tag{20.71}$$

where ν_{ref} and ν_i are the frequencies at which NMR absorption occurs for the reference nucleus and for nucleus i.

One finds that δ for the protons of a given kind of chemical group differs only slightly from compound to compound. Some typical proton δ values are

RCH_2R'	RCH_3	RNH_2	ROH	$CH_3C(O)R$	OCH_3	ArH	RC(O)H	RCOOH
1.1–1.5	0.8–1.2	1–4	1–6	2–3	3–4	6–9	9–11	10–13

where R and Ar are aliphatic and aromatic groups. Chemical shifts are affected by intermolecular interactions, so one usually observes the proton NMR spectrum of an organic compound in a dilute solution of an inert solvent, most commonly $CDCl_3$. The large δ range for alcohols is due to hydrogen bonding, the extent of which varies with the alcohol concentration.

NMR is an invaluable tool for structure determination.

Spin–Spin Coupling

Proton NMR spectra are more complex than so far indicated, because of the existence of nuclear **spin–spin coupling.** Each nucleus with spin $I \neq 0$ has a nuclear magnetic moment, and the magnetic field of this magnetic moment can affect the magnetic field experienced by a neighboring nucleus, thereby slightly changing the frequency at which the neighboring nucleus will undergo NMR absorption. Because of the rapid molecular rotation in liquids and gases, the direct nuclear spin–spin interaction averages to zero. However, there is an additional, indirect interaction between the nuclear spins that is transmitted through the bonding electrons. This interaction is unaffected by molecular rotation and causes splitting of the NMR peaks. The magnitude of the indirect spin–spin interaction depends on the number of bonds between the nuclei involved. *For protons separated by four or more bonds, this spin–spin interaction is usually negligible.* The magnitude of the spin–spin interaction between nuclei i and k is proportional to a quantity J_{ik}, called the **spin–spin coupling constant;** J_{ik} has units of frequency.

Some typical proton–proton J values in hertz are:

HC—CH	C=CH$_2$	*cis*-HC=CH	*trans*-HC=CH	HCOH	HCC(O)H
5 to 9	−3 to +3	5 to 12	12 to 19	5 to 10	1 to 3

The nonequivalent protons in CH_3CH_2Br are separated by three bonds (H—C_a, C_a—C_b, and C_b—H), and J for these protons is 7.2 Hz. The nonequivalent protons in CH_3OCH_2Br are separated by four bonds, and J is negligible for them.

The correct derivation of the NMR energy levels and frequencies allowing for spin–spin coupling requires a complicated quantum-mechanical treatment (often best done on a computer). Fortunately, for many compounds, a simple, approximate treatment called **first-order** analysis allows the spectrum to be accurately calculated. This approximation is valid provided both the following conditions hold:

1. The differences between NMR resonance frequencies of chemically different sets of protons are all much larger than the spin–spin coupling constants between nonequivalent protons.
2. There is only one coupling constant between any two sets of chemically equivalent spin-$\frac{1}{2}$ nuclei.

The table of δ values following Eq. (20.71) shows that in many cases the difference $\delta_i - \delta_j$ for chemically nonequivalent protons is equal to or greater than 1. Suppose this difference equals 1.5. From Eq. (20.68), the difference between the NMR absorption frequencies of protons i and j is $\nu_i - \nu_j = \gamma_p B_0(\sigma_j - \sigma_i)/2\pi$. But (20.70) gives $\delta_i - \delta_j = 10^6(\sigma_j - \sigma_i)$, so $\nu_i - \nu_j = \gamma_p B_0 10^{-6}(\delta_i - \delta_j)/2\pi$. Equation (20.69) gives $B_0 = 2\pi\nu_{spec}/\gamma_p$ (note that $\sigma_i \ll 1$), so

$$\nu_i - \nu_j = 10^{-6}\nu_{spec}(\delta_i - \delta_j) \tag{20.72}$$

For $\delta_i - \delta_j = 1.5$ and $\nu_{spec} = 100$ MHz, Eq. (20.72) gives $\nu_i - \nu_j = 150$ Hz. This is substantially greater than J_{ij}, which is typically 10 Hz for protons. Hence condition 1 is met in many organic compounds. However, in many cases condition 1 is not met. For example, in CHR=CHR′, the protons have δ_i very close to δ_j, and the first-order treatment cannot be used for a 100-MHz spectrometer.

We shall illustrate the first-order treatment by applying it to CH_3CH_2OH. Consider first the CH_3 protons. They are separated by four bonds from the OH proton, and so the spin–spin interaction between these two groups is negligible. Since the methyl protons are separated by three bonds from the CH_2 protons, the spin–spin interaction between CH_2 and CH_3 protons splits the CH_3 peak. One can prove from quantum mechanics that, when the first-order treatment applies, *the spin–spin interactions between equivalent protons do not affect the spectrum.* Therefore we can ignore the spin–spin interactions between one methyl proton and another. Only the CH_2 protons affect the CH_3 peak.

Since $I = \frac{1}{2}$ for a proton, each CH_2 proton can have $M_I = +\frac{1}{2}$ or $-\frac{1}{2}$. Let up and down arrows symbolize these proton spin states. ($M_I\hbar$ is the proton spin-angular-momentum component in the z direction—the direction of the applied field B_0.) The two CH_2 protons can have the following possible nuclear-spin alignments:

$$\begin{array}{cccc} \uparrow\uparrow & \uparrow\downarrow & \downarrow\uparrow & \downarrow\downarrow \\ (a) & (b) & (c) & (d) \end{array} \qquad (20.73)$$

Since the two CH_2 protons are indistinguishable, one actually takes symmetric and antisymmetric linear combinations of (b) and (c). States (a), (d), and the symmetric linear combination of (b) and (c) are analogous to the electron spin functions (18.40); the antisymmetric linear combination of (b) and (c) is analogous to (18.41). The two linear combinations of (b) and (c) each have a total M_I of 0. State (a) has a total M_I of 1. State (d) has a total M_I of -1.

In a sample of ethanol, 25% of the molecules will have the CH_2 proton spins aligned as in (a), 50% as in (b) or (c), and 25% as in (d). Alignments (b) and (c) do not affect the magnetic field experienced by the CH_3 protons, whereas alignments (a) and (d) either increase or decrease this field and so either increase or decrease the NMR absorption frequency of the CH_3 protons. The CH_2 protons therefore split the CH_3 NMR absorption peak into a triplet (Fig. 20.44). It turns out that *the frequency spacing between the lines of the triplet equals the coupling constant* J_{CH_2,CH_3} *between the* CH_2 *and* CH_3 *protons and is independent of the applied field* B_0. [Although one may vary the magnetic field and keep the frequency fixed, observed splittings in teslas are converted to hertz by multiplying by $\gamma_p/2\pi$; Eq. (20.67).] The preceding discussion shows that the intensity ratios of the members of the triplet are 1:2:1. These ratios deviate slightly from the experimental result in Fig. 20.44 because an approximate treatment is being used.

Now consider the CH_2 peak. The possible alignments of the CH_3 proton spins are

$$\begin{array}{cccccccc} \uparrow\uparrow\uparrow & \uparrow\uparrow\downarrow & \uparrow\downarrow\uparrow & \downarrow\uparrow\uparrow & \uparrow\downarrow\downarrow & \downarrow\uparrow\downarrow & \downarrow\downarrow\uparrow & \downarrow\downarrow\downarrow \\ (a) & (b) & (c) & (d) & (e) & (f) & (g) & (h) \end{array}$$

States (b), (c), and (d) have the same total M_I. States (e), (f), and (g) have the same total M_I. The CH_3 protons therefore act to split the CH_2 absorption into a quartet with 1:3:3:1 intensity ratios. The CH_2 protons are separated by three bonds from the OH proton, so we must also consider the effect of the OH proton. A trace of H_3O^+ or OH^-

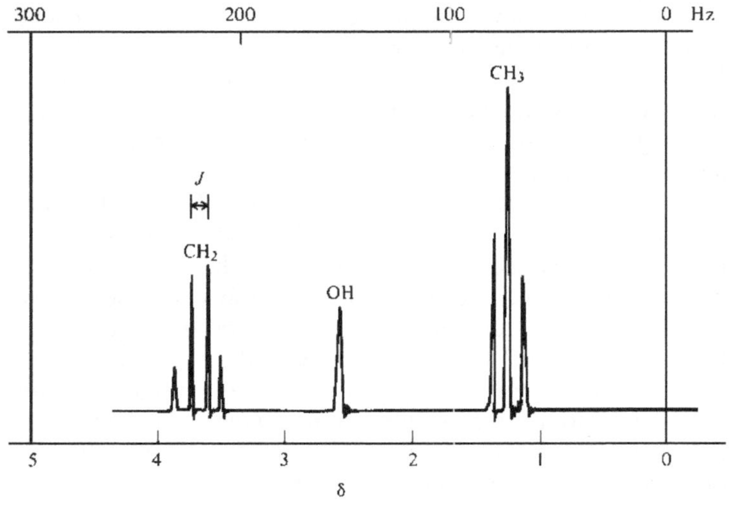

Figure 20.44

The high-resolution 60-MHz proton NMR spectrum of a dilute solution of CH_3CH_2OH in CCl_4 with a trace of acid. The different position of the OH peak compared with that in Fig. 20.43 is explained by hydrogen bonding in pure liquid ethanol. The relation between δ and the frequency-shift scale at the top is given by Eq. (20.72). J is the spin–spin coupling constant between CH_2 and CH_3 protons.

(including that coming from H_2O) will catalyze a rapid exchange of the OH protons between different ethanol molecules. This exchange eliminates the spin–spin interaction between the CH_2 protons and the OH proton, and the CH_2 peak remains a quartet with spacings equal to those in the CH_3 triplet. In pure ethanol, this exchange does not occur, and the OH proton acts to split each member of the CH_2 quartet into a doublet (corresponding to the OH proton spin states $\uparrow$ and $\downarrow$); the CH_2 absorption becomes an octet (eight lines) for pure ethanol. These eight lines are so closely spaced that it may be hard to resolve all of them.

In ethanol containing a trace of acid or base, the OH proton NMR peak is a singlet. In pure ethanol, the OH absorption is split into a triplet by the CH_2 protons.

We have seen that two equivalent protons act to split the absorption peak of a set of adjacent protons into three lines, and three equivalent protons act to split such a peak into four lines. In general, one finds that *n equivalent protons act to split the absorption peak of a set of adjacent protons into n + 1 lines, provided the spectrum is first-order.*

What about spin–spin splittings from nuclei other than ^1H? Since ^{12}C, ^{16}O, and ^{32}S each have $I = 0$, these nuclei don't split proton NMR peaks. ^{14}N has $I = 1$; ^{35}Cl, ^{37}Cl, ^{79}Br, and ^{81}Br each have $I = \frac{3}{2}$; ^{127}I has $I = \frac{5}{2}$. It turns out that nuclei with $I > \frac{1}{2}$ generally don't split proton NMR peaks. ^{19}F has $I = \frac{1}{2}$ and does split proton NMR peaks.

For large organic molecules, the chances are good that two or more nonequivalent sets of protons will have similar δ values, making the first-order treatment invalid. The spectrum becomes very complicated and is hard to interpret. To overcome this difficulty, one can use a spectrometer with higher values of B_0 and ν_{spec}, which are proportional to each other [Eq. (20.69)]. Note from (20.72) that $\nu_i - \nu_j$ increases as ν_{spec} increases, so at sufficiently high ν_{spec} we have $\nu_i - \nu_j \gg J_{ij}$ (condition 1), and the first-order treatment applies. Values of B_0 for commercially available research NMR spectrometers range from 4.7 to 21.6 T, corresponding to proton ν_{spec} values ranging from 200 to 920 MHz. Research NMR spectrometers are Fourier-transform instruments. High-frequency, high-field NMR spectrometers use an electromagnet whose wires are made superconductors by being cooled to 4 K by liquid helium. In addition to simplifying the spectrum, an increase in B_0 increases the signal strength (Prob. 20.71), so smaller amounts of sample can be studied.

Fourier-Transform NMR Spectroscopy

The NMR spectrum of a molecule contains lines at several frequencies, and it takes about 10^3 s to scan through the spectrum using a CW spectrometer. ^{13}C NMR spectra of organic compounds provide information on the "backbone" of organic compounds. The isotope ^{13}C is present in only 1% natural abundance, which makes the ^{13}C NMR absorption signals very weak. One can scan the spectrum repeatedly and feed the results into a computer that adds the results of successive scans, thereby enhancing the signal. However, a sufficiently large number of scans takes several days and is impractical.

To overcome this difficulty, one uses **Fourier-transform** (FT) NMR spectroscopy. Here, instead of continuously exposing the sample to rf radiation while B_0 or ν is slowly varied, B_0 is kept fixed and the sample is irradiated with a very short pulse of high-power radiation from an rf transmitter whose frequency is ν_{trans}, where ν_{trans} is fixed at a value in the range of the NMR frequencies for the kind of nucleus being studied. For ^{13}C in a field of 10 T, the NMR absorption frequency is given by (20.67) as 107.1 MHz, so ν_{trans} is taken as this value. The pulse lasts for several microseconds, so it contains only a limited number of cycles of rf radiation. Because of this, one can show mathematically by a technique called Fourier analysis that the pulse of rf radiation is equivalent to a mixture of all frequencies of radiation. However, only frequencies reasonably close to ν_{trans} have significant amplitudes in the mixture, so the pulse briefly exposes the sample to a band of frequencies centered about ν_{trans}.

Chemically nonequivalent ^{13}C nuclei absorb at slightly different NMR frequencies, all in the region near 107.1 MHz, and the pulse will excite all the ^{13}C spins. After the pulse ends, one observes the signal in the detector coil for about 1 s, thereby obtaining the signal as a function of time. This signal-versus-time function is called the **free-induction decay** (FID).

For the simplest case where all the ^{13}C nuclei are chemically equivalent and ν_{trans} equals the NMR absorption frequency of these nuclei, absorption from the pulse occurs at only one frequency, and the FID function is a simple exponential decay with time, as the excited nuclei return to the ground state to reestablish the equilibrium distribution of spins (a process called **relaxation**). When several nonequivalent ^{13}C's are present in a molecule, absorption from the pulse occurs at several frequencies and the FID shows a very complicated appearance containing oscillations superimposed on an exponential decay. One can show that by taking the mathematical Fourier transform (which is a certain integral) of the FID of signal intensity versus time, one obtains the usual NMR spectrum of signal intensity versus absorption frequency. (Recall that a similar procedure is done in FT-IR; the path difference δ in Sec. 20.9 is a function of time, so one transforms from signal versus t to signal versus ν in FT-IR.) A computer built into the NMR spectrometer very rapidly does the Fourier transformation of the FID to give the absorption spectrum.

Because one needs to observe the FID for only about 1 s in order to obtain the spectrum, an FT-NMR spectrometer is much faster than a CW instrument. By adding the FIDs of many successive pulses and then performing the Fourier transformation, the spectrum's signal-to-noise ratio is increased and the ^{13}C NMR spectrum is readily observed in spite of the low abundance of ^{13}C.

Pulsed FT NMR is not restricted to ^{13}C studies, and all high-quality commercial NMR spectrometers use this technique to increase sensitivity. Moreover, by using a complex sequence of pulses instead of a single pulse to produce the FID, one can obtain information that aids in assigning the signals in the spectra to the various proton and ^{13}C nuclei, and one can enhance the signals in the spectrum; see chap. 8 of *Friebolin.*

Double Resonance

In a double-resonance experiment, the sample is simultaneously exposed to rf radiation of two different frequencies, one frequency being used to observe radiation absorption and the second frequency to produce a perturbation that affects the spectrum. For example, in observing natural-abundance ^{13}C spectra in organic compounds, in addition to applying a pulse of rf radiation that covers the frequency range of the ^{13}C absorptions, one also usually applies continuous strong rf radiation whose frequencies cover the range of the proton absorption frequencies. The result is to remove the spin–spin coupling between the ^{1}H and ^{13}C nuclei (a process called **decoupling**), so the ^{1}H spins don't split the ^{13}C absorption lines. (Many other double-resonance techniques exist; see *Günther.*) Since the probability that two adjacent C nuclei are both ^{13}C is very small, there is no ^{13}C-^{13}C spin–spin splitting in natural-abundance ^{13}C NMR. *With no spin–spin splitting, the* ^{13}C *natural abundance spectrum contains one line for each set of nonequivalent carbons.* In ^{13}C NMR, the reference compound is $(CH_3)_4Si$ (TMS), and the ^{13}C chemical shifts in organic compounds usually lie in the range of δ values from -10 to 230. As with protons, the δ value is characteristic of the kind of carbon being observed. For example, δ for the C=O carbon in ketones is usually between 200 and 225. The combination of proton and ^{13}C NMR is an extremely powerful method of structure determination.

Dynamic NMR

Suppose we have a single nucleus moving back and forth at a frequency ν_{exch} between two environments 1 and 2 in which the magnetic field experienced by the nucleus differs; let ν_1 and ν_2 be the NMR absorption frequencies for the nucleus in environments 1 and 2. It can be shown that if the exchange frequency satisfies $\nu_{exch} \gg |\nu_1 - \nu_2|$, the NMR spectrum shows a single line at a frequency between ν_1 and ν_2, whereas if $\nu_{exch} \ll |\nu_1 - \nu_2|$, the spectrum shows a line at ν_1 and another line at ν_2. The rate of exchange varies with temperature, so by studying the temperature dependence of the NMR spectrum, one can obtain rate constants for the exchange reaction. First-order reactions with rate constants in the range 10^3 to 10^{-1} s^{-1} are most easily studied by NMR.

An example is dimethylformamide (Fig. 20.45). For this molecule, we can write a resonance structure with a double bond between C and N, and the partial double-bond character of this bond produces a substantial barrier to internal rotation. At room temperature, the 60-MHz proton NMR spectrum shows a line at $\delta = 8.0$ due to the CHO proton and shows two lines due to the methyl protons, one at $\delta = 2.79$ and one at 2.94. The presence of two methyl absorption lines shows that at 25°C the exchange rate of the two groups of methyl protons (labeled *a* and *b*) is far less than $|\nu_2 - \nu_1|$, and so the two sets of CH_3 protons are nonequivalent. (The *a* and *b* protons are separated by 4 bonds and do not split each other.) As the temperature is raised, the two methyl lines gradually coalesce, forming a single line at 120°C if a 60-MHz spectrometer is used. For 120°C and above, the internal rotation is so fast that $\nu_{exch} \gg |\nu_1 - \nu_2|$.

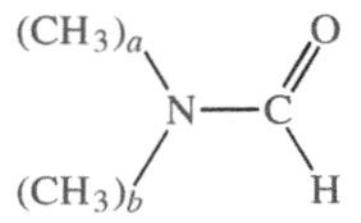

Figure 20.45

Dimethylformamide. Because of the partial double-bond character of the N—CO bond, the molecule is nearly planar except for the methyl hydrogens.

To obtain the rate constant at various temperatures, one must carry out a complicated quantum-mechanical analysis of the spectra for the intermediate temperatures, to find what value of *k* yields the observed spectrum at each *T*. Knowing *k* as a function of *T*, one can calculate the activation energy, which is the barrier to internal rotation, which is found to be 23 kcal/mol in dimethylformamide.

The temperature dependences of NMR spectra have been used to study the rates of internal rotation in substituted ethanes, proton exchange between alcohols and water, ring inversions, and inversion at nitrogen atoms; see *Friebolin*.

The Nuclear Overhauser Effect

Suppose the following double-resonance experiment is performed. The proton NMR spectrum of a molecule is recorded (using either a CW or FT spectrometer) while the sample is continuously irradiated with rf radiation of frequency ν_S that is the NMR absorption frequency of a specific set (which we call set S) of chemically equivalent protons in the molecule. One then finds that the intensities of all lines that are due to protons that are close to the set-S protons in the molecule are changed as compared with a spectrum taken without continuous radiation at ν_S. The radiation at ν_S changes the energy-level population distribution of the set-S protons, and the magnetic-dipole–magnetic-dipole interaction between the set-S protons and nearby protons changes the population distributions of the nearby protons, thereby changing the intensities of their NMR lines. This intensity change is the **nuclear Overhauser effect (NOE).** The magnitude of the NOE is usually proportional to $1/r^6$, where *r* is the distance between the set-S protons and the protons producing the line whose intensity is changed. The NOE is negligible for $r > 4$ Å. The NOE can be used to help assign spectra and to find internuclear distances in a molecule.

For example, $(CH_3)_2NCHO$ (Fig. 20.45) at 25°C shows CH_3 absorption lines at $\delta = 2.79$ and 2.94. To determine which line goes with which set of CH_3 protons, one can use the NOE. When continuous rf radiation at the $\delta = 2.94$ frequency is applied, the intensity of the CHO proton line is increased by 18%, whereas rf radiation at $\delta = 2.79$ produces a 2% decrease in the CHO line. Hence the $\delta = 2.94$ protons must be closer to (that is, cis to) the CHO proton.

Two-Dimensional NMR Spectroscopy

In Fig. 20.46, the rectangles denote rf pulses applied to an NMR sample. The duration of each pulse is very short and is greatly exaggerated in the figure. Figure 20.46*a* shows a single rf pulse applied to the system, followed by observation of the FID as a function of time *t*. Fourier transformation of this function of *t* gives the NMR spectrum as a function of frequency ν. This is a one-dimensional (1D) NMR spectrum. In Fig. 20.46*b*, a pulse is applied, and then after a time t_1, a second pulse is applied, and then the FID is observed as a function of time t_2 up to a time $t_{2,\max}$. Fourier transformation of this function of t_2 gives the NMR spectrum as a function of frequency. The spectrum's appearance will be influenced by the first pulse. We have obtained the spectrum as a function of a single variable, so this is still a 1D spectrum.

Now suppose we repeat the experiment in Fig. 20.46*b*, except that we use a slightly longer time t_1 between the two pulses; we then repeat the experiment using successively larger values of the interval t_1 between the two pulses. By collecting all the FIDs from the successive experiments, we get an FID that is now a function $f(t_1, t_2)$ of two variables, t_1 and t_2. If we now do a Fourier transformation of $f(t_1, t_2)$ by integrating over both t_1 and t_2, we will get an NMR spectrum that will be a function of two frequencies ν_1 and ν_2. This spectrum is a **two-dimensional (2D) NMR spectrum.** The frequency ν_2 has the same meaning as the frequency in 1D NMR. The significance of the frequency ν_1 depends on the nature of the pulses used in the experiments. Hundreds of different pulse patterns have been used in 2D NMR. Figures 20.46*c* and *d* show two other pulse patterns. In *c*, the time interval Δ remains fixed and t_1 is varied. In *d*, the second pulse is twice as long as the first.

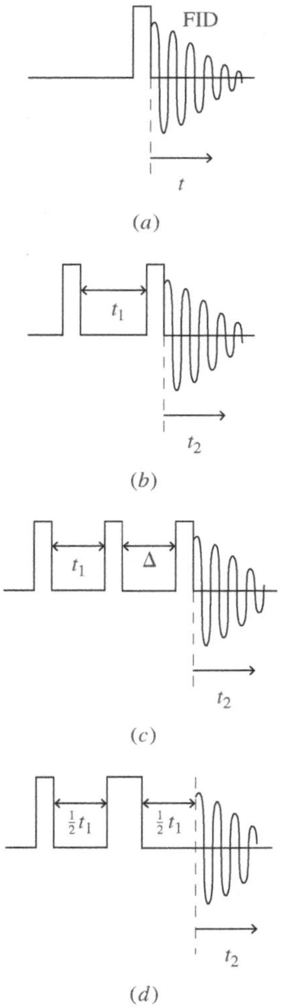

Figure 20.46

Some NMR pulse sequences.

The proton NMR spectra of large molecules such as steroids, oligosaccharides, proteins, and nucleic acids contain many overlapping lines, and the lines are extremely hard to assign to specific protons, even using very high-field spectrometers. By using 2D NMR, one can resolve very complicated spectra and determine the structure and conformation of compounds for which the 1D spectrum is hopelessly complex. The development of 2D NMR in the 1980s produced a revolutionary increase in the power of NMR to deduce the structure of large molecules. See W. R. Croasmun and R. M. K. Carlson (eds.), *Two-Dimensional NMR Spectroscopy*, 2nd ed., Wiley-VCH, 1994; J. Schraml and J. M. Bellama, *Two-Dimensional NMR Spectroscopy*, Wiley, 1988.

For a protein whose sequence of amino acids is known, application of 2D (and 3D) NMR and the NOE enables one to determine the distances between various pairs of protons. From this information, one can deduce the three-dimensional structure (conformation) of the protein in aqueous solution. (See J. N. S. Evans, *Biomolecular NMR Spectroscopy*, Oxford, 1995; J. Cavanagh et al., *Protein NMR Spectroscopy*, 2nd ed., Elsevier Academic, 2007.) NMR structures for over 6000 proteins have been determined. [Protein structures are tabulated in the Protein Data Bank (www.rcsb.org/pdb/).] NMR spectroscopy enables structures of proteins with molecular weights up to 100000 to be determined. Methods to extend NMR structure determination to proteins with higher molecular weights are being developed [A. G. Tzakos et al., *Annu. Rev. Biophys. Biomol. Struct.*, **35,** 319 (2006)].

NMR of Solids

As ordinarily observed, the NMR spectra of solids are of little value because the direct spin–spin interactions (which are averaged to zero in liquids, as noted earlier in this section), lead to broad, featureless absorptions with little information. However, by using special techniques, such as rapidly spinning the sample about an axis making a certain angle with B_0 [magic-angle spinning (MAS)], one can eliminate these spin–spin interactions and get high-resolution NMR spectra from solids. High-resolution solid-state NMR is used to study the structure and dynamics of solid polymers, biopolymers such as proteins, catalysts, molecules adsorbed on catalysts, coal, ceramics, glasses, etc.

(see E. O. Stejskal and J. D. Memory, *High Resolution NMR in the Solid State,* Oxford, 1994).

Magnetic Resonance Imaging

NMR spectroscopy can be used to form cross-sectional images of body parts in living subjects by displaying the intensity of a particular NMR transition (for example, protons in water) as a function of the coordinates in the cross-sectional plane. This technique of magnetic-resonance imaging (MRI) is widely used to diagnose diseases such as cancer. See I. L. Pykett, *Scientific American,* May 1982, p. 78; E. R. Andrews, *Acc. Chem. Res.,* **16,** 114 (1983). For discussions on the controversy over the apportionment of credit for developing MRI, see B. H. Kevles, *Naked to the Bone,* Rutgers, 1997, chap. 8; *Physics Today,* Dec. 1997, pp. 100–102.

Classical Description of NMR Spectroscopy. NMR spectroscopy is a quantum-mechanical phenomenon, and a fully correct description of it must be a quantum-mechanical one. The quantum mechanics of many aspects of NMR is difficult. Hence, a classical description of NMR is widely used. Consider a nucleus with spin $I = \frac{1}{2}$ and with positive g_N. In the presence of an applied magnetic field $\mathbf{B}_0$ along the z axis, the angle θ the nuclear-spin vector $\mathbf{I}$ makes with the z axis satisfies $\cos\theta = I_z/|\mathbf{I}| = M_I\hbar/[I(I+1)]^{1/2}\hbar = \pm\frac{1}{2}/(3/4)^{1/2} = \pm 1/3^{1/2}$, and $\theta = 54.7°$ or $125.3°$ (Fig. 20.40). As noted in Fig. 20.41*a*, the $M_I = +\frac{1}{2}$ orientation has lower energy. Since the magnetic moment $\mathbf{m}$ is proportional to $\mathbf{I}$, these are also the possible angles between $\mathbf{m}$ and $\mathbf{B}_0$. In classical mechanics, the interaction (20.58) between $\mathbf{m}$ and $\mathbf{B}_0$ produces a torque on $\mathbf{m}$ that makes $\mathbf{m}$ revolve around the field direction keeping its angle θ with $\mathbf{B}_0$ fixed, thereby sweeping out a conical surface. This motion is called **precession.** The precession frequency (called the *Larmor frequency*) equals $(\gamma/2\pi)\mathbf{B}_0$, where γ is the magnetogyric ratio of the nucleus.

For a collection of identical spin-$\frac{1}{2}$ nuclei in $\mathbf{B}_0$, slightly more than half the nuclei will have their $\mathbf{m}$ vectors on the cone with $\theta = 54.7°$, which has the lower energy. The magnetic moments $\mathbf{m}_i$ of the nuclei add vectorially to give a total magnetic moment $\Sigma_i\,\mathbf{m}_i$, and the quantity $\Sigma_i\,\mathbf{m}_i/V$, where V is the system's volume, is called the **magnetization M.** Because the nuclear spins have random components in the x and y directions, the components of $\Sigma_i\,\mathbf{m}_i$ in the x and y directions are zero and $M_x = 0 = M_y$. Thus $\mathbf{M}$ lies along the z direction, the direction of $\mathbf{B}_0$. Note that $M_I = +\frac{1}{2}$ nuclei and $M_I = -\frac{1}{2}$ nuclei give opposite contributions to M_z, and M_z is nonzero because there is a slight excess of low-energy $M_I = +\frac{1}{2}$ nuclei at equilibrium. The magnitude of M_z is proportional to the population difference between the two states.

In pulse FT NMR spectroscopy, the magnetic field $\mathbf{B}_1$ of the pulse of electromagnetic radiation is perpendicular to the direction of $\mathbf{B}_0$ and lies in the xy plane. (In most branches of spectroscopy, the electric field of the electromagnetic radiation interacts with the electric charges of the molecule to produce the transition. In NMR, the interaction of the magnetic field of the radiation with the nuclear magnetic moments produces the transition.) The presence of the field $\mathbf{B}_1$ causes the magnetization vector $\mathbf{M}$ to precess about the direction of $\mathbf{B}_1$ at the Larmor frequency $(\gamma/2\pi)B_1$, thereby rotating $\mathbf{M}$ away from the z axis and toward the xy plane. The amount of rotation is measured by the angle θ between $\mathbf{M}$ and the z axis. The amount of rotation is proportional to the duration t_{pulse} of the pulse; $\theta = bt_{\text{pulse}}$, where b is a constant. The rotation frequency is $(\gamma/2\pi)B_1$ and the period of this rotation is the reciprocal of the frequency, namely, $2\pi/\gamma B_1$. After one period, θ will equal 2π. Hence $\theta = bt_{\text{pulse}}$ becomes $2\pi = b2\pi/\gamma B_1$ and $b = \gamma B_1$. Thus

$$\theta = \gamma B_1 t_{\text{pulse}}$$

In FT NMR, one usually chooses the pulse duration to make $\theta = 90°$ (a 90° pulse) so that the pulse moves $\mathbf{M}$ into the xy plane.

The situation $M_z = 0$ means that there are equal numbers of $M_I = +\frac{1}{2}$ and $M_I = -\frac{1}{2}$ spins. The pulse causes both absorption and stimulated emission of radiation. Because of a net absorption of radiation, the 90° pulse has excited enough low-energy $M_I = +\frac{1}{2}$ nuclei to the $M_I = -\frac{1}{2}$ state to equalize the populations of the states (a situation called *saturation*).

After the pulse, the spin system moves back toward equilibrium (a process called **relaxation**). The rate at which M_z moves back toward its equilibrium value is proportional to the quantity $1/T_1$, where T_1 is called the **spin–lattice relaxation time** or the **longitudinal relaxation time.** (The longitudinal direction is the z direction.) M_z moving back to its equilibrium value corresponds to the $M_I = \pm\frac{1}{2}$ populations moving back to equilibrium by having some high-energy spins make transitions to the low-energy $M_I = +\frac{1}{2}$ state. For the relatively low frequency of NMR transitions, it turns out that spontaneous emission of radiation is too slow to contribute significantly to this process of reestablishing the population equilibrium. Instead, spin-population relaxation occurs because interactions between the nuclear spins and their surroundings (called the "lattice" because this relaxation was first studied in crystalline solids) produce nonradiative transitions that transfer energy from the high-energy spins to molecular translational and rotational energies. The rate at which the component M_{xy} in the xy plane changes back to its equilibrium value of zero is proportional to the quantity $1/T_2$, where T_2 is the **spin–spin** (or **transverse**) **relaxation time.**

The classical vector model of NMR is useful for understanding many aspects of NMR. However, because NMR is a quantum-mechanical phenomenon, there are many NMR multiple-pulse experiments that cannot be correctly treated with the classical model and a complicated quantum-mechanical treatment is needed.

"There can be little doubt that the spectroscopic tool that has done the most for chemistry is NMR" [E. B. Wilson, *Ann. Rev. Phys. Chem.*, **30,** 1 (1979)].

20.13 ELECTRON-SPIN-RESONANCE SPECTROSCOPY

In **electron-spin-resonance** (ESR) **spectroscopy** [also called **electron paramagnetic resonance** (EPR) **spectroscopy**], one observes transitions between the quantum-mechanical energy levels of an unpaired electron spin magnetic moment in an external magnetic field. Most ground-state molecules have all electron spins paired, and such molecules show no ESR spectrum. One observes ESR spectra from free radicals such as H, CH_3, $(C_6H_5)_3C$, and $C_6H_6^-$, from transition-metal ions with unpaired electrons, and from excited triplet states of organic compounds. The sample may be solid, liquid, or gaseous.

An electron has spin quantum numbers $s = \frac{1}{2}$ and $m_s = +\frac{1}{2}$ and $-\frac{1}{2}$. Relativistic quantum mechanics and experiment show that the g value of a free electron is $g_e = 2.0023$. Analogous to the equation $\mathbf{m} = (g_N e/2m_p)\mathbf{I}$ [Eq. (20.62)] for a nuclear spin, the magnetic dipole moment of a free electron is

$$\mathbf{m}_e = (-g_e e/2m_e)\mathbf{S}, \qquad g_e \approx 2 \tag{20.74}$$

where $-e$, m_e, and $\mathbf{S}$ are the electron charge, mass, and spin-angular-momentum vector. The magnitude of $\mathbf{S}$ is $[s(s+1)]^{1/2}\hbar = \frac{1}{2}\sqrt{3}\hbar$. An electron spin magnetic moment has two energy levels in an applied magnetic field, corresponding to the two orientations $m_s = +\frac{1}{2}$ and $m_s = -\frac{1}{2}$. The magnetic field B experienced by an unpaired electron in a molecular species differs somewhat from the applied field B_0, and we write $B = B_0(1 - \sigma)$, where σ is a shielding constant. The energy levels are [Eqs. (20.58) and (20.74)]

$$E = -\mathbf{m}_e \cdot \mathbf{B} = (g_e e/2m_e)\mathbf{S} \cdot \mathbf{B} = (g_e e/2m_e)B_0(1 - \sigma)S_z$$

$$= (g_e e/2m_e)B_0(1 - \sigma)(\pm\tfrac{1}{2}\hbar)$$

$$E = \pm\tfrac{1}{2}g_e\mu_B(1 - \sigma)B_0$$

where the **Bohr magneton** μ_B is

$$\mu_B \equiv e\hbar/2m_e = 9.274 \times 10^{-24}\ \text{J/T} \tag{20.75}$$

(Don't confuse the electron mass m_e with the electron magnetic moment $\mathbf{m}_e$.) The energy-level separation is $g_e\mu_B(1 - \sigma)B_0$ and equating this to $h\nu$, we get the ESR transition frequency as

$$\nu = g_e\mu_B(1 - \sigma)B_0/h = g\mu_B B_0/h \tag{20.76}$$

where the **g factor** for the molecular species is $g \equiv g_e(1 - \sigma)$. For organic radicals, g factors are close to the free-electron value $g_e = 2.0023$. For transition-metal ions, g can differ greatly from g_e. For $B_0 = 1$ T and $g = 2$, Eq. (20.76) gives the ESR frequency as 28 GHz, which is in the microwave region. ESR frequencies are much higher than NMR frequencies because the electron mass is 1/1836 the proton mass. Figure 20.47 is a block diagram of an ESR spectrometer. The klystron generates a fixed frequency of microwave radiation, which is propagated along the waveguide. The magnetic field applied to the sample is varied over a range to generate the spectrum.

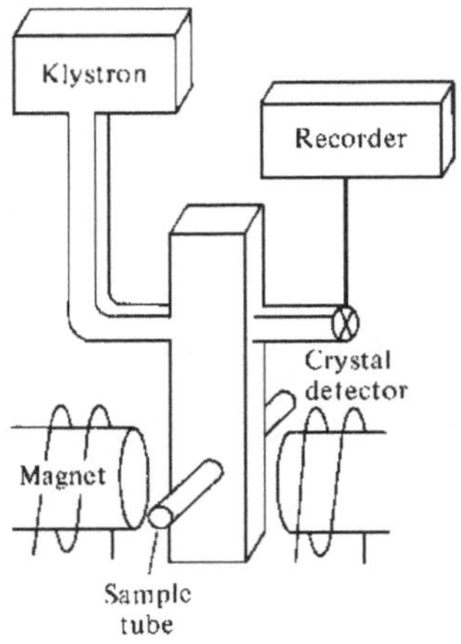

Figure 20.47

An ESR spectrometer.

Interaction between the electron spin magnetic moment and the nuclear spin magnetic moments splits the ESR absorption peak into several lines (*hyperfine splitting*). Both protons and electrons have spin $\frac{1}{2}$. Just as a proton NMR absorption peak is split into $n + 1$ peaks by n adjacent equivalent protons, the ESR absorption peak of a radical with a single unpaired electron is split into $n + 1$ peaks by n equivalent protons. For example, the ESR spectrum of the methyl radical $\cdot CH_3$ consists of 4 lines (Fig. 20.48). If the unpaired electron interacts with one set of m equivalent protons and a second set of n equivalent protons, the ESR spectrum has $(n + 1)(m + 1)$ lines. For example, the $\cdot CH_2CH_3$ ESR spectrum has 12 lines. The ^{12}C nuclei have $I = 0$ and don't split the ESR lines.

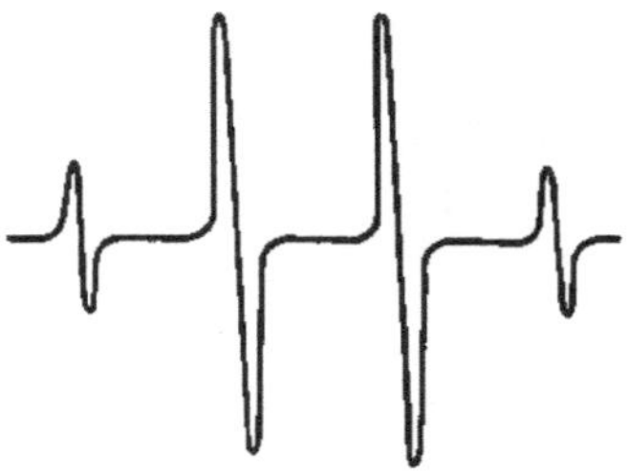

Figure 20.48

The ESR spectrum of the methyl radical. ESR spectrometers usually display the derivative (slope) of the absorption line. This slope is positive for the left half of the line and negative for the right half.

For the n-butyl radical (Fig. 20.49), the H's on the α carbon and the H's on the β carbon produce substantial ESR splitting; the H's on the γ carbon produce a very small splitting, which may or may not be resolved, depending on the quality of the spectrometer used. The H's on the δ carbon produce a negligible splitting. The n-butyl ESR spectrum will thus show either 3(3) = 9 lines or 3(3)(3) = 27 lines, depending on the spectrometer resolution. In radicals such as $C_6H_6^+$ and $C_6H_6^-$ formed from conjugated ring compounds, the odd electron is delocalized over the whole molecule and all the H's contribute to splitting the ESR lines.

ESR spectroscopy can detect free-radical reaction intermediates.

One can use ESR spectroscopy to obtain information on biological molecules by bonding an organic free radical to the macromolecule under study, a procedure called *spin labeling*. (See *Chang,* chap. 6; *Campbell and Dwek,* chap. 7.)

δ γ β α
$CH_3CH_2CH_2CH_2\cdot$

Figure 20.49

The n-butyl radical.

20.14 OPTICAL ROTATORY DISPERSION AND CIRCULAR DICHROISM

A molecule that rotates the plane of polarization of plane-polarized light (Fig. 20.1) is said to be **optically active.** Plane-polarized light is produced by passing unpolarized light through a polarizing crystal. A molecule that is not superimposable on its mirror image is optically active (provided the molecule and its mirror image cannot be interconverted by rotation about a bond with a low rotational barrier).

The **angle of optical rotation** α is the angle through which the light's electric-field vector has been rotated by passing through the sample. Most commonly, the optically active substance is in solution. α is proportional to the sample length l and to the concentration of optically active material in the solution. To get a quantity characteristic of the optically active substance, one defines the **specific rotation** $[\alpha]_\lambda$ of the optically active substance B as

$$[\alpha]_\lambda \equiv \frac{\alpha_\lambda}{[\rho_B/(g/cm^3)](l/dm)} \tag{20.77}$$

where ρ_B, the mass concentration of B, is the mass of B per unit volume in the solution [Eq. (9.2)] and l is the path length of the light through the sample. For a pure sample, ρ_B becomes the density of pure B. The optical rotation α and the specific rotation $[\alpha]$ depend on the radiation's wavelength (as indicated by the λ subscript) and depend on the temperature and solvent. If the polarization plane is rotated to the right (clockwise) as viewed looking toward the oncoming beam, the substance is *dextrorotatory* and $[\alpha]_\lambda$ is defined as positive. If the rotation is counterclockwise, the substance is *levorotatory* and $[\alpha]_\lambda$ is negative.

A plot of $[\alpha]_\lambda$ versus λ gives the **optical rotatory dispersion** (ORD) spectrum of the substance. $[\alpha]_\lambda$ changes very rapidly in a wavelength region where the optically active substance has an electronic absorption, and the ORD spectrum is of most interest in such regions—the UV and the visible.

Related to ORD is circular dichroism. For an optically active substance, the molar absorption coefficient ε_λ (Sec. 20.2) for left-circularly polarized light differs (typically by 0.01 to 0.1%) from ε_λ for right-circularly polarized light. **Circularly polarized** light is light in which the electric-field direction rotates about the direction of propagation as one moves along the wave, making one complete rotation in one wavelength of the light. The polarization is left or right depending on whether the field vector at a fixed position rotates counterclockwise or clockwise with time, respectively, when viewed from in front of the oncoming beam. Circularly polarized light can be produced by special devices.

The **circular dichroism** (CD) spectrum of a substance is a plot of $\Delta\varepsilon \equiv \varepsilon_L - \varepsilon_R$ versus light wavelength λ, where ε_L and ε_R are molar absorption coefficients [Eq. (20.10)] for left- and right-circularly polarized light. The circular dichroism $\Delta\varepsilon$ is nonzero only in regions of absorption and is most commonly measured in the UV and visible regions, the regions of electronic absorption. (Instead of $\Delta\varepsilon$, the *molar ellipticity* $[\theta] \equiv 3298\Delta\varepsilon$ is sometimes plotted.)

Many biological molecules are optically active. The ORD and CD spectra of proteins and nucleic acids are sensitive to the conformations of these macromolecules and can be used to follow changes in conformation as functions of temperature, binding of ligands, etc. Currently, CD spectra are used much more often than ORD.

By comparing the UV CD spectrum of a protein with CD spectra of proteins with known conformations one can (using one of many available computer programs) estimate with a fair degree of accuracy the percentages of α helix, parallel β sheet, antiparallel β sheet, etc., in the protein.

CD spectra are used to study the kinetics of protein folding and unfolding and have been used to study conformational changes in prions, infectious proteins believed responsible for such neurodegenerative diseases as mad-cow disease and Creutzfeldt–Jakob disease. The A, B, and Z forms of DNA can be distinguished from one another using their CD spectra.

CD spectroscopy is also used to measure $\Delta\varepsilon$ for IR absorption bands. By comparing a molecule's observed vibrational CD (VCD) spectrum with the spectrum predicted by a density-functional calculation (Sec. 19.10) for an assumed configuration, one can determine the absolute configuration of chiral molecules with up to 200 atoms.

For more on circular dichroism, see N. Berova and R. Woody (eds.), *Circular Dichroism*, 2nd ed., Wiley, 2000; G. D. Fasman (ed.), *Circular Dichroism and the Conformational Analysis of Biomolecules,* Plenum, 1996.

Measurement of the difference in Raman scattering intensity for right- and left-circularly polarized light as a function of Raman shift gives the branch of spectroscopy called *Raman optical activity* (ROA). ROA gives information about molecular configuration and conformation.

20.15 PHOTOCHEMISTRY

Photochemistry

Photochemistry is the study of chemical reactions produced by light. Absorption of a photon of light may raise a molecule to an excited electronic state, where it will be more likely to react than in the ground electronic state. In **photochemical reactions,** the activation energy is supplied by absorption of light. In contrast, the reactions studied in Chapter 16 are **thermal reactions,** in which the activation energy is supplied by intermolecular collisions.

The energy of a photon is $E_{\text{photon}} = h\nu = hc/\lambda$. The energy of one mole of photons is $N_A h\nu$. We find the following values of photon wavelength, energy, and molar energy:

λ/nm	200 (UV)	400 (violet)	700 (red)	1000 (IR)
E_{photon}/eV	6.2	3.1	1.8	1.2
$N_A h\nu$/(kJ/mol)	598	299	171	120

(20.78)

Since it usually takes at least $1\frac{1}{2}$ or 2 eV to put a molecule into an excited electronic state, photochemical reactions are initiated by UV or visible light.

Ordinarily, the number of photons absorbed equals the number of molecules making a transition to an excited electronic state. This is the **Stark–Einstein law** of photochemistry.

In exceptional circumstances, this law is violated. A high-power laser beam provides a very high density of photons (Prob. 20.89). There is some probability that a molecule will be hit almost simultaneously by two laser-beam photons, producing a transition in which a single molecule absorbs two photons at once. In rare cases, a single photon can excite two molecules in contact with each other. Liquid O_2 is light blue because of absorption of 630-nm (red) light; each photon absorbed excites two colliding O_2 molecules to the lowest-lying O_2 excited electronic state [E. A. Ogryzlo, *J. Chem. Educ.*, **42,** 647 (1965)].

Photochemical reactions are of tremendous biological importance. Most plant and animal life on earth depends on photosynthesis, a process in which green plants synthesize carbohydrates from CO_2 and water:

$$6CO_2 + 6H_2O \rightarrow C_6H_{12}O_6(\text{glucose}) + 6O_2 \qquad (20.79)$$

The reverse of this reaction provides energy for plants and animals. For (20.79), $\Delta G°$ is 688 kcal/mol, so the equilibrium lies far to the left in the absence of light. The presence of light and of the green pigment chlorophyll makes reaction (20.79) possible. Chlorophyll contains a conjugated ring system that allows it to absorb visible radiation. The main absorption peaks of chlorophyll are at 450 nm (blue) and 650 nm (red). Photosynthesis requires eight photons per molecule of CO_2 consumed.

The process of vision depends on photochemical reactions, such as the dissociation of the retinal pigment rhodopsin after it absorbs visible light. Other important photochemical reactions are the formation of ozone from O_2 in the earth's stratosphere, the formation of photochemical smog from automobile exhausts, the reactions in film photography, and the formation of vitamin D and skin cancer by sunlight.

Photochemical reactions are more selective than thermal reactions. By using monochromatic light, we can excite one particular species in a mixture to a higher electronic state. (The monochromaticity, high power, and tunability of lasers make them ideal sources for photochemical studies.) In contrast, heating a sample increases the translational, rotational, and vibrational energies of all species. Organic chemists use photochemical reactions as a tool in syntheses.

Certain chemical reactions yield products in excited electronic states. Decay of these excited states may then produce emission of light, a process called **chemiluminescence.** Fireflies and many deep-sea fish show chemiluminescence. In a sense, chemiluminescence is the reverse of a photochemical reaction.

Consequences of Light Absorption

Let B* and B_0 denote a B molecule in an excited electronic state and in the ground electronic state, respectively. The initial absorption of radiation is $B_0 + h\nu \rightarrow B^*$. In most cases, the ground electronic state is a singlet with all electron spins paired. The selection rule $\Delta S = 0$ (Sec. 20.11) then shows that the excited electronic state B* is also a singlet.

Following light absorption, many things can happen.

The B* molecule is usually produced in an excited vibrational level. Intermolecular collisions (especially collisions with the solvent if the reaction is in solution) can transfer this extra vibrational energy to other molecules, causing B* to lose most of its vibrational energy and attain an equilibrium population of vibrational levels, a process called **vibrational relaxation.**

The B* molecule can lose its electronic energy by spontaneously emitting a photon, thereby falling to a lower singlet state, which may be the ground electronic state: $B^* \rightarrow B_0 + h\nu$. Spontaneous emission of radiation by an electronic transition in which the total electronic spin doesn't change ($\Delta S = 0$) is called **fluorescence.** Fluorescence is favored in low-pressure gases, where the time between collisions is relatively long. A typical lifetime of an excited singlet electronic state is 10^{-8} s in the absence of collisions.

The B* molecule can transfer its electronic excitation energy to another molecule during a collision, thereby returning to the ground electronic state, a process called **radiationless deactivation:** $B^* + C \rightarrow B_0 + C$, where B_0 and C on the right have extra translational, rotational, and vibrational energies.

The B* molecule (especially after undergoing vibrational relaxation) can make a radiationless transition to a different excited electronic state: $B^* \rightarrow B^{*\prime}$. Conservation of energy requires that B* and $B^{*\prime}$ have the same energy. Generally, the molecule $B^{*\prime}$ has a lower electronic energy and a higher vibrational energy than B*.

If B* and $B^{*\prime}$ are both singlet states (or both triplet states), then the radiationless process $B^* \rightarrow B^{*\prime}$ is called **internal conversion.** If B* is a singlet electronic state and $B^{*\prime}$ is a triplet (or vice versa), then $B^* \rightarrow B^{*\prime}$ is called **intersystem crossing.** Recall that a triplet state has two unpaired electrons and total electronic spin quantum number $S = 1$.

Suppose $B^{*\prime}$ is a triplet electronic state. It can lose its electronic excitation energy and return to the ground electronic state during an intermolecular collision or by intersystem crossing to form B_0 in a high vibrational energy level. In addition, $B^{*\prime}$ can emit a photon and fall to the singlet ground state B_0. Emission of radiation with $\Delta S \neq 0$ is called **phosphorescence.** Phosphorescence violates the selection rule $\Delta S = 0$ and has a very low probability of occurring. The lifetime of the lowest-lying excited triplet electronic state is typically 10^{-3} to 1 s in the absence of collisions.

[The term **luminescence** refers to any emission of light by electronically excited species, and includes fluorescence and phosphorescence (which are emissions that follow electronic excitation by absorption of light), chemiluminescence, luminescence following collisions with electrons (as in gas-discharge tubes or from television screens), etc.]

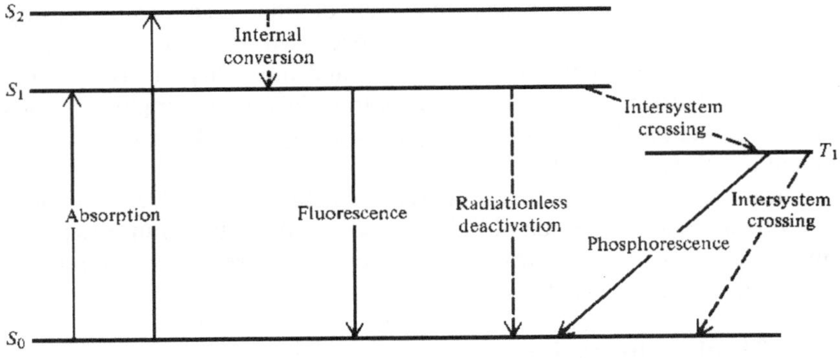

Figure 20.50

Photophysical processes. Dashed arrows indicate radiationless transitions. S_0 is the ground (singlet) electronic state. S_1 and S_2 are the lowest two excited singlet electronic states. T_1 is the lowest triplet electronic state. For simplicity, vibration–rotation levels are omitted.

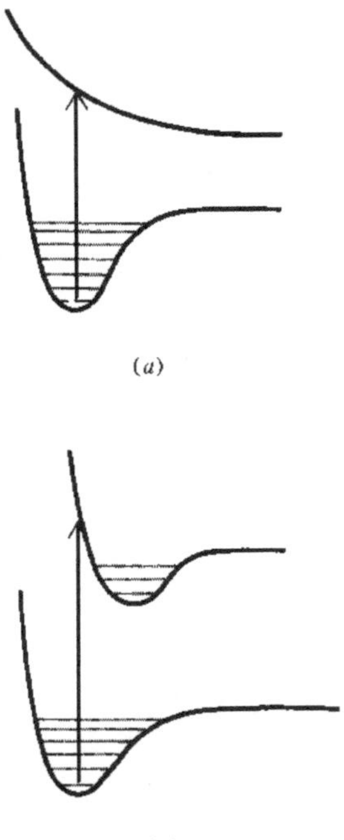

Figure 20.51

Electronic absorptions in a diatomic molecule that always lead to dissociation.

Figure 20.50 summarizes the preceding processes.

Besides the above physical processes, absorption of light can cause several kinds of chemical processes.

Since B* is often formed in a high vibrational level, the B* molecule may have enough vibrational energy to dissociate: B* → R + S. The decomposition products R and S may react further, especially if they are free radicals. If B* is a diatomic molecule with vibrational energy exceeding the dissociation energy D_e of the excited electronic state, then dissociation occurs in the time it takes one molecular vibration to occur, 10^{-13} s. For a polyatomic molecule with enough vibrational energy to break a bond, dissociation may take a while to occur. There are many vibrational modes, and it requires time for vibrational energy to flow into the bond to be broken. Excitation of a diatomic molecule to a repulsive electronic state [one with no minimum in the $E_e(R)$ curve] always causes dissociation. Excitation of a diatomic molecule to a bound excited electronic state with a minimum in the $E_e(R)$ curve causes dissociation if the vibrational energy of the excited molecule exceeds D_e. See Fig. 20.51. Sometimes a molecule undergoes internal conversion from a bound excited state to a repulsive excited electronic state, which then dissociates.

The vibrationally excited B* molecule may isomerize: B* → P. Many cis–trans isomerizations can be carried out photochemically.

The B* molecule may collide with a C molecule, the excitation energy of B* providing the activation energy for a bimolecular chemical reaction: B* + C → R + S.

The B* molecule may collide with an unexcited B or C molecule to form the excimer (BB)* or the exciplex (BC)* (Sec. 20.2), species stable only in an excited electronic state. This is especially common in solutions of aromatic hydrocarbons. The excimer or exciplex may then undergo fluorescence [(BB)* → 2B + $h\nu$ or (BC)* → $h\nu$ + B + C] or nonradiative decay to 2B or B + C.

The B* molecule may transfer its energy in a collision to another species D, which then undergoes a chemical reaction. Thus, B* + D → B + D*, followed by D* + E → products; alternatively, B* + D → B + P + R. This process is **photosensitization.** The species B functions as a photochemical catalyst. An example is photosynthesis, where the photosensitizer is chlorophyll.

All these chemical processes can be preceded by internal conversion or intersystem crossing, B* → B*′, so that it is B*′ that reacts.

The many possible chemical and physical processes make it hard to deduce the precise sequence of events in a photochemical reaction.

Photochemical Kinetics

A common setup for kinetic study of a photochemical reaction exposes the sample to a continuous beam of nearly monochromatic radiation. Of course, only radiation that is absorbed is effective in producing reaction. For example, exposing acetaldehyde to

400-nm radiation will have no effect, since radiation with a wavelength less than 350 nm is required to excite acetaldehyde to a higher electronic level. According to the Beer–Lambert law (20.11), the intensity I of radiation varies over the length of the reaction cell. Convection currents (and perhaps stirring) are usually sufficient to maintain a near-uniform concentration of reactants over the cell length, despite the variation in I.

As in any kinetics experiment, one follows the concentration of a reactant or product as a function of time. In addition, one measures the rate of absorption of light energy by comparing the energies reaching radiation detectors (such as photoelectric cells) after the beam passes through two side-by-side cells, one filled with the reaction mixture and one empty (or filled with solvent only).

The initial step in a photochemical reaction is

$$(1) \quad B + h\nu \rightarrow B^* \tag{20.80}$$

For the elementary process (20.80), the reaction rate is $r_1 \equiv d[B^*]/dt$, where $[B^*]$ is the molar concentration of B^*. From the Stark–Einstein law, r_1 equals $\mathcal{I}_a$, where $\mathcal{I}_a$ is defined as the number of moles of photons absorbed per second and per unit volume; $r_1 = \mathcal{I}_a$. We assume that B is the only species absorbing radiation. Let the reaction cell have length l, cross-sectional area $\mathcal{A}$, and volume $V = \mathcal{A}l$. Let I_0 and I_l be the intensities of the monochromatic beam as it enters the cell and as it leaves the cell, respectively. The intensity I is the energy that falls on unit cross-sectional area per unit time, so the radiation energy incident per second on the cell is $I_0\mathcal{A}$ and the energy emerging per second is $I_l\mathcal{A}$. The energy absorbed per second in the cell is $I_0\mathcal{A} - I_l\mathcal{A}$. Dividing by the energy $N_A h\nu$ per mole of photons and by the cell volume, we get $\mathcal{I}_a$, the moles of photons absorbed per unit volume per second:

$$r_1 = \mathcal{I}_a = \frac{I_0\mathcal{A} - I_l\mathcal{A}}{VN_A h\nu} = \frac{I_0}{lN_A h\nu}(1 - e^{-\alpha[B]l}) \tag{20.81}$$

where the Beer–Lambert law (20.11) was used. In (20.81), $\alpha = 2.303\varepsilon$, where ε is the molar absorption coefficient of B at the wavelength used in the experiment.

The **quantum yield** Φ_X of a photochemical reaction is the number of moles of product X formed divided by the number of moles of photons absorbed. Division of numerator and denominator in this definition by volume and time gives

$$\Phi_X = \frac{d[X]/dt}{\mathcal{I}_a} \tag{20.82}$$

Quantum yields vary from 0 to 10^6. Quantum yields less than 1 are due to deactivation of B^* molecules by the various physical processes discussed above and to recombination of fragments of dissociation. The quantum yield of the photochemical reaction $H_2 + Cl_2 \rightarrow 2HCl$ with 400-nm radiation is typically 10^5. Absorption of light by Cl_2 puts it into an excited electronic state that "immediately" dissociates into Cl atoms. The Cl atoms then start a chain reaction (Sec. 16.13), yielding many, many HCl molecules for each Cl atom formed.

An example of photochemical kinetics is the dimerization of anthracene ($C_{14}H_{10}$), which occurs when a solution of anthracene in benzene is irradiated with UV light. A simplified version of the accepted mechanism is:

$$
\begin{array}{lll}
(1) & A + h\nu \rightarrow A^* & r_1 = \mathcal{I}_a \\
(2) & A^* + A \rightarrow A_2 & r_2 = k_2[A^*][A] \\
(3) & A^* \rightarrow A + h\nu' & r_3 = k_3[A^*] \\
(4) & A_2 \rightarrow 2A & r_4 = k_4[A_2]
\end{array}
$$

where A is anthracene. Step (1) is absorption of a photon by anthracene to raise it to an excited electronic state; Eq. (20.81) gives $r_1 = \mathcal{I}_a$. Step (2) is dimerization. Step (3) is fluorescence. Step (4) is a unimolecular decomposition of the dimer.

The rate r for the overall reaction $2A \rightarrow A_2$ is

$$r = d[A_2]/dt = k_2[A^*][A] - k_4[A_2] \tag{20.83}$$

Use of the steady-state approximation for the reactive intermediate A* gives

$$d[A^*]/dt = 0 = \mathcal{I}_a - k_2[A][A^*] - k_3[A^*] \tag{20.84}$$

which gives $[A^*] = \mathcal{I}_a/(k_2[A] + k_3)$. Substitution in (20.83) gives

$$r = \frac{k_2[A]\mathcal{I}_a}{k_2[A] + k_3} - k_4[A_2] \tag{20.85}$$

Note that $\mathcal{I}_a$ depends on [A] in a complicated way [Eq. (20.81) with B replaced by A].

The quantum yield is given by (20.82) as

$$\Phi_{A_2} = \frac{d[A_2]/dt}{\mathcal{I}_a} = \frac{r}{\mathcal{I}_a} = \frac{k_2[A]}{k_2[A] + k_3} - \frac{k_4}{\mathcal{I}_a}[A_2] \tag{20.86}$$

If $k_4 = 0$ (no reverse reaction) and $k_3 = 0$ (no fluorescence), then Φ becomes 1. The first fraction on the right can be written as $k_2/(k_2 + k_3/[A])$; an increase in [A] increases Φ, since it increases r_2 (dimerization) compared with r_3 (fluorescence). This is the observed behavior. A typical Φ for this reaction is 0.2.

Instead of dealing with the individual rates of all the physical and chemical processes that follow absorption of radiation, one often adopts the simplifying approach of writing the initial step of the reaction as

$$\text{(I)} \quad B + h\nu \rightarrow R + S \qquad r_1 = \phi\mathcal{I}_a$$

Here, R and S are the first chemically different species formed following the absorption of radiation by B. Step I really summarizes several processes, namely, absorption of radiation by B to give B*, deactivation of B* by collisions and fluorescence, decomposition (or isomerization) of B* to R and S, and recombination of R and S immediately after their formation (recall the cage effect). The quantity ϕ, called the **primary quantum yield,** varies between 0 and 1. The greater the degree of collisional and fluorescent deactivation of B*, the smaller ϕ is. For absorption by gas-phase diatomic molecules that leads to dissociation, the dissociation is so rapid that deactivation is usually negligible and $\phi \approx 1$.

The Photostationary State

When a system containing a chemical reaction in equilibrium is placed in a beam of radiation that is absorbed by one of the reactants, the rate of the forward reaction is changed, thereby throwing the system out of equilibrium. Eventually a state will be reached in which the forward and reverse rates are again equal. This state will have a composition different from that of the original equilibrium state and is a **photostationary state.** It is a steady state (Sec. 1.2) rather than an equilibrium state, because removal of the system from its surroundings (the radiation beam) will alter the system's properties. An important photostationary state is the ozone layer in the earth's stratosphere (Sec. 16.16).

20.16 GROUP THEORY

Molecular symmetry elements and operations were discussed in Sec. 20.5. The full application of molecular symmetry uses the mathematics of group theory. This section gives an introduction to group theory and omits most proofs. For fuller details, see

Cotton or *Schonland*. Much of the material of this section is abstract, and is not as easy to read as a mystery thriller (but might well be mystifying the first time you read it).

Groups

Let $A, B, C, \ldots$ be a collection of entities, all of which are different from one another. The entities $A, B, C, \ldots$ might or might not be numbers. Let the symbol $*$ denote a specific rule for combining any two of the entities $A, B, C, \ldots$ to yield a third entity called the **product** of the two entities. For example, the equation $B*F = M$ says that M is the product of B and F. The rule for combining entities can be any well-defined rule, not necessarily ordinary multiplication. The entities $A, B, C, \ldots$ are said to form a **group** under the rule of combination $*$ if the following four conditions are satisfied: (*a*) closure, (*b*) associativity, (*c*) the existence of an identity entity, (*d*) the existence of an inverse for each entity. The meanings of these four conditions will be defined in the example that follows. The entities $A, B, C, \ldots$ that constitute the group are called the **elements** or **members** of the group.

Consider all the integers (whole numbers), positive, negative, and zero. Let the rule of combination be ordinary addition, so that $B*F$ becomes $B + F$.

The **closure** requirement means that if B and F are any two elements of the group (including the case where B and F are the same element), then the product $B*F$ is an element of the group. Since the sum of two integers is always an integer, the closure requirement is satisfied in this example.

The **associativity** requirement means that the rule of combination has the property that $(B*F)*J = B*(F*J)$, for all elements of the group. Since $(B + F) + J = B + (F + J)$, associativity holds. [Associativity should not be taken for granted. Is $(B/F)/J$ equal to $B/(F/J)$?]

The **identity element** I is a particular element of the group that has the property that $B*I = I*B = B$ for every element B in the group. For our example, the identity element is the integer zero. Since $B + 0 = 0 + B = B$, the requirement that an identity element exist is met.

The **inverse** B^{-1} of an element B has the property that $B*B^{-1} = B^{-1}*B = I$, where I is the identity element. If every element of the group has an inverse that is an element of the group, then the inverse requirement is met. For our example of integers, the inverse of B is $-B$ ($B^{-1} = -B$), since $B + (-B) = (-B) + B = 0$.

Since the four requirements are met, the set of all integers forms a group under the rule of combination of addition. The number of elements in a group is called its **order.** The group of integers under addition is of infinite order.

Note that there is no requirement that $B*D$ equal $D*B$. A group for which $B*D = D*B$ for all possible products is called **commutative** or **Abelian.**

From here on, the symbol $*$ for the rule of combination will be omitted and the product of the group elements B and D will be written as BD.

Symmetry Point Groups

We now show that the symmetry operations $\hat{A}, \hat{B}, \hat{C}, \ldots$ of a molecule (Sec. 20.5) form a group with the rule of combination for $\hat{B}$ and $\hat{F}$ being "take the product of the symmetry operations $\hat{B}$ and $\hat{F}$."

The **product** $\hat{B}\hat{F}$ of the symmetry operations $\hat{B}$ and $\hat{F}$ means we first apply the operation $\hat{F}$ to the molecule and we then apply $\hat{B}$ to the result found by applying $\hat{F}$. For example, consider the product $\hat{C}_2(z)\hat{\sigma}(xz)$ in the H_2O molecule. The $\hat{\sigma}(xz)$ operation (which is a reflection in the xz plane) interchanges the two hydrogens (Fig. 20.52). When the $\hat{C}_2(z)$ rotation is applied to the result, the two hydrogens are again interchanged. Since $\hat{C}_2(z)\hat{\sigma}(xz)$ returns all the atoms in H_2O to their original locations, one might think that $\hat{C}_2(z)\hat{\sigma}(xz)$ equals the identity operation $\hat{E}$, but this conclusion is too hasty. (Recall from Sec. 20.5 that the identity operation $\hat{E}$ does nothing.) The symmetry

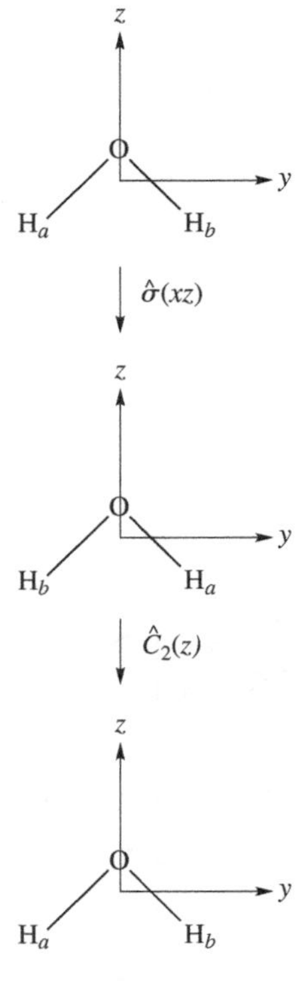

Figure 20.52
Effect of $\hat{C}_2(z)\hat{\sigma}(xz)$ on H_2O.

operation $\hat{\sigma}(yz)$, which is a reflection in the molecular plane, also leaves all atoms of H_2O unmoved. Symmetry operations are transformations of points in three-dimensional space. To see whether $\hat{C}_2(z)\hat{\sigma}(xz)$ equals $\hat{E}$ or $\hat{\sigma}(yz)$, we must consider what happens to a point at the three-dimensional location (x, y, z). The reflection $\hat{\sigma}(xz)$ in the xz plane leaves the x and z coordinates of any point unchanged and changes the y coordinate to its negative. The $\hat{C}_2(z)$ rotation about the z axis leaves the z coordinate of any point unchanged and sends the x and y coordinates to their respective negatives:

$$(x, y, z) \xrightarrow{\hat{\sigma}(xz)} (x, -y, z) \xrightarrow{\hat{C}_2(z)} (-x, y, z)$$

Thus the net effect of $\hat{C}_2(z)\hat{\sigma}(xz)$ is to send the x coordinate to its negative. This is what $\hat{\sigma}(yz)$ does, so $\hat{C}_2(z)\hat{\sigma}(xz) = \hat{\sigma}(yz)$. Note in Fig. 20.52, that, by convention, *the coordinate axes do not move when a symmetry operation is applied to the points in space.*

If $\hat{B}\hat{F} = \hat{F}\hat{B}$, the symmetry operations $\hat{B}$ and $\hat{F}$ are said to **commute.** Symmetry operations do not always commute (Prob. 20.97).

If $\hat{B}$ and $\hat{G}$ are symmetry operations of a molecule, each leaves the molecule in a position indistinguishable from the original position. Hence successive performance of these operations must leave the molecule in a position indistinguishable from the original, and the product $\hat{B}\hat{G}$ must be a symmetry operation. Thus the closure requirement is met.

Multiplication of symmetry operations can be shown to be associative, and the associativity requirement is met.

The identity element of a molecular symmetry group is the symmetry operation $\hat{E}$, the identity operation, which does nothing.

Clearly, each symmetry operation has an inverse, which undoes the effect of the operation. For example, the inverse of a $\hat{C}_4$ rotation is a $\hat{C}_4^3$ rotation about the same axis, since a 270° counterclockwise rotation ($\hat{C}_4^3$) is the same as a 90° clockwise rotation. A reflection is its own inverse.

Since the four requirements are met, the set of symmetry operations of a molecule is a group. The symmetry groups of molecules are called **point groups,** since each symmetry operation leaves the point that is the molecular center of mass unmoved. A molecule can be classified as belonging to one of a number of possible point groups, depending on what symmetry elements are present.

Commonly Occurring Point Groups

A molecule whose symmetry elements are a C_n axis (where n can be 2 or 3 or 4 or . . .) and n planes of symmetry that each contain the C_n axis belongs to the point group C_{nv}. The v stands for "vertical." A **vertical** symmetry plane (symbol σ_v) is one that contains the highest-order axis of symmetry of the molecule. The H_2O molecule has as its symmetry elements a C_2 axis and two planes of symmetry that contain this axis (Figs. 20.17 and 20.18), so its point group is C_{2v}. The symmetry operations of H_2O are $\hat{E}$, $\hat{C}_2(z)$, $\hat{\sigma}(xz)$, and $\hat{\sigma}(yz)$; the order of the group C_{2v} is 4. The point group of NH_3 is C_{3v}, whose symmetry operations are $\hat{C}_3$, $\hat{C}_3^2$, $\hat{E}$, $\hat{\sigma}_a$, $\hat{\sigma}_b$, and $\hat{\sigma}_c$; the order is 6.

A molecule whose only symmetry element is a plane of symmetry belongs to the group C_s. Examples are the bent molecule HOCl, where the symmetry plane is the molecular plane, and the tetrahedral molecule $CHFBr_2$, where the symmetry plane contains the nuclei H, C, and F.

Molecules whose point group is D_{nh} have a C_n symmetry axis, n C_2 symmetry axes perpendicular to the C_n axis, a horizontal symmetry plane σ_h perpendicular to the C_n axis, n vertical symmetry planes containing the C_n axis, and a center of symmetry if n is even; the C_n axis is also an S_n axis. An example is benzene (Fig. 20.53), whose point group is D_{6h} (Prob. 20.101).

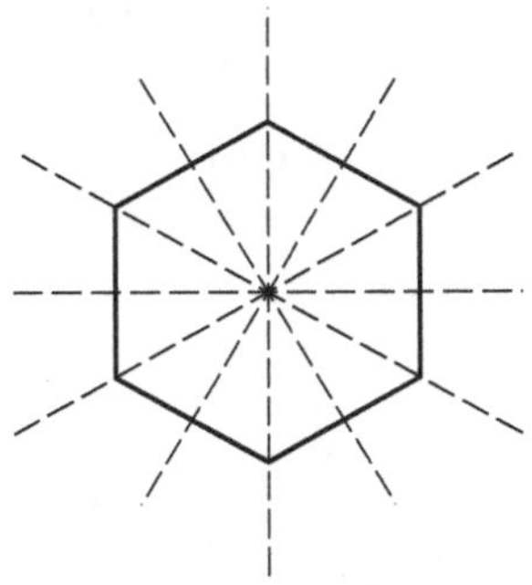

Figure 20.53

C_2 axes in C_6H_6.

The tetrahedral molecule CH_4 belongs to point group T_d (Prob. 20.102) and the octahedral molecule SF_6 belongs to O_h.

A molecule with no symmetry elements belongs to group C_1, since a $\hat{C}_1$ rotation equals the identity operation $\hat{E}$. The order of C_1 is 1.

In a linear molecule with no center of symmetry (for example, HF or OCS), the molecular axis is a C_∞ axis, since rotation about this axis by any angle is a symmetry operation. Also, any plane that contains the molecular axis is a symmetry plane, and there are an infinite number of such vertical symmetry planes. Hence linear molecules with no center of symmetry belong to group $C_{\infty v}$. In a linear molecule with a center of symmetry (for example, H_2 or CO_2), the molecular axis is a C_∞ axis, there are an infinite number of vertical symmetry planes, and also a horizontal symmetry plane perpendicular to the molecular axis. A centrosymmetric linear molecule has point group $D_{\infty h}$.

Some other point groups are considered in Probs. 20.104–20.106.

Multiplication Tables

The **multiplication table** of a group contains all possible products of members of the group. For example, the multiplication table of C_{2v}, the point group of H_2O, is shown in Table 20.2. The elements of the group are listed at the top of the table and at the far left of the table. Each of the 16 entries within the table is the product of the element at the far left of its row and at the top of its column. For example, the entry $\hat{\sigma}(xz)$ that lies in the column headed $\hat{\sigma}(yz)$ (the rightmost column) means that $\hat{C}_2(z)\hat{\sigma}(yz) = \hat{\sigma}(xz)$.

Each row (and each column) of a group multiplication table must contain each group element once and only once. To prove this, suppose the element F appeared twice in some row. Then we would have $DR = F$ and $DJ = F$, where R and J are two group elements. We have $DR = DJ$. Multiplication by D^{-1} on the left gives $D^{-1}DR = D^{-1}DJ$ and $IR = IJ$ (where I is the identity element), which becomes $R = J$. However, all elements of a group are different from one another, so R cannot equal J. Thus it is impossible for an element to appear twice in the same row of a group multiplication table.

The C_{2v} multiplication table is easily filled in as follows. The first row and first column of entries are easily found using the fact that $\hat{E}\hat{R} = \hat{R}$ and $\hat{S}\hat{E} = \hat{S}$, where $\hat{E}$ is the identity operation. The diagonal entries equal $\hat{E}$, since $\hat{C}_2^2 = \hat{E}$ and $\hat{\sigma}^2 = \hat{E}$. We saw (Fig. 20.52 and the associated discussion) that $\hat{C}_2(z)\hat{\sigma}(xz) = \hat{\sigma}(yz)$. All the remaining entries can then be filled in using the theorem that each element appears once in each row and once in each column. Note that C_{2v} is a commutative group.

Matrices

A **matrix** is a rectangular array of numbers (called the **elements** of the matrix). Matrices obey certain rules of combination. As well as being important in group theory, matrices play a key role in quantum-chemistry calculations. The most efficient

TABLE 20.2

Multiplication Table of C_{2v}

	$\hat{E}$	$\hat{C}_2(z)$	$\hat{\sigma}(xz)$	$\hat{\sigma}(yz)$
$\hat{E}$	$\hat{E}$	$\hat{C}_2(z)$	$\hat{\sigma}(xz)$	$\hat{\sigma}(yz)$
$\hat{C}_2(z)$	$\hat{C}_2(z)$	$\hat{E}$	$\hat{\sigma}(yz)$	$\hat{\sigma}(xz)$
$\hat{\sigma}(xz)$	$\hat{\sigma}(xz)$	$\hat{\sigma}(yz)$	$\hat{E}$	$\hat{C}_2(z)$
$\hat{\sigma}(yz)$	$\hat{\sigma}(yz)$	$\hat{\sigma}(xz)$	$\hat{C}_2(z)$	$\hat{E}$

way to solve the Hartree–Fock equations (18.57) and the Kohn–Sham equations (19.41) is by using matrices.

A matrix with m rows and n columns is called an m by n matrix and contains mn matrix elements. If **B** is a matrix, the notation b_{jk} symbolizes the element in row j and column k. Two matrices are said to be **equal** if they have the same number of rows, the same number of columns, and have all corresponding matrix elements equal to each other. The matrix equation $\mathbf{S} = \mathbf{T}$ is equivalent to the mn scalar equations $s_{jk} = t_{jk}$, where j goes from 1 to m and k goes from 1 to n.

Let **A** and **B** be 2 by 2 matrices. Let **C** denote the matrix product **AB**. The product **AB** is defined as a 2 by 2 matrix whose elements are found as

$$\mathbf{AB} = \begin{pmatrix} a_{11} & a_{12} \\ a_{21} & a_{22} \end{pmatrix}\begin{pmatrix} b_{11} & b_{12} \\ b_{21} & b_{22} \end{pmatrix} = \begin{pmatrix} a_{11}b_{11} + a_{12}b_{21} & a_{11}b_{12} + a_{12}b_{22} \\ a_{21}b_{11} + a_{22}b_{21} & a_{21}b_{12} + a_{22}b_{22} \end{pmatrix}$$

$$= \begin{pmatrix} c_{11} & c_{12} \\ c_{21} & c_{22} \end{pmatrix} = \mathbf{C} \tag{20.87}$$

The element $c_{11} = a_{11}b_{11} + a_{12}b_{21}$ of **C** is found by adding the products of corresponding elements of row 1 of **A** and column 1 of **B**. The element c_{12} is found from row 1 of **A** and column 2 of **B**. The element c_{21} is found from row 2 of **A** and column 1 of **B**.

EXAMPLE 20.5 Matrix multiplication

Find **AB** if

$$\mathbf{A} = \begin{pmatrix} 1 & 5 \\ 3 & 4 \end{pmatrix} \qquad \mathbf{B} = \begin{pmatrix} -1 & 6 \\ 3 & 2 \end{pmatrix} \tag{20.88}$$

We have

$$\mathbf{AB} = \begin{pmatrix} 1 & 5 \\ 3 & 4 \end{pmatrix}\begin{pmatrix} -1 & 6 \\ 3 & 2 \end{pmatrix} = \begin{pmatrix} 1(-1) + 5(3) & 1(6) + 5(2) \\ 3(-1) + 4(3) & 3(6) + 4(2) \end{pmatrix} = \begin{pmatrix} 14 & 16 \\ 9 & 26 \end{pmatrix}$$

Exercise

Find **BA**. Does **BA** equal **AB**? (*Answer:* The first-row elements are 17 and 19; the second-row elements are 9 and 23.)

As we saw in Example 20.5, matrix multiplication is not commutative; that is, **AB** and **BA** need not be equal.

If $\mathbf{T} = \mathbf{RS}$, where **R** and **S** are square matrices of the same size, the element in row j and column k of **T** is found by multiplying corresponding elements of row j of **R** and column k of **S** and adding the products (Prob. 20.110).

Addition of matrices is defined in Prob. 20.107.

A matrix that has one column is called a **column matrix** or a **column vector.** Let **V** be a 2 by 1 column vector. Multiplication of a square matrix by a column vector is illustrated by

$$\mathbf{AV} = \begin{pmatrix} a_{11} & a_{12} \\ a_{21} & a_{22} \end{pmatrix}\begin{pmatrix} v_{11} \\ v_{21} \end{pmatrix} = \begin{pmatrix} a_{11}v_{11} + a_{12}v_{21} \\ a_{21}v_{11} + a_{22}v_{21} \end{pmatrix} \tag{20.89}$$

AV is a column vector.

A **square** matrix has the same number of rows as columns. The **order** of a square matrix equals the number of rows. The elements $f_{11}, f_{22}, \ldots, f_{nn}$ (which go from the upper left corner to the lower right corner) of the square matrix **F** are said to lie on its **principal diagonal** and are called the **diagonal elements** of **F**. Elements not on the principal diagonal of a square matrix are **off-diagonal elements.** A **diagonal matrix** is a square matrix all of whose off-diagonal elements are equal to zero.

A **unit matrix** is a diagonal matrix each of whose diagonal elements equals 1. For example, $\begin{pmatrix} 1 & 0 \\ 0 & 1 \end{pmatrix}$ is a unit matrix of order 2. A unit matrix is denoted by the symbol **I**. Multiplication of a square matrix **C** by a unit matrix of the same order as **C** does not change **C**. Thus, **CI** = **IC** = **C**.

The **inverse** of the square matrix **B** is a square matrix that satisfies $\mathbf{B}^{-1}\mathbf{B} = \mathbf{B}\mathbf{B}^{-1} = \mathbf{I}$, where $\mathbf{B}^{-1}$ denotes the inverse and **I** is the unit matrix of the same order as **B**. A square matrix has an inverse if and only if the determinant of its matrix elements is nonzero. When the determinant of the matrix is nonzero, the matrix is said to be **nonsingular.**

A **block-diagonal matrix** is a square matrix whose nonzero elements all lie on square blocks centered on the principal diagonal. For example, if

$$\mathbf{M} = \begin{pmatrix} -1 & 5 & 0 \\ 3 & 8 & 0 \\ 0 & 0 & 7 \end{pmatrix} \equiv \begin{pmatrix} \mathbf{M}_1 & \mathbf{0} \\ \mathbf{0} & \mathbf{M}_2 \end{pmatrix} \qquad \mathbf{N} = \begin{pmatrix} 2 & 9 & 0 \\ 4 & 1 & 0 \\ 0 & 0 & 6 \end{pmatrix} \equiv \begin{pmatrix} \mathbf{N}_1 & \mathbf{0} \\ \mathbf{0} & \mathbf{N}_2 \end{pmatrix} \tag{20.90}$$

then **M** and **N** are block-diagonal matrices. **M** contains the 2 by 2 block $\mathbf{M}_1 = \begin{pmatrix} -1 & 5 \\ 3 & 8 \end{pmatrix}$ and the 1 by 1 block $\mathbf{M}_2 = (7)$. The product **MN** of the two block-diagonal matrices **M** and **N** is found to be a block-diagonal matrix whose nonzero blocks are the product of corresponding blocks of **M** and **N** (Prob. 20.111). That is,

$$\mathbf{MN} = \begin{pmatrix} \mathbf{M}_1\mathbf{N}_1 & \mathbf{0} \\ \mathbf{0} & \mathbf{M}_2\mathbf{N}_2 \end{pmatrix} \tag{20.91}$$

If **P** = **MN**, then **P** is a block-diagonal matrix whose blocks are [Eq. (20.91)] $\mathbf{P}_1 = \mathbf{M}_1\mathbf{N}_1$ and $\mathbf{P}_2 = \mathbf{M}_2\mathbf{N}_2$. That is, *corresponding blocks of two block-diagonal matrices* **M** *and* **N** *multiply the same way as* **M** *and* **N**, if **M** and **N** have the same block-diagonal form.

A diagonal matrix is a special case of a block-diagonal matrix that has its blocks all 1 by 1.

Representations

A symmetry operation moves each point in space to a new location. For example, the $\hat{\sigma}(xz)$ reflection of point group C_{2v} moves the point originally at (x, y, z) to $(x, -y, z)$. If we use a prime to denote the new location, we have $x' = x$, $y' = -y$, and $z' = z$. These three equations are equivalent to the matrix equation

$$\begin{pmatrix} x' \\ y' \\ z' \end{pmatrix} = \begin{pmatrix} 1 & 0 & 0 \\ 0 & -1 & 0 \\ 0 & 0 & 1 \end{pmatrix} \begin{pmatrix} x \\ y \\ z \end{pmatrix} \tag{20.92}$$

The effect of $\hat{C}_2(z)$ is to move the point at (x, y, z) to $(-x, -y, z)$, and we can write a matrix equation to express this. Each of the symmetry operations of C_{2v} is thus described by a matrix. The matrices that correspond to the symmetry operations are

$$\hat{E}: \begin{pmatrix} 1 & 0 & 0 \\ 0 & 1 & 0 \\ 0 & 0 & 1 \end{pmatrix} \quad \hat{C}_2(z): \begin{pmatrix} -1 & 0 & 0 \\ 0 & -1 & 0 \\ 0 & 0 & 1 \end{pmatrix} \quad \hat{\sigma}(xz): \begin{pmatrix} 1 & 0 & 0 \\ 0 & -1 & 0 \\ 0 & 0 & 1 \end{pmatrix} \quad \hat{\sigma}(yz): \begin{pmatrix} -1 & 0 & 0 \\ 0 & 1 & 0 \\ 0 & 0 & 1 \end{pmatrix} \tag{20.93}$$

It isn't hard to show (Prob. 20.112) that these four matrices multiply the same way as the corresponding symmetry operations multiply. That is, if $\hat{R}\hat{T} = \hat{W}$, where

$\hat{R}$, $\hat{T}$, and $\hat{W}$ are symmetry operations of C_{2v} and if **R**, **T**, and **W** are the matrices in (20.93) that correspond to these symmetry operations, then **RT** = **W**.

A set of nonnull square matrices that multiply the same way the corresponding members of a point group multiply is said to be a **representation** of the group. (A *nonnull* matrix is one with at least one nonzero matrix element.) The matrices that constitute the representation are called the **matrix representatives** of the point-group operations. The order of the matrices in a representation is called the **dimension** of the representation. The matrices in (20.93) are a three-dimensional representation of C_{2v}.

Because the four matrices in the representation (20.93) are in diagonal form, which is a special case of block-diagonal form, the italicized statement after (20.91) shows that corresponding diagonal elements of these matrices must multiply in the same way as the 3 by 3 matrices in (20.93). Therefore, the diagonal elements give us three one-dimensional representations of C_{2v}. From (20.93), these one-dimensional representations are

$\hat{E}$	$\hat{C}_2(z)$	$\hat{\sigma}(xz)$	$\hat{\sigma}(yz)$	
(1)	(−1)	(1)	(−1)	
(1)	(−1)	(−1)	(1)	(20.94)
(1)	(1)	(1)	(1)	

For example, Table 20.2 gives $\hat{C}_2(z)\hat{\sigma}(xz) = \hat{\sigma}(yz)$. Taking the corresponding matrix representatives in the topmost representation in (20.94), we have $(-1)(1) = (-1)$, which is a valid matrix equation. (Matrices of order 1 multiply the same way ordinary numbers multiply.)

Whenever the matrices of a representation are in block-diagonal form, we can break these matrices into smaller matrices that give us representations of lower dimension, and the original representation is said to be a **reducible representation.**

If the matrices **A**, **B**, **C**, . . . form a representation of a group, it can be proved (Prob. 20.114) that the matrices $\mathbf{P}^{-1}\mathbf{AP}$, $\mathbf{P}^{-1}\mathbf{BP}$, $\mathbf{P}^{-1}\mathbf{CP}$, . . . are a representation of the group; here, **P** is any nonsingular square matrix whose order is the same as that of **A**, **B**, **C**, ... The transformation of **A** to $\mathbf{P}^{-1}\mathbf{AP}$ is called a **similarity transformation.** Two representations whose matrices are related by a similarity transformation are said to be **equivalent** to each other.

If the matrices **A**, **B**, **C**, . . . are not in block-diagonal form but there exists some matrix **P** such that $\mathbf{P}^{-1}\mathbf{AP}$, $\mathbf{P}^{-1}\mathbf{BP}$, $\mathbf{P}^{-1}\mathbf{CP}$, . . . are all in the same block-diagonal form, then the representation **A**, **B**, **C**, . . . is also called a **reducible representation.** If the matrix representatives are not in block-diagonal form and no similarity transformation exists that will put them into block-diagonal form, the representation is **irreducible.** All one-dimensional representations are considered to be irreducible.

The members of a group can be divided into **classes.** If A, B, C, D, . . . are the elements of a group, the elements that belong to the same class as B are found by taking $A^{-1}BA$, $B^{-1}BB$, $C^{-1}BC$, $D^{-1}BD$, ... One can show that an element of a group cannot belong to two different classes. For a commutative group, we have $A^{-1}BA = BA^{-1}A = BI = B$. Hence each element of a commutative group is in a class by itself. The group C_{2v} is commutative (Table 20.2) and has four members. Hence it has four classes.

A theorem of group theory states that *the number of nonequivalent irreducible representations of a group is equal to the number of classes of the group.* Since C_{2v} has four classes, it has four nonequivalent irreducible representations. We found three of them in (20.94).

Another theorem states that *if $d_1, d_2, \ldots, d_c$ are the dimensions of the nonequivalent irreducible representations of a group whose order is h, then*

$$d_1^2 + d_2^2 + \cdots + d_c^2 = h \qquad (20.95)$$

(c is the number of classes.) The group C_{2v} has $h = 4$ and has four nonequivalent irreducible representations. We found three one-dimensional irreducible representations of C_{2v} in (20.94). Therefore, (20.95) gives us $1^2 + 1^2 + 1^2 + d_4^2 = 4$, and so $d_4 = 1$. The fourth irreducible representation of C_{2v} is found to be (Prob. 20.116)

$\hat{E}$	$\hat{C}_2(z)$	$\hat{\sigma}(xz)$	$\hat{\sigma}(yz)$
(1)	(1)	(−1)	(−1)

(20.96)

Character Tables

The sum of the diagonal elements of a square matrix is called the **trace** of the matrix. For example, the trace of **A** in (20.88) is $1 + 4 = 5$. The traces of the matrices of a representation of a group are called the **characters** of the representation. For example, the characters of the reducible C_{2v} representation in (20.93) are 3, −1, 1, and 1. For most applications of group theory to quantum mechanics, one needs only the characters of representations and not the full matrices. A table of the sets of characters for the nonequivalent irreducible representations of a group gives the **character table** of the group. For C_{2v}, the nonequivalent irreducible representations are all one-dimensional, and the nonequivalent irreducible representations are (20.94) and (20.96). The C_{2v} character table is given in Table 20.3.

One-dimensional irreducible representations are labeled A or B, according to whether the character of the highest-order symmetry axis is +1 or −1, respectively. The numerical subscripts distinguish different irreducible representations. The one-dimensional representation whose characters are all equal to 1 exists for every point group and is called the **totally symmetric representation.** The letters x, y, and z will be explained later. [For molecules with a center of symmetry, each irreducible representation is labeled with a g or u subscript (gerade or ungerade) according to whether the character of $\hat{i}$ in that representation is positive or negative, respectively.]

The character table of group C_{3v}, the point group of NH_3 is also given in Table 20.3. The symmetry operations are $\hat{C}_3$, $\hat{C}_3^2$, $\hat{E}$, $\hat{\sigma}_a$, $\hat{\sigma}_b$, and $\hat{\sigma}_c$. One finds that C_{3v} has three classes. $\hat{E}$ is in a class by itself (this is always true; see Prob. 20.117). A second class consists of $\hat{C}_3$ and $\hat{C}_3^2$. The final class consists of $\hat{\sigma}_a$, $\hat{\sigma}_b$, and $\hat{\sigma}_c$. Note that the members of a class are closely related symmetry operations. Group theory shows that *symmetry operations in the same class have the same characters in a given representation.* Rather than listing each symmetry operation separately, the top row of a character table lists members of the same class together. Thus, $2\hat{C}_3(z)$ in the character table stands for $\hat{C}_3$ and $\hat{C}_3^2$, whose characters are equal; $3\hat{\sigma}_v$ in the character table stands for the three symmetry reflections.

Since C_{3v} has three classes, it has three nonequivalent irreducible representations. The letter E denotes a two-dimensional irreducible representation. Because the trace of a 2 by 2 unit matrix is $1 + 1 = 2$, the character of the identity operation $\hat{E}$ is 2 in a

TABLE 20.3

Character Tables of C_{2v} and C_{3v}

C_{2v}	$\hat{E}$	$\hat{C}_2(z)$	$\hat{\sigma}(xz)$	$\hat{\sigma}(yz)$	
A_1	1	1	1	1	z
A_2	1	1	−1	−1	
B_1	1	−1	1	−1	x
B_2	1	−1	−1	1	y

C_{3v}	$\hat{E}$	$2\hat{C}_3(z)$	$3\hat{\sigma}_v$	
A_1	1	1	1	z
A_2	1	1	−1	
E	2	−1	0	(x, y)

two-dimensional representation. (Some groups have three-dimensional irreducible representations and these are denoted by either T or F, according to the preference of the person making up the table. G and H denote four- and five-dimensional irreducible representations, respectively.) For C_{3v}, Eq. (20.95) becomes $1^2 + 1^2 + 2^2 = 6$.

Basis Functions

When a symmetry operation is applied to the points in three-dimensional space, the point originally at (x, y, z) is moved to the location (x', y', z'). One finds that $x' = d_{11}x + d_{12}y + d_{13}z$, where the d's are constants, with similar equations holding for y' and z'. Thus x' is a *linear combination* of the functions x, y and z [where this term is defined after Eq. (18.23)]. Let $\mathbf{r}$ and $\mathbf{r}'$ be column vectors whose elements are x, y, z and x', y', z', respectively. Because each of the primed coordinates is a linear combination of the unprimed coordinates, it follows from the rule for matrix multiplication that $\mathbf{r}' = \mathbf{D}\mathbf{r}$, where $\mathbf{D}$ is a square matrix of order three whose elements are d_{11}, d_{12}, etc. An example is (20.92). (The matrix $\mathbf{D}$ is not necessarily diagonal.) For each of the symmetry operations of a point group, we have a matrix that relates $\mathbf{r}'$ to $\mathbf{r}$. For example, for the C_{2v} operations, these matrices are given in (20.93).

We noted that the $\mathbf{D}$ matrices in (20.93) form a representation of C_{2v}. These $\mathbf{D}$ matrices consist of numbers that tell how the functions x, y, and z are transformed into linear combinations of x, y, and z by the group's symmetry operations. The functions x, y, and z are called *basis functions* for the representation that consists of the $\mathbf{D}$ matrices. In general, one can show that *whenever each of the set of functions $f_1, f_2, \ldots, f_n$ is transformed into a linear combination of $f_1, f_2, \ldots, f_n$ by the group's symmetry operations, then the matrices of coefficients of the linear combinations form a representation of the point group;* the functions $f_1, f_2, \ldots, f_n$ are called **basis functions** for this representation. (For this result to be valid, the functions $f_1, f_2, \ldots, f_n$ must be linearly independent, meaning none of them can be expressed as a linear combination of the other functions in the set.) Note that *the dimension of the representation whose basis functions are $f_1, f_2, \ldots, f_n$ equals n, the number of functions in the basis.*

For the group C_{2v}, the basis functions x, y, and z give matrices in (20.93) that are diagonal, so the function x alone is transformed into a linear combination of x (namely, $+x$ or $-x$), and x by itself is the basis for a representation of C_{2v}. From (20.93), the first representation in (20.94), and Table 20.3, this representation is B_1. Hence the function x is put on the same line as B_1 in the character table. Similarly for y and z. For the point group C_{3v}, one finds that the group operations transform z into itself and transform each of x and y into linear combinations of x and y. The italicized statements in the last paragraph thus show that z is the basis for a one-dimensional representation of C_{3v} and the set of functions x and y is the basis for a two-dimensional representation. This is the meaning of (x, y) on the E line of the C_{3v} character table.

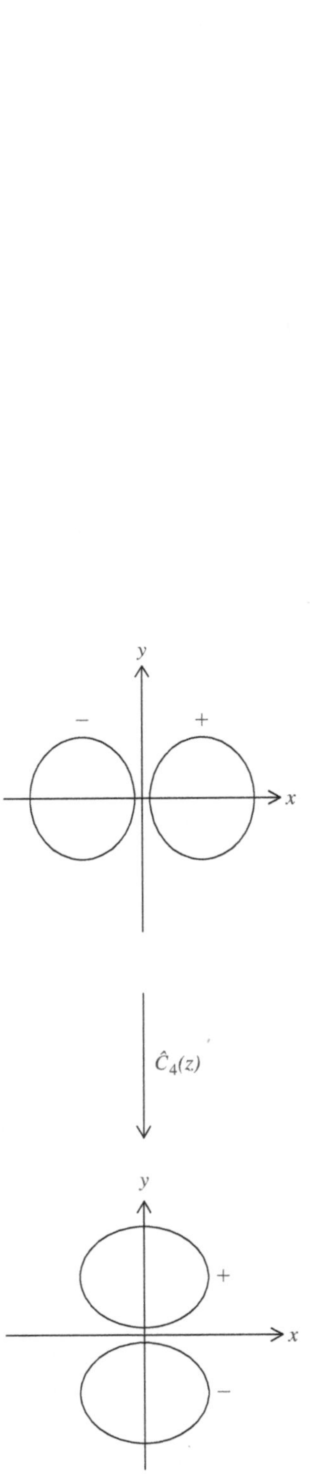

Figure 20.54

Effect of $\hat{C}_4(z)$ on $2p_x$.

One can consider the effects of symmetry operations on other functions besides x, y, and z. For example, consider a $2p_x$ AO, which has the form $2p_x = cxe^{-b(x^2+y^2+z^2)^{1/2}}$ [Eq. (18.26)], where c and b are constants. We saw that $\hat{C}_2(z)(x, y, z) \rightarrow (-x, -y, z)$, so $x' = -x$, $y' = -y$, and $z' = z$. Replacing x with $-x'$ and y with $-y'$ in $2p_x$, we get $-cx'e^{-b(x'^2+y'^2+z'^2)^{1/2}}$ so the transformed orbital is the negative of a $2p_x$ AO. Pictorially, a $2p_x$ AO has positive and negative lobes centered on the x axis, and a 180° $\hat{C}_2(z)$ rotation about the z axis interchanges the lobes and transforms $2p_x$ into $-2p_x$. A 90° counterclockwise $\hat{C}_4(z)$ rotation about the z axis transforms the $2p_x$ AO into the $2p_y$ AO (Fig. 20.54).

Group Theory and Quantum Mechanics

Let $\hat{H}$ be the electronic Hamiltonian operator in the molecular electronic Schrödinger equation. A symmetry operation of a molecule sends equivalent nuclei into one another.

It therefore follows that each molecular symmetry operation commutes with $\hat{H}$ (commutation is defined at the end of Sec. 18.4). For a nondegenerate level, it follows from a certain theorem of quantum mechanics that the molecular electronic wave function must be an eigenfunction of each of the molecular symmetry operations. For a degenerate energy level, every linear combination of the wave functions of the level is an eigenfunction of $\hat{H}$ (as noted in Sec. 18.3), and one can show that when a symmetry operation commutes with $\hat{H}$, the application of that operation to the wave function of an n-fold degenerate level transforms the wave function into a linear combination of the n wave functions of the level. Therefore (by the italicized statements in the last subsection) the linearly independent wave functions of a degenerate energy level form a basis for a representation of the molecular point group. This representation might be reducible or irreducible. For reasons discussed elsewhere (I. N. Levine, *Molecular Spectroscopy,* Wiley, 1975, pp. 413–415), this representation is very unlikely to be reducible, and we shall assume it to be irreducible.

Thus, the wave functions of each electronic energy level form a basis for an irreducible representation of the molecular point group. This means that the symmetry operations transform each of the wave functions of a level into a linear combination of the wave functions of the level, where the coefficients in the linear combinations form matrices that give an irreducible representation of the point group. We can classify each electronic energy level according to one of the possible irreducible representations of the molecule's point group. Since the number of functions in the basis for a representation equals the dimension of the representation, the degeneracy of an energy level equals the dimension of the irreducible representation for which the wave functions of that level form a basis. In discussing the symmetry of wave functions, the term **symmetry species** is often used instead of irreducible representation.

For example, each electronic state of H_2O can be classified as belonging to one of the irreducible representations (symmetry species) A_1, A_2, B_1, or B_2 of the point group C_{2v}. Recall that atomic terms (^1S, ^3P, etc.) are specified by giving the spin multiplicity $2S + 1$ (where S is the total spin angular momentum quantum number) as a left superscript on the letter ($S, P, D, \ldots$) specifying the total electronic orbital angular momentum. A molecular electronic term is specified by giving the spin multiplicity as a left superscript on the irreducible representation. Thus, for H_2O, one has electronic terms such as $^1A_1$, $^3A_1$, $^1B_1$, etc.

For most molecules, the irreducible representation (symmetry species) of the ground electronic state is the totally symmetric one and the electron spins are all paired to give a singlet state. The ground electronic state of H_2O is $^1A_1$.

One can show that each canonical MO of a molecule can be classified according to one of the irreducible representations (symmetry species) of the molecular point group. Lowercase letters are used for the symmetry species of MOs. The MOs belonging to a given symmetry species are numbered in order of increasing orbital energy. For example, the lowest a_1 MOs of H_2O are labeled $1a_1, 2a_1, 3a_1, \ldots$. The **electron configuration** of a molecule is specified by giving the number of electrons in each shell, where a **shell** is a set of MOs with the same energy. (Recall that an atomic electron configuration such as $1s^2 2s^2 2p^4$ gives the number of electrons in each subshell, where a subshell is a set of AOs having the same energy.)

For H_2O, one finds the ground-electronic-state electron configuration to be $(1a_1)^2(2a_1)^2(1b_2)^2(3a_1)^2(1b_1)^2$. The lowest MO $1a_1$ is essentially the same as the $1s$ AO on the oxygen atom. This $1s$ AO is unaffected by each of the four molecular symmetry operations and so belongs to the totally symmetric symmetry species A_1. The $1b_1$ MO is a lone-pair $2p_x$ AO on oxygen, where the x axis is perpendicular to the molecular plane (Fig. 20.52). This AO is sent into its negative by $\hat{C}_2(z)$ and by $\hat{\sigma}(yz)$, and is unaffected by $\hat{\sigma}(xz)$ and $\hat{E}$, which corresponds to B_1 in the character table in Table 20.3. Other H_2O MOs are considered in Prob. 20.119.

Just as a given atomic electron configuration can give rise to several terms having different energies (Fig. 18.13), so can a given molecular electron configuration. For example, the excited electron configuration $(1a_1)^2(2a_1)^2(1b_2)^2(3a_1)^2(1b_1)(4a_1)$ of H_2O gives rise to the terms $^3B_1$ and $^1B_1$.

For NH_3 (point group C_{3v}), the ground electronic state is a $^1A_1$ state with the electron configuration $(1a_1)^2(2a_1)^2(1e)^4(3a_1)^2$. The E symmetry species of C_{3v} is two-dimensional, so there are two degenerate $1e$ MOs. These two degenerate MOs constitute a shell that holds four electrons, two in each MO. For CH_4, the ground electron configuration is $(1a_1)^2(2a_1)^2(1t_2)^6$, where the T_2 symmetry species has dimension 3, so there are three degenerate $1t_2$ MOs. The $1a_1$ MO is essentially a $1s$ AO on C and the other four occupied MOs are bonding delocalized canonical MOs (Sec. 19.6).

Since the Schrödinger equation omits electron spin, the degeneracy specified by the dimension of an irreducible representation does not include spin degeneracy, and this degeneracy is called *orbital degeneracy*.

Now consider molecular vibrations. One can show that each normal vibrational mode of a molecule belongs to one of the irreducible representations of the molecular point group. For example, application of each of the four C_{2v} symmetry operations to the H_2O normal mode ν_1 in Fig. 20.27 leaves the vectors of this mode unchanged. Hence the symmetry species of ν_1 is the totally symmetric species a_1. Likewise, ν_2 has species a_1. For ν_3, a $\hat{C}_2(z)$ rotation changes each vector to its negative and $\hat{\sigma}(xz)$ also changes each vector to its negative (the molecular plane is the *yz* plane). ν_3 is unchanged by $\hat{E}$ and by $\hat{\sigma}(yz)$. Thus the symmetry species of ν_3 is b_2.

Group theory enables one to do a lot more than just classify electronic states, MOs, and normal modes according to their symmetry species and degeneracies. Using group theory, one can predict the symmetry species of the normal modes of vibration of a molecule, and one can predict the symmetry species of the MOs formed from the basis AOs. Moreover, solution of the Hartree–Fock equation (18.57) is considerably simplified by group theory, since it enables one to deal separately with MOs of different symmetry species. Group theory is invaluable in deriving selection rules to find the allowed transitions in vibrational spectra and in electronic spectra.

Recall that a reducible representation can be put into block-diagonal form by a similarity transformation, with the blocks corresponding to irreducible representations of the point group. One says that the reducible representation is the **direct sum** of the irreducible representations that occur in the blocks. For example, if Γ is the reducible representation whose matrices are those in (20.93), then $\Gamma = B_1 \oplus B_2 \oplus A_1$, where $\oplus$ denotes the direct sum.

Many of the applications of group theory in quantum chemistry and spectroscopy depend on starting with a reducible representation and finding which irreducible representations it is the direct sum of. Let Γ (capital gamma) be a reducible representation and let $a_{i\Gamma}$ denote the number of times the irreducible representation i occurs in Γ. For example, for the reducible representation (20.93), $a_{i\Gamma}$ is 1 for A_1, is 1 for B_1, is 1 for B_2, and is 0 for A_2. Let $\chi_\Gamma(\hat{R})$ and $\chi_i(\hat{R})$ denote the characters of the symmetry operation $\hat{R}$ in the representation Γ and in the irreducible representation i. For example, for (20.93), $\chi_\Gamma(\hat{E}) = 3$. (χ is the Greek letter chi.) Group theory shows that

$$a_{i\Gamma} = \frac{1}{h} \sum_{\hat{R}} \chi_i^*(\hat{R}) \chi_\Gamma(\hat{R}) \tag{20.97}$$

where h is the order of the group and the sum goes over the h different symmetry operations of the group. (The complex conjugate occurs because certain irreducible representations have complex characters.) For example, for Γ being the representation

(20.93), the characters are 3, −1, 1, and 1, and Eq. (20.97) gives the number of times that B_1 occurs in Γ as $a_{B_1\Gamma} = \frac{1}{4}[(1)(3) + (-1)(-1) + (1)(1) + (-1)(1)] = 1$. Although this result is obvious by inspection of (20.93), it is not so obvious when the representation's matrices are not in block-diagonal form or when we know only the characters of the representation.

20.17 SUMMARY

Electromagnetic radiation (light) consists of oscillating electric and magnetic fields. Spectroscopy studies the interaction of light and matter, especially absorption and emission of radiation by matter. The energy $h\nu$ of the photon absorbed or emitted in a transition is equal to the energy difference between the stationary states m and n involved in the transition: $h\nu = |E_m - E_n|$. For a radiative transition to have significant probability of occurring, the transition-moment integral (20.5) must be nonzero. The selection rules give the allowed transitions for a given kind of system. The Beer–Lambert law (20.11) relates the fraction of radiation in a small wavelength range absorbed by a sample to the molar absorption coefficient ε.

The energy of a molecule is (to a good approximation) the sum of translational, rotational, vibrational, and electronic energies. The spacings between levels increase in the order: translational, rotational, vibrational, electronic.

For a diatomic molecule, the rotational energy is $E_{\text{rot}} \approx J(J+1)\hbar^2/2I_e = B_e hJ(J+1)$, where $J = 0, 1, 2, \ldots$, the equilibrium moment of inertia is $I_e = \mu R_e^2$ (where μ is the reduced mass of the two atoms), and B_e is the equilibrium rotational constant. Because vibration affects the average bond distances, B_e is replaced by B_v [Eq. (20.32)] when the molecule is in vibrational state v. For polyatomic molecules, the rotational-energy expression depends on whether the molecule is a spherical, symmetric, or asymmetric top.

The vibrational energy of a diatomic molecule is approximately that of a harmonic oscillator: $E_{\text{vib}} \approx (v + \frac{1}{2})h\nu_e$, where $v = 0, 1, 2, \ldots$ and the equilibrium vibration frequency is $\nu_e = (1/2\pi)(k_e/\mu)^{1/2}$. Anharmonicity adds a term proportional to $-(v + \frac{1}{2})^2$ to E_{vib} and causes the vibrational levels to converge as v increases (Fig. 20.9). For an $\mathcal{N}$-atom molecule, the vibrational energy is approximately the sum of $3\mathcal{N} - 6$ (or $3\mathcal{N} - 5$ if the molecule is linear) harmonic-oscillator energies, one for each normal vibrational mode [Eq. (20.48)].

The pure-rotational spectrum lies in the microwave (or far-IR) region and is studied in microwave spectroscopy. Molecular geometries are obtainable from the pure-rotational spectra of molecules with nonzero dipole moments. For a diatomic molecule, the radiative rotational selection rule is $\Delta J = \pm 1$.

The vibration–rotation spectrum lies in the infrared and consists of a series of bands. Each line in a band corresponds to a different rotational change. A normal mode is IR-active if it changes the dipole moment. Analysis of the IR spectrum yields the IR-active vibrational frequencies and (provided the rotational lines are resolved) the moments of inertia, from which the molecular structure can be found.

Pure-rotational and vibration–rotation transitions can also be observed in Raman spectroscopy. Here, rather than absorbing a photon, a molecule scatters a photon, exchanging energy with it in the process.

Transitions of valence electrons to higher levels produce absorption in the UV and visible regions.

In NMR spectroscopy, one observes transitions of nuclear spin magnetic moments in an applied magnetic field. The local field at a nucleus is influenced by the electronic environment, so nonequivalent nuclei in a molecule undergo NMR transitions at different frequencies. Moreover, because of interactions between nuclear spins within a molecule, the NMR absorption lines are split. Provided the spectrum is first-order, n equivalent protons act to split the absorption peak of a set of adjacent protons into $n + 1$ peaks. Spin–spin splitting is usually negligible for protons separated by more than three bonds.

In ESR spectroscopy, one observes transitions between the energy levels of an unpaired electron spin magnetic moment in an applied magnetic field.

Photochemistry studies chemical reactions produced by absorption of light.

The symmetry operations of a molecule form a mathematical group. Matrices that multiply the same way as the members of a group form a representation of the group. The character table of a group lists the characters of the nonequivalent irreducible representations of the group, where the characters are the traces of the matrices of the representations. Molecular wave functions, MOs, and normal modes can be classified according to their irreducible representations (symmetry species).

Important kinds of calculations discussed in this chapter include

- Calculation of absorption and emission frequencies and wavelengths from quantum-mechanical energy levels using $E_{\text{upper}} - E_{\text{lower}} = h\nu = hc/\lambda$ and the selection rules for the system.
- Evaluation of the transition-moment integral $\int \psi_m^*(\Sigma_i Q_i \mathbf{r}_i)\psi_n \, d\tau$ to find the selection rules for the allowed transitions.
- Use of the Beer–Lambert law $I = I_0 10^{-\varepsilon cl}$ to find the transmittance I/I_0 at a particular λ.
- Use of $E_{\text{rot}} = J(J + 1)\hbar^2/2I$ and $I = \mu R_e^2$ to calculate rotational energy levels and bond distances in diatomic molecules.
- Use of $E_{\text{vib}} \approx (v + \frac{1}{2})h\nu$ and $\nu = (1/2\pi)(k/\mu)^{1/2}$ to find force constants of diatomic molecules.
- Calculation of relative populations of energy levels using the Boltzmann distribution law $N_i/N_j = (g_i/g_j)e^{-(E_i-E_j)/kT}$, where N_i, N_j and g_i, g_j are the populations and degeneracies of energy levels i and j.
- Calculation of the NMR transition frequency ν.
- Calculation of the specific optical rotation $[\alpha]$.
- Reduction of a reducible representation to the direct sum of irreducible representations.

FURTHER READING AND DATA SOURCES

Straughan and Walker; Harmony; Chang; Campbell and Dwek; Brand, Speakman, and Tyler; Herzberg; Gordy; Gordy and Cook; Mann and Akitt; Günther; Sugden and Kenney; Long; Hollas; Graybeal; Friebolin; Bovey.

Molecular structures: *Landolt–Börnstein,* New Series, Group II, vols. 7, 15, and 21, *Structure Data of Free Polyatomic Molecules; Herzberg,* vol. III, app. VI.

Molecular spectroscopic constants: *Herzberg,* vol. III, app. VI; K. P. Huber and G. Herzberg, *Constants of Diatomic Molecules,* Van Nostrand Reinhold, 1979; *Landolt–Börnstein,* New Series, Group II, vols. 4, 6, 14a, 14b, 19, and 20.

Molecular dipole moments: R. D. Nelson et al., Selected Values of Electric Dipole Moments for Molecules in the Gas Phase, *Natl. Bur. Stand. U.S. Publ.* NSRDS-NBS 10, 1967; A. L. McClellan, *Tables of Experimental Dipole Moments,* vol. 1, Freeman, 1963, vol. 2, Rahara Enterprises, 1974, vol. 3, Rahara Enterprises, 1989.

Collections of spectra: W. M. Simons (ed.), *The Sadtler Handbook of Infrared Spectra,* Sadtler, 1978; C. J. Pouchert, *The Aldrich Library of Infrared Spectra,* 3d ed., Aldrich, 1981; W. Simons (ed.), *The Sadtler Handbook of Proton NMR Spectra,*

Sadtler, 1978; C. J. Pouchert, *The Aldrich Library of NMR Spectra,* 2d ed., Aldrich, 1983. The Integrated Spectral Data Base System for Organic Compounds (SDBS) at riodb01.ibase.aist.go.jp/sdbs/ contains thousands of ^{1}H and ^{13}C NMR, IR, Raman, and ESR spectra, and also mass spectra.

PROBLEMS

Section 20.1

20.1 True or false? (*a*) Electromagnetic radiation always travels at speed c. (*b*) In an electromagnetic wave traveling in the y direction, the electric field vector **E** always points in the same direction at each location along the wave. (*c*) The fields **E** and **B** are perpendicular to the direction of travel of an electromagnetic wave and are perpendicular to each other. (*d*) Microwaves have higher frequency than visible light. (*e*) An infrared photon has a lower energy than an ultraviolet photon.

20.2 Find the frequency, wavelength, and wavenumber of light with photons of energy 1.00 eV per photon.

20.3 Find the speed, frequency, and wavelength of sodium D light in water at 25°C. For data, see the material following Eq. (20.4).

Section 20.2

20.4 True or false? (*a*) When a molecule absorbs a photon and makes a transition to a stationary state of energy E_k, the absorption frequency ν satisfies $h\nu = E_k$. (*b*) When a molecule emits a photon of frequency ν, it undergoes an energy change given by $\Delta E = h\nu$. (*c*) When a molecule absorbs a photon of frequency ν, it undergoes an energy change given by $\Delta E = h\nu$. (*d*) The longer the wavelength of a transition, the smaller the energy difference between the two levels involved in the transition. (*e*) Exposing a molecule in state n to electromagnetic radiation of frequency $\nu = (E_n - E_m)/h$ will increase the probability that the molecule will make a transition to the lower state m with emission of a photon of frequency ν.

20.5 Give the SI units of (*a*) frequency; (*b*) wavenumber; (*c*) speed; (*d*) absorbance; (*e*) molar absorption coefficient.

20.6 Verify Eq. (20.6) for the particle-in-a-box transition moment.

20.7 Use the harmonic-oscillator selection rule $\Delta v = \pm 1$ to find the frequency or frequencies of light absorbed by a harmonic oscillator with vibrational frequency ν_{vib}.

20.8 Use Fig. 17.18 and Table 14.1 to evaluate the transition-moment integral (20.5) for each of the following pairs of states of a charged harmonic oscillator: (*a*) $v = 0$ and $v = 1$; (*b*) $v = 0$ and $v = 2$; (*c*) $v = 0$ and $v = 3$. Are the results consistent with the selection rule $\Delta v = \pm 1$?

20.9 For a certain quantum-mechanical system, the wavelength for an absorption transition from level A to level C is 485 nm and the wavelength for an absorption transition from level B to level C is 884 nm. Find λ for a transition between levels A and B.

20.10 A hypothetical quantum-mechanical system has the energy levels $E = bn(n + 2)$, where $n = 1, 2, 3, \ldots$ and b is a positive constant. The selection rule for radiative transitions is $\Delta n = \pm 2$. For a collection of such systems distributed among many energy levels, the lowest-frequency absorption transition is observed to be at 80 GHz. Find the next lowest absorption frequency.

20.11 (*a*) For an electron confined to a one-dimensional box of length 2.00 Å, calculate the three lowest possible absorption frequencies for transitions that start from the ground stationary state. (*b*) Repeat (*a*) without assuming that the initial state is the ground state.

20.12 A hypothetical quantum-mechanical system has the energy levels $E = an(n + 4)$, where $n = 0, 1, 2, \ldots$ and a is a positive constant. The selection rule for radiative transitions is $\Delta n = \pm 3$. Find the formula for the allowed absorption frequencies in terms of n_{lower}, a, and h.

20.13 A hypothetical quantum-mechanical system has the energy levels $E = AK(K + 3)$; $K = 1, 2, 3, \ldots$; $A > 0$. The radiative-transition selection rule is $\Delta K = \pm 1$. The $K = 2$ to 3 transition occurs at 60 GHz. A transition at 135 GHz is observed. What levels is this transition between?

20.14 For each of the absorbance values 0.1, 1, 2, and 10, calculate the transmittance and the percentage of radiation absorbed.

20.15 Ethylene has a UV absorption peak at 162 nm with $\varepsilon = 1.0 \times 10^4$ dm^3 mol^{-1} cm^{-1}. Calculate the transmittance of 162-nm radiation through a sample of ethylene gas at 25°C and 10 torr for a cell length of (*a*) 1.0 cm; (*b*) 10 cm.

20.16 Methanol has a UV absorption peak at 184 nm with $\varepsilon = 150$ dm^3 mol^{-1} cm^{-1}. Calculate the transmittance of 184-nm radiation through a 0.0010 mol dm^{-3} solution of methanol in a nonabsorbing solvent for a cell length of (*a*) 1.0 cm; (*b*) 10 cm.

20.17 A certain solution of the enzyme lysozyme (molecular weight 14600) in D_2O with a mass concentration of 80 mg/cm^3 in a 0.10-mm-long absorption cell is found to have a transmittance of 8.3% for infrared radiation of wavelength 6000 nm. Find the absorbance of the solution and the molar absorption coefficient of lysozyme at this wavelength.

20.18 A solution of 2.00 g of a compound transmits 60.0% of the 430-nm light incident on a 3.00-cm-long cell. What percent of 430-nm light will be transmitted by a solution of 4.00 g of this compound in the same cell?

20.19 At 330 nm, the ion $Fe(CN)_6^{3-}(aq)$ has $\varepsilon = 800$ dm^3 mol^{-1} cm^{-1}, and $Fe(CN)_6^{4-}(aq)$ has $\varepsilon = 320$ dm^3 mol^{-1} cm^{-1}. The reduction of $Fe(CN)_6^{3-}$ to $Fe(CN)_6^{4-}$ is being followed spectrophotometrically in 1.00-cm-long cell. The solution has an initial $Fe(CN)_6^{3-}$ concentration of 1.00×10^{-3} mol dm^{-3} and no $Fe(CN)_6^{4-}$. After 340 s, the absorbance is 0.701. Calculate the percent of $Fe(CN)_6^{3-}$ that has reacted.

Section 20.3

20.20 True or false? (*a*) The spacings between adjacent low-lying molecular translational, rotational, and vibrational levels satisfy $\Delta\varepsilon_{tr} < \Delta\varepsilon_{rot} < \Delta\varepsilon_{vib}$. (*b*) At room temperature, many rotational levels of gas-phase molecules are substantially populated. (*c*) At room temperature, many vibrational levels of $O_2(g)$ are substantially populated. (*d*) The vibrational energy levels of a diatomic molecule are accurately given by the harmonic-oscillator expression $(v + \frac{1}{2})h\nu$. (*e*) A bound electronic state of a diatomic molecule has a finite number of vibrational levels. (*f*) As the vibrational quantum number increases, the spacing between adjacent vibrational levels of a diatomic molecule decreases. (*g*) $D_0 > D_e$. (*h*) As the rotational quantum number J increases, the spacing between adjacent rotational levels of a diatomic molecule increases.

Section 20.4

20.21 True or false? (*a*) Diatomic-molecule vibration–rotation absorption bands always have $\Delta v = 1$. (*b*) For diatomic-molecule pure-rotational absorption spectra, only $\Delta J = 1$ lines occur. (*c*) Because only $\Delta J = 1$ is allowed in pure-rotational absorption spectra of diatomic molecules, a diatomic-molecule pure-rotational spectrum contains only one line.

20.22 Use data in Table 20.1 in Sec. 20.4 to calculate D_0 for the ground electronic state of (*a*) $^{14}N_2$; (*b*) $^{12}C^{16}O$.

20.23 (*a*) Explain why D_e and k_e for $D^{35}Cl$ are essentially the same as D_e and k_e for $H^{35}Cl$ but D_0 for these two species differs. (*b*) Use data in Table 20.1 in Sec. 20.4 to calculate D_0 for each of these two species. Neglect the difference in $\tilde{\nu}_e x_e$ for the two species.

20.24 As noted before Eq. (20.20), the vibrational coordinate x for a diatomic molecule equals $R - R_e$. We can estimate the typical departure A of a diatomic-molecule bond length from its equilibrium value due to zero-point vibration by setting the harmonic-oscillator ground-state vibrational energy equal to the classical-mechanical expression for the maximum potential energy of a harmonic oscillator. Show that this gives $A = (h/4\pi^2\nu_e\mu)^{1/2}$. Use Table 20.1 in Sec. 20.4 to calculate this typical departure for $H^{35}Cl$ and for $^{14}N_2$.

20.25 If the $J = 2$ to 3 rotational transition for a diatomic molecule occurs at $\lambda = 2.00$ cm, find λ for the $J = 6$ to 7 transition of this molecule.

20.26 The $J = 0 \rightarrow 1$, $v = 0 \rightarrow 0$ transition for $^1H^{79}Br$ occurs at 500.7216 GHz, and that for $^1H^{81}Br$ occurs at 500.5658 GHz. (*a*) Calculate the bond distance R_0 in each of these molecules. Use a table inside the back cover. Neglect centrifugal distortion. (*b*) Predict the $J = 1 \rightarrow 2$, $v = 0 \rightarrow 0$ transition frequency for $^1H^{79}Br$. (*c*) Predict the $J = 0 \rightarrow 1$, $v = 0 \rightarrow 0$ transition frequency for $^2H^{79}Br$. [Actually, each pure-rotational transition of these species is split into several lines because of the electric quadrupole moments of the ^{79}Br and ^{81}Br nuclei. The frequencies given are for the centers of the $J = 0 \rightarrow 1$ transitions.]

20.27 The $J = 2 \rightarrow 3$ pure-rotational transition for the ground vibrational state of $^{39}K^{37}Cl$ occurs at 22410 MHz. Neglecting centrifugal distortion, predict the frequency of the $J = 0 \rightarrow 1$ pure-rotational transition of (*a*) $^{39}K^{37}Cl$; (*b*) $^{39}K^{35}Cl$.

20.28 Verify that neglect of vibration–rotation interaction gives Eq. (20.38) for $\tilde{\nu}_P$.

20.29 Verify Eqs. (20.39) and (20.40) for *R*- and *P*-branch wavenumbers.

20.30 For $^{16}O_2$, $\tilde{\nu}_e = 1580$ cm^{-1}. Find k_e for $^{16}O_2$.

20.31 (*a*) From the IR data following Eq. (20.36), calculate $\tilde{\nu}_e$ and $\tilde{\nu}_e x_e$ for $^1H^{35}Cl$. (*b*) Use the results of (*a*) to predict $\tilde{\nu}_{origin}$ for the $v = 0 \rightarrow 6$ transition of this molecule.

20.32 In the $v = 0 \rightarrow 1$ band of the $^{12}C^{16}O$ IR spectrum, the four lines closest to the band origin lie at 2150.858, 2147.084, 2139.427, and 2135.548 cm^{-1}, where the band origin is between the second and third of these lines. (*a*) Give the initial and final J values for each of these lines without looking at figures in the text. Give your reasoning. (*b*) Find $\tilde{\alpha}_e$, $\tilde{B}_e$, and $\tilde{\nu}_{origin}$. A good way to do this is to use the Excel Solver to minimize the sum of the squares of deviations of calculated from observed wavenumbers. (*c*) Find R_e for $^{12}C^{16}O$.

20.33 For each of the $J = 1$ through $J = 6$ rotational levels of the ground vibrational state of $^1H^{35}Cl$, use data in Table 20.1 to calculate the energy-level population ratio N_J/N_0 at 300 K, where N_0 is the $J = 0$ population.

20.34 The absorption frequency for a transition from energy level a to level b has a frequency 1.54×10^{12} Hz. Find the energy-level population ratio N_b/N_a at 300 K if (*a*) levels a and b are nondegenerate; (*b*) the level degeneracies are 3 for a and 5 for b; (*c*) repeat (*a*) and (*b*) at 1000 K.

20.35 Match each of the symbols B_e, α_e, D, $\nu_e x_e$ with one of these terms: anharmonicity, vibration–rotation interaction, centrifugal distortion, rotational constant.

20.36 (*a*) For O_2, O_2^+, and O_2^-, which species has the largest k_e in the ground electronic state and which has the smallest k_e? (*b*) For N_2 and N_2^+, which species has the larger ν_e in the ground electronic state? (*c*) For N_2 and O_2, which molecule has the larger k_e in the ground electronic state? (*d*) For Li_2 and Na_2, which molecule has the greater rotational energy in the $J = 1$ level?

Section 20.5

20.37 True or false? (*a*) $\hat{C}_4^2 = \hat{C}_2$; (*b*) $\hat{C}_3^4 = \hat{C}_3$; (*c*) $\hat{\sigma}^3 = \hat{\sigma}$; (*d*) $\hat{S}_3^3 = \hat{E}$.

20.38 List all the symmetry elements present in (*a*) H_2S; (*b*) CF_3Cl; (*c*) XeF_4; (*d*) PCl_5; (*e*) IF_5; (*f*) *p*-dibromobenzene; (*g*) HCl; (*h*) CO_2. State clearly the location of each symmetry element.

20.39 List all the symmetry operations for (*a*) H_2S; (*b*) CF_3Cl.

20.40 The inversion operation $\hat{\imath}$ moves a nucleus at x, y, z to $-x, -y, -z$. What effect does each of the following operations have on a nucleus at x, y, z? (*a*) A $\hat{C}_2$ rotation about the z axis; (*b*) a reflection in the xy plane; (*c*) an $\hat{S}_2$ rotation about the z axis. From the answer to (*c*), what statement can be made about the $\hat{S}_2$ operation?

Section 20.6

20.41 Without doing any calculations, describe as fully as you can the locations of the principal axes of inertia of (*a*) BF_3; (*b*) H_2O; (*c*) CO_2.

20.42 Classify each of the following as a spherical, symmetric, or asymmetric top: (*a*) SF_6; (*b*) IF_5; (*c*) H_2S; (*d*) PF_3; (*e*) benzene; (*f*) CO_2; (*g*) $^{35}ClC^{37}Cl_3$; (*h*) BF_3.

20.43 Each bond length in BF_3 is 1.313 Å. Calculate the principal moments of inertia of $^{11}B^{19}F_3$.

20.44 For PCl_5, the two axial bond lengths are 2.12 Å and the three equatorial bond lengths are 2.02 Å. Calculate I_a, I_b, and I_c for $^{31}P^{35}Cl_5$.

20.45 For CF_3I, the rotational constants are $\tilde{A} = 0.1910\ \text{cm}^{-1}$ and $\tilde{B} = 0.05081\ \text{cm}^{-1}$. (*a*) Calculate E_{rot}/h for the $J = 0$ and $J = 1$ rotational levels. (*b*) Calculate the two lowest microwave absorption frequencies.

20.46 The R_0 bond lengths in the linear molecule OCS are $R_{\text{OC}} = 1.160$ Å and $R_{\text{CS}} = 1.560$ Å. (*a*) The z coordinate of the center of mass (com) of a set of particles with masses m_i and z coordinates z_i is $z_{\text{com}} = (\sum_i m_i z_i)/(\sum_i m_i)$. Find the position of the center of mass of $^{16}O^{12}C^{32}S$. (*b*) Find the moment of inertia of $^{16}O^{12}C^{32}S$ about an axis through the center of mass and perpendicular to the molecular axis. (*c*) Find the three lowest microwave absorption frequencies of $^{16}O^{12}C^{32}S$.

20.47 Analysis of the infrared spectrum of $^{12}C^{16}O_2$ gives the rotational constant as $\tilde{B}_0 = 0.39021\ \text{cm}^{-1}$. Find the CO bond length in CO_2.

Section 20.8

20.48 Give the number of normal vibrational modes of (*a*) SO_2; (*b*) C_2F_2; (*c*) CCl_4.

20.49 H_2O vapor has an IR absorption band at $\tilde{\nu}_{\text{origin}} = 7252\ \text{cm}^{-1}$. The lower vibrational level for this band is 000. What are the possibilities for the upper vibrational level?

20.50 Use data in Sec. 20.8 to calculate the zero-point vibrational energy of (*a*) CO_2; (*b*) H_2O.

20.51 Two of the BF_3 normal modes of vibration are described below. State whether each mode is IR-active or IR-inactive. (*a*) Simultaneous stretching of each bond; (*b*) each atom moving perpendicular to the molecular plane, with B moving in the opposite direction as each F.

Section 20.9

20.52 State which of each of the following pairs of vibrations has the higher vibrational frequency. (*a*) C=C stretching, C≡C stretching; (*b*) C—H stretching, C—D stretching; (*c*) C—H stretching, CH_2 bending.

20.53 C—H stretching vibrations in organic compounds occur near 2900 cm^{-1}. Near what wavenumber would C—D stretching vibrations occur?

20.54 From the CH and C=O stretching frequencies listed in Sec. 20.9, estimate the force constants for stretching vibrations of these bonds.

Section 20.10

20.55 True or false? (*a*) The Raman shift of a given Raman spectral line does not change if the frequency ν_0 of the incident radiation is changed. (*b*) Stokes lines have positive Raman shifts. (*c*) The selection rules for pure-rotational Raman transitions are the same as for ordinary pure-rotational absorption transitions. (*d*) A molecule with no electric dipole moment will not show a pure-rotational absorption spectrum but may have pure-rotational lines in its Raman spectrum. (*e*) Every molecule has a vibrational Raman spectrum. (*f*) In vibrational Raman spectra, Stokes lines are more intense than anti-Stokes lines. (*g*) A normal mode that is IR-inactive must be Raman-active.

20.56 (*a*) Derive the formula for the Raman shifts of the pure-rotational Raman lines of a linear molecule. What is the spacing between adjacent pure-rotational Raman lines? (*b*) The pure-rotational Raman spectrum of $^{14}N_2$ shows a spacing of 7.99 cm^{-1} between adjacent rotational lines. Find the bond distance in N_2. (*c*) What is the spacing between the unshifted line at ν_0 and each of the pure-rotational linear-molecule lines closest to ν_0? (*d*) If 540.8-nm radiation from an argon laser is used as the exciting radiation, find the wavelengths of the two pure-rotational Raman $^{14}N_2$ lines nearest the unshifted line.

Section 20.11

20.57 True or false? (*a*) Transitions between electronic states occuring with absorption or emission of radiation always have $\Delta S = 0$. (*b*) Most of the radiation emitted from a fluorophore is usually at longer wavelengths than the radiation exciting the fluorescence. (*c*) Excited electronic states of a molecule usually have equilibrium geometries quite close to that of the ground electronic state.

20.58 Calculate the wavelength of the series limit of the Balmer lines of the hydrogen-atom spectrum.

20.59 Calculate the wavelengths of the first three lines in the Paschen series of the hydrogen-atom spectrum.

20.60 Calculate the wavelength of the $n = 2 \rightarrow 1$ transition in Li^{2+}.

20.61 The Schumann–Runge bands of O_2 are due to a transition between the ground electronic state and an excited electronic state designated as the B state. The O_2 ground electronic state dissociates to two ground-state O atoms. The O_2 B state dissociates to one ground-state O atom and an O atom in an excited state 1.970 eV above the O ground state. The $v' = 0$ to $v'' = 0$ band of the Schumann–Runge bands is at 202.60 nm, and the bands converge to a continuous absorption beginning at 175.05 nm. Find D_0 of the O_2 ground state and of the B state.

20.62 For $^{12}C^{16}O$ molecules in a certain jet, the $J = 1$ to $J = 0$ population ratio is 0.181. Find the rotational temperature of the $^{12}C^{16}O$ molecules. See Table 20.1.

Section 20.12

20.63 True or false? (*a*) The proton spin state with $M_I = +\frac{1}{2}$ has lower energy in a magnetic field than that with $M_I = -\frac{1}{2}$. (*b*) The proton spin states $M_I = +\frac{1}{2}$ and $M_I = -\frac{1}{2}$ have the same energy in the absence of a magnetic field. (*c*) For all nuclei with $I = \frac{1}{2}$, the spin state with $M_I = +\frac{1}{2}$ has lower energy in a magnetic field than that with $M_I = -\frac{1}{2}$. (*d*) A nucleus such as ^{12}C or ^{16}O that has $I = 0$ has no NMR spectrum. (*e*) γ is a constant that has the same value for every nucleus. (*f*) g_N has the same value for every nucleus. (*g*) The nuclear magneton μ_N has the same value for every nucleus. (*h*) The chemical shift δ_i does not change when B_0 and ν_{spec} change. (*i*) The frequency shift $\nu_i - \nu_{ref}$ does not change when B_0 changes. (*j*) J_{ij} does not change when B_0 changes. (*k*) The first-order treatment holds accurately for virtually all molecules studied in NMR spectroscopy. (*l*) A pulse FT-NMR spectrometer uses a fixed magnetic field B_0.

20.64 Calculate the force on an electron moving at 3.0×10^8 cm/s through a magnetic field of 1.5 T if the angle between the electron's velocity vector and the magnetic field is (*a*) 0°; (*b*) 45°; (*c*) 90°; (*d*) 180°.

20.65 Calculate the magnetic dipole moment of a particle with charge 2.0×10^{-16} C moving on a circle of radius 25 Å with speed 2.0×10^7 cm/s.

20.66 (*a*) Verify Eq. (20.63) for γ. (*b*) Verify the $\gamma/2\pi$ value for ^1H given after (20.67).

20.67 The nucleus ^{11}B has $I = \frac{3}{2}$ and $g_N = 1.792$. Calculate the energy levels of a ^{11}B nucleus in a magnetic field of (*a*) 1.50 T; (*b*) 3.00 T.

20.68 Use data in Prob. 20.67 to find the NMR absorption frequency of ^{11}B in a magnetic field of (*a*) 1.50 T; (*b*) 2.00 T.

20.69 For an NMR spectrometer whose proton absorption frequency is 600 MHz, find the ^{13}C absorption frequency.

20.70 Calculate the value of B in a proton magnetic resonance spectrometer that has ν_{spec} equal to (*a*) 60 MHz; (*b*) 300 MHz.

20.71 (*a*) Calculate the ratio of the populations of the two nuclear-spin energy levels of a proton in a field of 1.41 T (the field in a 60-MHz spectrometer) at 25°C. (*b*) Explain why increasing the applied field B_0 increases the NMR absorption line strengths.

20.72 For an applied field of 7.05 T (the field used in a 300-MHz spectrometer), calculate the difference in NMR absorption frequencies for two protons whose δ values differ by 1.0.

20.73 (*a*) Sketch the proton NMR spectrum of acetaldehyde, CH_3CHO, for a 60-MHz spectrometer. Include both δ and ν scales. Estimate δ and J from tables in this chapter. (*b*) Repeat (*a*) for a 300-MHz spectrometer.

20.74 It is conventional to plot NMR spectra with the chemical shift δ_i increasing to the left. State whether each of the following increases to the left or to the right in a plotted NMR spectrum. (*a*) The absorption frequency ν_i for a fixed magnetic field B_0. (*b*) The frequency shift $\nu_i - \nu_{ref}$. (*c*) The shielding constant σ_i. (*d*) The magnetic field $B_{0,i}$ at which absorption occurs in a fixed-frequency NMR spectrometer.

20.75 For each of the following, state how many proton NMR peaks occur, the relative intensity of each peak, and whether each peak is a singlet, doublet, triplet, etc.: (*a*) benzene; (*b*) CH_3F; (*c*) C_2H_6; (*d*) $CH_3CH_2OCH_2CH_3$; (*e*) $(CH_3)_2CHBr$; (*f*) methyl acetate; (*g*) $CH_2{=}CHBr$ in a magnetic field large enough to produce a first-order spectrum; (*h*) C_2H_5CHO.

20.76 For each of the following molecules, state whether requirement 2 (only one coupling constant between any two sets of chemically equivalent spin-$\frac{1}{2}$ nuclei) is met: (*a*) $CH_2{=}CF_2$; (*b*) $CH_2{=}C{=}CF_2$.

20.77 Repeat Prob. 20.75 for the natural-abundance ^{13}C NMR spectra observed with proton–^{13}C spin–spin splittings eliminated by double resonance.

20.78 Give the number of lines in the spin-decoupled ^{13}C NMR spectrum of (*a*) *o*-xylene $[C_6H_4(CH_3)_2]$; (*b*) *m*-xylene; (*c*) *p*-xylene.

20.79 Use Fig. 20.44 to estimate J for the CH_2 and CH_3 protons in ethanol.

20.80 Suppose the proton NMR spectrum of CH_3CH_2OH is observed using a 300-MHz spectrometer and a 600-MHz spectrometer. State whether each of the following quantities is the same or different for the two spectrometers. If different, by what factor does the quantity change when one goes from 300 to 600 MHz? (*a*) $\delta_{CH_3} - \delta_{CH_2}$; (*b*) $\nu_{CH_3} - \nu_{CH_2}$; (*c*) J_{CH_2,CH_3}.

20.81 Verify Eq. (20.71). Use the fact that $\sigma \ll 1$.

20.82 (*a*) For a first-order process like the dimethylformamide internal rotation where there is no spin–spin coupling between the protons being interchanged and where the two singlet lines have equal intensities, one can show that the rate constant k_c at the temperature at which the lines coalesce satisfies $k_c = \pi|\nu_2 - \nu_1|/2^{1/2}$, where $\nu_2 - \nu_1$ is the frequency difference between the lines in the absence of interchange. Use data in the text to find k_c (at 120°C) for proton interchange in dimethylformamide. (*b*) If the spectrometer frequency ν_{spec} is increased, will the coalescence temperature for this process increase, decrease, or remain the same?

Section 20.13

20.83 Calculate the ESR frequency in a field of 2.500 T for $\cdot CH_3$, which has $g = 2.0026$.

20.84 How many lines will the ESR spectrum of the naphthalene negative ion $C_{10}H_8^-$ have?

20.85 Give the number of lines in the ESR spectrum of each of the following; assume that only protons on the α and β carbons cause splitting: (*a*) $(CH_3)_2CH\cdot$; (*b*) $CH_3CH_2CH_2CH_2\cdot$; (*c*) $(CH_3)_3C\cdot$; (*d*) $(CH_3)_2CHCH_2\cdot$.

Section 20.14

20.86 A solution of the amino acid L-lysine in water at 20°C containing 6.50 g of solute per 100 mL in a 2.00-dm-long tube has an observed optical rotation of +1.90° for 589.3-nm light. Find the specific rotation at this wavelength.

20.87 For a freshly prepared aqueous solution of α-D-glucose, $[\alpha]_D^{20} = +112.2°$, where the subscript and superscript on the specific rotation indicate sodium D (589.3 nm) light and 20°C. For a freshly prepared aqueous solution of β-D-glucose, $[\alpha]_D^{20} = +17.5°$. As time goes by, $[\alpha]_D^{20}$ for each solution changes and reaches the limiting value 52.7°, which is for the equilibrium mixture of α- and β-D-glucose. Find the percentage of α-D-glucose present at equilibrium at 20°C in water.

Section 20.15

20.88 Verify the photon energy and molar-energy calculations in (20.78).

20.89 (*a*) Show that in a light beam whose intensity is I, the number of photons per unit volume equals $I/h\nu c'$, where c' is the speed of light in the medium through which the beam is passing. (*Hint:* Start with the definition of I.) (*b*) A pulsed argon laser operating at 488 nm can readily produce a focused beam with $I = 10^{15}$ W/m^2, where W stands for watts. If this beam passes through water with refractive index 1.34, find the number of photons per unit volume in the water and compare with the number of water molecules per unit volume.

20.90 In a certain photochemical reaction using 464-nm radiation, the incident-light power was 0.00155 W and the system absorbed 74.4% of the incident light; 6.80×10^{-6} mole of product was produced during an exposure of 110 s. Find the quantum yield.

20.91 The photochemical decomposition of HI proceeds by the mechanism

$$HI + h\nu \rightarrow H + I$$
$$H + HI \rightarrow H_2 + I$$
$$I + I + M \rightarrow I_2 + M$$

where the rate of the first step is $\phi \mathcal{I}_a$ with $\phi = 1$. (*a*) Show that $-d[HI]/dt = 2\mathcal{I}_a$. Hence the quantum yield with respect to HI is 2. (*b*) How many HI molecules will be decomposed when 1.00 kcal of 250-nm radiation is absorbed?

20.92 In the anthracene dimerization, the rate of the forward reaction is negligible in the absence of radiation, but let us assume a forward bimolecular reaction with rate constant k_5 in the absence of radiation:

$$(5) \quad 2A \rightarrow A_2 \qquad r_5 = k_5[A]^2$$

in addition to reactions (1) to (4) preceding Eq. (20.83). With inclusion of step 5, express the reaction rate r in terms of [A], $[A_2]$, and $\mathcal{I}_a$. Set $r = 0$ to obtain the photostationary-state A_2 concentration. Compare with the equilibrium A_2 concentration in the absence of radiation.

Section 20.16

20.93 Which of the following are groups? (*a*) The numbers 1 and −1 with the rule of combination being ordinary multiplication. (*b*) The set of all integers, positive, negative, and zero, with multiplication as the rule of combination. (*c*) The numbers 1, 0, and −1 with the rule of combination being addition.

20.94 Which of these arithmetical operations are associative? (*a*) Addition. (*b*) Subtraction. (*c*) Multiplication. (*d*) Division.

20.95 (*a*) A group of order two has two elements A and I, where I is the identity element. Give the multiplication table of the group. (*b*) For a group of order three with the elements A, B, and I, find all possible forms of the multiplication table.

20.96 True or false? (*a*) The symmetry elements of a molecule are the members (elements) of the molecule's symmetry point group. (*b*) The symmetry operations of a molecule are the members (elements) of the molecule's symmetry point group.

20.97 Draw the octahedral molecule SF_6 with the z axis passing through two trans fluorines and with the x and y axes each bisecting FSF angles. Then number the F atoms. By applying operations to SF_6, find whether the operations in each of the following pairs commute with each other: (*a*) $\hat{C}_4(z)$ and $\hat{C}_2(x)$; (*b*) $\hat{C}_4(z)$ and $\hat{\sigma}(xz)$; (*c*) $\hat{i}$ and $\hat{\sigma}(xy)$.

20.98 Give the inverse of each of the following operations: (*a*) $\hat{E}$; (*b*) $\hat{\sigma}$; (*c*) $\hat{i}$; (*d*) $\hat{C}_5$; (*e*) $\hat{C}_5^2$; (*f*) $\hat{S}_3$.

20.99 True or false? (*a*) $\hat{C}_6^3 = \hat{C}_2$; (*b*) $\hat{S}_6^3 = \hat{S}_2$; (*c*) $\hat{S}_2 = \hat{i}$; (*d*) $\hat{S}_1 = \hat{\sigma}$.

20.100 For PCl_3, (*a*) list all the symmetry elements; (*b*) list all the symmetry operations.

20.101 For benzene, (*a*) list all the symmetry elements; (*b*) list all the symmetry operations.

20.102 For CH_4, list all the symmetry elements. State clearly where each symmetry element is located.

20.103 Find the point group of each of the following molecules: (*a*) CF_4; (*b*) HCN; (*c*) N_2; (*d*) BF_3; (*e*) SF_6; (*f*) IF_5; (*g*) XeF_4; (*h*) PCl_3; (*i*) PCl_5; (*j*) C_2H_4; (*k*) $CHCl_3$; (*l*) SF_5Br; (*m*) CHFClBr; (*n*) 1-fluoro-2-chlorobenzene.

20.104 The group C_n (where n can be 2, 3, 4, . . .) has a C_n axis as its only symmetry element. Give the symmetry operations of this group and give the order of this group.

20.105 The group C_{nh} (where n can be 2, 3, 4, . . .) contains a C_n axis, a σ_h plane of symmetry perpendicular to the axis, and a center of symmetry if n is even, but has no C_2 axes perpendicular to the C_n axis. For C_{nh}, (*a*) Is the C_n axis also an S_n axis? (*b*) When is the C_n axis also a C_2 axis? (*c*) Give the point group of each of these molecules: *cis*-dichloroethylene; *trans*-dichloroethylene; 1,1-dichloroethylene.

20.106 The group D_{nd} (n = 2, 3, 4, . . .) has a C_n axis, n C_2 axes perpendicular to the C_n axis, and n vertical planes of symmetry (called diagonal planes σ_d) that each contain the C_n axis and that bisect the angles between adjacent C_2 axes. The C_n axis is also an S_n axis. (A common error is to overlook the C_2 axes of a D_{nd} molecule.) Give the point group of each of these molecules: the staggered conformation of C_2H_6; allene. (When you see a molecule with two equal halves staggered with respect to each other, think D_{nd}.)

20.107 (*a*) The matrix sum $\mathbf{F} = \mathbf{A} + \mathbf{B}$ (which exists if **A** and **B** have the same number of rows and have the same number of columns) is the matrix each of whose elements is the sum of corresponding elements of **A** and **B**; that is, $f_{ij} = a_{ij} + b_{ij}$. Find the sum of **A** and **B** in (20.88). (*b*) The matrix $k\mathbf{A}$, where k is a scalar, is the matrix each of whose elements is k times the corresponding element of **A**. Find 3**A**, where **A** is given in (20.88).

20.108 If $\mathbf{A} = \begin{pmatrix} 0.2 & 4 \\ -1 & 3 \end{pmatrix}$ and $\mathbf{B} = \begin{pmatrix} 4 & 1 \\ 5 & 8 \end{pmatrix}$, find **AB** and **BA**.

20.109 If $\mathbf{A} = \begin{pmatrix} 1 & 2 \\ 3 & 4 \end{pmatrix}$ and $\mathbf{W} = \begin{pmatrix} 2 \\ 6 \end{pmatrix}$, find **AW**.

20.110 If $\mathbf{T} = \mathbf{RS}$, express the matrix element t_{mn} in terms of a certain sum of matrix elements of **R** and **S**.

20.111 Verify the block-diagonal equation (20.91) for the matrices in (20.90).

20.112 Verify that the matrices in (20.93) multiply the same way as the corresponding symmetry operations multiply. To save time, note that the product of two diagonal matrices is a diagonal matrix whose elements are the products of corresponding elements of the matrices being multiplied.

20.113 Find the matrix that gives the effect of each of these operations on the point (x, y, z); (*a*) $\hat{C}_4(z)$ (assume the rotation is counterclockwise when viewed from the positive z axis); (*b*) $\hat{i}$; (*c*) $\hat{\sigma}(xy)$.

20.114 (*a*) If $\mathbf{AC} = \mathbf{F}$, prove that $(\mathbf{P}^{-1}\mathbf{AP})(\mathbf{P}^{-1}\mathbf{CP}) = \mathbf{P}^{-1}\mathbf{FP}$. (Matrix multiplication can be shown to be associative.) (*b*) Explain why if **A, B, C**, . . . form a representation of a group then $\mathbf{P}^{-1}\mathbf{AP}$, $\mathbf{P}^{-1}\mathbf{BP}$, $\mathbf{P}^{-1}\mathbf{CP}$, . . . are a representation of the group.

20.115 Use (20.95) to prove that all the nonequivalent irreducible representations of a commutative group are one-dimensional.

20.116 Verify that (20.96) is a representation of C_{2v}.

20.117 Prove that the identity operation $\hat{E}$ is always in a class by itself.

20.118 Into what function is a $2p_y$ AO transformed by each of these operations (take all rotations as counterclockwise when viewed from the positive side of the rotation axis): (*a*) $\hat{C}_2(y)$; (*b*) $\hat{C}_2(z)$; (*c*) $\hat{C}_4(y)$; (*d*) $\hat{C}_4(z)$; (*e*) $\hat{\sigma}(yz)$; (*f*) $\hat{\sigma}(xz)$; (*g*) $\hat{i}$; (*h*) $\hat{S}_4(y)$.

20.119 Some occupied MOs of the ground electronic state of H_2O are described in this problem. For each MO, examine the symmetry behavior of the oxygen AOs and of the combination of hydrogen AOs, and then give the symmetry species (irreducible representation) of the MO. The z axis is the C_2 axis and the molecular plane is the yz plane. (*a*) This MO contains substantial contributions from the AOs O2s, O2p_z, and $H_1 1s + H_2 1s$. (*b*) This MO has substantial contributions from O2p_y and from $H_1 1s - H_2 1s$.

20.120 Give the symmetry species of each of the following normal modes of vibration. (*a*) In H_2CO, a mode in which all atoms move perpendicular to the molecular yz plane with the carbon moving in the opposite direction as the other three atoms. (*b*) In NH_3, a mode in which the H atoms vibrate in the bond directions while the N atom vibrates along the C_3 axis (the z axis) with the vertical (z) component of motion of the three hydrogens being in the opposite direction as the z component of the N motion.

20.121 Use (20.97) to verify that the representation (20.93) is the direct sum of A_1, B_1, and B_2.

20.122 A certain representation of C_{2v} has the following characters:

$\hat{E}$	$\hat{C}_2(z)$	$\hat{\sigma}_v(xz)$	$\hat{\sigma}_v(yz)$
9	-1	1	3

(*a*) What is the dimension of this representation? (*b*) Express this representation as the direct sum of irreducible representations.

20.123 A certain representation of C_{3v} has the following characters:

$\hat{E}$	$2\hat{C}_3$	$3\hat{\sigma}_v$
293	-118	9

Express this representation as the direct sum of irreducible representations. [*Hint:* The sum in (20.97) is over the symmetry operations, not over the classes.]

20.124 Give the characters for a C_{3v} representation Γ for which $\Gamma = 4A_1 \oplus A_2 \oplus 6E$. (A similarity transformation does not change the trace of a matrix.)

General

20.125 Consider the molecules N_2, HBr, CO_2, H_2S, CH_4, CH_3Cl, and C_6H_6. (*a*) Which have pure-rotational absorption spectra? (*b*) Which have vibration–rotation absorption spectra? (*c*) Which have pure-rotational Raman spectra?

20.126 For H_2 and D_2, state which has the greater value of each of the following. In some cases, the value is the same for both. Neglect deviations from the Born–Oppenheimer approximation. (*a*) k_e; (*b*) ν_e; (*c*) I_e; (*d*) B_e; (*e*) D_e; (*f*) D_0; (*g*) number of bound vibrational levels; (*h*) fraction of molecules having $v = 0$ at 2000 K; (*i*) fraction of molecules having $J = 0$ at 300 K.

20.127 An electron has 2 possible values for its m_s quantum number, so (18.39) shows $2 \times 2 = 4$ two-electron spin functions. The functions $\alpha(1)\alpha(2)$ and $\beta(1)\beta(2)$ in (18.39) that have the same m_s value are symmetric and are acceptable since they do not distinguish between the identical electrons. The other two functions must be combined into the symmetric and antisymmetric combinations shown in (18.40) and (18.41). For a nucleus with spin quantum number I, there are $2I + 1$ values of M_I, so instead of the 2×2 functions in (18.39), we start with $(2I + 1)^2$ two-nuclei spin functions. Use the same procedure used with the electron-spin functions to derive the numbers of symmetric and antisymmetric nuclear-spin functions given in Sec. 20.3.

20.128 The abbreviation for a unit that is named after a person is capitalized. (*a*) Name at least 10 SI units named after persons. (*b*) Name 4 non-SI units named after persons.

20.129 True or false? (*a*) Linear molecules are symmetric tops. (*b*) A molecule with zero dipole moment cannot change its rotational state. (*c*) A molecule whose dipole moment is zero must have a center of symmetry. (*d*) Whenever a molecule goes from one energy level to another, it emits or absorbs a photon whose energy is equal to the energy difference between the levels. (*e*) An asymmetric top cannot have any axes of symmetry. (*f*) The Raman shift of a given Raman-spectrum line is independent of the value of the exciting frequency ν_0. (*g*) The rotational energy of every molecule is given by $BhJ(J + 1)$ provided centrifugal distortion is neglected. (*h*) The vibrational levels of a given electronic state of a diatomic molecule are unequally spaced. (*i*) For a system in thermal equilibrium, a state with a higher energy than another state always has a smaller population than the lower-energy state. (*j*) For a system in thermal equilibrium, an energy level with a higher energy always has a smaller population than the lower-energy level. (*k*) We all have moments of inertia.

Chapter 20

20.1 **(a)** F; **(b)** F; **(c)** T; **(d)** F; **(e)** T.

20.2 $E = (1.00 \text{ eV})(1.602 \times 10^{-19} \text{ J/1 eV}) = 1.602 \times 10^{-19} \text{ J} = h\nu$, so
$\nu = (1.602 \times 10^{-19} \text{ J})/(6.626 \times 10^{-34} \text{ J s}) = 2.42 \times 10^{14} \text{ s}^{-1}$.
$\lambda = c/\nu = (2.998 \times 10^{10} \text{ cm/s})/(2.42 \times 10^{14}\text{/s}) = 1.24 \times 10^{-4}$ cm.
$\sigma = 1/\lambda = (1.24 \times 10^{-4} \text{ cm})^{-1} = 8.06 \times 10^{3} \text{ cm}^{-1}$.

20.3 $c_{H_2O} = c/n_{H_2O} = (2.998 \times 10^{10} \text{ cm/s})/1.33 = 2.25 \times 10^{10}$ cm/s. The frequency is unchanged in water and $\lambda_{H_2O} = c_{H_2O}/\nu = c_{H_2O}\lambda_{vac}/c = \lambda_{vac}/n_{H_2O} =$ (589 nm)/1.33 = 443 nm. $\nu = \nu_{vac} = c/\lambda_{vac} = (2.9979 \times 10^{8} \text{ m/s})/(589 \times 10^{-9} \text{ m})$ $= 5.09 \times 10^{14}$ Hz.

20.4 **(a)** F; **(b)** F; **(c)** T; **(d)** T; **(e)** T.

20.5 **(a)** s^{-1} or Hz; **(b)** m^{-1}; **(c)** m/s; **(d)** no units; **(e)** m^2/mol.

20.6 Using identities before Eq. (20.6), we get
$\mu_{mn} = (2Q/a)\frac{1}{2}\int_0^a \{x \cos[(m-n)\pi x/a] - x\cos[(m+n)\pi x/a]\}\, dx =$

$$\frac{Q}{a}\left[\frac{a^2}{(m-n)^2\pi^2}\cos\frac{(m-n)\pi x}{a} + \frac{xa}{(m-n)\pi}\sin\frac{(m-n)\pi x}{a} - \frac{a^2}{(m+n)^2\pi^2}\cos\frac{(m+n)\pi x}{a} - \frac{xa}{(m+n)\pi}\sin\frac{(m+n)\pi x}{a}\right]\Bigg|_0^a$$

which reduces to Eq. (20.6), since $\sin j\pi = 0$ for j an integer and $\cos 0 = 1$.

20.7 $\nu_{light} = |\Delta E|/h = [(\upsilon_2 + \frac{1}{2})h\nu_{vib} - (\upsilon_1 + \frac{1}{2})h\nu_{vib}]/h = (\upsilon_2 - \upsilon_1)\nu_{vib} = \nu_{vib}$.

20.8 **(a)** $\hat{\mu} = Qx$, since this is a one-particle, one-dimensional system.
$Q\int \psi_0^* x\psi_1\, dx = Q(\alpha/\pi)^{1/4}(4\alpha^3/\pi)^{1/4}\int_{-\infty}^{\infty} x^2 e^{-\alpha x^2}\, dx =$

$Q\alpha(2/\pi)^{1/2}2(2!)\pi^{1/2}/2^3 1!\alpha^{3/2} = Q/(2\alpha)^{1/2}$, where integrals 1 and 3 of Table 14.1 were used.

(b) $Q(\alpha/\pi)^{1/4}(\alpha/4\pi)^{1/4} \int_{-\infty}^{\infty} (2\alpha x^3 - x)e^{-\alpha x^2}\, dx = 0$, where integral 4 was used.

(c) $Q(\alpha/\pi)^{1/4}(\alpha/9\pi)^{1/4} \int_{-\infty}^{\infty} (2\alpha^{3/2}x^4 - 3\alpha^{1/2}x^2)e^{-\alpha x^2}\, dx =$ $Q(\alpha/3\pi)^{1/2}[2\alpha^{3/2}2(4!)\pi^{1/2}/2^5 2!\alpha^{5/2} - 3\alpha^{1/2}2(2!)\pi^{1/2}/2^3 1!\alpha^{3/2}] = 0$. The results (a)–(c) are consistent with $\Delta\upsilon = \pm 1$.

20.9 $E_C - E_A = hc/\lambda_{AC}$. $E_C - E_B = hc/\lambda_{BC}$. $E_B - E_A = hc/\lambda_{AB} =$ $(E_C - E_A) - (E_C - E_B) = hc/\lambda_{AC} - hc/\lambda_{BC}$, so $1/\lambda_{AB} = 1/\lambda_{AC} - 1/\lambda_{BC} =$ $(485\text{ nm})^{-1} - (884\text{ nm})^{-1} = 0.000931\text{ nm}^{-1}$ and $\lambda_{AB} = 1075$ nm.

20.10 Since E increases as $n^2 + n$, the spacing between levels increases as n increases. Therefore the lowest-frequency absorption is due to the transition from $n = 1$ to $n = 3$. We have $\nu_{\text{lowest}} = 80\text{ GHz} = |\Delta E|/h = [b(3)(5) - b(1)(3)]/h$ $= 12b/h$ and $b = (80\text{ GHz})h/12$. The next-lowest absorption frequency is that from $n = 2$ to $n = 4$ and its frequency is $\nu = |\Delta E|/h = [4(6)b - 2(4)b]/h =$ $16b/h = 16(80\text{ GHz})h(12)^{-1}/h = (16/12)(80\text{ GHz}) = 107$ GHz.

20.11 (a) $\nu = |\Delta E|/h = (h^2/8ma^2)(n_2^2 - n_1^2)/h = (h/8ma^2)(n_2^2 - n_1^2) =$ $(6.626 \times 10^{-34}\text{ J s})(n_2^2 - n_1^2)/8(9.109 \times 10^{-31}\text{ kg})(2.00 \times 10^{-10}\text{ m})^2 =$ $(2.273 \times 10^{15}/\text{s})(n_2^2 - n_1^2)$. The selection rule is that Δn is odd, so the lowest frequencies result from $n = 1 \to 2$, $n = 1 \to 4$, and $n = 1 \to 6$. The frequencies are $(2.273 \times 10^{15}/\text{s})(4 - 1) = 6.82 \times 10^{15}$ Hz, 3.41×10^{16} Hz, and 7.96×10^{16} Hz.

(b) The smallest values of $n_2^2 - n_1^2$ with $n_2 - n_1$ odd are for $n = 1 \to 2$, $n = 2 \to 3$, and $n = 3 \to 4$. We get 6.82×10^{15} Hz, 1.14×10^{16} Hz, and 1.59×10^{16} Hz.

20.12 $\nu = (E_{\text{upper}} - E_{\text{lower}})/h = [an_{\text{upper}}(n_{\text{upper}} + 4) - an_{\text{lower}}(n_{\text{lower}} + 4)]/h =$ $a[(n_{\text{lower}} + 3)(n_{\text{lower}} + 3 + 4) - n_{\text{lower}}(n_{\text{lower}} + 4)]/h = (6n_{\text{lower}} + 21)a/h$, where $n_{\text{lower}} = 0, 1, 2, \ldots$

20.13 Using u and ℓ for upper and lower, we have $\nu = h^{-1}A[K_u(K_u + 3) - K_\ell(K_\ell + 3)] = Ah^{-1}[(K_\ell + 1)(K_\ell + 4) - K_\ell(K_\ell + 3)] =$

$2Ah^{-1}(K_\ell + 2)$, where $K_\ell = 1, 2, 3, \ldots$. Then 60 GHz = $2Ah^{-1}(4)$; $A = 7.5h$ GHz; $\nu = (15\text{ GHz})(K_\ell + 2)$. 135 GHz = (15 GHz)($K_\ell$ + 2) and $K_\ell = 7$.

20.14 $T = 10^{-A}$. For $A = 0.1$, $T = 10^{-0.1} = 0.79$ and 21% is absorbed. For $A = 1$, $T = 0.10$ and 90% is absorbed. For $A = 10$, $T = 10^{-10}$ and 99.99999999% is absorbed.

20.15 $c = n/V = P/RT$ and $T = I_\lambda/I_{\lambda,0} = 10^{-\varepsilon_\lambda c_B \ell} = 10^{-\varepsilon_\lambda P_B \ell / RT}$.

(a) $\varepsilon_\lambda P_B \ell /RT = (10^4\text{ dm}^3/\text{mol-cm})[(10/760)\text{ atm}](1.0\text{ cm})/ (0.08206\text{ dm}^3\text{-atm/mol-K})(298\text{ K}) = 5.3_8$. $T = 10^{-5.38} = 4._2 \times 10^{-6}$.

(b) $T = 10^{-53.8} = 1._6 \times 10^{-54}$.

20.16 $T = I_\lambda/I_{\lambda,0} = 10^{-\varepsilon_\lambda c_B \ell}$.

(a) $\varepsilon_\lambda c_B \ell = (150\text{ dm}^3/\text{mol-cm})(10^{-3}\text{ mol/dm}^3)(1.0\text{ cm}) = 0.15$ and $T = 10^{-0.15} = 0.71$.

(b) $T = 10^{-1.5} = 0.03_2$.

20.17 $A = \log (I_0/I) = \log (T^{-1}) = \log (1/0.083) = 1.08_1$. $c = (0.080\text{ g/cm}^3)(1\text{ mol}/14600\text{ g})(10^3\text{ cm}^3/1\text{ dm}^3) = 0.0055\text{ mol/dm}^3$. $T = I/I_0 = 10^{-\varepsilon cl}$ and $\varepsilon = -(1/cl)\log T = -(0.0055\text{ mol/dm}^3)^{-1}(0.010\text{ cm})^{-1}\log 0.083 = 2.0 \times 10^4\text{ dm}^3\text{ mol}^{-1}\text{ cm}^{-1}$.

20.18 $A_2/A_1 = [\log (I_{\lambda,0}/I_\lambda)_2]/[\log (I_{\lambda,0}/I_\lambda)_1] = \varepsilon_\lambda c_{B,2} l_2/\varepsilon_\lambda c_{B,1} l_1 = c_{B,2}/c_{B,1} = 2$. So $\log (I_{\lambda,0}/I_\lambda)_2 = 2 \log (1/0.60) = 0.444$ and $I_{\lambda,0}/I_\lambda = 2.78$. We have $T = 1/2.78 = 0.36$ and 36% of the light is transmitted.

20.19 Let the subscripts 3 and 4 denote $Fe(CN)_6^{3-}$ and $Fe(CN)_6^{4-}$, respectively. Use of $A = (\varepsilon_3 c_3 + \varepsilon_4 c_4)l$ and $c_3 + c_4 = 1.00 \times 10^{-3}\text{ mol/dm}^3$ gives $0.701 = [(800\text{ dm}^3/\text{mol-cm})c_3 + (320\text{ dm}^3/\text{mol-cm})(0.00100\text{ mol/dm}^3 - c_3)](1.00\text{ cm})$. We find $c_3 = 7.94 \times 10^{-4}\text{ mol/dm}^3$. Then $c_3/c_{3,0} = 0.000794/0.00100 = 0.794$. The % reacted is 20.6%.

20.20 **(a)** T; **(b)** T; **(c)** F; **(d)** F; **(e)** T; **(f)** T; **(g)** F; **(h)** T.

20.21 **(a)** F; **(b)** T; **(c)** F.

20.22 Division of Eq. (20.25) by hc gives $D_0/hc = D_e/hc - \frac{1}{2}\tilde{\nu}_e + \frac{1}{4}\tilde{\nu}_e x_e$.

(a) $D_0/hc = [79890 - \frac{1}{2}(2359) + \frac{1}{4}(14)]\text{ cm}^{-1} = 78714\text{ cm}^{-1}$;
$D_0 = (78714\text{ cm}^{-1})(100\text{ cm/m})(6.6261 \times 10^{-34}\text{ J s})(2.9979 \times 10^8\text{ m/s}) = 1.5636 \times 10^{-18}\text{ J} = 9.759\text{ eV}$.

(b) $D_0/hc = [90544 - \frac{1}{2}(2170) + \frac{1}{4}(13)]\text{ cm}^{-1} = 89462\text{ cm}^{-1}$;
$D_0 = 1.7771 \times 10^{-18}\text{ J} = 11.092\text{ eV}$.

20.23 **(a)** D_e is the depth of the electronic energy curve E_e and k_e equals $E_e''(R_e)$. In the Born–Oppenheimer approximation, $E_e(R)$ is found by solving the electronic Schrödinger equation (19.7) in which the nuclei are fixed; the nuclear masses do not occur in (19.7) and (19.6). Hence, $E_e(R)$ is the same for $^2H^{35}Cl$ and $^1H^{35}Cl$, which have the same nuclear charges. From Eq. (20.25), D_0 differs from D_e by the zero-point vibrational energy. The vibrational frequency ν_e equals $(1/2\pi)(k/\mu)^{1/2}$. The reduced mass μ differs substantially for $^2H^{35}Cl$ and $^1H^{35}Cl$, so their ν_e's differ and their D_0's differ.

(b) For $^1H^{35}Cl$, $D_0/hc = D_e/hc - \frac{1}{2}\tilde{\nu}_e + \frac{1}{4}\tilde{\nu}_e x_e =$
$[37240 - \frac{1}{2}(2990.9) + \frac{1}{4}(52.8)]\text{ cm}^{-1} = 35758\text{ cm}^{-1}$;
$D_0 = 7.103_1 \times 10^{-19}\text{ J} = 4.433_4\text{ eV}$. From Eqs. (20.23) and (17.79):
$\nu_{e,\text{DCl}}/\nu_{e,\text{HCl}} = (\mu_{\text{HCl}}/\mu_{\text{DCl}})^{1/2}$; $\mu_{\text{HCl}} = 1.00782(34.969)(\text{g/mol})/35.977N_A = (0.97959\text{ g/mol})/N_A$; $\mu_{\text{DCl}} = (1.9044\text{ g/mol})/N_A$. So $\tilde{\nu}_{e,\text{DCl}} = (2990.9\text{ cm}^{-1})(0.97959/1.9044)^{1/2} = 2145.1\text{ cm}^{-1}$. Also, $D_{e,\text{DCl}} = D_{e,\text{HCl}}$. For $^2H^{35}Cl$ we then have $D_0/hc = [37240 - \frac{1}{2}(2145) + \frac{1}{4}(53)]\text{ cm}^{-1} = 36181\text{ cm}^{-1}$ and $D_0 = 7.187 \times 10^{-19}\text{ J} = 4.486\text{ eV}$ (where we neglected the change in $\nu_e x_e$).

20.24 $\frac{1}{2}h\nu_e = \frac{1}{2}kA^2$. But $\nu_e = (1/2\pi)(k/\mu)^{1/2}$, so $k = 4\pi^2\mu\nu_e^2$; $\frac{1}{2}h\nu_e = 2\pi^2\mu\nu_e^2 A^2$ and $A = (h/4\pi^2\mu\nu_e)^{1/2}$. For $H^{35}Cl$, $\mu = (1.0)(35.0)\text{g}/(36.0)(6.02 \times 10^{23}) = 1.6 \times 10^{-24}\text{ g}$ and $\nu_e = (2.998 \times 10^{10}\text{ cm/s})(2991\text{ cm}^{-1}) = 9.0 \times 10^{13}\text{ s}^{-1}$, so $A = [(6.63 \times 10^{-34}\text{ J s})/4\pi^2(1.6 \times 10^{-27}\text{ kg})(9.0 \times 10^{13}/\text{s})]^{1/2} = 1.1 \times 10^{-11}\text{ m} =$

0.11 Å. For $^{14}N_2$, $\mu = 14(14)\text{g}/28N_A = 1.16 \times 10^{-23}$ g, $\nu_e = 7.1 \times 10^{13}\ \text{s}^{-1}$ and $A = 0.045$ Å.

20.25 Use of (20.31) gives $\lambda = c/\nu = c/2(J+1)B$, so $2.00\ \text{cm} = c/2(3)B$ and $B = c/(12.0\ \text{cm})$. We then have $\lambda = c/2(J+1)c(12.0\ \text{cm})^{-1} = (6.0\ \text{cm})/(J+1)$. $\lambda_{6\to7} = (6.0\ \text{cm})/7 = 0.86$ cm.

20.26 From Eqs. (20.31), (20.33), and (20.15): $\nu_{J\to J+1} = 2(J+1)B_0 = 2B_0 = 2h/8\pi^2 I_0 = h/4\pi^2\mu R_0^2$ and $R_0 = (h/\mu\nu)^{1/2}/2\pi$.

(a) Use of the table of atomic masses gives $\mu = m_1 m_2/(m_1 + m_2) = (1.0078250)(78.91834)\text{g}/(79.92616_5)(6.02214 \times 10^{23}) = 1.652431 \times 10^{-24}$ g for $^1H^{79}Br$ and $\mu = 1.652945 \times 10^{-24}$ g for $^1H^{81}Br$. For $^1H^{79}Br$, $R_0 =$ $[(6.62607 \times 10^{-34}\ \text{J s})/(1.652431 \times 10^{-27}\ \text{kg})(500.7216 \times 10^9/\text{s})]^{1/2}/2\pi =$ 1.424257 Å. For $^1H^{81}Br$, we get $R_0 = 1.424257$ Å.

(b) For $J = 1$ to 2, $\nu = 2(J+1)B_0 = 4B_0 = 2\nu_{J=0\to1} = 2(500.7216\ \text{GHz}) =$ 1001.4432 GHz with centrifugal distortion neglected.

(c) For the $J = 0$ to 1 transition, $\nu_{DBr}/\nu_{HBr} = 2B_{0,DBr}/2B_{0,HBr} = \mu_{HBR}/\mu_{DBr}$, since R_0 is essentially unchanged on isotopic substitution. $\mu_{DBr} = (2.014102)(78.91834)\text{g}/(80.93244_2)(6.02214 \times 10^{23}) =$ 3.261264×10^{-24} g. $\mu_{HBr}/\mu_{DBr} = 1.652431/3.261264 = 0.506684$ and $\nu_{DBr} = 0.506684(500.7216\ \text{GHz}) = 253.708$ GHz.

20.27 From Eq. (20.31), $\nu_{J\to J+1} = 2B_0(J+1)$, so $B_0(^{39}K^{37}Cl) = (22410\ \text{MHz})/2(3) =$ 3735 MHz.

(a) $\nu = 2(3735\ \text{MHz})1 = 7470$ MHz.

(b) The reasoning in Prob. 20.23a shows that R_e for the two isotopic species is the same; further, R_0 should differ only very slightly for the two species (see, for example, Prob. 20.26). Equations (20.33) and (20.15) then give $B_0(^{39}K^{35}Cl)/B_0(^{39}K^{37}Cl) = I_0(^{39}K^{37}Cl)/I_0(^{39}K^{35}Cl) =$ $\mu(^{39}K^{37}Cl)/\mu(^{39}K^{35}Cl) = 18.9693/18.4292 = 1.02931$, where the reduced masses were found from $m_A m_B/(m_A + m_B)$. Then $B_0(^{39}K^{35}Cl) =$ $1.02931(3735\ \text{MHz}) = 3844\tfrac{1}{2}$ MHz and $\nu_{J=0\to1} = 2(3844\tfrac{1}{2}\ \text{MHz}) =$ 7689 MHz for $^{39}K^{35}Cl$.

20.28 The equation preceding (20.37) is $\tilde{\nu} = \tilde{\nu}_{\text{origin}} + \tilde{B}_e J'(J' + 1) - \tilde{B}_e J''(J'' + 1)$. For P branch lines, $J' = J'' - 1$ and $J'(J' + 1) - J''(J'' + 1) = (J'' - 1)J'' - J''(J'' + 1) = -2J''$, so $\tilde{\nu}_P = \tilde{\nu}_{\text{origin}} - 2\tilde{B}_e J''$.

20.29 Eq. (20.39) multiplied out gives $\tilde{\nu}_R = \tilde{\nu}_{\text{origin}} + 2\tilde{B}_e(J'' + 1) - \tilde{\alpha}_e \upsilon'(J''^2 + 3J'' + 2) + \tilde{\alpha}_e \upsilon''(J''^2 + J'') - \tilde{\alpha}_e(J'' + 1)$. With centrifugal distortion neglected, Eq. (20.26) gives $\tilde{\nu}_R = (E_{J''+1,\upsilon'} - E_{J'',\upsilon''})/hc =$ $\tilde{\nu}_e(\upsilon' + \frac{1}{2}) - \tilde{\nu}_e x_e(\upsilon' + \frac{1}{2})^2 + \tilde{B}_e(J'' + 1)(J'' + 2) -$ $\tilde{\alpha}_e(\upsilon' + \frac{1}{2})(J'' + 1)(J'' + 2) - \tilde{\nu}_e(\upsilon'' + \frac{1}{2}) + \tilde{\nu}_e x_e(\upsilon'' + \frac{1}{2})^2 - \tilde{B}_e J''(J'' + 1) +$ $\tilde{\alpha}_e(\upsilon'' + \frac{1}{2})J''(J'' + 1) = \tilde{\nu}_e(\upsilon' - \upsilon'') - \tilde{\nu}_e x_e(\upsilon'^2 + \upsilon' - \upsilon''^2 - \upsilon'') +$ $\tilde{B}_e(2J'' + 2) + \tilde{\alpha}_e \upsilon''(J''^2 + J'') - \tilde{\alpha}_e \upsilon'(J''^2 + 3J'' + 2) - \tilde{\alpha}_e(J'' + 1)$, which Eq. (20.34) shows to be the same as the above multiplied-out form of Eq. (20.39), verifying (20.39). When multiplied out (20.40) is $\tilde{\nu}_P = \tilde{\nu}_{\text{origin}} - 2B_e J'' + \tilde{\alpha}_e \upsilon'(J'' - J''^2) + \tilde{\alpha}_e \upsilon''(J'' + J''^2) + \tilde{\alpha}_e J''$. Then $\tilde{\nu}_P = (E_{J''-1,\upsilon'} - E_{J'',\upsilon''})/hc$ and use of (20.26) leads to the multiplied-out form of (20.40).

20.30 $\nu_e = (1/2\pi)(k_e/\mu)^{1/2}$ and $k_e = 4\pi^2\nu_e^2\mu = 4\pi^2c^2\tilde{\nu}_e^2\mu = 4\pi^2(2.9979 \times 10^8\ \text{m/s})^2 \times (158000\ \text{m}^{-1})^2(0.01599491\ \text{kg})/2(6.0221 \times 10^{23}) = 1176\ \text{N/m}$, where we used $\mu = m_A m_A/(m_A + m_A) = m_A/2$.

20.31 **(a)** From Eq. (20.34), $\tilde{\nu}_{\text{origin}}(0 \to 1) = \tilde{\nu}_e - 2\tilde{\nu}_e x_e$ and $\tilde{\nu}_{\text{origin}}(0 \to 2) = 2\tilde{\nu}_e - 6\tilde{\nu}_e x_e$. Hence $3\tilde{\nu}_{\text{origin}}(0 \to 1) - \tilde{\nu}_{\text{origin}}(0 \to 2) = \tilde{\nu}_e = 3(2886.0\ \text{cm}^{-1}) - 5668.0\ \text{cm}^{-1} = 2990.0\ \text{cm}^{-1}$. Then $\tilde{\nu}_e x_e = \frac{1}{2}[\tilde{\nu}_e - \tilde{\nu}_{\text{origin}}(0 \to 1)] = \frac{1}{2}(2990.0 - 2886.0)\text{cm}^{-1} = 52.0\ \text{cm}^{-1}$.

(b) From (20.34), $\tilde{\nu}_{\text{origin}}(0 \to 6) = 6(2990.0\ \text{cm}^{-1}) - 42(52.0\ \text{cm}^{-1}) = 15756\ \text{cm}^{-1}$.

20.32 **(a)** The distance of a line from the band origin is the change in rotational energy for the line's transition. The selection rule is $\Delta J = \pm 1$ and the spacings between rotational levels increase as J increases. Hence the two lines closest to the band origin involve the $J = 0$ and 1 levels. The 2139

cm^{-1} line is lower in frequency and energy than the band origin and so must be the $J = 1 \to 0$ line, where the rotational energy is decreasing. The 2147 cm^{-1} line has $J = 0 \to 1$. The 2151 cm^{-1} line has $J = 1 \to 2$. The 2135.5 cm^{-1} line has $J = 2 \to 1$.

(b) Let *a*, *b*, *c*, *d* be the four given wavenumbers in order of increasing wavenumber. From (20.40) and (20.39), we have

$$a = \tilde{\nu}_{\text{origin}} - (2\tilde{B}_e - 2\tilde{\alpha}_e)2 - 4\tilde{\alpha}_e = \tilde{\nu}_{\text{origin}} - 4\tilde{B}_e$$

$$b = \tilde{\nu}_{\text{origin}} - (2\tilde{B}_e - 2\tilde{\alpha}_e)1 - \tilde{\alpha}_e = \tilde{\nu}_{\text{origin}} - 2\tilde{B}_e + \tilde{\alpha}_e$$

$$c = \tilde{\nu}_{\text{origin}} + (2\tilde{B}_e - 2\tilde{\alpha}_e)1 - \tilde{\alpha}_e = \tilde{\nu}_{\text{origin}} + 2\tilde{B}_e - 3\tilde{\alpha}_e$$

$$d = \tilde{\nu}_{\text{origin}} + (2\tilde{B}_e - 2\tilde{\alpha}_e)2 - 4\tilde{\alpha}_e = \tilde{\nu}_{\text{origin}} + 4\tilde{B}_e - 8\tilde{\alpha}_e$$

Three spreadsheet cells are designated for $\tilde{\nu}_{\text{origin}}$, $\tilde{B}_e$, and $\tilde{\alpha}_e$. The initial guess for $\tilde{\nu}_{\text{origin}}$ can be taken as the average of *b* and *c*, namely 2143.25 cm^{-1}. The initial guesses for $\tilde{B}_e$ and $\tilde{\alpha}_e$ can be taken as zero. The four formulas for *a, b, c,* and *d* are entered into four cells and the squares of the deviations of the *a, b, c,* and *d* formula values from the observed values are calculated, and summed. The Solver is set up to minimize the sum of the squares of the deviations by varying $\tilde{\nu}_{\text{origin}}$, $\tilde{B}_e$, and $\tilde{\alpha}_e$ subject to the constraints that $\tilde{B}_e$ and $\tilde{\alpha}_e$ be positive. An excellent fit to the observed lines is found with $\tilde{\nu}_{\text{origin}} = 2143.2695\ \text{cm}^{-1}$, $\tilde{B}_e = 1.93023\ \text{cm}^{-1}$, and $\tilde{\alpha}_e = 0.016411$.

(c) From (20.16) and (20.15), $B_e = h/8\pi^2\mu R_e^2$ and $R_e = (h/8\pi^2\mu B_e)^{1/2}$.
$\mu = [(12)(15.994915)/27.994915]\text{g}/(6.02214 \times 10^{23}) = 1.138500 \times 10^{-23}$ g.
$B_e = \tilde{B}_e c = (1.93023\ \text{cm}^{-1})(2.997925 \times 10^{10}\ \text{cm/s}) = 5.78668 \times 10^{10}\ \text{s}^{-1}$.
$R_e = [(6.62607 \times 10^{-34}\ \text{J s})/8\pi^2(1.138500 \times 10^{-26}\ \text{kg})(5.78668 \times 10^{10}\ \text{s}^{-1})]^{1/2}$ $= 1.12863 \times 10^{-10}\ \text{m} = 1.12863$ Å, which is smaller than R_0 in Example 20.3.

20.33 With centrifugal distortion neglected, the $\upsilon = 0$ vibration-rotation levels are given by Eq. (20.26) as $E_{\text{vib-rot}} = \frac{1}{2}h\nu_e - \frac{1}{4}h\nu_e x_e + hB_e J(J+1) - \frac{1}{2}h\alpha_e J(J+1)$. For $J = 0$, $E_{\text{vib-rot}}(0) = \frac{1}{2}h\nu_e - \frac{1}{4}h\nu_e x_e$. We have $E_{\text{vib-rot}}(J) - E_{\text{vib-rot}}(0) =$ $h(B_e - \frac{1}{2}\alpha_e)J(J+1) = (\tilde{B}_e - \frac{1}{2}\tilde{\alpha}_e)hcJ(J+1)$. $(\tilde{B}_e - \frac{1}{2}\tilde{\alpha}_e)hc/kT =$ $[10.594 - ½(0.31)]\ \text{cm}^{-1}\ (100\ \text{cm/m})(6.6261 \times 10^{-34}\ \text{J s})(2.9979 \times 10^{8}\ \text{m/s})/$

$(1.3807 \times 10^{-23}$ J/K)(300 K) = 0.050064. The degeneracy of each level is $2J + 1$, so the Boltzmann distribution law gives the relative populations as $N_J/N_0 = (2J+1)e^{-\Delta E/kT} = (2J+1)\exp[-(\tilde{B}_e - \frac{1}{2}\tilde{\alpha}_e)hcJ(J+1)/kT] = (2J+1)e^{-0.050064J(J+1)}$. We find

J	1	2	3	4	5	6
N_J/N_0	2.714	3.703	3.839	3.307	2.450	1.588

20.34 **(a)** $E_b - E_a = h\nu_{a\to b} = (6.626 \times 10^{-34}$ J s)$(1.54 \times 10^{12}$ s$^{-1}) = 9.64 \times 10^{-18}$ J. $N_b/N_a = (g_b/g_a)\exp[-(E_b - E_a)/kT] = (g_b/g_a)\exp[-(9.64 \times 10^{-18}$ J)/(1.3806 $\times 10^{-23}$ J/K)$T] = (g_b/g_a)\exp[-(6.98 \times 10^5$ K)/$T]$. $N_b/N_a = \exp[-(6.98 \times 10^5$ K)/(300 K)] $= \exp(-2326.7)$. $\ln(N_b/N_a) = -2326.7 = 2.3026 \log(N_b/N_a)$ and $N_b/N_a \approx 10^{-1010.5} = 0.3 \times 10^{-1010}$, where (1.71) was used, and where more significant figures than are justified were used. Since N_b/N_a is far less than Avogadro's number, we can say that $N_b/N_a = 0$. (See also the discussion after Example 6.6.)

(b) $g_b/g_a = 5/3$, so $N_b/N_a \approx (5/3)(0.3 \times 10^{-1010}) = 0.5 \times 10^{-1010}$. As in (a), $N_b/N_a = 0$.

(c) $N_b/N_a = \exp[-(6.98 \times 10^5$ K)/(1000 K)] $= \exp(-698)$. $\ln(N_b/N_a) = -698 = 2.3026 \log(N_b/N_a)$ and $N_b/N_a \approx 10^{-303.1} = 0.8 \times 10^{-303}$. $N_b/N_a \approx (5/3)(0.8 \times 10^{-303}) = 1.3 \times 10^{-303}$. As in (a), $N_b/N_a = 0$.

20.35 B_e—rotational constant; α_e—vibration–rotation interaction; D—centrifugal distortion; $\nu_e x_e$—anharmonicity.

20.36 **(a)** From Fig. 19.18, O_2 has 4 more bonding electrons than antibonding electrons, O_2^+ has 5 net bonding electrons, and O_2^- has 3 net bonding electrons. Therefore O_2^+ has the strongest bond and the largest k_e and O_2^- has the smallest k_e.

(b) Use of Fig. 19.15 shows that N_2 has 6 net bonding electrons and N_2^+ has 5 net bonding electrons, so N_2 has the stronger bond, the larger k_e and the larger ν_e since $\nu_e = (1/2\pi)(k_e/\mu)^{1/2}$.

(c) N_2 has a triple bond and O_2 a double bond. The N_2 bond is stronger and N_2 has the larger k_e.

(d) $E_{rot} = J(J+1)\hbar^2/2I$, where $I = \mu R_e^2$. An Na atom is heavier and larger than an Li atom, so Na_2 has the larger μ and the larger R_e. So $E_{rot,J=1}$ is greater for Li_2.

20.37 **(a)** T; **(b)** T; **(c)** T; **(d)** F.

20.38 **(a)** A C_2 axis and two symmetry planes.

(b) A C_3 axis (through the CCl bond) and three symmetry planes (each one containing C, Cl, and one F).

(c) The molecule is square planar. A C_4 axis perpendicular to the molecular plane; an S_4 axis and a C_2 axis, each coincident with the C_4 axis; a symmetry plane coincident with the molecular plane; four symmetry planes perpendicular to the molecular plane; four C_2 axes in the molecular plane (two pass through pairs of opposite F's and two lie between the F's); a center of symmetry.

(d) The structure is trigonal bipyramidal. A C_3 axis; an S_3 axis coincident with the C_3 axis; a (horizontal) plane of symmetry containing the equatorial Cl's; three planes of symmetry that each contain the two axial Cl's; three C_2 axes, each lying in the horizontal symmetry plane.

(e) The VSEPR method shows the structure is a square-based pyramid. A C_4 axis and four symmetry planes.

(f) A C_2 axis perpendicular to the molecular plane; a center of symmetry; two C_2 axes in the molecular plane; three symmetry planes—one coincident with the molecular plane and two perpendicular to it. See Prob. 20.40c.

(g) A C_∞ axis through the nuclei and an infinite number of symmetry planes that each contain the C_∞ axis.

(h) A C_∞ axis (which is also an S_∞ axis), an infinite number of symmetry planes through this axis, a symmetry plane perpendicular to this axis, a center of symmetry, an infinite number of C_2 axes perpendicular to the molecular axis.

20.39 **(a)** The symmetry elements (Prob. 20.38a) are a C_2 axis and two symmetry planes, which we call σ_a and σ_b. The symmetry operations are $\hat{E}$, $\hat{C}_2$, $\hat{\sigma}_a$, $\hat{\sigma}_b$.

(b) $\hat{E}$, $\hat{C}_3$, $\hat{C}_3^2$, $\hat{\sigma}_a$, $\hat{\sigma}_b$, $\hat{\sigma}_c$.

20.40 **(a)** Moves a nucleus at x, y, z to $-x, -y, z$.

(b) From x, y, z, to $x, y, -z$.

(c) The $\hat{S}_2$ rotation about the z axis consists of a $\hat{C}_2$ rotation about z followed by reflection in the xy plane. From the answers to (a) and (b) this moves a point at x, y, z to $-x, -y, -z$. We see that $\hat{S}_2 = \hat{i}$.

20.41 **(a)** The three principal axes intersect at the B nucleus (which is the center of mass). One principal axis is perpendicular to the molecular plane (and coincides with the C_3 axis). The other two principal axes lie in the molecular plane; one of these can be taken to coincide with a BF bond, and the other is perpendicular to this one. (As in XeF_4, the orientation of the principal axes is not unique.)

(b) The three principal axes intersect at the center of mass, which lies on the C_2 axis. One principal axis coincides with the C_2 axis. The second lies in the molecular plane and is perpendicular to the C_2 axis. The third is perpendicular to the molecular plane.

(c) The three principal axes intersect at the C nucleus. One principal axis coincides with the molecular axis (which is a C_∞ axis); the other two can be taken as any two axes through the C that are perpendicular to the molecular axis and perpendicular to each other.

29.42 **(a)** SF_6 has more than one noncoincident C_4 axis and is a spherical top.

(b) IF_5 (which is a square-based pyramid) has one C_4 axis and is a symmetric top.

(c) One C_2 axis. Asymmetric top.

(d) One C_3 axis. Symmetric top.

(e) One C_6 axis. Symmetric top.

(f) One C_∞ axis. Symmetric top.

(g) One C_3 axis. Symmetric top.

(h) Symmetric.

20.43 The principal axes intersect at the center of mass, which is the B nucleus. One principal axis is the C_3 axis. For this axis, $I_c = \Sigma_i m_i r_{i,c}^2 =$ $3(18.998 \text{ amu})(1.313 \text{ Å})^2 = 98.3 \text{ amu Å}^2$. The other two principal axes lie in the molecular plane and we can take one of them to coincide with a B–F bond. Hence, $I_a = \Sigma_i m_i r_{i,a}^2 = 2(18.998 \text{ amu})[(1.313 \text{ Å})(\sin 60°)]^2 = 49.1 \text{ amu Å}^2$. With one C_3 axis, BF_3 is a symmetric top, so $I_b = I_a = 49.1 \text{ amu Å}^2$. I_b can also be calculated by taking distances from the F atoms to an in-plane line through B and perpendicular to a B–F bond: $I_b =$ $(19.0 \text{ amu})(1.313 \text{ Å})^2 + 2[(\cos 60°)(1.313 \text{ Å})]^2(19.0 \text{ amu}) = 49.1 \text{ amu Å}^2$.

20.44 One principal axis of this symmetric top is the C_3 axis through the axial bonds. For this axis, $I = 3(34.97 \text{ amu})(2.02 \text{ Å})^2 = 428 \text{ amu Å}^2$. Another principal axis can be taken to coincide with an equatorial P–Cl bond, and for this axis $I =$ $2(34.97 \text{ amu})(2.12 \text{ Å})^2 + 2(34.97 \text{ amu})[(2.02 \text{ Å})(\sin 60°)]^2 = 528 \text{ amu Å}^2$. Since this is a symmetric top, the moments of inertia about the principal axes that are perpendicular to the C_3 axis are equal, and the third principal moment is 528 amu Å^2.

20.45 The molecule is a symmetric top with $I_a \neq I_b = I_c$. From Eq. (20.45), $E_{rot}/h =$ $BJ(J+1) + (A-B)K^2 = [\tilde{B}J(J+1) + (\tilde{A}-\tilde{B})K^2]c$.

(a) For $J = 0$, $K = 0$ and $E_{rot}/h = 0$. For $J = 1$, $K = -1, 0, 1$. For $J = 1$ and $K = 0$, $E_{rot}/h = 2\tilde{B}c = 2(0.05081 \text{ cm}^{-1})(2.9979 \times 10^{10} \text{ cm/s}) =$ $3.046 \times 10^9 \text{ s}^{-1}$. For $J = 1$ and $|K| = 1$, $E_{rot}/h = [2\tilde{B} + (\tilde{A} - \tilde{B})]c =$ $(\tilde{B} + \tilde{A})c = (0.2418 \text{ cm}^{-1})c = 7.249 \times 10^9 \text{ s}^{-1}$.

(b) From Eq. (20.47), $\nu = 2B(J+1) = 2\tilde{B}c(J+1) = 2\tilde{B}c, 4\tilde{B}c, \ldots =$ $3.046 \times 10^9 \text{ s}^{-1}, 6.093 \times 10^9 \text{ s}^{-1}, \ldots = 3046 \text{ MHz}, 6093 \text{ MHz}$.

20.46 (a) Let the molecule lie on the positive half of the z axis with the origin at the oxygen nucleus. Then $z_{com} = [12(1.160 \text{ Å}) + 31.972071(2.720 \text{ Å})]/$ $59.966986 = 1.682 \text{ Å}$.

(b) $I_0 = \Sigma_i m_i r_i^2 = [(15.994915)(1.682)^2 + 12(1.682 - 1.160)^2 + 31.972071(2.720 - 1.682)^2](\text{g Å})^2/(6.02214 \times 10^{23}) = 1.377_7 \times 10^{-22}$ g Å^2 $= 1.377_7 \times 10^{-45}$ kg m^2.

(c) $\nu_{J\to J+1} = 2(J+1)B_0$. $B_0 = h/8\pi^2 I_0 = (6.62608 \times 10^{-34}$ J s$)/8\pi^2(1.377_7 \times 10^{-45}$ kg m$^2) = 6.091_2 \times 10^9$ s^{-1}. $\nu_{0\to 1} = 2B_0 = 12.18$ GHz; $\nu_{1\to 2} = 4B_0 = 24.36$ GHz; $\nu_{2\to 3} = 6B_0 = 36.55$ GHz.

20.47 $B_0 = \tilde{B}_0 c = (0.39021$ cm$^{-1})(2.99792 \times 10^{10}$ cm/s$) = 1.16982 \times 10^{10}$ s^{-1}. $I_0 = h/8\pi^2 B_0 = (6.62607 \times 10^{-34}$ J s$)/8\pi^2(1.16982 \times 10^{10}$ s$^{-1}) = 7.1738 \times 10^{-46}$ kg m^2. The center of mass is at the carbon atom and the principal axes pass through this atom. If d is the CO bond length, then $I_0 = \Sigma_i m_i r_i^2 = m_O d^2 + m_O d^2 = 2m_O d^2$, so $d = (I_0/2m_O)^{1/2} =$
$[(7.1738 \times 10^{-46}$ kg m$^2)(6.02214 \times 10^{23})/2(15.994915 \times 10^{-3})kg]^{1/2} =$
1.162×10^{-10} m = 1.62 Å.

20.48 **(a)** From the VSEPR method, SO_2 is nonlinear; $3\mathcal{N} - 6 = 9 - 6 = 3$.
(b) Linear. $3\mathcal{N} - 5 = 7$. **(c)** $3\mathcal{N} - 6 = 9$.

20.49 We look for sets of integers i, j, k such that $3657i + 1595j + 3756k$ is slightly greater than 7252. Systematic trial and error (best done by first setting $j = 0$ and looking for i and k values that fit, then setting $j = 1$ and looking for i and k, then setting $j = 2$, etc.) gives the $\upsilon_1'\upsilon_2'\upsilon_3'$ possibilities for the 7252 cm^{-1} band as (calculated frequencies in parentheses) 200 (7314), 101 (7413), 002 (7512).

20.50 **(a)** $\frac{1}{2}\Sigma_i h\nu_i = \frac{1}{2}hc\,\Sigma_i \tilde{\nu}_i = \frac{1}{2}hc(1340 + 667 + 667 + 2349)$ cm^{-1} $=$ 4.99×10^{-20} J = 0.311 eV.
(b) $\frac{1}{2}hc(3657 + 1595 + 3756)$ cm^{-1} $= 8.95 \times 10^{-20}$ J = 0.558 eV.

20.51 **(a)** Inactive, since the dipole moment remains unchanged.
(b) Active, since the dipole moment changes.

20.52 $\nu = (1/2\pi)\sqrt{k/\mu}$.

(a) The C≡C bond is stronger and has the greater k and the greater ν.

(b) C—H has the smaller μ and the greater ν.

(c) Bending vibrations are generally lower-frequency than stretching, so C—H stretching has the greater ν.

20.53 $\tilde{\nu} = \nu/c = (1/2\pi)(k/\mu)^{1/2}/c$. Isotopic substitution does not affect the electrons and hence doesn't affect k. We have $\mu = m_1m_2/(m_1 + m_2) \approx m_1m_2/m_2 = m_1$, where $m_1 = m_H$ (or m_D) and m_2 is the mass of the rest of the molecule, and we used $m_2 >> m_1$. Therefore $\mu_{CD} \approx 2\mu_{CH}$ and $\tilde{\nu}_{CD} \approx \tilde{\nu}_{CH}/2^{1/2} = (2900\ \text{cm}^{-1})/2^{1/2}$ $= 2050\ \text{cm}^{-1}$.

20.54 $\nu = (1/2\pi)(k/\mu)^{1/2}$ and $k = 4\pi^2\nu^2\mu = 4\pi^2\tilde{\nu}^2c^2\mu$. $\mu_{CH} = 12(1)\ \text{g}/13(6.02 \times 10^{23})$ $= 1.53 \times 10^{-24}$ g, and $\mu_{CO} = 12(16)\ \text{g}/28(6.02 \times 10^{23}) = 1.14 \times 10^{-23}$ g. So $k_{CH} = 4\pi^2(3000\ \text{cm}^{-1})^2(100\ \text{cm/m})^2(3.00 \times 10^8\ \text{m/s})^2(1.53 \times 10^{-27}\ \text{kg}) =$ 489 N/m. Also, $k_{CO} = 4\pi^2(1750\ \text{cm}^{-1})^2(100\ \text{cm/m})^2(3.00 \times 10^8\ \text{m/s})^2 \times$ $(1.14 \times 10^{-26}\ \text{kg}) = 1240$ N/m.

20.55 **(a)** T; **(b)** T; **(c)** F; **(d)** T (provided it is not a spherical top); **(e)** T. **(f)** T; **(g)** F.

20.56 **(a)** The rotational levels of a linear molecule are $E_{rot} = BhJ(J+1)$ [Eq. (20.45) with $K = 0$] and the pure rotational Raman selection rule is $\Delta J = \pm 2$. So $\nu_0 - \nu_{scat} = \Delta E/h = \pm(Bh/h)[(J+2)(J+3) - J(J+1)] =$ $\pm(4J+6)B$, where $J = 0, 1, 2, \ldots$. The spacing between adjacent lines is $[4(J+1)+6]B - (4J+6)B = 4B$.

(b) $4\tilde{B} = 7.99\ \text{cm}^{-1}$ and $\tilde{B} = 1.998\ \text{cm}^{-1} = 199.8\ \text{m}^{-1}$. We have $\tilde{B} = B/c =$ $h/8\pi^2Ic = h/8\pi^2c\mu R^2$ and $R = (h/8\pi^2c\mu\tilde{B})^{1/2}$. Also, $\mu = m_1m_2/(m_1+m_2) =$ $m_1^2/2m_1 = m_1/2$ and $R =$

$$\left[\frac{2(6.626 \times 10^{-34}\ \text{J s})(6.022 \times 10^{23}\ \text{mol}^{-1})}{8\pi^2(2.998 \times 10^8\ \text{m/s})(0.01401\ \text{kg/mol})(199.8\ \text{m}^{-1})}\right]^{1/2} =$$

1.098×10^{-10} m = 1.098 Å (as in Table 20.1).

(c) The lowest J is 0 and $\nu_0 - \nu_{scat} = [4(0) + 6]B = 6B$.

(d) $\nu_0 = c/\lambda_0 = (2.9979 \times 10^8\ \text{m/s})/(540.8 \times 10^{-9}\ \text{m}) = 5.543_5 \times 10^{14}\ \text{s}^{-1}$. $\nu_0 - \nu_{scat} = \pm 6B = \pm 6\tilde{B}c = \pm 6(199.8\ \text{m}^{-1})(2.998 \times 10^8\ \text{m/s}) =$ $\pm 0.00359 \times 10^{14}\ \text{s}^{-1}$. $\nu_{scat} = 5.543_5 \times 10^{14}\ \text{s}^{-1} \pm 0.00359 \times 10^{14}\ \text{s}^{-1} =$

$5.547_1 \times 10^{14}$ s^{-1} and $5.539_9 \times 10^{14}$ s^{-1}. Then $\lambda_{scat} = c/\nu_{scat} = 540.4_5$ nm and 541.1_5 nm.

20.57 **(a)** F; **(b)** T; **(c)** F.

20.58 For the Balmer series, $n_b = 2$ in Eq. (20.50); for the series limit, $n_a = \infty$ and $1/\lambda = R/4 = (109678 \text{ cm}^{-1})/4$. Then $\lambda = 3.647 \times 10^{-5}$ cm = 364.7 nm.

20.59 For the Paschen series, $n_b = 3$ in Eq. (20.50); the first three lines have $n_a =$ 4, 5, and 6. So $1/\lambda = R(1/9 - 1/16)$, $R(1/9 - 1/25)$, $R(1/9 - 1/36)$. We get $\lambda =$ 1.8756×10^{-4} cm, 1.2822×10^{-4} cm, 1.0941×10^{-4} cm.

20.60 Li^{2+} is a hydrogenlike atom and Eq. (18.14) gives the energy levels. The Li nucleus is substantially heavier than the H nucleus, so we can take μ equal to the electron mass. So $E = -9[2\pi^2 me^4/(4\pi\varepsilon_0)^2 n^2 h^2]$ and $1/\lambda = \nu/c = |\Delta E|/hc =$ $9[2\pi^2 me^4/(4\pi\varepsilon_0)^2 ch^3](1/n_b^2 - 1/n_a^2) = 9(109736 \text{ cm}^{-1})(1/1 - 1/4) =$ 7.4072×10^5 cm^{-1}. Then $\lambda = 1.3500 \times 10^{-6}$ cm.

20.61 Using the notation of Fig. 20.38 and Eq. (20.52), we have D_0 of the B state as $D_0' = h\nu_{cont} - h\nu_{00} = (6.6261 \times 10^{-34} \text{ J s})(2.998 \times 10^8 \text{ m/s}) \times$ $[(1750.5 \times 10^{-10} \text{ m})^{-1} - (2026.0 \times 10^{-10} \text{ m})^{-1}] = 1.543 \times 10^{-19} \text{ J} = 0.963$ eV. D_0 of the ground state is given by (20.52) as $D_0'' = (6.6261 \times 10^{-34} \text{ J s})(2.9979 \times 10^8 \text{ m/s})(1750.5 \times 10^{-10} \text{ m})^{-1} \times$ $(1 \text{ eV})/(1.60218 \times 10^{-19} \text{ J}) - 1.970 \text{ eV} = 5.11$ eV.

20.62 $N_{J=1}/N_{J=0} = 0.181 = (3/1)e^{-\Delta\varepsilon/kT}$ and $\Delta\varepsilon/kT = -\ln(0.181/3) = 2.81$. So $T =$ $0.356\Delta\varepsilon/k$. From (20.15), (20.33), and (20.32), $\Delta\varepsilon = 1(2)\hbar^2/2I - 0 = h^2/4\pi^2 I =$ $2\tilde{B}_0 hc = 2(\tilde{B}_e - \frac{1}{2}\tilde{\alpha}_e)hc = 2(1.931 - 0.009) \text{ cm}^{-1} (1 \text{ cm}/0.01 \text{ m}) \times$ $(6.626 \times 10^{-34} \text{ J s})(2.9979 \times 10^8 \text{ m/s}) = 7.63_6 \times 10^{-23}$ J. $T = 0.356\,\Delta\varepsilon/k =$ $0.356(7.63_6 \times 10^{-23} \text{ J})/(1.3807 \times 10^{-23} \text{ J/K}) = 1.97$ K.

20.63 **(a)** T; **(b)** T; **(c)** F; **(d)** T; **(e)** F; **(f)** F; **(g)** T; **(h)** T; **(i)** F; **(j)** T; **(k)** F; **(l)** T.

20.64 $F = BQ\upsilon \sin\theta = (1.5\text{ T})(1.60 \times 10^{-19}\text{ C})(3.0 \times 10^{6}\text{ m/s}) \sin\theta =$ $(7.2 \times 10^{-13}\text{ N}) \sin\theta$.

(a) $F = 0$. **(b)** $F = (7.2 \times 10^{-13}\text{ N}) \sin 45° = 5.1 \times 10^{-13}$ N.

(c) 7.2×10^{-13} N. **(d)** 0.

20.65 From the paragraph before (20.61), $|\mathbf{m}| = Q\upsilon r/2 =$ $(2.0 \times 10^{-16}\text{ C})(2.0 \times 10^{5}\text{ m/s})(25 \times 10^{-10}\text{ m})/2 = 5.0 \times 10^{-20}$ J/T.

20.66 **(a)** $e/2m_p = (1.602176 \times 10^{-19}\text{ C})/2(1.672622 \times 10^{-27}\text{ kg}) = 4.78942 \times 10^{7}$ Hz/T, where (20.54) was used.

(b) $\gamma/2\pi = (4.78942 \times 10^{7}\text{ Hz/T})5.58569/2\pi = 42.5775 \times 10^{6}$ Hz/T.

20.67 **(a)** Equations (20.65) and (20.63) give $E = -\gamma\hbar BM_I =$ $-1.792(4.7894 \times 10^{7}\text{ s}^{-1}/\text{T})1.792(6.626 \times 10^{-34}\text{ J s})(2\pi)^{-1}(1.50\text{ T})M_I =$ $-(1.35_8 \times 10^{-26}\text{ J})M_I$. Since $I = 3/2$, $M_I = 3/2, 1/2, -1/2$, and $-3/2$. The levels are -2.04×10^{-26} J, -0.679×10^{-26} J, 0.679×10^{-26} J, 2.04×10^{-26} J.

(b) B is doubled, so the four energies in (a) are each doubled..

20.68 **(a)** From (20.67) and (20.63), $\nu = |\gamma|B/2\pi =$ $(4.7894 \times 10^{7}\text{ s}^{-1}/\text{T})1.792(1.50\text{ T})/2\pi = 2.049 \times 10^{7}/\text{s} = 20.49$ MHz.

(b) 27.32 MHz.

20.69 From (20.67), $\nu_A/\nu_B = |\gamma_A|/|\gamma_B|$ and $\nu(^{13}\text{C}) = (10.708/42.577)(600\text{ MHz}) =$ 151 MHz.

20.70 **(a)** Equations (20.67) and (20.63) give $B = 2\pi\nu/|\gamma| =$ $2\pi(60 \times 10^{6}\text{ s}^{-1})/[5.5857(4.7894 \times 10^{7}\text{ s}^{-1}/\text{T})] = 1.41$ T.

(b) $(300/60)(1.41\text{ T}) = 7.05$ T.

20.71 **(a)** $M_I = +½$ and $-½$. The energy separation is given by (20.65) as $\Delta E =$ $|\gamma|\hbar B = (4.7894 \times 10^{7}\text{ s}^{-1}/\text{T})5.5857(6.626 \times 10^{-34}\text{ J s})(2\pi)^{-1}(1.41\text{ T}) =$ 3.98×10^{-26} J. The levels are nondegenerate and the population ratio is

$e^{-\Delta E/kT} = \exp[-(3.98 \times 10^{-26}\text{ J})/(1.381 \times 10^{-23}\text{ J/K})(298\text{ K})] =$
$\exp(-0.00000967) = 0.9999903.$

(b) An increase in B increases the separation between energy levels, thereby producing a greater population difference between the initial and final states. Hence the absorption intensity increases.

20.72 From Eq. (20.72), $\nu_i - \nu_j = 10^{-6}(300 \times 10^6\text{ Hz})(1.0) = 300\text{ Hz}.$

20.73 From the tables in the text, δ is 2 to 3 for the CH_3 protons and is 9 to 11 for the CHO proton; J is 1 to 3 Hz. The CH_3 doublet has three times the total intensity of the CHO quartet. Thus (all splittings are about 2 Hz)

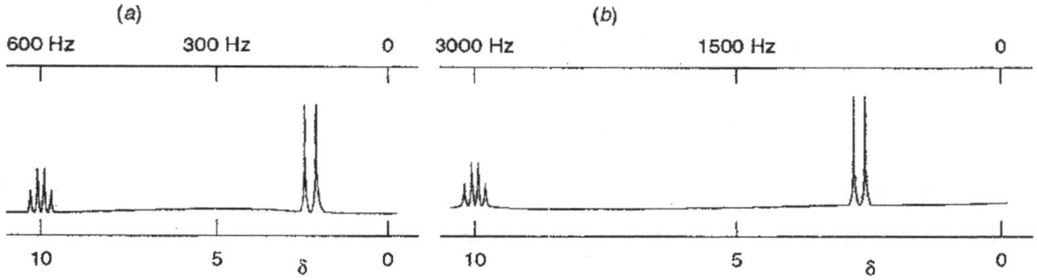

20.74 (a) From (20.70) and (20.68), increasing δ_i means decreasing σ_i and increasing ν_i, so ν_i increases to the left.

(b) To the left.

(c) To the right.

(d) From (20.70) and (20.69), increasing δ_i means decreasing σ_i and decreasing $B_{0,i}$, so $B_{0,i}$ increases to the right.

20.75 (a) One singlet peak.

(b) One proton NMR peak that is split into a doublet by the F.

(c) One singlet.

(d) The four methylene protons give a quartet of relative intensity 2; the six methyl protons give a triplet of relative intensity 3.

(e) The $(CH_3)_2$ protons give a doublet of relative intensity 6; the CH proton gives a septet of relative intensity 1.

(f) Two equal-intensity singlet peaks. The CH_3 groups are not equivalent and don't split each other.

(g) Three quartet peaks of equal intensity. In each of the three quartets, the 4 lines have equal intensity but only two of the 3 spacings are equal.

(h) The CH_3 protons give a triplet of relative intensity 3; the CHO proton gives a triplet of relative intensity 1; the CH_2 protons give an octet of relative intensity 2.

20.76 **(a)** This molecule has two different proton–^{19}F coupling constants: one for the ^{1}H and ^{19}F nuclei that are cis to each other, and a different coupling constant for ^{1}H and ^{19}F nuclei that are trans to each other.

(b) Here, the F atoms lie in a plane that is perpendicular to the plane of the CH_2 group, and there is only one ^{1}H–^{19}F coupling constant.

20.77 All peaks except in (b) are singlets.

(a) One peak.

(b) One peak that is split into a doublet by the F.

(c) One peak.

(d) Two peaks of equal intensity.

(e) Two peaks with 2:1 intensity ratio.

(f) Three equal-intensity peaks.

(g) Two peaks of equal intensity.

(h) Three equal-intensity peaks.

20.78 **(a)** 4 (there are 4 kinds of carbons); **(b)** 5; **(c)** 3.

20.79 In Fig. 20.44, 100 Hz corresponds to a length of 28 mm and J corresponds to a length of $2\frac{1}{4}$ mm, so $J = (2\frac{1}{4}/28)(100 \text{ Hz}) = 8$ Hz.

20.80 **(a)** Unchanged; see Eq. (20.70) and the following paragraph.

(b) Each ν and the difference between the ν's is multiplied by 2; see Eqs. (20.68) and (20.72).

(c) Unchanged, as noted after (20.73).

20.81 From Eq. (20.69), $\sigma_i = 1 - 2\pi\nu_{spec}/|\gamma_i|B_{0,i}$. so $\delta_i = 10^6(\sigma_{ref} - \sigma_i) = 10^6(-2\pi\nu_{spec}/|\gamma_i|)(1/B_{0,ref} - 1/B_{0,i}) = 10^6 B_0(B_{0,ref} - B_{0,i})/B_{0,ref}\, B_{0,i}$. Since $\sigma_i << 1$, we have $B_{0,i} \approx 2\pi\nu_{spec}/|\gamma_i| = B_0$, and $\delta_i = 10^6(B_{0,ref} - B_{0,i})/B_{0,ref}$.

20.82 **(a)** For a 60-MHz spectrometer, coalescence occurs at 120°C and the formula gives k_c at 120°C as $k_c = \pi|\nu_2 - \nu_1|/2^{1/2}$. From (20.72), $\nu_2 - \nu_1 = 10^{-6}(60\text{ MHz})(2.94 - 2.79) = 9.0\text{ Hz}$ and $k_c = \pi(9.0\text{ s}^{-1})/2^{1/2} = 20\text{ s}^{-1}$.

(b) Eq. (20.72) and the fact that δ_i is independent of ν_{spec} show that an increase in ν_{spec} increases $\nu_i - \nu_j$. Coalescence occurs when $\nu_{exch} >> |\nu_1 - \nu_2|$, and the increase in $|\nu_1 - \nu_2|$ and the fact that ν_{exch} increases as T increases mean that the coalescence temperature will increase when ν_{spec} is increased.

20.83 $\nu = g\beta_e B_0/h = 2.0026(9.274 \times 10^{-24}\text{ J/T})(2.50\text{ T})/(6.6261 \times 10^{-34}\text{ J s}) = 7.01 \times 10^{10}\text{ s}^{-1} = 70.1\text{ GHz}$.

20.84 There is one set of 4 equivalent protons and a second set of 4 equivalent protons, so there are 5(5) = 25 lines.

20.85 **(a)** 2(7) = 14; **(b)** 3(3) = 9; **(c)** 10; **(d)** 3(2) = 6.

20.86 $[\alpha] = \alpha/[\rho_B/(\text{g/cm}^3)](l/\text{dm}) = 1.90°/(0.0650)(2.00) = 14.6°$.

20.87 The observed α of the mixture is the sum of the α's of the α and β isomers: $\alpha = \alpha_\alpha + \alpha_\beta = [\alpha]_\alpha \rho_\alpha l(\text{cm}^3/\text{dm g}) + [\alpha]_\beta \rho_\beta l(\text{cm}^3/\text{dm g})$ (1). The total solute mass m is $m = m_\alpha + m_\beta$ and division by the solution's volume gives $\rho = \rho_\alpha + \rho_\beta$. Division of Eq. (1) by $\rho l(\text{cm}^3/\text{dm g})$ gives $[\alpha] = [\alpha]_\alpha w_\alpha + [\alpha]_\beta w_\beta$ (2), where $w_\alpha = \rho_\alpha/(\rho_\alpha + \rho_\beta) = m_\alpha/(m_\alpha + m_\beta)$ and w_β are the mass fractions (and also the mole fractions) of the α and β isomers. Equation (2) gives $52.7° = w_\alpha 112.2° + (1 - w_\alpha)17.5°$ and $w_\alpha = 0.372$, or 37.2% α-D-glucose.

20.88 $E = h\nu = hc/\lambda$. The first entry in (20.78) is $E = (6.626 \times 10^{-34}\text{ J s}) \times (2.998 \times 10^8\text{ m/s})/(200 \times 10^{-9}\text{ m}) = 9.93 \times 10^{-19}\text{ J} = 6.20\text{ eV}$; etc. The first entry in the following line is $N_A h\nu = N_A hc/\lambda = (6.022 \times 10^{23}/\text{mol}) \times (6.626 \times 10^{-34}\text{ J s})(2.998 \times 10^8\text{ m/s})/(200 \times 10^{-9}\text{ m}) = 5.98 \times 10^5$ J/mol; etc.

20.89 **(a)** I is the energy per unit time that falls on unit cross-sectional area perpendicular to the beam. In time t, the beam travels a distance $c't$ and the photons that pass through cross-sectional area A in time t are contained in a volume $Ac't$. Each photon has energy $h\nu$. If N is the number of photons in the volume $Ac't$, the energy of the photons in this volume is $Nh\nu$. The intensity equals this energy divided by the time t and the cross-sectional area A: so $I = Nh\nu/tA$. So $N = ItA/h\nu$. Then the number of photons per unit volume is $N/V = N/Ac't = (ItA/h\nu)/Ac't = I/h\nu c'$.

(b) $c' = c/1.34 = 2.24 \times 10^8$ m/s. $\nu = c/(488 \times 10^{-9}\text{ m}) = 6.14 \times 10^{14}\text{ s}^{-1}$. $N/V = I/h\nu c' = (10^{15}\text{ W/m}^2)/[(6.63 \times 10^{-34}\text{ J s})(6.1 \times 10^{14}\text{ s}^{-1})(2.2 \times 10^8\text{ m/s})] = 1 \times 10^{25}\text{ photons/m}^3 = 1 \times 10^{19}\text{ photons/cm}^3$. 18 g of water contains 6×10^{23} molecules in a volume of 18 cm^3, and the number of molecules per cm^3 is 3×10^{22}.

20.90 The energy absorbed is $0.744(0.00155\text{ J/s})(110\text{ s}) = 0.126_9$ J. The number of moles of photons absorbed is $(0.1269\text{ J})/(N_A hc/\lambda) = 4.92 \times 10^{-7}$ mole. So $\Phi = (6.80 \times 10^{-6}\text{ mole})/(4.92 \times 10^{-7}\text{ mole}) = 13.8$.

20.91 **(a)** $d[\text{HI}]/dt = -\phi\mathcal{I}_a - k_2[\text{H}][\text{HI}]$. The steady-state approximation for H gives $d[\text{H}]/dt = 0 = \phi\mathcal{I}_a - k_2[\text{H}][\text{HI}]$ and $k_2[\text{H}][\text{HI}] = \phi\mathcal{I}_a$. So $d[\text{HI}]/dt = -2\phi\mathcal{I}_a = -2\mathcal{I}_a$, since $\phi \approx 1$.

(b) The number N of photons absorbed satisfies $4184\text{ J} = Nh\nu = Nhc/\lambda$ and $N = (4184\text{ J})(250 \times 10^{-9}\text{ m})/(6.626 \times 10^{-34}\text{ J s})(2.998 \times 10^8\text{ m/s}) = 5.27 \times 10^{21}$. The number of HI molecules decomposed is $2(5.27 \times 10^{21}) = 1.05 \times 10^{22}$.

20.92 With the inclusion of reaction (5), Eq. (20.83) becomes $r = k_2[\text{A}^*][\text{A}] - k_4[\text{A}_2] + k_5[\text{A}]^2$. Reaction (5) does not involve A*, so (20.84) still holds and we have $r = k_2[\text{A}]\mathcal{I}_a/(k_2[\text{A}] + k_3) - k_4[\text{A}_2] + k_5[\text{A}]^2$. For the photostationary state, $r = 0$ and we get $[\text{A}_2] = k_5[\text{A}]^2/k_4 + k_2[\text{A}]\mathcal{I}_a/(k_4k_2[\text{A}] + k_3k_4)$. In the absence of radiation, $\mathcal{I}_a = 0$ and the last equation becomes $[\text{A}_2] = k_5[\text{A}]^2/k_4$. The concentration of A_2 is greater in the presence of radiation.

20.93 **(a)** Yes; **(b)** no; **(c)** no. (1 + 1 is not a member of the set.)

20.94 **(a)** Yes; **(b)** no; **(c)** yes; **(d)** no.

20.95 **(a)** We have $AI = IA = A$, since I is the identity element. Since each group element appears exactly once in each row of the multiplication table, we must have $AA = I$. The multiplication table is thus

	I	***A***
I	I	A
A	A	I

(b) Using the property of the identity element, we start with

	I	***A***	***B***
I	I	A	B
A	A	w	x
B	B	y	z

where w, x, y, and z are to be determined. w cannot equal A because this would put two A's in row 2. So w must either equal I or B. If we put $w = I$, then the theorem that each group element appears exactly once in each row means that $x = B$. But with $x = B$, we have B appearing twice in column three, which is not allowed. Hence $w = B$. Filling in the rest of the table, we get

	I	***A***	***B***
I	I	A	B
A	A	B	I
B	B	I	A

This table shows that each element has an inverse, and using the table, one finds that associativity is satisfied, so this is the multiplication table of a group.

20.96 **(a)** F; **(b)** T.

20.97 As in Fig. 20.52, the coordinate axes do not move when a symmetry operation is applied.

(a) They do not commute. **(b)** They do not commute. **(c)** They commute.

20.98 **(a)** $\hat{E}$; **(b)** $\hat{\sigma}$; **(c)** $\hat{i}$; **(d)** $\hat{C}_5^4$; **(e)** $\hat{C}_5^3$; **(f)** $\hat{S}_3^5$. (Since $\hat{S}_3^2$ involves two reflections, which amounts to no reflections, it is not the inverse of $\hat{S}_3$.)

20.99 **(a)** T. **(b)** T. **(c)** T. (If we call the rotation axis the z axis, then the 180° rotation part of $\hat{S}_2$ converts the x and y coordinates to their negatives and the reflection converts the z coordinate to its negative.) **(d)** T.

20.100 **(a)** The structure is trigonal pyramidal. A C_3 axis and three planes of symmetry (each of which contains one of the bonds).

(b) $\hat{C}_3$, $\hat{C}_3^2$, $\hat{E}$, $\hat{\sigma}_a$, $\hat{\sigma}_b$, $\hat{\sigma}_c$.

20.101 Symmetry elements: A C_6 axis that is also an S_6 axis, an S_3 axis and a C_2 axis. A center of symmetry. Seven symmetry planes, six perpendicular to the molecular plane and one coincident with it. Six C_2 axes lying in the molecular plane. Symmetry operations: $\hat{E}$, $\hat{C}_6$, $\hat{C}_6^2$, $\hat{C}_6^3$, $\hat{C}_6^4$, $\hat{C}_6^5$, $\hat{S}_6$, $\hat{S}_6^5$, $\hat{S}_3$, $\hat{S}_3^2$, $\hat{i}$, six $\hat{C}_2$ rotations, and seven $\hat{\sigma}$ reflections, for a total of 24 operations. (The operation $\hat{S}_6^3$ is the same as $\hat{i}$; see the answer to Prob. 20.99c.)

20.102 Four C_3 axes (one along each bond). Three S_4 axes (which are also C_2 axes), one through each pair of opposite faces of the cube in Fig. 20.20. Six symmetry planes, each of which contains two bonds.

20.103 **(a)** T_d; **(b)** $C_{\infty v}$; **(c)** $D_{\infty h}$; **(d)** D_{3h}; **(e)** O_h; **(f)**; C_{4v} **(g)** D_{4h}; **(h)** C_{3v}; **(i)** D_{3h}; **(j)** D_{2h}; **(k)** C_{3v}; **(l)** C_{4v}; **(m)** C_1; **(n)** C_s.

20.104 $\hat{C}_n, \hat{C}_n^2, \hat{C}_n^3, \ldots, \hat{C}_n^{n-1}, \hat{E}$. Order n.

20.105 (a) Yes, since the product $\hat{C}_n\sigma_h$ is a symmetry operation.

(b) When n is even, since $\hat{C}_n^{n/2} = \hat{C}_2$ is then a symmetry operation.

(c) C_{2v}; C_{2h}; C_{2v}.

20.106 D_{3d}; D_{2d}.

20.107 (a) $\begin{pmatrix} 0 & 11 \\ 6 & 6 \end{pmatrix}$ **(b)** $\begin{pmatrix} 3 & 15 \\ 9 & 12 \end{pmatrix}$

20.108 $\mathbf{AB} = \begin{pmatrix} 0.2(4)+4(5) & 0.2(1)+4(8) \\ -1(4)+3(5) & -1(1)+3(8) \end{pmatrix} = \begin{pmatrix} 20.8 & 32.2 \\ 11 & 23 \end{pmatrix}$ $\mathbf{BA} = \begin{pmatrix} -0.2 & 19 \\ -7 & 44 \end{pmatrix}$

20.109 $\begin{pmatrix} 1 & 2 \\ 3 & 4 \end{pmatrix}\begin{pmatrix} 2 \\ 6 \end{pmatrix} = \begin{pmatrix} 1(2)+2(6) \\ 3(2)+4(6) \end{pmatrix} = \begin{pmatrix} 14 \\ 30 \end{pmatrix}$

20.110 Let k be the number of columns in **R** and the number of rows in **S**. (These quantities must be equal or the matrix product is not defined.) The element t_{mn} is calculated using row m of **R** and column n of **S**. So $t_{mn} = \sum_{i=1}^{k} r_{mi}s_{in}$.

20.111 $\mathbf{M}_1\mathbf{N}_1 = \begin{pmatrix} -1 & 5 \\ 3 & 8 \end{pmatrix}\begin{pmatrix} 2 & 9 \\ 4 & 1 \end{pmatrix} = \begin{pmatrix} 18 & -4 \\ 38 & 35 \end{pmatrix}$ $\mathbf{M}_2\mathbf{N}_2 = (7)(6) = (42)$

$$\mathbf{MN} = \begin{pmatrix} -1 & 5 & 0 \\ 3 & 8 & 0 \\ 0 & 0 & 7 \end{pmatrix}\begin{pmatrix} 2 & 9 & 0 \\ 4 & 1 & 0 \\ 0 & 0 & 6 \end{pmatrix} = \begin{pmatrix} 18 & -4 & 0 \\ 38 & 35 & 0 \\ 0 & 0 & 42 \end{pmatrix}$$

20.112 For example, the product of the $\hat{\sigma}(xz)$ and $\hat{\sigma}(yz)$ matrices (in either order) is

$$\begin{pmatrix} 1 & 0 & 0 \\ 0 & -1 & 0 \\ 0 & 0 & 1 \end{pmatrix}\begin{pmatrix} -1 & 0 & 0 \\ 0 & 1 & 0 \\ 0 & 0 & 1 \end{pmatrix} = \begin{pmatrix} -1 & 0 & 0 \\ 0 & -1 & 0 \\ 0 & 0 & 1 \end{pmatrix},$$ which is the matrix that corresponds to $\hat{C}_2(z)$, in agreement with the product in Table 20.2.

20.113 (a) $\hat{C}_4(z)$ moves the point at (x, y, z) to $(-y, x, z)$ and Eq. (20.92) becomes

$$\begin{pmatrix} x' \\ y' \\ z' \end{pmatrix} = \begin{pmatrix} 0 & -1 & 0 \\ 1 & 0 & 0 \\ 0 & 0 & 1 \end{pmatrix}\begin{pmatrix} x \\ y \\ z \end{pmatrix}$$

(b) $\hat{i}$ moves the point at (x, y, z) to $(-x, -y, -z)$ and the matrix that does this is $\begin{pmatrix} -1 & 0 & 0 \\ 0 & -1 & 0 \\ 0 & 0 & -1 \end{pmatrix}$.

(c) $\hat{\sigma}(xy)$ moves the point at (x, y, z) to $(x, y, -z)$ and the matrix is

$$\begin{pmatrix} 1 & 0 & 0 \\ 0 & 1 & 0 \\ 0 & 0 & -1 \end{pmatrix}.$$

20.114 (a) Using associativity, we have $(\mathbf{P}^{-1}\mathbf{AP})(\mathbf{P}^{-1}\mathbf{CP}) = (\mathbf{P}^{-1}\mathbf{A})(\mathbf{PP}^{-1})(\mathbf{CP}) = (\mathbf{P}^{-1}\mathbf{A})(\mathbf{I})(\mathbf{CP}) = (\mathbf{P}^{-1}\mathbf{A})(\mathbf{CP}) = \mathbf{P}^{-1}\mathbf{ACP} = \mathbf{P}^{-1}\mathbf{FP}$, since multiplication by a unit matrix has no effect.

(b) The result of part (a) of this problem shows that the transformed matrices $\mathbf{P}^{-1}\mathbf{AP}$, $\mathbf{P}^{-1}\mathbf{BP}$,... multiply the same way as the original matrices **A, B**,...; since the original matrices multiply the same way as the group members, so do the transformed matrices.

20.115 As noted a couple of paragraphs before (20.95), each element of a commutative group is in a class by itself, so the number of classes c equals the number of elements h in a commutative group. There are thus h terms on the left side of (20.95). The smallest possible value of each term on the left of (20.95) is 1, and if any representation had a dimension greater than 1, then the value of the left side of (20.95) would exceed h and (20.95) would not hold. So all irreducible representations of a commutative group are one-dimensional.

20.116 For example, the products of the matrices corresponding to $\hat{C}_2(z)$ and $\hat{\sigma}(xz)$ are $(1)(-1) = (-1)$ and $(-1)(1) = (-1)$, and (-1) is the matrix corresponding to $\hat{\sigma}(yz)$, which according to Table 20.2 is the correct product. The other products are verified similarly.

20.117 To find the elements in the same class as E, we form the products $A^{-1}EA$, $B^{-1}EA$,... . But $A^{-1}EA = A^{-1}A = E$ for every element A, so the identity element E is in a class by itself.

20.118 **(a)** $2p_y$; **(b)** $-2p_y$, since the positive and negative lobes are interchanged; **(c)** $2p_y$; **(d)** $-2p_x$; **(e)** $2p_y$; **(f)** $-2p_y$; **(g)** $-2p_y$; **(h)** $-2p_y$.

20.119 **(a)** The orbitals $O2s$, $O2p_z$, and $H_11s + H_21s$ are each unchanged by each of the four symmetry operations and the C_{2v} character table (Table 20.3) gives the symmetry species as a_1.

(c) The $O2p_y$ orbital is unchanged by $\hat{E}$ and by $\hat{\sigma}(yz)$ and is transformed to $-O2p_y$ by $\hat{C}_2(z)$ and by $\hat{\sigma}(xz)$ and $H_11s - H_21s$ shows the same behavior. The character table gives the symmetry species as b_2.

20.120 **(a)** We examine the effects of each symmetry operation on the vectors showing the normal-mode motions of the atoms. The point group is C_{2v} and the z axis coincides with the C_2 axis through the C and O atoms. The $\hat{C}_2(z)$ operation reverses the direction of a vector pointing in the $+x$ or $-x$ direction (and also interchanges the vectors on the H atoms), so for this normal mode the $\hat{C}_2(z)$ character is -1. For $\hat{\sigma}(yz)$, the direction of a vector pointing in the $+x$ or $-x$ direction is reversed and the character is

-1. The $\hat{\sigma}(xz)$ operation interchanges the vectors in the H atoms but leaves the directions of vectors pointing in the $\pm x$ direction unchanged and the character is $+1$. The character table (Table 20.2) gives the symmetry species as b_1.

(b) The point group is C_{3v}. The N-atom vibration vector points along the z axis and is unaffected by each of the six symmetry operations listed in the character table (Table 20.3). The H-atom vibration vectors point along the bonds and although the symmetry operations may interchange or permute these vectors, the direction of each vector on each H is unchanged by each symmetry operation. Thus the characters are all $+1$ for this mode and the mode has species a_1.

20.121 The representation Γ consists of the matrices in (20.93), and the point group is C_{2v} with order $h = 4$. The irreducible-representation characters χ_i are taken from Table 20.3. The characters χ_Γ of Γ are found by taking the traces (the sums of the diagonal elements) of the matrices in (20.93); these characters are $3, -1, 1$, and 1 for $\hat{E}$, $\hat{C}_2(z)$, $\hat{\sigma}(xz)$, and $\hat{\sigma}(yz)$, respectively. Eq. (20.97) gives $a_{A_1(20.93)} = (1/4)[1(3) + 1(-1) + 1(1) + 1(1)] = 1$,

$a_{A_2(20.93)} = (1/4)[1(3) + 1(-1) + (-1)(1) + (-1)(1)] = 0$

$a_{B_1(20.93)} = (1/4)[1(3) + (-1)(-1) + 1(1) + (-1)(1)] = 1$

$a_{B_2(20.93)} = (1/4)[1(3) + (-1)(-1) + (-1)(1) + 1(1)] = 1$ so $\Gamma_{(20.93)} = A_1 \oplus B_1 \oplus B_2$.

20.122 (a) $\hat{E}$ is represented by a unit matrix whose order equals the dimension of the representation. The trace of a unit matrix of order n is n. Hence this representation has dimension 9.

(b) We use Eq. (20.97) with the χ_i values taken from the character table in Table 20.3. We have $a_{A_1\Gamma} = (1/4)[1(9) + 1(-1) + 1(1) + 1(3)] = 3$,

$a_{A_2\Gamma} = (1/4)[1(9) + 1(-1) + (-1)(1) + (-1)(3)] = 1$

$a_{B_1\Gamma} = (1/4)[1(9) + (-1)(-1) + 1(1) + (-1)(3)] = 2$

$a_{B_2\Gamma} = (1/4)[1(9) + (-1)(-1) + (-1)(1) + 1(3)] = 3$

so $\Gamma = 3A_1 \oplus A_2 \oplus 2B_1 \oplus 3B_2$.

20.123 We use Eq. (20.97) with the χ_i values taken from the character table in Table 20.3. The sum in (20.97) is over the h symmetry operations. The characters in

the C_{3v} character table are listed for each class, and if we take the sum in (20.97) to be over the classes, we must multiply each term in the sum by the number of operations in that class, so as to include each of the h operations in the sum. We have $a_{A_1\Gamma} = (1/6)[1(293) + 2(1)(-118) + 3(1)9] = 14$,
$a_{A_2\Gamma} = (1/6)[1(293) + 2(1)(-118) + 3(-1)9] = 5$
$a_{E\Gamma} = (1/6)[2(293) + 2(-1)(-118) + 3(0)9] = 137$. So $\Gamma = 14A_1 \oplus 5A_2 \oplus 137E$.

20.124 The trace of a matrix such as **M** in (20.90) that is in block-diagonal form is clearly equal to the sum of the traces of each block. (For example, the trace of **M** is $-1 + 8 + 7 = 14$ and the traces of $\mathbf{M}_1$ and $\mathbf{M}_2$ are $8 - 1 = 7$ and 7.) Hence the trace of a matrix of a reducible representation equals the sums of the traces of all the irreducible-representation matrices that it is the direct sum of. So $\chi_\Gamma(\hat{E}) = 4(1) + 1 + 6(2) = 17$, $\chi_\Gamma(\hat{C}_3) = 4(1) + 1 + 6(-1) = -1$,
$\chi_\Gamma(\hat{\sigma}_v) = 4(1) + (-1) + 6(0) = 3$.

20.125 (a) HBr, H_2S, CH_3Cl, which have nonzero dipole moments.

(b) HBr, CO_2, H_2S, CH_4, CH_3Cl, C_6H_6, which have vibrations that change the dipole moment.

(c) N_2, HBr, CO_2, H_2S, CH_3Cl, and C_6H_6, which are not spherical tops.

20.126 (a) $k_e = E_e''(R_e)$. The function $E_e(R)$ is found by solving the electronic Schrödinger equation, which is the same for H_2 and D_2, so $E_e(R)$ is the same for H_2 and D_2 and this makes k_e, D_e, and R_e the same for H_2 and D_2. (This answer neglects very slight deviations from the Born–Oppenheimer approximation that make E_e, k_e, R_e, and D_e differ extremely slightly for H_2 and D_2.)

(b) $\nu_e = (1/2\pi)(k_e/\mu)^{1/2}$. Since μ is greater for D_2 and k_e is essentially the same, ν_e is smaller for D_2.

(c) $I_e = \mu R_e^2$. Since μ is substantially greater for D_2 and R_e is essentially the same for the two species, I_e is greater for D_2.

(d) $B_e = h/8\pi^2 I_e$. Since I_e is greater for D_2, B_e is greater for H_2.

(e) The same, since $E_e(R)$ is the same for the two.

(f) $D_0 \cong D_e - \frac{1}{2}h\nu_e$. Since ν_e is smaller for D_2 and D_e is essentially the same for the two, D_0 is greater for D_2.

(g) Since ν_e is smaller for D_2 and D_e is the same, D_2 has more bound vibrational levels.

(h) Since ν_e is smaller for D_2, the separation between vibrational levels is smaller for D_2, and more D_2 molecules are in excited vibrational levels at a given T, so the fraction in $\upsilon = 0$ is greater for H_2.

(i) Since I_e is greater for D_2, the separation between D_2 rotational levels is smaller for D_2 and the fraction in $J = 0$ is greater for H_2.

20.127 Since there are $2I + 1$ values of M_I, there are $2I + 1$ two-electron spin functions where electron 1 and electron 2 have the same spin function and so are symmetric functions [similar to $\alpha(1)\alpha(2)$]. The remaining $(2I+1)^2 - (2I+1)$ two-electron spin functions are neither symmetric nor antisymmetric, but have forms like $\alpha(1)\beta(2)$ and $\alpha(2)\beta(1)$. These functions must be combined to form symmetric and antisymmetric functions of form similar to $2^{-1/2}[\alpha(1)\beta(2) + \alpha(2)\beta(1)]$ and $2^{-1/2}[\alpha(1)\beta(2) - \alpha(2)\beta(1)]$; the total number of such combined functions will equal the number of uncombined functions used, and so will be $(2I+1)^2 - (2I+1)$. Half of these will be symmetric and half antisymmetric. Thus there will be $\frac{1}{2}[(2I+1)^2 - (2I+1)] = I(2I+1)$ combined symmetric functions and $\frac{1}{2}[(2I+1)^2 - (2I+1)] = I(2I+1)$ combined antisymmetric spin functions. Adding in the $2I + 1$ uncombined symmetric spin functions, we get a total of $I(2I+1) + 2I + 1 = (I+1)(2I+1)$ symmetric two-electron spin functions. And we have $I(2I+1)$ antisymmetric two-electron spin functions.

20.128 (a) joule, newton, watt, pascal, hertz, coulomb, volt, tesla, ohm, kelvin, ampere, siemens;

(b) poise, debye, angstrom, svedberg, dalton, torr, bohr, hartree.

20.129 (a) T.

(b) F. It cannot change its rotational state by absorption or emission of radiation, but can change its rotational state during collisions.

(c) F. Counterexamples are CH_4 and BF_3.

(d) F. The energy might be transferred to another molecule in a collision.

(e) F. A counterexample is H_2O.

(f) T.

(g) F. This formula is only for linear and spherical-top molecules.

(h) T.

(i) T.

(j) F.

(k) This question is too silly to answer.